Studies on Decapoda and Copepoda in Memory of Michael Türkay

CRUSTACEANA MONOGRAPHS constitutes a series of books on carcinology in its widest sense. Contributions are handled by the Series Editor(s) and may be submitted through the office of KONINKLIJKE BRILL Academic Publishers N.V., P.O. Box 9000, NL-2300 PA Leiden, The Netherlands.

Published in this series:

CRM 001 - *Stephan G. Bullard*	Larvae of anomuran and brachyuran crabs of North Carolina
CRM 002 - *Spyros Sfenthourakis et al. (eds.)*	The biology of terrestrial isopods, V
CRM 003 - *Tomislav Karanovic*	Subterranean Copepoda from arid Western Australia
CRM 004 - *Katsushi Sakai*	Callianassoidea of the world (Decapoda, Thalassinidea)
CRM 005 - *Kim Larsen*	Deep-sea Tanaidacea from the Gulf of Mexico
CRM 006 - *Katsushi Sakai*	Upogebiidae of the world (Decapoda, Thalassinidea)
CRM 007 - *Ivana Karanovic*	Candoninae (Ostracoda) from the Pilbara region in Western Australia
CRM 008 - *Frank D. Ferrari & Hans-Uwe Dahms*	Post-embryonic development of the Copepoda
CRM 009 - *Tomislav Karanovic*	Marine interstitial Poecilostomatoida and Cyclopoida (Copepoda) of Australia
CRM 010 - *Carrie E. Schweitzer et al.*	Systematic list of fossil decapod crustacean species
CRM 011 - *Peter Castro et al. (eds.)*	Studies on Brachyura: a homage to Danièle Guinot
CRM 012 - *Patricio R. De los Ríos-Escalante*	Crustacean zooplankton communities in Chilean inland waters
CRM 013 - *Katsushi Sakai*	Axioidea of the world and a reconsideration of the Callianassoidea (Decapoda, Thalassinidea, Callianassida)
CRM 014 - *Charles H.J.M. Fransen et al. (eds.)*	Studies on Malacostraca: Lipke Bijdeley Holthuis Memorial Volume
CRM 015 - *Akira Asakura et al. (eds.)*	New Frontiers in Crustacean Biology: Proceedings of the TCS Summer Meeting, Tokyo, 20-24 September 2009
CRM 016 - *Danielle Defaye et al. (eds.)*	Studies on Freshwater Copepoda: a Volume in Honour of Bernard Dussart
CRM 017 - *Hironori Komatsu et al. (eds.)*	Studies on Eumalacostraca: a homage to Masatsune Takeda
CRM 018 - *Masahiro Dojiri & Ju-Shey Ho*	Systematics of the Caligidae, copepods parasitic on fishes
CRM 019 - *Darren C.J. Yeo et al. (eds.)*	Advances in Freshwater Decapod systematics and biology
CRM 020 - *Finn Viehberg et al. (eds.)*	The Recent and Fossil meet Kempf Database Ostracoda Festschrift Eugen Karl Kempf — Proceedings of the 15th International German Ostracodologists' Meeting
CRM 021 - *L.A.M. Neethling & A. Avenant-Oldewage*	Branchiura — A compendium of the geographical distribution and a summary of their biology

Authors' addresses:

C. Magalhães, Instituto National de Pesquisas da Amazônia, Av. André Araújo, 2936, 69.067-375 Manaus, AM, Brazil; e-mail: celiomag@inpa.gov.br

First published as a Special Issue of Crustaceana. *When citing from this book, refer to* Crustaceana *90 (7-10): 771-1288.*

Cover: Lithoscaptus tuerkayi sp. nov., drawing by Inge van Noortwijk in the paper by Sancia van der Meij; see fig. 1A on page 260 of this volume.

Studies on Decapoda and Copepoda in Memory of Michael Türkay

By

Célio Magalhães, Carola Becker, Peter Davie, Sven Klimpel,
Pedro Martínez-Arbizu and Moritz Sonnewald (Editors)

CRUSTACEANA MONOGRAPHS, 22

BRILL

LEIDEN • BOSTON

Library of Congress Cataloging-in-Publication Data
The Library of Congress Cataloging-in-Publication Data is available from the Publisher.

Authors' addresses (continues from p. ii):
C. Becker, Queen's University Marine Laboratory, 12-13 The Strand, Portaferry, Northern Ireland, U.K.; e-mail: c.becker@qub.ac.uk
P.J.F. Davie, Queensland Museum, P.O. Box 3300, South Brisbane, QLD, Australia; e-mail: peter.davie@qm.qld.gov.au
S. Klimpel, Biodiversity and Climate Research Centre (BiK-F), Goethe-University, Senckenberganlage 25, D-60325 Frankfurt am Main, Germany; e-mail: sven.klimpel@senckenberg.de
P. Martínez-Arbizu, German Center for Marine Biodiversity Research (DZMB), Forschungsinstitut Senckenberg, Suedstrand 44, D-26382 Wilhelmshaven, Germany; e-mail: pmartinez@senckenberg.de
M. Sonnewald, Senckenberg Research Institute and Natural History Museum, Marine Zoology – Crustacean Section, Senckenberganlage 25, D-60325 Frankfurt am Main, Germany; e-mail: moritz.sonnewald@senckenberg.de

The contents of this volume were originally published in 2017 in *Crustaceana* volume 90, issue 7-10.

This book is printed on acid-free paper.

ISBN13: 978 90 04 36273 4
E-ISBN: 978 90 04 36643 5

PRINTED IN THE NETHERLANDS

CONTENTS

First published as a Special Issue of *Crustaceana*. The page numbers in the above Table of Contents refer to the bracketed page numbers in this volume. The other page numbers are the page numbers in *Crustaceana* 90/7-10. When citing from this book, refer to *Crustaceana* 90 (2017) 771-1288 and the page numbers without brackets.

[When citing this volume, please refer to Crustaceana 90 (2017) 771-1288]

PREFACE

The sudden and untimely death of Prof. Dr. Michael Türkay in September 2015 came as a shock to his many friends and colleagues all over the world. Only the year before, he and his team at Senckenberg had skilfully organized the Eighth International Crustacean Conference in Frankfurt am Main, and he had seemed his normal vibrant and hospitable self. Thus, within a week of receiving the sad news of his passing, a group of friends and colleagues conferred through e-mail, and all agreed that a Memorial Issue of a scientific periodical should be published to honour Michael's great contribution to the study of Crustacea. In fact two such issues were agreed upon. Célio Magalhães became the guest editor for a "Nauplius" tribute, and this has seen a number of papers published electronically as each was ready (Nauplius, **24** (2016) and **25** (2017)). The other is the present special issue of "Crustaceana". Peter Davie first took on the coordinating role for the Crustaceana volume, and sent out a large number of special invitations to Michael's friends and collaborators over his lifetime. Unfortunately, Peter was forced to withdraw from most of the active editing, and Célio Magalhães took over as the Coordinating Editor (on top of his duties with Nauplius). Many colleagues accepted their invitations, and Célio (with the help of the other guest editors) was faced with the huge task of reviewing, accepting, and editing the series of contributions in a most professional and commendable way. Here we present no less than 31 contributed papers, plus the special obituary tribute to Michael's life which begins the issue. We think that together they make a most worthy homage to one of the great modern figures of carcinology, Michael Türkay. We are all both thankful and proud to have had the opportunity to be involved in this process, and we take great pleasure in presenting this compilation of scientific work on Crustacea to Michael's widow, Heide, and their two sons, Stephan and Christoph. Heide was an integral part of Michael's career and successes, and also deserves accolades for the support she always gave her husband in his pursuit of the science to which we are all so much dedicated.

Guest Editors
Célio Magalhães
Carola Becker
Peter J. F. Davie
Sven Klimpel

Pedro Martínez-Arbizu
Moritz Sonnewald
On behalf of the regular Board of Editors
Charles H. J. M. Fransen
Peter K. L. Ng
Christoph D. Schubart
J. Carel von Vaupel Klein

[When citing this volume, please refer to Crustaceana 90 (2017) 771-1288]

Michael Türkay Memorial Issue

MICHAEL TÜRKAY
(3 APRIL 1948–9 SEPTEMBER 2015)

BY

PETER DAVIE[1,3]) and CAROLA BECKER[2,4])

[1]) Queensland Museum, P.O. Box 3300, South Brisbane, QLD, Australia

[2]) Queen's University Marine Laboratory, 12-13 The Strand, Portaferry, Northern Ireland, U.K.

Michael Türkay on board R.V. "Senckenberg" during a February 2010 winter cruise to the Dogger Bank. Photo by Forschungsinstitut Senckenberg.

Michael Türkay was an extraordinary man. He was a larger-than-life presence on the world stage as a marine biologist, carcinologist, educator, curator and museum champion. And most importantly an honourable and loyal friend to so many — as the French would say an "homme sérieux".

[3]) Corresponding author; e-mail: Peter.Davie@qm.qld.gov.au

[4]) e-mail: c.becker@qub.ac.uk

He was a true world citizen, and although the rest of the world placed him as a "German", Michael himself identified much more as a proud "Frankfurter", and this somehow symbolized his character much better. Frankfurt was confirmed as an Imperial Free City of the Holy Roman Empire following the Peace of Westphalia in 1648, and later, a Free City of the German Confederation following the Congress of Vienna in 1816. Frankfurt was also home to one of the greatest figures of European intellectual life, Johann Wolfgang von Goethe, a figure greatly admired by Michael (Peters & Türkay, 1999). Goethe, on what was to be his last visit home in 1815, declared: "A free spirit befits a free city... It befits Frankfurt to shine in all directions and to be active in all directions." No one could have personified this great ideal more than Michael Türkay.

CREATING THE MAN — CHILDHOOD, SCHOOLING, FAMILY

Michael was born in Frankfurt am Main, Germany on 3 April 1948, the eldest son of Dr. Karnik Türkay and his wife Bettina ("Betty"). Betty was born and raised in Frankfurt, the daughter of a strongly committed unionist and social democrat, whose opposition to the rise of the Nazi Party sadly culminated in a long period of imprisonment. She undertook her tertiary studies at the Johann Wolfgang Goethe University in Frankfurt (as Michael would, many years later), specializing in the study of oriental philology (the structure, historical development, and relationships of languages). Later, her young son Michael would also be imbued with this same passion for language. Michael's father, Dr. Karnik Türkay was a specialist in obstetrics and gynaecology.

Large parts of Frankfurt were destroyed during the heavy bombings of World War II, and on 22 March 1944, a British attack destroyed virtually the entire Old City, killing over 1000 people. Thus, the baby Michael was born into a devastated city, and a war ravaged country.

As they grew up, Michael's family had many seaside holidays to the shores of the Mediterranean. Karnik would take Michael and his little brother, Armand, on fishing trips, and they would spend many happy hours fishing, and also diving for crustaceans, seashells, and any other marine animals they could find. Armand well remembers he and Michael catching crabs by tying small brass coins (with a hole punched in them) to lengths of string, and then dangling them into cavities along concrete jetties. Small crabs would grab the coins in their claws and the little boys would carefully pull them out by the dozens. However, in truth, Michael's first fascinations were not for his later beloved crustaceans, but for molluscs (especially bivalves), fish, and beetles (he apparently had a vast collection of Coleoptera).

It was in Grade 8 at school that a biology teacher helped change the course of young Michael's life. Already a keen insect collector, his brother Armand remembers Michael coming home spouting many strange words that he explained were the Latin names for the animals he had showed his teacher. Michael was to later credit this teacher with the change in him from a random collector of "shiny things", into a truly systematic collector with a taxonomist's drive for discovery. He and Armand from then on used to collect with a purpose, and started to assemble a great collection of butterflies and beetles, as well as many other insects. Dried and pinned, but with no display cases, these insects took the place of the books on all the shelves around the Türkay's family home.

Michael was 16 years old when he first darkened the doors of the Senckenberg Museum. He was firmly clutching a box of Mediterranean seashells under his arm, and hoping for someone to help him to identify them! He was initially introduced to Dr. Adolf Zilch, the curator of the Malacology Section, but Dr. Zilch's primary interest was in land snails, and it appears he failed to enthuse the young man. Instead, Michael was introduced to the curator of the Crustacea Section, Dr. Richard Bott, who had been a school teacher before coming to Senckenberg, and perhaps this earlier vocation helped Richard to recognize and nurture Michael's great potential. Michael always talked of Dr. Bott with great respect, and wrote an obituary for him after his death (Türkay, 1975).

Michael's tertiary education began with his enrolment at the Johann Wolfgang Goethe University (at present: Goethe University Frankfurt) in October 1967, situated immediately adjacent to the Senckenberg Museum itself. The students of this university played a major role in the German student movement that culminated in the violent protests across Germany during the Easter of 1968 ("68er-Bewegung" ("68 Movement")). Although the movement was "suppressed" and calm followed, it indeed heralded the beginning of great changes in German society, with the increasing recognition of the rights of women, new ideas on education and child-raising, and a blossoming of art and cultural activities. Michael was definitely not a "student activist", and in fact their activities annoyed him because they interfered with his studies, but this period did herald an exciting new future for Germany, and Michael certainly enjoyed the changes and new possibilities that were to come.

He undertook studies majoring in biology and chemistry with the initial aim of becoming a teacher, and he received his Pre-Diploma (equivalent to a Bachelor degree) in April 1970. His Main Diploma (equivalent to a Master of Science degree) followed three years later on 7 November 1973. He had majored in zoology, accompanied by some botany and organic chemistry, but as a portent of his future, his thesis topic was on decapod crustaceans from various R.V. "Meteor" expeditions in the eastern Atlantic.

He finally completed his Doctorate, also through the Johann Wolfgang Goethe University, around 10 years later, on 27 June 1983, but during this period he had begun full-time work at the Museum to which he would dedicate his life. His thesis was entitled "Morphology and taxonomy of the Gecarcinidae. A contribution towards the comparative morphology of Brachyura (Crustacea: Decapoda)", and the subsequent publications were a major step-forward in our understanding of the taxonomy of this iconic group of land crabs.

Behind every great man or woman there is usually a great life-partner, and Michael's wife Heide was of incalculable support to him throughout his career. They first met as fellow university students, both enrolling in biology for the winter semester of 1967/68. Heide was particularly interested in fish at that time, and this group was the subject of her Diploma Thesis. Kindred spirits, they married in 1978, and though Heide was pleased to play the supporting family role in the relationship, she has nevertheless kept her interest in biology throughout her life. Heide also developed a great personal knowledge on the diversity of plants, and Michael always conceded her "the last word" when it came to the identification of the plants they found in their travels. Two sons were born to the happy couple, Stephan in 1984 and Christoph in 1986.

Michael and Heide brought up their family in a lovely old home in Dreieich, a quiet town just south of Frankfurt, which was a wonderful sanctuary from the bustle of city life. Many colleagues were invited home for an excellent meal, and convivial company stretching late into the night discussing everything from philosophy, history, world politics, food, and the intricacies of regional differences in German wine-making.

Although they were thoroughly devoted to each other, Heide must have had periods of great forbearance trying to manage her young family during Michael's many long absences travelling and collecting, and his long hours working late at the

Michael and Heide Türkay in 2008.

office. But she loved and cared for him dearly, and always gave him the support to allow him to be who he was. Michael naturally hoped his children ("mein Schatz" called loudly when they were in trouble), would follow in his scientific footsteps. He initially found it difficult when they chose their own divergent paths in life, but as they began to succeed in their own right he could not have been a prouder father.

THE SENCKENBERGISCHE NATURFORSCHENDE GESELLSCHAFT AND ITS MUSEUM

Johann Wolfgang von Goethe is credited with the initiative to form the Senckenbergische Naturforschende Gesellschaft (SNG) (Senckenberg Nature Research Society), more recently known as the Senckenberg Gesellschaft für Naturforschung. The purpose of this society was to conduct research in the natural sciences and popularize the results for the citizens of Frankfurt. The society was first enabled through an endowment from the physician, naturalist, botanist and philanthropist Johann Christian Senckenberg (1707-1772), after whom it was named. It was originally formed by a group of 17 wealthy merchants who, together, had the money and influence to both found and support institutions for science and culture (Sakurai, 2013). The most important and enduring of these was the Senckenberg Museum, with which the society is still inextricably entwined. The Museum was opened in 1821 "to host collections already existing in Frankfurt and to be brought together in the future", collecting and preparing museum specimens as a "common treasure" (instead of simply purchasing them) (Tuerkay & Scholz, 2014). Indeed it was the vast collections donated by the great Frankfurt-born naturalist and explorer Eduard Rüppell, made during his major zoological expeditions across Africa, that established the Senckenberg Museum as an equal amongst the great European museums in London, Paris, Berlin and Vienna.

Michael Türkay played an important role in the SNG, being a member of its board of directors from 2007 to 2013. But it was to the Society's great museum, and its wonderful historical traditions, to which Michael devoted his entire working life. The Society valued Michael to such an extent that, as one of its "grand" members, it sponsored a "Festkolloquium for Professor Michael Türkay on the occasion of his 60th birthday", with the theme "Marine Biodiversity Research". More than a hundred guests from Germany and abroad accepted an invitation to honour their friend and colleague. In his introductory address the President of the Society, Wolfgang Strutz, paid tribute to Michael's active involvement in the organization and management of the Senckenberg institute, and to his commitment to the Senckenberg School ("Senckenberg-Schule"), the only school in Germany dedicated to training technicians for natural history museums and research institutes.

Talks were given by some of Michael's long-time friends and luminaries in the study of marine biology. These included Peter Davie (Queensland Museum,

Brisbane, QLD, Australia) on the history and future of carcinology; Dirk Brandis (Zoological Museum, University of Kiel, Kiel, Germany), a former PhD and postdoctoral student of Michael's; Hjalmar Thiel (Hamburg, Germany), one of the pioneers of German deep-sea biodiversity research, and recipient of the prestigious Cretzschmar medal for "Senckenberg and biological deep-sea research"; Karsten Reise, Director of the Wattenmeerstation, Sylt (Wadden Sea Station) (Alfred Wegener Institute), who collaborated with Michael for many years studying North Sea fauna and ecology; and finally, Klaus Hausmann (Freie Universität, Berlin), a protozoan specialist but also a frequent shipmate of Michael's on expeditions aboard the R.V. "Meteor".

This SNG Festschrift honoured Michael as an exceptional scientist, a passionate academic teacher, and a "Vollblut-Senckenberger" ("Thoroughbred Senckenbergian"), who, over his decades of service, had significantly shaped the history of the Museum. Dr. Strutz described Michael as "ein unendlich sympathischer Mensch" (a thoroughly likeable person), to the applause of all present.

At the end of this wonderful event, the SNG presented Michael with its official birthday gift, a detailed model of the research vessel "Senckenberg".

Michael is presented with a model of the R.V. "Senckenberg" by the President of SNG, Wolfgang Strutz (left), and the Director of Senckenberg, Volker Mossbrugger (centre), as a gift on the occasion of the Festschrift celebrating his 60th birthday.

MICHAEL'S CAREER AT THE SENCKENBERG

A week after Michael's death, the "Frankfurt Neue Presse" (in the online version of that newspaper) published a short obituary announcing his passing (Janovic, 2015). Heading the article was the exclamation "Senckenberg ohne Michael Türkay — das ist eigentlich undenkbar" (Senckenberg without Michael Türkay — that is virtually unthinkable). So long, and so prominent, had Michael's association with Senckenberg been, that the identities of the man and the institution had become inextricably linked!

Michael had volunteered at the Museum ever since 1964, initially coming in after school whenever he could, and then at every opportunity through all his years at the university. In fact, by the time of Richard Bott's death in January 1974, he already had firmly established his credibility as a crustacean expert with 16 published scientific papers to his credit. It was thus a natural progression that, with Dr. Bott gone, the then 26-year-old Michael should be formally employed to work in the Crustacea Section. Initially this was only as a "first grade assistant", with a very low salary. He was finally elevated to full curator status, and head of the Crustacean Section in November 1976. His talents continued to be recognized, and in 1989 he became head of the Department for Invertebrate Zoology (then named "Zoologie II"), and in 2000 the head of that reorganized department, then newly named as the "Department for Marine Zoology".

Always ready and willing to take on a heavy load of the Senckenberg research institute's administrative needs, his competence in this role earned him much respect from his colleagues. Thus, from July 1995 to January 2002, he was appointed as Deputy Director of Science, and then Director of Science from January 2002 to June 2007. Following this he was made Personnel Manager from 2007-2010, and also from 1 July 2007 to 1 July 2013 the Deputy Director General, as well as a full member of the board of directors of the Senckenberg Research Society.

Michael received his "Habilitation" from the Goethe University in 2001 (the highest academic qualification issued by a German university, and acknowledging eligibility as a Professor), leading to his appointment as Senckenberg Lead Researcher in 2002. He was made a board member of the Senckenberg directorate from 2007 to 2013, and an extracurricular Professor in 2008. Amongst all his other duties he also found time to be the Headmaster of the Senckenberg School from 1996 to 2014!

He formally retired on 30 June 2013, but kept his position in the Crustacean Section. His dream was to get back to working on his crabs once his formal responsibilities had been lifted, but this dream was tragically cut short.

While he found huge satisfaction in immersing himself in his own research, he strongly and selflessly felt that, in the end, he would be able to achieve far more for the Museum, and for biological science as a whole, by being able to have a strong influence on institutional decision-making — his wonderful legacy is proof of this!

Collection, databasing and web-based resources

Michael was always a great supporter of implementing computer technology to give ready access to the vast amounts of data associated with museum collections. Databasing the collections was the single-most important thing museums could do to keep themselves relevant into the 21st century. In this regard, Michael championed the in-house development of "SeSam — Senckenberg Collection Management Database", at some significant cost. Its central database design provided a joint data-pool for systematics, geographical data and literature. SeSam has now been in use for many years to manage the Senckenberg's zoological, botanical and fossil collection material, and its internet interactivity has also allowed web-based interrogation of their collection data by scientists and the interested public worldwide. SeSam has been proven a great success, it has also been used by a number of other museums, including the Museum für Naturkunde Berlin and the Zoological Museum of the Christian-Albrechts University Kiel.

Michael was also a significant contributor of decapod crustacean information to the European Register of Marine Species (ERMS), a major database initiative founded in 1998 by a grant from the European Unions's Marine Science and Technology Programme. With the idea of extending this to worldwide coverage, in 2004 Michael along with Peter Davie, Danièle Guinot and Peter Ng applied to GBIF for funding for a long-imagined project called "DECACAT — Electronic catalogue of the names of decapod crustaceans". It was intended to assemble an international working group of experts to compile the first complete list of the modern decapod fauna, and was to include references to original descriptions and broad data on geographic occurrence. The Senckenberg was to be the organizing hub for the project. Unfortunately, the applications for both that year and the next were declined, but at least the now well-known WoRMS (World Register of Marine Species) initiative, that grew out of ERMS, is finally bringing this dream to fruition, and the large amount of data that Michael had already entered is now available on this new platform.

Display

Michael always recognized the importance of good displays for promoting popular goodwill towards museums, as well as being important tools for the education of students and the general public. Crustaceans have thus been well represented in the museum cases, and Michael was heavily involved in an interesting exhibition on hydrothermal vents — an area of special personal interest. His involvement in deep sea research also saw his active support during 2008-2009, of the major display "Deep sea — discover, explore, experience". This comprised 40 original specimens and 35 models of deep-sea animals, and was built in partnership with the Museum of Natural History in Basel. A central element of the exhibition was the encounter of a giant squid with a seven metre long sperm whale especially modelled at the Senckenberg. The exhibition was a major success, attracting nearly 200 000 visitors to the Museum — not least due to Michael's ability to fascinate and attract the public with his many media interviews, public talks, and guided tours. He passed on all his knowledge and enthusiasm not just to the people in Frankfurt, but also in Basel, Dresden and Berlin, where the display was also to travel.

TÜRKAY AS A TEACHER

Michael was always passionate about passing on his knowledge, and in exciting future generations about the natural world. A naturally inquisitive and very intelligent man, Michael had an exceptionally broad biological knowledge, from taxonomy and systematics through to functional morphology, and evolutionary biology to ecology and zoogeography. He was like an encyclopaedia — the type of traditional naturalist that would fit Goethe's concept of a "Universalgelehrter" (polymath). But he also had the uncommon gift of being able to explain complicated facts and ideas in a clear, understandable way. His students often had the impression that there was no question he could not somehow answer, and many were excited and inspired to continue their own journeys of learning.

Recognizing that museums need more than just research scientists to function properly, Michael became a driving force behind the unique Senckenberg School. Inaugurated in 1973, its mission is to train natural history museum technicians. Michael not only gave the zoology lectures, but also further developed the curriculum to better teach subjects in botany, palaeontology, taxidermy, conservation techniques, databasing, and many more. For him this was not just a job, but a vocation. He subsequently became the headmaster of the Senckenberg School from 1996 to 2014. The current Collection Manager of the Senckenberg Research Institute, Andreas Allspach, was actually one of his first students at this school, and many of his other graduates became highly sought after by natural history museums across Germany.

One of his major achievements was to promote co-teaching between Goethe University and Senckenberg so as to enable students to gain insights into the important role of museums in biodiversity research, especially through taxonomic work on museum collections. Michael regularly taught courses on arthropods and molluscs to undergraduate students at the University, but also conducted seminars and courses on aspects of invertebrate comparative zoology, especially the morphology and systematics of crustaceans and molluscs. He would sometimes joke disparagingly about taxonomic approaches that simply counted differences and similarities ("Borstenzähler" ("counters of setae")) — when Michael looked at a morphological structure he always strove to understand its function, its evolutionary significance, and its ecological adaptive value. Michael's approach to biological education emphasized a broad integrative approach that stressed not only the taxonomy, anatomy, physiology and ecology of the animals themselves, but also their cultural significance. For some species, this meant their importance as food and how to effectively manage their fishery, but for others, their use in medicine, their potential to yield bioactive compounds, their industrial and commercial uses, their symbolism to native peoples, and their place in art were all highlighted.

It was always a great concern to Michael that the values of a traditional zoological education were slowly being eroded and degraded by new genetic approaches that appear to offer quick and easy answers, whilst ignoring whole organisms and their ecological role. He hated the idea that biology students could become so focussed on laboratory and computer processes that they would no longer understand Darwin's inspiring words: "There is grandeur in this view of life…from so simple a beginning endless forms most beautiful and most wonderful have been, and are being, evolved."

Perhaps closest to Michael's heart were the courses he taught on marine biology, a field not typically embraced by a University located far from the coast. He would often, with a wink and a twinkle in his eye, refer to his home city as Frankfurt am Meer (Frankfurt on the Sea) instead of Frankfurt am Main (Frankfurt on the river Main). He would thrill his students with his knowledge of topics as diverse as hydrothermal vent biology, giant deep sea squids (*Architeuthis*), or the adaptations and impacts of wood-boring bivalves (*Teredo*).

But Michael's pedagogy would not be confined to the lecture hall, and he strongly believed in the value of students interacting with the living world.

In 1972, the 24-year-old Michael attended a conference at the Ruđer Bošković Institute in Rovinj, Croatia (then Yugoslavia). This led the young man to establish close contacts with his Rovinj colleagues and, in particular, he formed a friendship with Prof. Zdravko Števčić that was to last a lifetime. Professors Kurt Fiedler and Christian Winter led student excursions every two years to Rovinj, and Michael took the opportunity to accompany them as a tutor. On the retirement of Kurt Fiedler, Michael stepped-up into co-organizer and leadership roles for this excursion, and in the process also formed a close friendship with Christian Winter. This excursion soon became a student's favourite and many participants would long remember those two weeks as a highlight of their time at university. The schedule included many field trips, by boat or by bus, to different marine habitats around the Northern Adriatic Sea, including rocky shores, seagrass beds, estuaries and marine caves. The days often started with students donning snorkelling gear before swarming out to collect specimens to bring back to the lab. Specimen identification sessions sometimes continued late into the night, but Michael's professional guidance and his captivating personality always ensured that students enjoyed themselves despite the long hours and hard work. While the focus was on marine life, Michael's holistic approach to the environment meant that excursions for bird watching and terrestrial field trips were also included. He felt it was important for the students to also gain at least some understanding of the plant and insect communities nurtured by the Mediterranean. Moreover, he was a font of knowledge about the cultural, social and political heritage of a location. Carola Becker has a special memory of the regular visit to a small chapel with a beautiful mosaic displaying marine animals. The chapel lies high on a hill, and as always, Michael would stride out alone leading a pack of students struggling to keep up with him under the hot midday sun. He would always reach the top of the hill first, but all this effort was not a show of superiority — instead he would turn around to welcome his exhausted and sweating group, holding a gigantic watermelon that he had secretly carried all the way up to offer as refreshment.

Starting in 1986, the student excursion to Rovinj, in the Adriatic, was annually alternated with a similar Marine Biological Student Excursion to the North Sea. Although mainly based at Senckenberg am Meer in Wilhelmshaven, the island of Wangerooge and the Biologische Anstalt Helgoland were also visited. A highlight of the trip was a one day excursion on the R.V. "Senckenberg" to the Jade Bight, with the day's catches sorted, identified, counted and measured. Again, Michael followed a very integrative approach to teaching students about the Wadden Sea and its surrounding areas. He included such topics as dyke protection and land use in his lectures, and somehow managed to take his students on an educational visit to the famous Jever brewery! These trips always ended with a barbecue on the beautiful island of Helgoland. Michael's ability to still identify a species of fish when filleted and wrapped in aluminium foil became legendary among students. Michael's cruises continued right up until the year of his death in 2015, and to each new crop of young students he showed the same happy and excited enthusiasm he always felt about the miracle of marine life.

The R.V. "Senckenberg" was launched in 1977, and from that time Michael used it annually for a cruise to the North Sea. Although many localities were visited, they were able to gather an extremely valuable long-term data-set of the Dogger Bank epibenthos extending over 21 years. The analysis of these data became the subject of a doctoral thesis by Michael's student, Moritz Sonnewald, and it has

Michael Türkay with students and colleagues recovering specimens from deep-sea bottom trawls on board R.V. "Meteor" during the expedition "Diversity of the Atlantic Benthos" in 2009 (from left to right: Carola Becker, Michael Türkay, Matthias Schneider, Karin Meißner).

revealed changes in faunal distribution that appear to reflect the impacts of climate change in the North Sea (Sonnewald, 2012; Sonnewald & Türkay, 2012).

In total, Michael supervised 27 diploma, bachelor and master theses, and 10 doctoral theses. Many successful European scientific careers were born under Michael's inspiration and supervision, including Michael Apel, who went on to become director of the Museum Mensch und Natur in Munich; Dirk Brandis, first a Curator at the Zoological Museum of the University of Kiel and now its Director, Moritz Sonnewald, now a Biologist at Senckenberg itself; Mathias Gutmann, now a Professor at the Karlsruhe Institute of Technology; and Carola Becker, currently a research scientist at the Queen's University Marine Laboratory, Portaferry, U.K.

Michael was highly regarded and admired by his students for his enthusiasm and sophistication. Students would secretly refer to him as a walking encyclopaedia, and had the deepest respect for him. Despite his broad and inexhaustible knowledge, Professor Dr. Türkay always remained an approachable and very human kind of teacher. He encouraged curiosity, welcomed any kind of question, and never belittled anyone for their lack of knowledge.

THE BON VIVANT

Michael lived his life to the full, and loved good food, good beer, good wine and good fellowship! In his younger days at least, he had a number of happy expeditions to Greece working with his long-time friends Anastassios Eleftheriou (Institute for Marine Biology, Heraklion, Crete) and Athanasios Koukouras (Aristotle University of Thessaloniki), who introduced Michael to the full delights of "Taverna-ology".

Michael also loved to cook, and took great pleasure in the study and practice of "applied carcinology", much to the pleasure of many colleagues. The courses in comparative zoology that he taught at the Goethe University always contained an obligatory "mussel festival", and the students on his annual trawling expeditions to the North Sea were also never left hungry after a successful catch. In fact his cooking skills were so good that in 1994 he was to partner with the world-renowned Austrian chef Eckart Witzigmann, and the equally famous pastry chef and food photographer Christian Teubner, to create a beautifully illustrated and mouth-watering German language cook-book entitled "Shrimp, lobster and crawfish. Food and kitchen practices of crustaceans. With the best recipes of Eckart Witzigmann" (Teubner & Türkay, 1994). This was followed in 1998 by an equally sumptuous molluscan sequel, "Shells and oysters" (Türkay et al., 1998).

Always proud of the local Frankfurt specialties, I have a vivid memory of Michael taking me and a visiting contingent of elderly Chinese carcinologists, including Chen Hui-Lan, to a restaurant in the "old town" that was renowned for its "Schweinshaxe" (pig knuckle), an enormous piece of meat that fills the plate, and is liberally surrounded with potato and Sauerkraut. I can still see the look of almost frightened awe on the faces of the gathered Chinese confronted with this gigantic meal. He was right though, it was delicious!

INTERNATIONAL OUTREACH

The Senckenberg crustacean collection being both wide in the scope of its modern collections, as well as of having great historical importance, attracted large numbers of taxonomic researchers from all around the world. But, the use of the collection was vastly enhanced by Michael's active programme of researcher invitations and financial support. Thus collaborations and friendships were forged with colleagues from as far afield as Russia, China, Japan, Iran and the Arabian region, Brazil, Singapore, Australia, and many other countries. Throughout his career, Michael was able to raise €2 330 000 of competitive external funds, mainly received from DFG (German Research Foundation) and BMBF (Federal Ministry of Education and Research, Bonn/Berlin, Germany). A good deal of this would no doubt have been used to support his collaborative research with the international community.

Michael thus became a conduit through whom many scientists could connect, even if modern politics made such associations otherwise difficult — neither political correctness nor religious beliefs were allowed to encroach on, or impede, the pursuit of scientific discovery. In Michael's laboratory all nationalities were one, all were friends at the bench, and all of us were enriched by the experience!

Michael loved, and had a great facility for, languages. He refused to let language barriers separate him from uniting with peoples in the love of biology and science. Besides his native German tongue, he was fluent in English, French, Italian, Armenian and Turkish; he also had a good grasp of Spanish, Portuguese, Greek and Serbo-Croatian. Wherever he went, Michael always tried to learn what he could, even to the extent of attempting to read Russian, Thai, Chinese and Japanese script.

South America and the New World

Michael Türkay's first three scientific publications dealt with marine brachyuran crabs (including seven new species and a subspecies) from the western coasts of Central and South America and from Venezuela in the north. So it was somewhat inevitable that he would eventually form a close relationship with South American colleagues (Magalhães, 2016). In the late 1970s he worked with Gilberto Rodríguez, a Venezuelan carcinologist who was then preparing his monograph on freshwater pseudothelphusid crabs. However, most papers on Neotropical freshwater crabs that were to follow were written with the Brazilian scientist Célio Magalhães, whom Michael first met in December 1984, when Célio visited Europe for the first time to examine collections of Amazonian freshwater decapods. Michael guided the young man towards the crabs that would later become the subject of his doctoral thesis, and with Michael's friendship and support, Célio was to return to Frankfurt many times. The two men collaborated on 11 papers over the next 30 years.

Russia

Michael formed a close alliance with Russian scientists, particularly following the fall of the Berlin wall. The Soviet regime had long had an active deep-sea research programme and Michael saw opportunities for synergies with his own research. He began to actively pursue cooperative research projects with prominent Russian carcinologists such as Rudolf N. Burukovsky and Nikolai A. Zarenkov, and then with their successors such as Vassily Spiridonov, Dmitry G. Zhadan and Ivan Marin. Spiridonov in particular made numerous visits to work in Michael's lab, and they became close friends while coauthoring several works together.

Michael also visited Russia on several occasions, and took part in field studies in the White Sea to study the shrimp *Crangon allmanni* Kinahan, 1860.

Japan

Michael had a special kinship with Dr. Katsushi Sakai and arranged for him to have extended stays at Senckenberg under a Humboldt-fellowship and other travel grants. Dr. Sakai became a fluent German speaker, and co-authored a total of 19 papers with Michael. These strong personal links finally led to Katsushi donating to the Senckenberg the large personal collection of Japanese crabs that had belonged to his father, the legendary Japanese carcinologist Tune Sakai. It also led to Michael, along with Peter Davie and Danièle Guinot, being invited to help Dr. Sakai revise Tune Sakai's "Crabs of Japan". This was published in 2004 in a CD-Rom format through Dr. Sakai's involvement with the Expertise Centre for Taxonomic Identification (ETI) based in the Netherlands (Sakai et al., 2004). Many will not know, but Michael was even bestowed a Japanese name.

鶴飼美蛙

Michael's Japanese name. The first two characters represent his first name and mean "beautiful frog", while the last two characters are for his family name and mean "crane raiser" (Kawai, 2015).

China

Michael had a long relationship with Chinese scientists and Senckenberg researchers played a significant role in the biological surveys of Hainan Island in 1990 and 1992. Subsequently a number of Chinese scientists were hosted in Frankfurt to study and describe the fauna. In particular, Chen Hui-Lian (Institute of Oceanology, Academia Sinica, Qingdao) visited several times, and they wrote three papers together, as did Yang Si-Liang from the Beijing Natural History Museum. Michael formed a close relationship with the late Prof. Liu Rui-yu, a senior member of the Chinese Academy of Sciences, a founder of the Chinese Crustacean Society, and a central figure in the development of their Institute of Oceanology. It was fitting that at a small special dinner hosted by Prof. Liu during the Seventh International Crustacean Congress in Qingdao in 2010, Michael was honoured to be seated to the right of the great man and feted by him as an illustrious guest.

The Middle East

In more recent years Michael took a closer interest in the Middle East, and forged ties with Iranian scientists. In particular, he procured financial support from

Michael Türkay seated with Prof. Liu Rui-yu at a meal during the Seventh International Crustacean Congress in Qingdao in 2010.

the DAAD (German Academic Exchange Service) to support his last doctoral student, Reza Naderloo from the University of Tehran (= Teheran), who is now forging a successful crustacean career.

Australia

I first saw Michael when he stood up to present a talk at the International Crustacean Conference held in Sydney in 1980. He took the podium to present an interesting paper on the systematic relationships of the endemic Australian ocypodoid crab, *Heloecius cordiformis* (H. Milne Edwards, 1837), but the very first thing he did was to let loose a wind-up, bright yellow, toy plastic crab, which scuttled across the stage with claws waving and eyes bobbing up-and-down — from that moment I knew we would be friends! In the week following the conference Michael visited the Queensland Museum, and he was delighted to accompany me to visit with my old mentor, Prof. W. "Bill" Stephenson, of the University of Queensland, who had done so much pioneering work on revising the taxonomy of the swimming crab family Portunidae. This was the first of several happy visits to Queensland, especially because his brother Armand and his family

Michael really enjoyed spending time in Australia, especially because the "Aussie" and German attitude to beer drinking is rather similar. This cartoon featuring a popular Queensland beer being held up by a "mud crab", *Scylla serrata* (Forskål, 1775), particularly appealed to him, combining, as it does, his favourite items of consumption.

had also made Brisbane their home. He always collected along the coast, travelling as far north as Cooktown. On one occasion he showed me his special technique for digging out ghost crabs (*Ocypode*) that I still regularly use. Michael last visited in 2006, as part of a German delegation hoping to initiate a formal "German-Australian cooperation on Biodiversity".

Michael also had a strong friendship with Diana ("Di") Jones from the Western Australian Museum, whom he sponsored for stays at the Senckenberg to progress studies on the taxonomy of the world's fiddler crabs (*Uca*), a project he had earlier begun with Chen Hui-Lian and Yang Si-Liang from China. This study finally resulted in the multi-authored generic revision of *Uca* (Shih et al., 2016) published the year after his death, and which owed much to Michael's accrued knowledge of this charismatic group of crabs.

CONFERENCES AT THE SENCKENBERG INSTITUTE

Michael organized and staged three important major crustacean meetings at the Senckenberg in Frankfurt, the first being the highly successful International Senckenberg-Symposium on Crustacea Decapoda, held from 18-22 October 1993. Due to his long association with freshwater crabs, he then hosted the 21st International Senckenberg Conference on Biology of Freshwater Decapods, Frankfurt am Main, 9 December 2010, which was again well attended. And finally in August 2014, just over one year before his death, he convened the largest scientific meeting for crustaceans, the 8th International Crustacean Congress. He also hosted the German National Crustacean Conference twice, once in 1985 and again in 2007.

THE RESEARCHER

Michael was only 19 and volunteering at the Senckenberg, when in 1967, with the encouragement and guidance of Dr. Bott, he published his first scientific paper — an account of the brachyurans from South America's west coast, that included the description of three new species (Türkay, 1967).

Although he published on a wide variety of topics and taxa, crabs remained his main preoccupation. Through Michael eyes, crabs were fascinating and beautiful creatures ("Krebse sind faszinierend und wunderschön"). During his studies of land crabs, Michael maintained specimens of gecarcinid species in tanks in his office. He would check on his crabs every day and was soon able to imitate their way of communicating via substrate vibrations by knocking rhythmically on the floor. If the female crabs responded, he would reward them with a male so that he could observe their courtship behaviour.

His research interests expanded year-by-year. His Curriculum Vitae lists: "Taxonomy, systematics, morphology, phylogeny, and zoogeography of decapod crustaceans. Main geographic areas: North Sea, Red Sea and Gulf of Aden (deep sea), and eastern Asian seas (China, Japan). Main systematic interest in brachyuran crabs with special emphasis on terrestrial and semiterrestrial taxa of the tropics and deep sea. Interest in marine ecology, especially of the North Sea, Red Sea and the deep sea of the world oceans."

One family of crabs that particularly fascinated Michael was the Pinnotheridae (pea crabs). He was intrigued that there is still so little known about these crabs, but the fact that these tiny, highly modified, symbiotic animals are found in bivalves, meant that he could sometimes conduct his research during dinner! Over his career he built up a vast collection complete with careful records on the hosts, and Carola Becker was delighted to be invited by Michael to undertake her Diploma Thesis in 2003, and subsequently her PhD project on the taxonomy, host-ecology and reproductive biology of pinnotherids. Together, they revised and re-described the European species (Becker & Türkay, 2010) and studied the peculiar adaptations of their reproductive systems (e.g., Becker et al., 2012). The last part of their pea crab study is published in this issue (Becker & Türkay, 2017).

Michael also always looked carefully for new morphological characters that might carry a strong phylogenetic signal, and was a strong advocate for the use of gastric mill structure. He and his co-workers, especially Katsushi Sakai and Yang Si-Liang, proved in their revision of the *Helice*/*Chasmagnathus* complex of shore crabs, that gastric mill morphological characters could be very useful at the generic level (Sakai et al., 2006).

Always at the centre of Michael's world-view were the organisms themselves. He firmly believed that any understanding of evolutionary processes, relationships,

or ecosystems, must begin with the study of the organisms and their functions. In this regard, Wolfgang F. Gutmann, a biologist who was contemporary with Michael at the Senkenberg, was to have a strong impact on Michael's scientific approach. During the 1970s and 1980s, Gutmann, with the support of a number of other Senckenberg colleagues, developed new, somewhat revolutionary ideas of evolution based on constructional morphology. This approach became known as the "Frankfurt Theory", "Organism-centred Theory", or more recently "engineering morphology". Gutmann's ideas were considered controversial, and perceived by many as anti-Darwinist. Türkay, however, considered Gutmann's concepts to be a useful bio-engineering approach for understanding an animal's morphology and body plan ("Körperkonstruktion") as a whole functioning organism (Gutmann & Türkay, 2001). Michael was a strong believer of Darwin's evolutionary theories, but was critical that an organism was not always seen as an integral coherent biomechanical entity — rather as a "hatrack" upon which characters could be exchangeably hung ("Kleiderständer an dem Merkmale austauschbar dranhängen"). By incorporating the laws of biomechanics, Michael regarded the Frankfurt Theory as a useful supplementary approach to interpreting evolutionary relationships — in much the same way as new evidence from other fields of study, such as genetics, are now being incorporated in our modern evolutionary understanding. He was to use "constructional morphology" in his own doctoral thesis, and ensured that the concepts were understood by the students that he supervised.

FIELD EXPEDITIONS AND RESEARCH CRUISES

Michael loved getting out of the office and getting his feet wet, and took every opportunity. He was interested in virtually all habitats, and wherever he went he would collect. He caught crabs from along the Caribbean coast of Colombia in 1978; from the coasts of Shikoku Island and Honshu Island, Japan, in 1979; from Hainan Island, China in 1991 and 1992; from the coast of Queensland, Australia in 1980, 1995 and 1999; and from Martinique in 2004.

But Michael's greatest love was his trips to sea. He found his "sea-legs" in 1975 when he boarded the R.V. "Meteor" for what would be a series of ten major cruises. This initial sea-going expedition reputedly saw him wreck his first Agassiz Trawl near Casablanca, but more happily resulted in a research paper covering Portuguese and Moroccan decapods. As mentioned earlier, every year from 1977, Michael would leave Wilhelmshaven aboard the R.V. "Senckenberg" to explore various localities in the North Sea.

His real passion, however, was for deep-sea research. This interest was largely fostered by his friendship with Prof. Hjalmar Thiel from Hamburg, and excited by the great German deep-sea "Valdivia" expeditions of 1898-1899 (Thiel & Türkay, 2002). Together, Michael and Hjalmar initiated a new era of German deep-water exploration. The work of Michael and his colleagues on abyssal bottom biotopes and communities, using deep-water trawls accompanied by photography and video, represented the first serious return to research on abyssal fauna since the Danish R.V. "Galathea" and Soviet R.V. "Vityaz" cruises of the 1950s and 1960s. Cruises to the Red Sea and the Gulf of Aden (R.V. "Sonne", 1977; R.V. "Valdivia", 1981; R.V. "Meteor", 1987) retraced the Austro-Hungarian "Pola" expedition of 1895-1898, uncovering unique deep-sea fauna, as well as novel information on deep-sea ecology and physical conditions. Michael was the organizer or co-organizer of five Atlantic expeditions using the R.V. "Meteor" (to the continental slope of Africa and adjacent seamounts in 1975, the abyssal basins off Angola in 2000, the mid-Atlantic ridge in 2004 and 2005, and the western Atlantic in 2009). The Mediterranean Sea also was the subject of six expeditions, including four deep-sea "Meteor" cruises (in 1978, 1993, 2006 and 2007).

Finally, Michael was fascinated by deep-water hydrothermal vent fauna and its ecology, and participated in cruises aboard the German R.V. "Sonne" to investigate these vents in numerous locations including: the North Fiji Basin in 1995; the Bismarck Archipelago, Papua New Guinea in 1998; the Pacific Antarctic ridge in the eastern Pacific in 2001; the Lau Basin, Tonga Arc, Louiseville Ridge in 2002; and in 2004, aboard R.V. "Meteor", he visited the Logachev hydrothermal fields on the mid-Atlantic Ridge.

PUBLICATIONS AND NEW TAXA

Michael Türkay published over 310 scientific, technical and popular science papers as author or co-author (see complete list in Sonnewald & Apel, 2016a, b), of which around half were reviewed scientific works in international journals. They covered a wide variety of subjects, but papers on crustacean taxonomy predominate. Decapods were his favourite research group, with crabs being his ultimate specialty. Altogether he described 71 species and 26 genera of decapod crustaceans, across a wide range of families (see list in Sonnewald & Apel, 2015). He also recognized two new subfamilies though neither are currently considered valid. Heloeciinae Türkay, 1983 was erected for an endemic genus and species of east Australian intertidal ocypodoid crab, but unfortunately the earlier name Heloeciacaea of H. Milne Edwards, 1852, has precedence; and Heliceinae Sakai, Türkay & Yang, 2006, is now regarded as a subjective junior synonym of Cyclograpsinae H. Milne Edwards, 1853 (see Ng et al., 2008).

Twenty papers were exclusively dedicated to New World decapods, with most of these concerned with South American freshwater crustaceans. Altogether, he described eight new genera, 23 new species, and one new subspecies of American and Antillean decapods belonging to 17 families (Magalhães, 2016).

In the original spirit of the SNG, a museum scientist should "popularize the results for the citizens of Frankfurt", and Michael embraced this role with enthusiasm. The Museum indeed has its own popular publication "Natur und Museum" (currently "Senckenberg Natur Forschung Museum"), and he was a strong supporter, contributing around 43 articles on a wide variety of subjects and museum activities.

"DO NOT GO GENTLE INTO THAT GOOD NIGHT!"

For so many of us Michael had such a strong life-force that it seemed inconceivable that he would be snatched away at such a young age. True he had had some health issues over the years, but no one would have guessed that being admitted to hospital for a painful back would have heralded a cancer prognosis. Despite this terrible news, Michael sailed on with his life, submitting to the awful chemotherapy treatments, and letting very few of his colleagues, or even close friends, know that there was anything seriously wrong. Despite his gregarious outgoing nature, in the end he was a very private man. Knowing the sea as well as he did, perhaps he was confident that he would be able to successfully navigate between his personal Scylla and Charybdis? Always one to look optimistically to the future, he was still on the e-mail with his colleagues, and making new plans for future projects, right up until the last few days of his life.

AN EPITAPH

The great English essayist and poet Joseph Addison, wrote in 1710: "*It is indeed wonderful to consider, that there should be a sort of learned men who are wholly employed in gathering together the refuse of nature, if I may call it so, and hoarding up in their chests and cabinets such creatures as others industriously avoid the sight of.*" Michael Türkay was such a traditional Museum man, but he was also so much more than this. Michael was a man of great intellect, and a man that epitomized the values that keep the museums of the world as relevant today as they were back in the great period of natural history exploration of the 18th and 19th century. Michael thrilled at finding new creatures, however humble, but he equally buzzed with the excitement of new ideas and theories to explore. He treasured the historical traditions of museology, but merged them seamlessly with computerization and new technologies that would help uncover new truths and popularize the natural world — especially to modern generations that have

Prof. Dr. Michael Türkay at sea, holding a *Cancer pagurus* Linnaeus, 1758, on a cruise aboard R.V. "Senckenberg" on the North Sea, July 2007. This photograph is well-known to many of us, and has appeared in print before, but it so well encapsulates Michael's work and personality that we reproduce it again. Michael himself was also very fond of it, and personally sent it to many of his colleagues and friends.

become more and more disconnected with nature, numbed by the exploitation of the environment, and largely look to the future with apathy. Michael was, and remains, an antidote of enthusiasm and hope, and the world is a poorer place without him.

ACKNOWLEDGEMENTS

This account of Michael's life is based not only on personal knowledge, but has also relied on the obituaries and tributes that have already been published by

Michael's friends since his death. In particular those of Vassily Spiridonov (2015), Dirk Brandis, Dieter Fiege and Ingrid Kröncke (2015), Tadashi Kawai (2015), Moritz Sonnewald and Michael Apel (2016) and Célio Magalhães (2016). The famous line "Do not go gentle into that good night." was written in a poem by the great Welsh poet Dylan Thomas. Michael's brother, Armand Türkay (now von Stein), shared some lovely memories of their early family life. And finally we are very grateful to Michael's wife Heide, and their two sons Stephan and Christoph, for checking the text and providing some personal information that has helped fill in the story of his life.

TAXA NAMED IN HONOUR OF MICHAEL TÜRKAY
(including those in the present issue)

Actumnus tsurukaii Takeda & Komatsu, 2017 (Brachyura: Pilumnidae) [this volume; name is spelled as *tsurukaii* after the Japanese translation of the family name Türkay: tsuru (= crane) and kai (= rearing)]
Ambilimbus tuerkayi Martinez Arbizu, 1999 (Copepoda: Erebonasteridae)
Amphicrossus tuerkayi Martinez Arbizu, 1999 (Copepoda: Erebonasteridae)
Ankerius grusocurare Naruse & Uyeno, 2017 (Brachyura: Aphanodactylidae) [this volume; name is the Roman spelling of Michael's Japanese name]
Atyaephyra tuerkayi Cristodoulou, Antoniou, Magoulas & Koukouras, 2012 (Caridea: Atyidae)
Bathymodiolus tangaroana tuerkayi Von Cosel & Jannsen, 2008 (Mollusca: Mytilidae)
Calappa tuerkayana Pastore, 1995 (Brachyura: Calappidae)
Calyptogena tuerkayi Krylova & Jannsen, 2006 (Mollusca: Vesicomyidae)
Collodes tuerkayi Santana & Tavares, 2017 (Brachyura: Inachoididae) [this volume]
Discoplax michalis Ng & Shih, 2015 (Brachyura: Gecarcinidae)
Divacuma tuerkayi Muhlenhardt-Siegel, 2003 (Cumacea: Diastylidae)
Enosteoides turkayi Osawa, 2016 (Anomura: Porcellanidae)
Ergasilus turkayi Marques, Clebsh, Córdova & Boeger, 2017 (Copepoda: Ergasilidae)
Euchirograpsus tuerkayi Crosnier, 2001 (Brachyura: Plagusiidae)
Eunoe tuerkayi Barnich & Fiege, 2003 (Polychaeta: Polynoidae)
Hymenopenaeus tuerkayi Crosnier, 1995 (Penaeidae: Solenoceridae)
Leptochela (*Leptochela*) *tuerkayi* Wang, Gan & Li, 2017 (Caridea: Pasiphaeidae) [this volume]
Lithodes turkayi Macpherson, 1987 (Anomura: Lithodidae)
Lithoscaptus tuerkayi Van der Meij, 2017 (Brachyura: Cryptochiridae)
Mantisgebia tuerkayi Sakai, 2011 (Gebiidea: Upogebiidae)
Michaelimenes Okuno, 2017 (Caridea: Palaemonidae)
Michaelthelphusa Magalhães, 2017 (Brachyura: Pseudothelphusidae) [this volume]
Michaelthelphusa tuerkayi Magalhães, 2017 (Brachyura: Pseudothelphusidae) [this volume]
Michalisquilla Van der Wal & Ahyong, 2017 (Stomatopoda: Squillidae)
Munida michaeli Komai, 2017 (Anomura: Munididae)
Munida tuerkayi Macpherson, Beuck, Roder & Voolstra, 2017 (Anomura: Munididae) [this volume]
Nematocarcinus tuerkayi Burukovsky, 2005 (Caridea: Nematocarcinidae)
Obliquogobius turkayi Goren, 1992 (Pisces: Gobiidae)
Pachycheles tuerkayi Werding & Hiller, 2017 (Anomura: Porcellanidae) [this volume]
Paraclimenaeus michaeli Odijk & Fransen, 2017 (Caridea: Palaemonidae) [this volume]
Parastacus tuerkayi Ribeiro, Huber, Schubart & Araujo, 2017 (Astacidea: Parastacidae)

Perisesarma tuerkayi Shahdadi, Davie & Schubart, 2017 (Brachyura: Sesarmidae)
Petrolisthes tuerkayi Naderloo & Apel, 2014 (Anomura: Porcellanidae)
Pterochirella tuerkayi Schulz, 1990 (Copepoda: Aetideidae)
Solitariopagurus tuerkayi McLaughlin, 1997 (Anomura: Paguridae)
Sundathelphusa miguelito Mendoza & Si, 2017 (Brachyura: Gecarcinucidae)
Sundathelphusa tuerkayi Ng & Anker, 2016 (Brachyura: Gecarcinucidae)
Tuerkaygourretia Sakai, 2017 (Axiidea: Gourretiidae) [this volume]
Tuerkayogebia Sakai, 1982 (Gebiidea: Upogebiidae)
Tuerkayogebia kiiensis (Sakai, 1971) (Gebiidea: Upogebiidae)
Typhlamphiascus tuerkayi Ma & Li, 2017 (Copepoda: Miraciidae) [this volume]
Uroptychus tuerkayi Baba & Macpherson, 2017 (Anomura: Chirostylidae) [this volume]
Uroptychus michaeli Ahyong & Baba, 2017 (Anomura: Chirostylidae) [this volume]

REFERENCES

BECKER, C. & M. TÜRKAY, 2010. Taxonomy and morphology of European pea crabs (Crustacea, Brachyura, Pinnotheridae). Journal of Natural History, **44**: 1555-1575.

BECKER, C. & M. TÜRKAY, 2017. Host specificity and feeding in European pea crabs (Brachyura; Pinnotheridae). Crustaceana, **90**: 819-844.

BECKER, C., M. TÜRKAY & D. BRANDIS, 2012. The male copulatory system of European pea crabs (Crustacea, Decapoda, Brachyura, Pinnotheridae). Journal of Morphology, **273**: 1306-1318.

BRANDIS, D., D. FIEGE & I. KRÖNCKE, 2015. In Memoriam Professor Dr. Michael Türkay. GfBS Newsletter, **31**: 21-23.

GUTMANN, M. & M. TÜRKAY, 2001. Bauplan und Konstruktion — Funktions- und konstruktionsmorphologische Grundlagen. In: P. JANICH, M. GUTMANN & K. PRIESS (eds.), Biodiversität. Wissenschaftliche Grundlagen und gesellschaftliche Relevanz. Wissenschaftsethik und Technikfolgenbeurteilung, **10**: 115-147. (Springer, Berlin).

JANOVIC, I., 2015. Senckenberg trauert um Meeresforscher Türkay [Senckenberg mourns the marine explorer Türkay]. Frankfurt Neue Presse, 16 September 2015, available online at http://www.fnp.de/lokales/frankfurt/Senckenberg-trauert-um-Meeresforscher-Tuerkay;art675,1591583.

KAWAI, T., 2015. IAA friend: Professor Dr. Michael Türkay (1948-2015). Crayfish News, **37**(3): 14-15.

MAGALHÃES, C., 2016. A tribute to Michael Türkay (3 April 1948 - 9 September 2015): contributions and legacy of a lifelong study on New World decapods, and personal impressions. Nauplius, **24**: e2016012, 17 pp. DOI:10.1590/2358-2936e2016012.

NG, P. K. L., D. GUINOT & P. J. F. DAVIE, 2008. Systema Brachyurorum: Part I. An annotated checklist of extant brachyuran crabs of the world. Raffles Bulletin of Zoology, (Supplement) **17**: 1-286.

PETERS, D. S. & M. TÜRKAY, 1999. Goethe und die Zoologie. In: F. STEININGER & A. KOSSATZ-POMPÉ (eds.), Quer durch Europa. Naturwissenschaftliche Reisen mit Johann Wolfgang von Goethe. Kleine Senckenberg-Reihe, **30**: 105-109. (Senckenberg Institut, Frankfurt am Main).

SAKAI, K., P. DAVIE, D. GUINOT & M. TÜRKAY, 2004. The crabs of Japan. (Version 1.0.) (ETI CD-ROM, University of Amsterdam, Amsterdam).

SAKAI, K., M. TÜRKAY & S.-L. YANG, 2006. Revision of the *Helice*/*Chasmagnathus*-complex (Crustacea: Decapoda: Brachyura). Abhandlungen der Senckenbergischen naturforschenden Gesellschaft, **565**: 1-76.

SAKURAI, A., 2013. Science and societies in Frankfurt am Main. (University of Pittsburgh Press, Pittsburgh, PA).

SHIH, H.-T., P. K. L. NG, P. J. F. DAVIE, C. D. SCHUBART, M. TÜRKAY, R. NADERLOO, D. JONES & M.-Y. LIU, 2016. Systematics of the family Ocypodidae Rafinesque, 1815 (Crustacea: Brachyura), based on phylogenetic relationships, with a reorganization of subfamily rankings and a review of the taxonomic status of *Uca* Leach, 1814, sensu lato and its subgenera. Raffles Bulletin of Zoology, **64**: 139-175.

SONNEWALD, M., 2012. Long-term monitoring of the Dogger Bank mega-epibenthos: analyses about changes in species composition, correlated with environmental coefficients. (Ph.D. Thesis, Johann-Wolfgang-Goethe-Universität, Frankfurt am Main).

SONNEWALD, M. & M. TÜRKAY, 2012. Long term research of the Dogger Bank epibenthos (North Sea): loss of biodiversity and changes in climate. In: H. KORN, K. KRAUS & J. STADLER (eds.), Proceedings of the European Conference on Biodiversity and Climate Change — Science, Practice and Policy. BfN-Skripten, Bonn, **310**: 28.

SPIRIDONOV, V. A., 2015. [Obituary: Professor Michael Tuerkay (3.04.1948-9.09.2015)]. Invertebrate Zoology, **12**: 215-220. [In Russian.]

TEUBNER, C. & M. TÜRKAY, 1994. Shrimps, Hummer und Langusten. Warenkunde und Küchenpraxis der Krustentiere. Mit den besten Rezepten von Eckart Witzigmann: 1-144. (Teubner, Füssen).

THIEL, H. & M. TÜRKAY, 2002. Carl Chun (1852-1914) and the early days of biological deep sea research in Germany. Historisch-Meereskundliches Jahrbuch, **9**: 101-136.

TÜRKAY, M., 1967. Neue Brachyuren von der Westküste Südamerikas (Crustacea, Decapoda). Senckenbergiana Biologica, **48**: 361-364.

TÜRKAY, M., 1975. Dr. phil. nat. Richard Bott (1902-1974). Leben und carcinologisches Werk. Crustaceana, **28**: 298-302.

TUERKAY, M. & J. SCHOLZ, 2014. The Senckenberg-Museum in Frankfurt: history and Japan-related collections. Siebold-Symposium October 2014 (8th Siebold Collections Working Conference). Abstract, available online at https://siebold-museum.byseum.de/magic/show_image.php?id=10768&download.

TÜRKAY, M., E. WITZIGMANN & C. TEUBNER, 1998. Muscheln & Austern. (Teubner, Füssen).

First received 10 July 2017.
Final version accepted 18 July 2017.

[When citing this volume, please refer to Crustaceana 90 (2017) 771-1288]

Michael Türkay Memorial Issue

UROPTYCHUS MICHAELI (DECAPODA, CHIROSTYLIDAE), A NEW SPECIES OF DEEP-WATER SQUAT LOBSTER FROM NORTH-WESTERN AUSTRALIA AND TAIWAN

BY

SHANE T. AHYONG[1,2,4]) and KEIJI BABA[3])

[1]) Marine Invertebrates, Australian Museum, 1 William St., Sydney, NSW 2010, Australia
[2]) School of Biological, Earth and Environmental Sciences, Kensington, NSW 2052, Australia
[3]) Faculty of Education, Kumamoto University, 2-40-1 Kurokami, Kumamoto 860-8555, Japan

ABSTRACT

Uroptychus michaeli sp. nov. is described from northwestern Australia and Taiwan. The new species closely resembles *U. nigricapillis*, to which northwestern Australian and some Taiwanese records had been previously referred. *Uroptychus michaeli* sp. nov. is readily distinguished from *U. nigricapillis* by the deeply excavate cervical groove on the carapace (versus shallow, weakly indicated), more elongate pereopods 2-4 in which the pereopod 2 merus is longer than the postorbital carapace length (versus shorter), and the proportionally longer pereopod 2 carpus, which is as long as or longer than half postorbital carapace length (versus less than half) and approximately twice the length of the dactylus (versus 1.2× or less).

ZUSAMMENFASSUNG

Hier wird *Uroptychus michaeli* sp. nov. aus Nordwest-Australien und Taiwan beschrieben. Die neue Art ähnelt stark *U. nigricapillis*, auf welche Nordwest-Australische und einige Taiwanesische Funde bislang verwiesen. *Uroptychus michaeli* kann leicht von *U. nigricapillis* durch seine tiefe (im Gegensatz zur flachen, schwach angedeuteten) Zervikalfurche auf dem Carapax unterschieden werden. Zusätzlich hat die Art längere Pereopoden 2-4, bei welchen der Merus des Pereopod 2 länger (im Gegensatz zu kürzer) als die postorbitale Carapaxlänge ist, sowie einen proportional längeren Carpus des Pereopod 2, welcher so lang wie – oder länger als die halbe postorbitale Carapaxlänge ist (im Gegensatz dazu: weniger als halb so lang) und ungefähr die doppelte Daktyluslänge aufweist (im Gegensatz zu 1,2-fach oder weniger).

INTRODUCTION

The Australian chirostylid squat lobsters of the genus *Uroptychus* Henderson, 1888 are known from only a handful of studies (Henderson, 1885, 1888; Haig,

[4]) Corresponding author; e-mail: shane.ahyong@austmus.gov.au

1974; Baba, 1986, 2000; Ahyong & Baba, 2004; Ahyong & Poore, 2004; McCallum & Poore, 2013). Forty-six of more than 120 known species of *Uroptychus* (Baba et al., 2008) are currently recorded from Australia. Of these, Ahyong & Baba (2004) tentatively reported an unusual specimen from northwestern Australia as *Uroptychus nigricapillis* Alcock, 1901, which differed from the type description and figures in features of the carapace and length of the walking legs. Subsequently, Baba et al. (2009) reported two forms of *U. nigricapillis* from Taiwan, a common form corresponding to *U. nigricapillis* sensu stricto, and a second form (comprising three specimens) corresponding to that from northwestern Australia. McCallum & Poore (2013) reported several specimens of the northwestern Australian form (as *Uroptychus nigricapillis*) from off Port Hedland, Western Australia, and since then, further specimens from Taiwan and northwestern Australia have become available to us for study. Based on reconsideration of *U. nigricapillis*, the northwestern Australian and atypical Taiwanese forms are determined to be new to science and named herein.

MATERIAL AND METHODS

Carapace length (cl) is measured along the midline from the tip of the rostrum to the mid-posterior margin of the carapace. Postorbital carapace length (pcl) is measured from the posterior margin of the orbit to the mid-posterior margin of the carapace. In the Description, measurements and meristic counts of the holotype are indicated in square brackets. Specimens are deposited in the Australian Museum, Sydney, NSW, Australia (AM), Museum Victoria, Melbourne, VIC, Australia (NMV), Natural Taiwan Ocean University, Taipei, Taiwan (NTOU) and Northern Territory Museum and Art Gallery, Darwin, NT, Australia (NTM).

TAXONOMY

Family CHIROSTYLIDAE Ortmann, 1892
Genus *Uroptychus* Henderson, 1888
Uroptychus michaeli sp. nov.
(fig. 1)

Uroptychus nigricapillis.— Ahyong & Baba, 2004: 60-62, fig. 2A-I.— Baba et al., 2008: 37-38 (part, North West Shelf record only).— Baba et al., 2009: 51-52, fig. 42 (part, specimens from stn CP28, CP170, CP268 only).— Poore et al., 2011: 329, pl. 7F.— McCallum & Poore, 2013: 165, fig. 12E. (Not *U. nigricapillis* Alcock, 1901.)

Type material.— Holotype: AM P67832, ovigerous female (cl 11.0 mm, pcl 7.6 mm), North West Shelf, 240 km NW of Port Hedland, Western Australia, 18°06′S 115°45′E, 500 m, RV "Soela", stn S02/82/31, coll. J. Paxton, 7 April 1982.

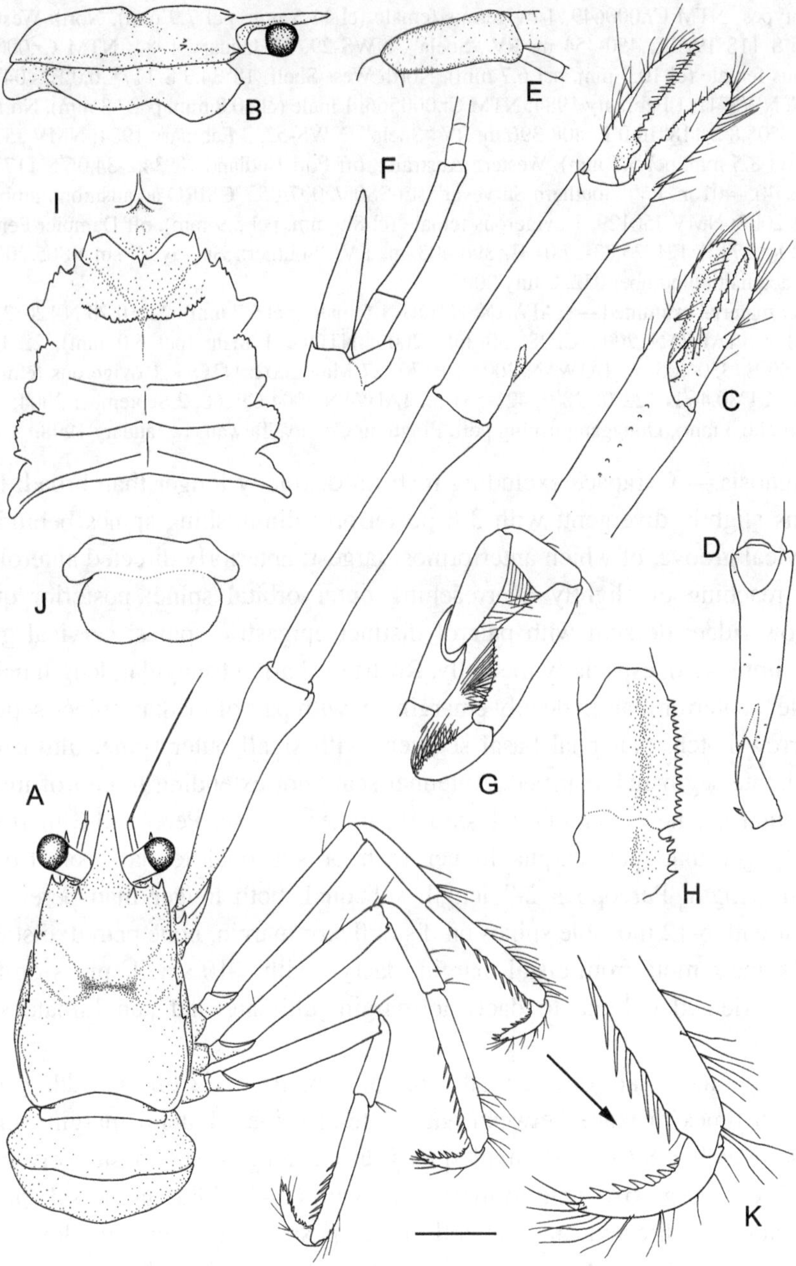

Fig. 1. *Uroptychus michaeli* sp. nov., ovigerous female holotype (cl 11.0 mm, pcl 7.6 mm), AM P67832. A, dorsal habitus; B, carapace, right lateral view; C, cheliped, distal ventral view; D, cheliped, proximal ventral view; E, right antennule, distal peduncular segment, lateral view; F, right antenna, ventral view; G, right maxilliped 3, lateral view; H, right maxilliped 3 crista dentata; I, sternal plastron; J, telson; K, right pereopod 2 dactylus and propodus. Scales: A-D = 3.0 mm; E, F, H = 0.7 mm; G, I, J = 1.5 mm; K = 1.2 mm. (A-D and F-I modified after Ahyong & Baba, 2004.)

Paratypes: NTM Cr.000649, 1 ovigerous female (cl 11.5 mm, pcl 7.9 mm), North West Shelf, 17°55.5′S 118°19.5′E, 450-454 m, RV "Soela", NWS-29, 27 January 1984; NTM Cr.000650, 1 ovigerous female (cl 10.0 mm, pcl 6.7 mm), North West Shelf, 18°34.3′S 117°30.0′E, 404 m, RV "Soela", NWS-47, 1 February 1984; NTM Cr.000566, 1 male (cl 10.8 mm, pcl 7.3 mm), North West Shelf, 18°05.8′S 118°10.0′E, 408-396 m, RV "Soela", NWS-52, 2 February 1984; NMV J57245, 1 female (cl 8.5 mm, pcl 5.7 mm), Western Australia, off Port Hedland 18°34.2-34.06′S 117°27.86-28.63′E, 405-401 m, RV "Southern Surveyer" stn SS05/2007/052, CSIRO acquisition number 043, 14 June 2007; NMV J56129, 1 ovigerous female (cl 8.9 mm, pcl 5.9 mm), off Dampier Peninsula, 14°51.20-50.71′S 121°25.83-27.01′E, 396-403 m, RV "Southern Surveyer" stn SS05/2007/144, CSIRO acquisition number 052, 2 July 2007.

Other material examined.— TAIWAN: NTOU, 1 female (pcl 6.7 mm), 22°10.70′N 120°27.90′E, 398-388 m, TAIWAN 2000 CP28, 30 July 2000; NTOU, 1 male (pcl 6.0 mm), 22°12.09′N 120°24.50′E, 330-405 m, TAIWAN 2002 CP170, 27 May 2002; NTOU, 1 ovigerous female (pcl 8.3 mm), 24°30.46′N 122°06.28′E, 421-531 m, TAIWAN 2004 CP268, 2 September 2004; NTOU, 1 male (pcl 6.3 mm), Donggang fishing port, Pingtung County, Taiwan, 14 January 1998.

Diagnosis.— Carapace excluding rostrum distinctly longer than broad; lateral margins slightly divergent; with 3-8 posteriorly diminishing spines behind base of cervical groove, of which anteriormost largest; anteriorly directed anterolateral spine, reaching or slightly overreaching outer orbital spine; posterior quarter with low ridge; dorsum with pair of distinct epigastric spines; cervical groove deeply impressed, especially medially. Rostrum sharply triangular, length half pcl. Sternite 3 anterior margin deeply emarginate, with pair of median spines separated by narrow notch. Antennal basal segment with small outer spine; ultimate and penultimate segments unarmed. Antennal scale not extending to end of ultimate peduncular segment. Pereopod 1 smooth, fingers setose. Pereopod 2 merus 1.1-1.2× longer than pcl; carpus longer than or subequal to half pcl, 1.6-1.8× dactylus length. Pereopods 2-3 length subequal, both longer than pereopod 4; propodi with 6-12 movable spines on distal flexor margin, none paired; distalmost flexor spine remote from distal margin; dactyli with 8-10 small spines on flexor margin, oriented oblique to dactylar margin, ultimate and penultimate spines similar.

Description.— Carapace: Length excluding rostrum 1.2 times width; greatest width 1.6 times distance between anterolateral spines. Lateral margins slightly divergent; with 3-8 (3 or 4) small spines diminishing in size posteriorly behind base of cervical groove, anteriormost spine with (1)-3 additional dorsal spinules; with anteriorly directed anterolateral spine; posterior quarter with low carina. Rostrum sharply triangular, length half pcl, internal angle 17-21°; margins straight, unarmed. Outer orbital angle produced to small spine extending to or falling short of level of anterolateral spines. Dorsum smooth, with pair of prominent epigastric spines, sometimes flanked laterally by spinule or small tubercle; cervical groove deep, distinct medially. Pterygostomian flap anteriorly rounded, with small anterior spine; surface unarmed.

Thoracic sternum: Excavated sternum with short anterior spine between bases of maxillipeds 1 and low tubercle anterior to sternite 3. Sternal plastron length 0.8-(0.9)× width, widening posteriorly. Sternite 3 (at base of maxilliped 3) depressed, anterior margin narrow, deeply emarginate, with pair of median spines separated by narrow notch; with distinct anterolateral tooth. Sternite 4 (at base of pereopod 1) with distinct anterolateral tooth extending anteriorly to level of base of emargination of sternite 3; margins dentate, irregular; demarcation between sternites 4 and 5 tuberculate to dentate.

Abdomen: Tergites glabrous, unarmed. Pleura 2-5 triangular, elongate, apices blunt, rounded. Telson length (0.6)-0.7× width; distal portion posteriorly emarginate, (1.7)-1.8× length of proximal portion.

Eye: Length 1.8 times cornea width; cornea moderately dilated, reaching to distal quarter of rostrum.

Antennule: Distal peduncular article length (2.6)-2.7× height.

Antenna: Basal segment with small outer spine. Peduncle extending to distal third of rostrum. Flagellum 16-18 segmented; about 3 times as long as peduncle, not reaching end of cheliped merus. Ultimate and penultimate segments unarmed; ultimate segment 2.0× length of penultimate segment. Antennal scale slightly wider than opposite peduncular segments, extending beyond midlength but not beyond distal 0.1 of ultimate peduncular segment.

Maxilliped 3: Dactylus, propodus, carpus and merus unarmed. Basis inner margin with 3-(4-5) denticles. Ischium crista dentata with (14-17)-20 denticles.

Pereopod 1 (chelipeds): Slender, cylindrical, 3.3-(3.5)× cl, 4.9-(5.1)-5.3× pcl; glabrous dorsally; fingers setose. Palm length about 4× width, 1.4-(1.6)× dactylus length; lateral margin with row of low granules. Fingers crossing distally; occlusal margins dentate, those of movable finger with obtuse process proximally; occlusal margin of fixed finger without distinct prominence. Carpus 1.1-(1.3)× longer than merus; glabrous; mesial surface with low, scattered granules. Merus length (1.2)-1.3× pcl; usually with several low granules on inner proximal margin. Ischium with short, slender dorsal spine.

Pereopods 2-4 (walking legs): Slender, mesiolaterally compressed, similar; sparsely setose proximally, becoming more strongly setose on distal half. Pereopods 2 and 3 length subequal, longer than pereopod 4. Meri unarmed except for short distoventral spine; length successively shorter posteriorly (though pereopods 2-3 merus length subequal); pereopod 2 merus 1.05-1.10× pereopod 3 merus length, both longer than pcl; pereopod 4 merus length 0.8 that of pereopods 2-3 merus; width subequal on pereopods 2-3, slightly narrower on pereopod 4; length-width ratio 7.1-(7.4)-7.7 on pereopod 2, 6.5-(6.8)-7.2 on pereopod 3, 5.4-(6.1) on pereopod 4; pereopod 2 merus (1.1)-1.2× pcl, (1.4)-1.5× propodus length; pereopod 3 merus 1.2-(1.3)× propodus length; pereopod 4 merus (1.1)× propodus length. Carpi unarmed, successively shorter posteriorly or subequal on pereopod 2 and 3; pereopod 2-3 carpus longer than or subequal to half pcl; carpus-propodus length ratio, (0.7)-0.8 on pereopod 2, (0.6)-0.7 on pereopod 3, (0.6) on pereopod 4; carpus-dactylus length ratio, 1.6-(1.9) on pereopod 2, 1.5-(1.7) on pereopod 3, (1.3)-1.5 on pereopod 4. Propodi slender, not expanded distally; pereopod 3 propodus slightly longer than pereopod 2 propodus, both longer than pereopod 4 propodus; flexor margin with (6-9)-12 movable spines along half to three-fourths of length of pereopod 2, (8)-9 along distal half to two-thirds length of pereopod 3, and (6)-8 along distal half of pereopod 4; distal (ultimate) movable spine single, relatively remote from dactylar articulation, closer to penultimate spine than dactylus. Dactyli strongly curved at proximal third; dactylus-propodus length ratio, 0.4 on pereopods 2-4; flexor margin with 8-10 elongate, triangular, well-spaced spines oriented obliquely to dactylar margin, ultimate and penultimate spines similar, antepenultimate subequally spaced between penultimate and fourth spines or slightly closer to fourth spine.

Eggs: 0.80 × 0.94 mm to 1.13 × 1.25 mm; about 50 carried.

Colour in life.— Overall pink-orange; body paler than pereopods, translucent; red intestinal tract visible through posterior carapace and abdomen (Baba et al., 2008, fig. 42; McCallum & Poore, 2013: fig. 12E; as *U. nigricapillis*).

Etymology.— We name this new species in honour of our late friend and colleague, Michael Türkay. The specific name thus is a noun in the genitive singular.

Remarks.— Ahyong & Baba (2004) first reported *Uroptychus michaeli* sp. nov. under the name *U. nigricapillis* from Western Australia, noting a number of features that differed from *U. nigricapillis* sensu stricto as described and figured by Alcock (1901: pl. 3 fig. 3, 3a) and Alcock & McArdle (1902: pl. 56, fig. 3) for the holotype collected at 1224 m from the Andaman Sea. *Uroptychus michaeli* sp. nov. differs from *U. nigricapillis* as follows: the walking legs of *U. michaeli* are proportionally longer, with the merus of the pereopod 2 longer than, instead of shorter than, the postorbital carapace length (pcl); the pereopod 2 carpus is as long as or longer than, rather than shorter than half pcl, and is significantly longer than the dactylus (1.6-1.8× versus about 1.2× length); the pereopod 3 merus length is subequal to that of pereopod 2, rather than less than 0.9 its length; the cervical groove is deeply impressed, particularly on the midline, rather than weakly indicated. *Uroptychus nigricapillis* is presently considered widespread in the Indo-West Pacific, from East Africa to Japan, itself requiring closer taxonomic scrutiny given its extensive apparent range (Baba et al., 2008, 2009).

With respect to the Australian fauna, specimens recently reported from eastern Australia as *U. gracilimanus* (Henderson, 1885) by Ahyong & Poore (2004) are referrable to *U. nigricapillis*, whereas previous records of *U. nigricapillis* from

Western Australia are referrable to *U. michaeli* sp. nov. As such, in Australia, *U. nigricapillis* is currently known only from the east coast, and *U. michaeli* from the west coast. Among other regional species, *Uroptychus michaeli* closely resembles *U. singularis* Baba & Lin, 2008 from Taiwan and *U. australis* (Henderson, 1885) from Australia, New Zealand and Indonesia, sharing a similar carapace shape and the presence of the epigastric spines. *Uroptychus michaeli* is readily distinguished from both *U. australis* and *U. singularis* by the distinct lateral branchial carapace spines (versus absent or at most a few tubercles) and the pereopod 2-4 dactyl spines, which, in the new species, are obliquely oriented rather than lying against the dactyl margin.

Distribution.— Northwestern Australia (North West Shelf) and Taiwan; 330-531 m.

ACKNOWLEDGEMENTS

We gratefully acknowledge Tin-Yam Chan (NTOU), Joanne Taylor (NMV) and Gavin Dally (NTM) for the loan of specimens, two anonymous reviewers for constructive comments on the manuscript, and Célio Magalhães for inviting this paper. This is a contribution from the Australian Museum Research Institute.

REFERENCES

AHYONG, S. T. & K. BABA, 2004. Chirostylidae from north-western Australia (Crustacea: Decapoda: Anomura). Memoirs of Museum Victoria, **61**: 57-64.

AHYONG, S. T. & G. C. B. POORE, 2004. The Chirostylidae of southern Australia (Crustacea: Decapoda: Anomura). Zootaxa, **436**: 1-88.

ALCOCK, A., 1901. Descriptive catalogue of the Indian Deep Sea Crustacea Decapoda Macrura and Anomura in the Indian Museum. Being a revised account of the deep-sea species collected by the Royal Indian Marine Survey Ship Investigator. (Trustees of the Indian Museum, Calcutta).

ALCOCK, A. & A. F. MCARDLE, 1902. Illustrations of the Zoology of the Royal Indian Marine Survey Steamer "Investigator", Crustacea. Part 10, pl. 56-67. (Trustees of the Indian Museum, Calcutta).

BABA, K., 1986. Two new anomuran Crustacea (Decapoda: Anomura) from North-West Australia. The Beagle, Occasional Papers of the Northern Territory Museum of Arts and Sciences, **3**: 1-5.

BABA, K., 2000. Two new species of chirostylids (Decapoda: Anomura: Chirostylidae) from Tasmania. Journal of Crustacean Biology, **20**(Special Number 2): 246-252.

BABA, K. & C.-W. LIN, 2008. Five new species of chirostylid crustaceans (Crustacea: Decapoda: Anomura: Chirostylidae) from Taiwan. Zootaxa, **1919**: 1-24.

BABA, K., E. MACPHERSON, C.-W. LIN & T.-Y. CHAN, 2009. Crustacean Fauna of Taiwan: squat lobsters (Chirostylidae and Galatheidae) (1st ed.). (National Science Council, Taipei).

BABA, K., E. MACPHERSON, G. C. B. POORE, S. T. AHYONG, A. BERMUDEZ, P. CABEZAS, C.-W. LIN, M. NIZINSKI, C. RODRIGUES & K. E. SCHNABEL, 2008. Catalogue of squat lobsters of the world (Crustacea: Decapoda: Anomura — families Chirostylidae, Galatheidae and Kiwaidae). Zootaxa, **1905**: 1-220.

HAIG, J., 1974. The anomuran crabs of Western Australia: their distribution in the Indian Ocean and adjacent seas. Journal of the Marine Biological Association of India, **14**: 443-451.

HENDERSON, J. R., 1885. Diagnoses of the new species of Galatheidea collected during the "Challenger" Expedition. Annals and Magazine of Natural History, series 5, **16**: 407-421.

HENDERSON, J. R., 1888. Report on the Anomura collected by H.M.S. Challenger during the years 1873-76. Report on the Scientific Results of the Voyage of H.M.S. Challenger during the years 1873-76, Zoology, **27**, i-vi + 1-221, pl. 1-21.

MCCALLUM, A. W. & G. C. B. POORE, 2013. Chirostylidae of Australia's western continental margin (Crustacea: Decapoda: Anomura), with the description of five new species. Zootaxa, **3664**: 149-175.

ORTMANN, A., 1892. Die Decapoden-Krebse des Strassburger Museums IV. Die Abtheilungen Galatheidea und Paguridea. Zoologischen Jahrbuchern, Abtheilung für Systematik, Geographie und Biologie der Tiere, **6**: 241-326, pl. 11, 12.

POORE, G. C. B., S. T. AHYONG & J. TAYLOR, 2011. The biology of squat lobsters. Crustacean Issues, **19**. (CSIRO Publishing, Melbourne, VIC).

First received 29 August 2016.
Final version accepted 19 October 2016.

[When citing this volume, please refer to Crustaceana 90 (2017) 771-1288]

Michael Türkay Memorial Issue

UROPTYCHUS TUERKAYI SP. NOV. (ANOMURA, CHIROSTYLIDAE), A NEW SQUAT LOBSTER FROM THE ATLANTIS-GREAT METEOR SEAMOUNT CHAIN IN THE EASTERN ATLANTIC

BY

KEIJI BABA[1,3]) and ENRIQUE MACPHERSON[2])

[1]) Kumamoto University, Faculty of Education, 2-40-1 Kurokami, Kumamoto 860-8555, Japan

[2]) Centro de Estudios Avanzados de Blanes (CSIC), C. acc. Cala San Francesc 14, E-17300 Blanes, Girona, Spain

ABSTRACT

A new species of chirostylid squat lobster, *Uroptychus tuerkayi* sp. nov., is described based upon material collected by the French "Seamount 2" project (1993) from the Atlantis-Great Meteor Seamount Chain south of the Azores Islands, at a depth of 340-730 m. *Uroptychus tuerkayi* resembles *U. maroccanus* Türkay, 1976 from the Moroccan coast, but it can be readily distinguished by the eyes being distinctly longer instead of as long as broad (globular in *U. maroccanus*), the antennal article 5 with a small instead of prominent distomesial spine, the anterolateral spine of the carapace slightly smaller than or subequal to, instead of much smaller than the lateral orbital spine, the pterygostomian flap anteriorly acuminate and not strongly produced to a spine as in *U. maroccanus*, and in having pereopod 1 with obsolescent instead of distinct spines on the merus and carpus. This is the sixth species of *Uroptychus* from the eastern Atlantic. A key to the eastern Atlantic species of *Uroptychus* is provided.

ZUSAMMENFASSUNG

Hier wird eine neue Art eines chirostyliden Furchenkrebses, *Uroptychus tuerkayi* sp. nov. beschrieben. Das Material stammt aus dem Französischen "Seamount 2" – Projekt (1993) von der Großen Meteor-Seegebirgskette südlich der Azoren im Atlantik aus 430-730 m Tiefe. *Uroptychus tuerkayi* ähnelt *U. maroccanus* Türkay, 1976 von der Küste Marokkos, kann jedoch gut durch die folgenden Merkmale unterschieden werden: die Augen sind erkennbar länger als breit (im Gegensatz zu: so lang wie breit – kugelförmig bei *U. maroccanus*), der fünfte Antennalfortsatz trägt einen kleinen (im Gegensatz zu einem markanten) Distomesialdorn, der Anterolateraldorn des Carapax ist ein wenig kleiner oder weniger als gleich so lang wie der laterale Orbitaldorn (im Gegensatz zu viel kleiner als letzterer), die Pterygostomialklappe ist am Vorderende zugespitzt und nicht als stark hervorstehender Dorn ausgebildet wie bei *U. maroccanus*. Überdies tragen Merus und Carpus des Pereopod 1 reduzierte anstatt deutliche Dornen. Dies ist die sechste *Uroptychus*-Art aus dem Ostatlantik. Ein Bestimmungsschlüssel für die Ostatlantischen *Uroptychus*-Arten ist hier enthalten.

[3]) Corresponding author; e-mail: kbaba.kumamoto@gmail.com

INTRODUCTION

The squat lobsters of the genus *Uroptychus* in the Atlantic Ocean are noticeably less speciose compared to those in the Indo-West Pacific, especially in the eastern Atlantic (Baba et al., 2008; Baba & Macpherson, 2012; Baba & Wicksten, 2015, in press). Only five species are known from the eastern Atlantic: *U. rubrovittatus* (A. Milne-Edwards, 1881); *U. concolor* (A. Milne-Edwards & Bouvier, 1894); *U. bouvieri* Caullery, 1896; *U. maroccanus* Türkay, 1976; and *U. cartesi* Baba & Macpherson, 2012.

While examining the collection in the Muséum national d'Histoire naturelle, Paris (MNHN), unidentified material belonging to *Uroptychus* has come to our attention. This was collected by the French "Seamount 2" project (chief: Philippe Bouchet; Gofas, 1993) from the Atlantis-Great Meteor Seamount Chain, south of the Azores Islands, at depths of 340-730 m. It resembles *U. maroccanus* Türkay, 1976 from the Moroccan coast in having a smooth, glabrous carapace with a row of tiny spines along the lateral margins, in shape of the sternum, and in spination of the distal articles of pereopods 2-4, but it is clearly different in the shape of the eye, antenna and pterygostomian flap, and in spination of pereopod 1. This material is herein described as *Uroptychus tuerkayi* sp. nov.

A key to species from the eastern Atlantic is provided, incorporating characters that were verified by examination of type and/or comparative material.

The terminology used follows Baba et al. (2011). The measurements of specimens indicate the postorbital carapace length, unless otherwise mentioned. In description of the species, the holotype characters are shown in square brackets where meristic variations are observed in the type series. The abbreviations used in the text include: Mxp = maxilliped; P1 = pereopod 1 (cheliped); P2-4 = pereopods 2-4 (walking legs 1-3).

TAXONOMIC ACCOUNT

Family CHIROSTYLIDAE Ortmann, 1892

Genus *Uroptychus* Henderson, 1888

Uroptychus tuerkayi sp. nov.

(figs. 1, 2)

Type material.— Holotype. GREAT METEOR BANK, "Seamount 2", R/V "Le Suroît", Stn DW152, 30°01.99′N 28°22.09′W-30°02.20′N 28°22.29′W, 470 m (465 m on label), bioclastic coarse sand with rich fauna: sponges, hydroids, mollusks, etc., 11 Jan. 1993: female (3.6 mm), MNHN-IU-2016-2939.

Paratypes. GREAT METEOR BANK, station data as for holotype: 2 males (3.4 mm, carapace broken in another male), 2 ovigerous females (3.3, 3.5 mm), 1 female (3.4 mm), MNHN-IU-2016-2940.— Stn CP146, 30°11.20′N 28°28.10′W-30°11.18′N 28°27.37′W, 420 m, 10 Jan. 1993, many

Fig. 1. *Uroptychus tuerkayi* sp. nov., A-C, E-I, holotype, female (3.6 mm), MNHN-IU-2016-2939; D, paratype, male (3.0 mm), MNHN-IU-2016-2941. A, Carapace and anterior part of abdomen, dorsal; B, same, lateral; C, lateral part of carapace, dorsal; D, same; E, sternal plastron, with excavated sternum and basal parts of Mxp1; F, telson; G, right antennal peduncle, anterior part of pterygostomian flap included, ventral; H, Mxp3, ventral; I, same, setae omitted, lateral. Scales = 1 mm.

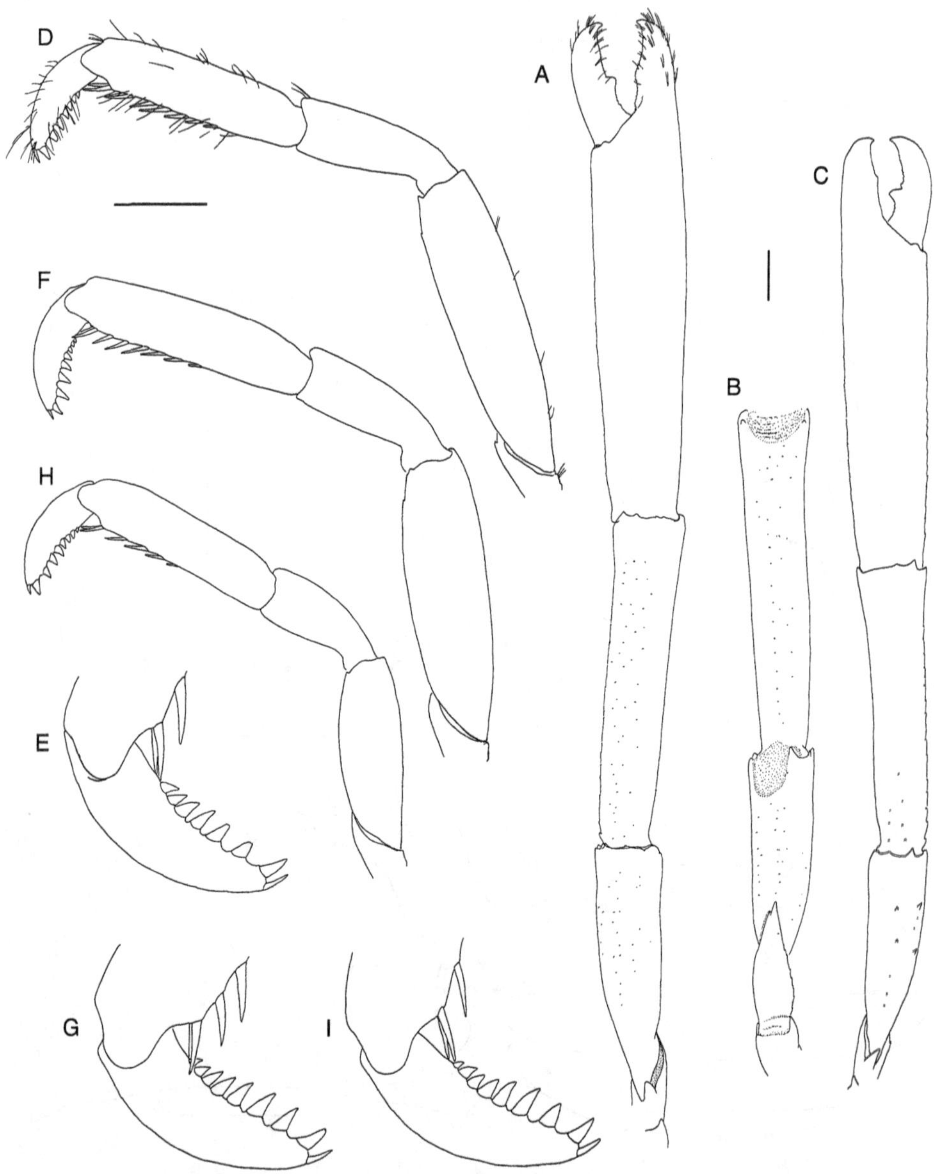

Fig. 2. *Uroptychus tuerkayi* sp. nov., A, B, D-I, holotype, female (3.6 mm), MNHN-IU-2016-2939; C, paratype, male (3.0 mm), MNHN-IU-2016-2941. A, Right P1, dorsal; B, same, distal part omitted, ventral; C, left P1, dorsal; D, left P2, lateral; E, same, distal part, setae omitted, lateral; F, left P3, lateral; G, same, distal part, lateral; H, left P4, lateral; I, same, distal part, lateral. Scales = 1 mm.

scleractinians, *Flabellum chunii* Marenzeller, 1904, some gorgonians: 1 male (3.7 mm), MNHN-IU-2016-530 & 1 ovigerous female (3.5 mm), MNHN-IU-2016-529.— Stn DW147, 30°11.16′N 28°27.15′W-30°11.16′N 28°27.03′W, 340-345 m, bioclastic sand and sponges, 10 Jan. 1993: 3 males (2.7-3.1 mm), 2 ovigerous females (3.0, 3.4 mm), 2 females (2.7, 2.7 mm), MNHN-IU-2016-2941.— Stn DW172, 30°05.10′N 28°41.50′W-30°04.77′N 28°41.54′W, 455 m, 14 Jan. 1993, bioclastic sand, molluscs, corals and foraminiferans: 1 male (3.8 mm), MNHN-IU-2016-528 & 1 ovigerous female (3.6 mm), MNHN-IU-2016-527.— Stn DW179, 30°00.60′N 28°42.30′W-30°00.92′N 28°42.02′W, 600-730 m, 15 Jan. 1993, bioclastic sand, indurated sediment blocks and a few pieces of basalt, some sponges, Stylasteridae (Hydrozoa), and fragments of *Cyathidium* (Crinoidea): 1 female (3.1 mm), MNHN-IU-2016-531. HYERES SEAMOUNT, Stn DW182, 31°23.23′N 28°53.46′W-31°22.93′N 28°53.59′W, 480 m, bioclastic sand and large sponges, epifauna with hydroids, gorgonians, sponges, 16 Jan. 1993: 2 males (2.7, 5.0 mm), 2 ovigerous females (4.0, 4.9 mm), 1 female (4.6 mm), MNHN-IU-2016-2942. ATLANTIS SEAMOUNT, Stn DW263, 34°25.89′N 30°32.49′W-34°26.14′N 30°32.78′W, 610-655 m, bioclastic sand and sponges, 3 Feb. 1993: 1 male (2.7 mm), 1 female (4.0 mm), MNHN-IU-2016-2943.

Etymology.— Named after our dear friend, the late Michael Türkay, who contributed much to the knowledge of crustacean biology. The specific name thus is a noun in the genitive singular.

Description.— Carapace: Broader than long (0.7-0.9 (0.8) as long as broad); greatest breadth 1.4-1.9 (1.8)× distance between anterolateral spines. Dorsal surface smooth, glabrous and unarmed, moderately convex from side to side (more or less flattish in large specimens), slightly convex from anterior to posterior, with or (without) feeble depression between gastric and cardiac regions; without distinct ridge along branchial margin. Lateral margins convex or straight divergent posteriorly to convex point at posterior fourth, with row of tiny spinules or (denticle-like tubercles); anterolateral spine subequal to or (smaller than) lateral orbital spine, barely or not reaching tip of that spine. Rostrum 1.4-1.6 (1.5)× as long as broad, directed straight forward horizontally or (slightly ventrally), or slightly upcurved distally, sharp triangular with interior angle of 20-25 (21)°; dorsal surface concave; lateral margin smooth and nearly straight; length (0.70)-0.75 that of remaining carapace, breadth about half carapace breadth at posterior carapace margin. Lateral orbital spine well developed, situated slightly anterior to anterolateral spine. Pterygostomian flap anteriorly angular and ending in acuminate tip without distinct spine, smooth on surface; greatest height at anterior half 1.5-(1.6)× that of posterior half (height measured between upper margin along linea anomurica and ridge along lower margin).

Thoracic sternum: Excavated sternum sharply triangular on anterior margin, surface sharply carinated in midline. Sternal plastron 0.7 as long as broad, lateral extremities divergent posteriorly, sternite 7 slightly broader than sternite 6. Sternite 3 weakly depressed from level of sternite 4 in ventral view; anterior margin moderately concave in broad V-shape with wedge-shaped median notch without distinct flanking spine, anterolateral angle irregularly rounded, obsolescently denticulate. Sternite 4 anterolateral margin irregular, obsolescently denticulate and convex, without spine at anterior end; posterolateral margin very short, less than 0.3 length of anterolateral margin. Anterolateral margins of sternite 5 strongly convex, 2× longer than posterolateral margin of sternite 4.

Abdomen: Barely setose. Somite 1 slightly convex from anterior to posterior. Somite 2 tergite 2.9-3.4 (3.1)× broader than long; pleural lateral margins moderately concave, divergent in dorsal view, posterolaterally angular with blunt end. Pleuron of somite 3 narrowed laterally with blunt end; those of somites 4 and 5 proportionately broad laterally, ending in rounded margin. Telson slightly less than half (0.44-0.46 (0.46)) as long as broad; lateral margins constricted between anterior and posterior plates (not constricted in small specimens of CL 2.7 mm); posterior plate as long as and 0.8 as broad as anterior plate, posterior margin slightly concave.

Eye: Relatively short (length (1.4)-1.6× breadth), slightly falling short of midlength of rostrum. Cornea not dilated, length (0.6)-0.8× that of remaining eyestalk.

Antennule and antenna: Ultimate article of antennular peduncle 2.5-3.6 (2.9)× longer than high. Antennal peduncle overreaching midlength of rostrum. Article 2 with small distolateral spine. Antennal scale 1.3-(1.8)× broader than article 5, (terminating at) or slightly overreaching midlength of article 5. Article 3 distomesially (rounded) or produced to small spine. Article 4 with small but distinct distomesial spine. Article 5 with distomesial spine smaller than that of article 4, length 1.3-1.6 (1.4)× that of article 4, breadth (0.5)-0.6 height of ultimate article of antennule. Flagellum consisting of 7-9 (8 or 9) segments, (terminating at) or slightly overreaching rostral tip, not reaching distal end of P1 merus.

Mxp: Mxp1 with bases broadly separated. Mxp3 basis with distal denticle on mesial ridge. Ischium with flexor margin sharply ridged, not rounded distally; crista dentata with 23-29 (25 (left) or 27 (right)) denticles. Merus barely setose, 1.8× longer than ischium, without distolateral spine; flexor margin not cristate but roundly ridged, without spine. Carpus unarmed.

P1: 5.7 (6.3)-6.5× (females), 6.0-6.6× (males) longer than carapace, barely setose except for fingers, feebly granulose except for fingers. Ischium with basally depressed dorsal spine, ventromesial margin with well-developed subterminal spine proximally followed by row of denticle-like small or obsolescent spines. Merus 1.3-1.5 (1.4)× longer than carapace; distal margin with 5 or 6 spines (mesial one of 2 dorsal spines and ventromesial spine larger, other spines low), dorsal surface with 2 small spines along mesial margin (distal larger, proximal one occasionally absent in large specimens, or both obsolete in large specimens including holotype) and row of several tiny spines or denticles in midline (obsolete in large specimens). Carpus subcylindrical, 1.2-1.4 (1.3)× longer than merus, distally bearing 2 tiny blunt ventral spines (1 mesial and 1 lateral), proximal portion with denticle-like dorsal and mesial spines distinct in small specimens, obsolescent in large specimens. Palm minutely granulose along mesial margin, somewhat narrowed proximally, 0.7 as high as broad, 3.6-4.3 (4.0)× (females), 3.3-4.4× (males) longer than broad; length 1.0-(1.1 or 1.2)× (females), 1.1-1.2× (males) that of carpus. Fingers distally strongly incurved and crossing when closed, somewhat gaping in males, not gaping in females, sparingly bearing relatively short setae; movable finger 0.30-0.38 (0.31) (females), 0.30-0.40 (males) as long as palm, opposable margin with prominent process slightly proximal to midpoint; opposable margin of fixed finger with low median eminence (lower in females than in males) proximally followed by longitudinal groove (obsolete in small specimens) accommodating opposite process of movable finger when closed.

P2-4: Sparsely setose. Meri strongly compressed mesiolaterally, successively shorter posteriorly (P3 merus 0.9 length of P2 merus, P4 merus 0.7 length of P3 merus), P2 merus (0.9)-1.0× length of carapace, (1.2)-1.3× length of P2 propodus; P3 merus (1.0)-1.1× length of P3 propodus; P4 merus 0.7-0.9 (0.8) length of P4 propodus; length-breadth ratio, 3.5-4.2 (3.8) on P2, (3.3)-4.0 on P3, 2.6-3.0 (2.7) on P4; dorsal crest with or without small distal spine, unarmed elsewhere, occasionally with a few obsolescent proximal protuberances on P2 only, ventrolateral margin distally ending in small spine occasionally followed by much smaller, often obsolescent spine proximal to it, ventromesial margin smooth and unarmed. Carpi subequal in length on P2 and P3, shortest on P4 (P4 carpus 0.8 length of P3 carpus); carpus-propodus length ratio, 0.6 on P2, 0.5-(0.6) on P3, 0.5 on P4. Propodi longer on P3 than on P2, subequal on P2 and P4 or (shortest on P4); flexor margin nearly straight, with pair of slender movable terminal spines preceded by 5 or (6) similar unpaired spines on P2, 4-6 (6) spines on P3, 3 or (4) spines on P4. Dactyli relatively stout, dactylus-carpus length ratio, (0.7)-0.8 on P2 and P3, 0.9 on P4; dactylus-propodus length ratio, 0.4 on P2-4; flexor margin feebly curving, terminating in slender spine preceded by 7 or 8 stout triangular spines proximally somewhat diminishing, all slightly obliquely directed.

Eggs: Number of eggs carried, 5; size 1.06 × 1.01 mm-0.97 × 1.06 mm (CL 3.3 mm), 0.94 × 1.17 mm-1.04 × 1.25 mm (CL 3.5 mm); 6, 0.90 × 1.00 mm-1.05 × 1.10 mm (CL 4.9 mm); 9 eggs, 0.90 × 1.08 mm-0.90 × 1.16 mm (CL 3.0 mm); 15 eggs, 1.05 × 0.92 mm-1.12 × 0.94 mm (CL 3.6 mm).

Remarks.— The present material agrees with the gross morphology originally described for *Uroptychus maroccanus* Türkay, 1976. Examination of the type material of *U. maroccanus* and of the material reported by Garcia Raso (1996) from the Moroccan coast and now kept in the Muséum national d'Histoire naturelle, Paris (figs. 3 and 4) disclosed additional morphological similarity in spination of the P2-4 propodi and dactyli, and in shape of the thoracic sternum including

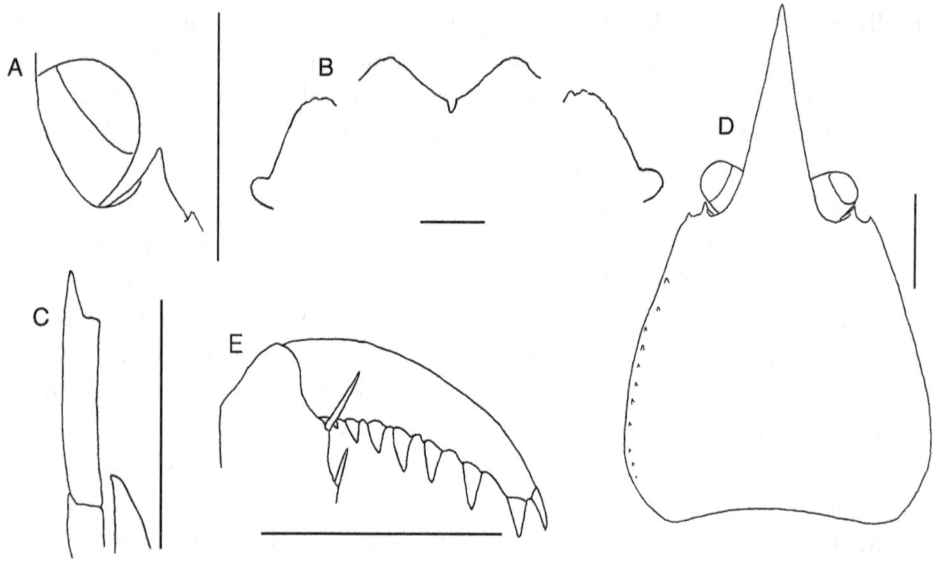

Fig. 3. *Uroptychus maroccanus* Türkay, 1976, A-C, holotype, male, SMF 6770; D, paratype, male (5.2 mm, rostrum included); E, paratype, ovigerous female (8.0 mm, rostrum included), SMF 6093. A, Eye and anterolateral part of carapace, dorsal; B, anterior part of sternal plastron; C, left antenna, proximal part omitted, ventral; D, carapace, small spines along right lateral margin omitted, dorsal; E, distal part of right P2, lateral. Scales = 1 mm.

a sharply carinated excavated sternum. However, the present material can be distinguished from *U. maroccanus* by the following differences:

- The eyestalks including cornea are distinctly longer than broad instead of being globular (about as long as broad).
- The anterolateral spine of the carapace is subequal to or somewhat smaller than instead of much smaller than the lateral orbital spine.
- The antennal article 5 bears a small instead of prominent distomesial spine, and the antennal scale reaches or slightly overreaches the midlength of the antennal article 5, instead of falling short of the midlength or terminating in the proximal fifth of article 5.
- The P1 spination in *U. tuerkayi* is weaker, with the merus being spineless in large specimens or bearing at most 2 small dorsal spines in small specimens, and with the carpus bearing at most a number of small denticle-like spines on the proximal part. In *U. maroccanus*, two rows (mesial and dorsal) of numerous, distinct spines are dorsally visible on the merus and carpus.
- The pterygostomian flap is anteriorly acuminate, not strongly produced to a spine as in *U. maroccanus*.

In addition, the P2-4 meri are unarmed instead of bearing a row of spines on the dorsal crest.

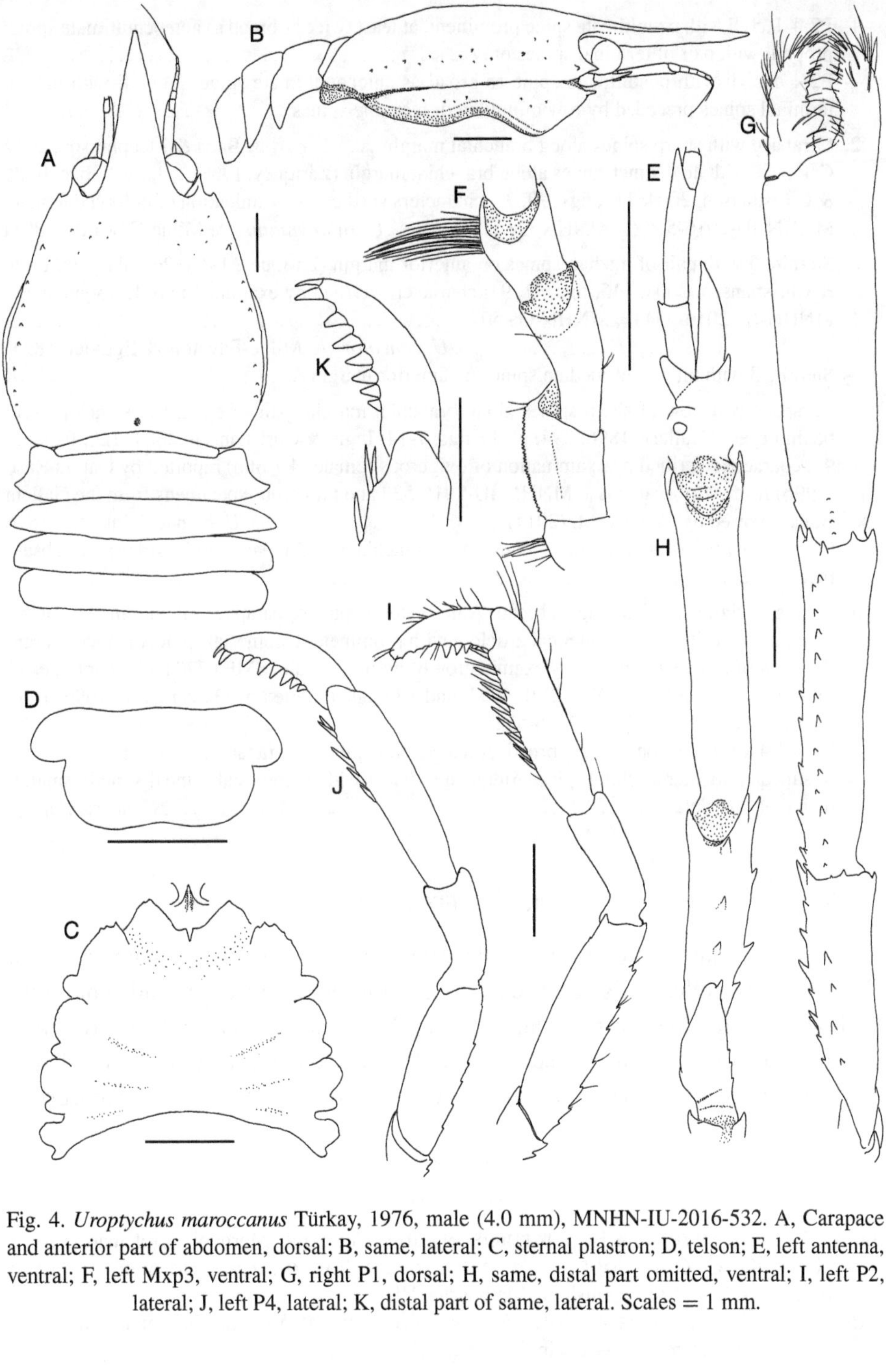

Fig. 4. *Uroptychus maroccanus* Türkay, 1976, male (4.0 mm), MNHN-IU-2016-532. A, Carapace and anterior part of abdomen, dorsal; B, same, lateral; C, sternal plastron; D, telson; E, left antenna, ventral; F, left Mxp3, ventral; G, right P1, dorsal; H, same, distal part omitted, ventral; I, left P2, lateral; J, left P4, lateral; K, distal part of same, lateral. Scales = 1 mm.

Key to Species of *Uroptychus* from the Eastern Atlantic Ocean

1. P2-4 dactyli with penultimate spine prominent, at least twice as broad as antepenultimate spine; propodi with pair of terminal spines only 2
– P2-4 dactyli with penultimate spine as broad as antepenultimate spine; propodi with pair of terminal spines preceded by row of unpaired, movable spines 3

2. Carapace with sharp spines along branchial margin .... *U. cartesi* Baba & Macpherson, 2012
– Carapace with no distinct spines along branchial margin (Zariquey, 1968: 254, fig. 93b, c; Ingle & Christiansen, 2004: 115, figs. 90, 93; characters verified by examination of holotype female, MNHN-IU-2016-535 (= MNHN-Ga 514))....... *U. rubrovittatus* (A. Milne-Edwards, 1881)

3. Sternite 3 with pair of median spines on anterior margin (Zariquey, 1968: 265, fig. 93a; Ingle & Christiansen, 2004: 115, figs. 91, 94; characters verified by examination of lectotype male, MNHN-IU-2016-534 (= MNHN-Ga 507))...
.... *U. concolor* (A. Milne-Edwards & Bouvier, 1894)
– Sternite 3 without pair of median spines on anterior margin 4

4. Carapace with row of sharp spines along branchial margin; pair of epigastric spines present behind eyes (Caullery, 1896: 394, pl. 17 figs. 7-14; Ingle & Christiansen, 2004: 120, figs. 92, 95; characters verified by examination of ovigerous female (4.7 mm) reported by García Raso (1996) from Moroccan coast, MNHN-IU-2016-533 and numerous specimens from the Galicia Bank reported by Cartes et al. (2014)) .... *U. bouvieri* Caullery, 1896
– Carapace with row of very small spines along branchial margin; pair of epigastric spines absent behind eyes 5

5. Eyes globular, nearly as long as broad. Anterolateral spine of carapace minute, much smaller than lateral orbital spine. Antennal article 5 with prominent distomesial spine, distinctly larger than that of article 4 (verified by examination of holotype male, SMF 6770 and 2 paratypes (1 male and 1 ovigerous female), SMF 6093, and 1 female reported by Garcia-Raso (1996) from Moroccan coast, MNHN-IU-2016-532)...................... *U. maroccanus* Türkay, 1976
– Eyes 1.4-1.6 times longer than broad. Anterolateral spine of carapace somewhat smaller than or subequal to lateral orbital spine. Antennal article 5 with distomesial spine distinctly smaller than that of article 4 .... *U. tuerkayi* n. sp.

ACKNOWLEDGEMENTS

We thank Laure Corbari and Paula Martin-Lefevre of the Muséum national d'Histoire naturelle, Paris for access to the museum collection and laboratory facilities, and for loan of material. The late M. Türkay is thanked for providing laboratory facilities during a visit of KB to the Senckenberg Institute in 1989, and we thank Shane T. Ahyong of the Australian Museum for reading the manuscript.

REFERENCES

Baba, K., S. T. Ahyong & E. Macpherson, 2011. Chapter 1. Morphology of marine squat lobsters. In: G. C. B. Poore, S. T. Ahyong & J. Taylor (eds.), The biology of squat lobsters: 1-37. (CSIRO Publishing, Melbourne, VIC).

Baba, K. & E. Macpherson, 2012. A new squat lobster (Crustacea: Decapoda: Anomura: Chirostylidae) from off NW Spain. Zootaxa, **3224**: 49-56.

BABA, K., E. MACPHERSON, G. C. B. POORE, S. T. AHYONG, A. BERMUDEZ, P. CABEZAS, C.-W. LIN, M. NIZINSKI, C. RODRIGUES & K. E. SCHNABEL, 2008. Catalogue of squat lobsters of the world (Crustacea: Decapoda: Anomura families Chirostylidae, Galatheidae and Kiwaidae). Zootaxa, **1905**: 1-220.

BABA, K. & M. WICKSTEN, 2015. *Uroptychus minutus* Benedict, 1902 and a closely related new species (Crustacea: Anomura: Chirostylidae) from the western Atlantic Ocean. Zootaxa, **3957**: 215-225.

BABA, K. & M. WICKSTEN, in press. *Uroptychus nitidus* (A. Milne Edwards, 1880) and related species (Crustacea: Decapoda: Anomura: Chirostylidae) from the western Atlantic. Zootaxa.

CARTES, J. E., V. PAPIOL, I. FRUTOS, E. MACPHERSON, C. GONZÁLEZ-POLA, A. PUNZÓN, X. VALEIRAS & A. SERRANO, 2014. Distribution and biogeographic trends of decapod assemblages from Galicia bank (NE Atlantic) at depths between 700 and 1800 m, with connexions to regional water masses. Deep-Sea Research II, **106**: 165-178.

CAULLERY, M., 1896. Crustacés schizopodes et décapodes. In: R. KOEHLER (ed.), Resultats scientifiques de la Campagne du Caudan dans le Golfe de Gascogne, août-septembre, 1895. Annales de l'Université de Lyon, **26**: 365-419, pls. 13-17.

D'UDEKEM D'ACOZ, C., 1999. Inventaire et distribution des crustacés décapodes de l'Atlantique nord-oriental, de la Méditerranée et des eaux continentales adjacentes au nord de 25°N. Patrimoines naturels (M.N.H.N./S.P.N.), **40**: 1-383.

GARCÍA RASO, J. E., 1996. Crustacea Decapoda (excl. Sergestidae) from Ibero-Maroccan waters. Results of Balgim-84 expedition. Bulletin of Marine Science, **58**: 730-752.

GOFAS, S., 1993. Mission Océanographique Seamount 2. Compte-rendu et liste des stations. (MNHN, unpublished report).

HENDERSON, J. R., 1888. Report on the Anomura collected by H.M.S. Challenger during the years 1873-76. Report on the scientific results of the Voyage of H.M.S. Challenger during the years 1873-76. Zoology, **27**: vi + 221 pp., 21 pls.

INGLE, R. W. & M. E. CHRISTIANSEN, 2004. Lobsters, mud shrimps and anomuran crabs. Keys and notes for the identification of the species. In: J. H. CROTHERS & P. J. HAYWORD (eds.), Synopses of the British Fauna (New Series), **55**: 1-271. (Linnean Society, London).

MILNE EDWARDS, A., 1881. Compte rendu sommaire d'une exploration zoologique faite dans l'Atlantique, a bord du navire le Travailleur. Comptes Rendus Hebdomadaires de Séances de l'Académie des Sciences, Paris, **93**: 931-936.

MILNE EDWARDS, A. & E. L. BOUVIER, 1894. Considérations générales sur la famille des Galathéides. Annales des Sciences Naturelles, Zoologie (ser. 7), **16**: 191-327.

ORTMANN, A. E., 1892. Die Decapoden-Krebse des Strassburger Museums, mit besonderer Berücksichtigung der von Herrn Dr. Döderlein bei Japan und bei den Liu-Kiu-Inseln gesammelten und zur Zeit im Strassburger Museum aufbewahrten Formen. IV. Die Abtheilungen Galatheidea und Paguridea. Zoologische Jahrbücher, Abtheilung für Systematik, Geographie und Biologie der Thiere, **6**: 241-326, pls. 11, 12.

TÜRKAY, M., 1976. Decapoda Reptantia von der portugiesischen und marokkanischen Küste Auswertung der Fahrten 8, 9c (1967), 19 (1970), 23 (1971) und 36 (1975) von F.S. "Meteor". "Meteor" Forschungs-Ergebnisse, Reihe D, **23**: 23-44.

ZARIQUIEY ALVAREZ, R., 1968. Crustáceos decápodos Ibéricos. Investigacion Pesquera, **32**: xv + 510 pp.

First received 20 August 2016.
Final version accepted 17 October 2016.

[When citing this volume, please refer to Crustaceana 90 (2017) 771-1288]

Michael Türkay Memorial Issue

HOST SPECIFICITY AND FEEDING IN EUROPEAN PEA CRABS (BRACHYURA, PINNOTHERIDAE)

BY

CAROLA BECKER[1,2,3]) and MICHAEL TÜRKAY[2,†])

[1]) Queen's University Belfast, Queen's Marine Laboratory, 12-13 The Strand, Portaferry BT22 1PF, U.K.

[2]) Senckenberg Research Institute and Natural History Museum, Senckenberganlage 25, D-60325 Frankfurt, Germany

ABSTRACT

Pinnotherids, or pea crabs, are symbionts of invertebrates used for shelter and as food source. Feeding strategies and morphological adaptations to food uptake are important to understand host relations and how the host specificity is determined. We herein re-examine the host range of pinnotherids based on long-term collections from different localities in European waters. Both species of *Pinnotheres* are restricted to bivalves. *Pinnotheres pisum* infests mussels, oysters, the noble pen shell *Pinna nobilis* and other bivalves. *Pinnotheres pectunculi* is symbiotic with *Glycymeris glycymeris* and several other species of venerids. *Nepinnotheres pinnotheres* infests ascidians and *Pinna nobilis*. Observations on feeding in *Pinnotheres* reveal how mucus strings are brushed from the host gills with a setal comb in the chelipeds, which we describe by using scanning electron microscopy. Our observations on feeding structures are discussed in relation to the host specificity of each species and compared to other pinnotherid taxa, taking additional factors of host choice into account.

Key words. — *Nepinnotheres*, *pisum*, *pectunculi*, behaviour, ecology, chelipeds, claws, scanning electron microscopy (SEM)

ZUSAMMENFASSUNG

Pinnotheriden oder Muschelwächter leben in Wirtsbeziehung mit einer Reihe Wirbelloser, die als Unterschlupf und Nahrungsquelle genutzt werden. Kenntnisse über die Nahrungsaufnahme und morphologische Anpassungen an das Fressverhalten sind von großer Bedeutung um die spezifische Wirtsbeziehung und das Wirtsspektrum einzelner Arten besser zu verstehen. In der vorliegenden Arbeit untersuchen wir das Wirtsspektrum von Pinnotheriden auf der Basis von Langzeituntersuchungen verschiedener Fundorte in europäischen Meeresgebieten. Die *Pinnotheres*-Arten sind ausschließlich in Muscheln zu finden. *Pinnotheres pisum* bewohnt Miesmuscheln, Austern und die große Steckmuschel, *Pinna nobilis*. *Pinnotheres pectunculi* kommt in *Glycymeris glycymeris* und mehreren

[3]) Corresponding author; e-mail: c.becker@qub.ac.uk

[†]) Deceased.

Muschelarten der Familie Veneridae vor. *Nepinnotheres pinnotheres* bewohnt solitäre Ascidien und *Pinna nobilis*. Beobachtungen des Fressverhaltens beider *Pinnotheres*-Arten im Wirt zeigen, dass die Krabbe den mit Nahrungspartikeln angereicherten Kiemenschleim der Muschel gewinnt, indem sie die Kiemen mit einem Borstenkamm an den Scheren abbürstet. Diesen Borstenkamm untersuchen und beschreiben wir mit Hilfe von Rasterelektronenmikroskopie. Unsere Beobachtungen zum Fressverhalten werden im Zusammenhang mit dem beobachteten Wirtsspektrum der untersuchten Arten und im Vergleich zu anderen Arten unter Berücksichtigung zusätzlicher Faktoren, die das Wirtsspektrum beeinflussen, diskutiert.

INTRODUCTION

Pinnotherid crabs are symbionts in a variety of invertebrates used as a permanent refuge by the breeding female and often as a food source. Members of the subfamily Pinnotherinae De Haan, 1833 are found inside the body cavities of bivalves (Campos, 2001), gastropods (Campos, 1990; Geiger & Martin, 1999), ascidians (Campos, 1996a), holothurians (Ng & Manning, 2003) and brachiopods (Feldmann et al., 1996), as ectosymbionts of sea urchins (Bell, 1988; Campos, 1990; George & Boone, 2003), or inside the tubes of sessile polychaetes (Bezerra et al., 2006). Several bivalve-infesting species were studied due to their impact on commercially exploited bivalve species (Berner, 1952; Bierbaum & Ferson, 1986; Bierbaum & Shumway, 1988; Navarte & Saiz, 2004; Trottier et al., 2012). Infestation with pinnotherids can cause mechanical injuries to gills (Christensen & McDermott, 1958; Bierbaum & Ferson, 1986), reduce filter efficiency (Sugiura et al., 1960), oxygen consumption (Bierbaum & Shumway, 1988) and decrease metabolism in bivalves (Mercado-Silva, 2005) and, therefore, growth (Kruczynski, 1972; Navarte & Saiz, 2004; Trottier et al., 2012) and the reproductive potential of the host (Berner, 1952; O'Beirn & Walker, 1999; Bologna & Heck Jr., 2000; Ocampo et al., 2014). Taking these damages to hosts into account, many pinnotherid species can be regarded as truly parasitic having an important commercial impact on aquaculture and fisheries of bivalves (Berner, 1952; Bierbaum & Ferson, 1986; Bierbaum & Shumway, 1988; Navarte & Saiz, 2004, Trottier et al., 2012).

Three species of pinnotherids included in Pinnotherinae have been recorded from European waters: *Nepinnotheres pinnotheres* (Linnaeus, 1758), *Pinnotheres pisum* (Linnaeus, 1767) and *Pinnotheres pectunculi* Hesse, 1872. Two additional species, *Pinnotheres ascidicola* Hesse, 1872 and *Pinnotheres marioni* Gourret, 1888, both originally described from ascidians, were later synonymized with *N. pinnotheres* (see Becker & Türkay, 2010). In 2011, a fourth species, *Afropinnotheres monodi* Manning, 1993, previously only known from the coast of West Africa, has been discovered in several new hosts of bivalves in the Gulf of Cádiz (Dulce Subida et al., 2011).

The adaptations of pinnotherids for feeding, which determine the host specificity, are hardly understood. Pinnotherids are difficult to collect and identify due to their small size and cryptic way of life inside hosts. We re-examined the host range of *N. pinnotheres*, *P. pisum* and *P. pectunculi* based on material from extensive collections of potential hosts conducted between 1983 and 2010 at different localities in European waters. We also provide findings from behavioural observations on feeding in the bivalve-inhabiting species *P. pisum* and *P. pectunculi*. The structures involved in feeding are studied by scanning electron microscopy (SEM) and compared to fine structures in *N. pinnotheres* in particular regarding the host specificity of each species. Our results are discussed in comparison with data in the literature for the studied species and other pinnotherid taxa, as well as in terms of factors other than feeding, which play a role in host choice.

MATERIAL AND METHODS

Sampling

North Sea.— Specimens of *Pinnotheres pisum* were collected from the same population of horse mussels, *Modiolus modiolus* (Linnaeus, 1758), annually from 1983 to 1992 (table I), 2003 and 2005 to 2010 (table II) in the Helgoland Trench of the German Bight south of Island Helgoland with RV "Senckenberg" using a beam trawl. GPS data of hauls ranged from 54°08.419′-54°08.599′N to 07°50.921′-07°53.431′E at depths of 50 to 55 m. A total of 259 specimens of *P. pisum* were collected from 576 specimens of *M. modiolus* (tables I, II). In 1985, two additional exemplars of *P. pisum* were collected from 1869 specimens of *Spisula solida* (Linnaeus, 1758) from the Loreley Bank east of Island Helgoland, in depths of 12 to 15 with RV "Senckenberg" using a ring dredge (table III).

The Dogger Bank of the North Sea was sampled annually in summer (July/August) and potential hosts were inspected for pea crabs from 2004 to 2010 with RV "Senckenberg" (table IV). The sampling grid consisted of 37 stations covering an area of approximately 17.000 km^2 with depths of 16 to 33 m (GPS of central Dogger Bank: 54°43′17.17″N 2°46′4.07″E). In January 2010 an additional winter cruise was conducted with RV "Heincke". A total of 27 specimens was extracted from *Mactra stultorum* (Linnaeus, 1758), *Gari fervensis* (Gmelin, 1791), *Donax vittatus* (Da Costa, 1778) and *Spisula elliptica* (Brown, 1827) (table IV).

Non-infested species from the Dogger Bank (with numbers of examined specimens in parentheses):

Bivalves: *Abra alba* (W. Wood, 1802) (7), *Acanthocardia echinata* (Linnaeus, 1758) (13), *Aequipecten opercularis* (Linnaeus, 1758) (45), *Chamelea* spp. (88), *Clausinella fasciata* (Da Costa, 1778) (4), *Corbula gibba* (Olivi, 1792) (5), *Dosinia*

TABLE I

Long-term study of *Pinnotheres pisum* (Linnaeus, 1767) in the horse mussel *Modiolus modiolus* (Linnaeus, 1758) in the Helgoland Trench of the German Bight, North Sea from 1983 to 1992

Month	*Modiolus* (*n*)	Infested (*n*)	Infestation rate (%)	Single ♀		Pair ♂♀		Single ♂	
				n	%	*n*	%	*n*	%
January	27	14	52	8	57	5	36	1	7
February	30	7	23	4	57	3	43	0	0
March	50	25	50	18	72	5	20	2	8
April	33	15	45	10	67	3	20	2	13
May	85	37	44	29 (3 ovi)	78	8 (1 ovi)	22	0	0
June	25	6	24	3 (2 ovi)	50	3 (1 ovi)	50	0	0
July	14	12	86	7 (7 ovi)	58	4 (4 ovi)	33	1	8
August	54	22	41	16 (6 ovi)	73	5	23	1	5
September	41	21	51	18	86	3	14	0	0
October	6	3	50	1	33	2	67	0	0
November	89	42	47	33	79	7	17	2	5
December	93	50	54	29	58	15	30	6	12
Total	**547**	**254**	**47**	**176**	**69**	**63**	**25**	**15**	**6**

Numbers of investigated hosts and pinnotherids, infestation rates and sex-ratios are shown. Proportions of ovigerous females in parentheses. *n*, number; ovi, ovigerous.

TABLE II
Sampling in the Helgoland Trench, North Sea from 2003 to 2010

Date	*Modiolus modiolus* (Linnaeus, 1758)	Infestation with *Pinnotheres pisum* (Linnaeus, 1767)	Infestation rate (%)
May 2003	14	5	36
August 2003	13	–	0
August 2005-2010	2	–	0

Annual summer samplings from 2005 to 2010 are combined. Each year 1-3 hauls were conducted.

spp. (8), *Ensis* spp. (59), *Nucula* cf. *nitidosa* Winckworth, 1930 (15), *Spisula solida* (3), *Spisula subtruncata* (Da Costa, 1778) (6), *Spisula* sp. (21), *Tapes rhomboides* (Pennant, 1777) (17), *Thracia* sp. (11), *Venerupis senegalensis* (Gmelin, 1791) (23).

Ascidians: *Ascidiella scabra* (Müller, 1776) (>1000).

Atlantic.— Specimens of *P. pisum*, *P. pectunculi* and *N. pinnotheres* were hand-collected from different hosts during fieldwork at the Roscoff Marine Station (Brittany, France) by collectors in 1990, 1991 and 2006, 2007 (table V). Additional specimens of *P. pisum* and *P. pectunculi* were collected in 1993, 1994 from different hosts and locations on the French northeastern Atlantic coast between Brest and Le Havre (table V). In April 2008, additional material of *P. pectunculi* in *Glycymeris glycymeris* (Linnaeus, 1758) was obtained from the diving service of the Roscoff Marine Station collected in waters around Roscoff (table V) and sent alive to the Senckenberg Research Institute for observations on behaviour of pinnotherids. Specimens of *P. pisum* were obtained from cultured *Mytilus edulis* Linnaeus, 1758 from Oléron (Bay of Biscay, France) bought on the fish market in 2005 and 2007 (René Laudigeois, Kleinmarkthalle Frankfurt, Germany).

List of examined but not infested species (with numbers of specimens examined in parentheses): [please see next page]

TABLE III
Pinnotheres pisum (Linnaeus, 1767) in *Spisula solida* (Linnaeus, 1758) from the Loreley Bank (German Bight, North Sea)

Date	Infestation (hosts)
May 1985	1♂♀ (787)
July 1985	– (329)
Aug. 1985	– (448)
Nov. 1985	– (305)
Total n/n	**1/1869**

Only one couple was found out of 1869 potential hosts. Numbers of investigated hosts in parentheses. n/n, number of infestations/number of investigated hosts; ♂♀, pair.

TABLE IV
Pinnotheres pisum (Linnaeus, 1767) in different bivalve species from the Dogger Bank, North Sea

Date	*Mactra stultorum* (Linnaeus, 1758)	*Gari fervensis* (Gmelin, 1791)	*Donax vittatus* (Da Costa, 1778)	*Spisula elliptica* (Brown, 1827)
2004 summer	1/9	0/3	–	–
2006 summer	3/14	2/8	0/1	–
2008 summer	6/8	1/4	–	0/2
2009 summer	5/15	0/1	0/2	–
2010 winter	1/20	0/10	1/38	1/11
2010 summer	4/34	0/11	1/56	0/99
Total n/n	**21/100**	**3/37**	**2/97**	**1/112**
Infestation rate (%)	**21**	**8**	**2**	**<1**

n/n, number of infestations/number investigated hosts.

Bivalves: *Acropagia crassa* (Pennant, 1777) (16), *Cerastoderma edule* (Linnaeus, 1758) (41), *Dosinia* sp. (1), *Mactra stultorum* (1), *Venerupis senegalensis* (15).

Ascidians: *Phallusia mammillata* (Cuvier, 1815) (124).

Mediterranean.— In the northern Adriatic Sea (Rovinj, Croatia), *P. pisum* and *N. pinnotheres* were obtained from different hosts in 2003, 2005, 2007 and 2009 (table VI), either hand-collected by scuba- and skin-diving in depths from 1 to 35 m or by beam trawl on trips with RV "Burin" from the Ruđer Bošković Institute on different sample sites ranging from 45°02′-45°07′N to 13°36′-13°40′E.

As an endangered species and under strict protection according to the European Council Directive 92/43/EEC (Annex IV) and the national laws of most Mediterranean countries, the noble pen shell *Pinna nobilis* Linnaeus, 1758 was not collected during fieldwork. A few specimens were however obtained as by-catch from local fishermen (table VII).

List of non-infested species (numbers of specimens examined in parentheses):

Bivalves: *Spondylus gaederopus* Linnaeus, 1758 (15), *Pecten jacobaeus* (Linnaeus, 1758) (7), *Chlamys* spp. (11), *Lima* sp. (5), *Venus verrucosa* Linnaeus, 1758 (4).

Ascidians: *Phallusia mammillata* (33), *Ascidia virginea* Müller, 1776 (3).

The material used for the present study is also listed in Becker & Türkay (2010) and is integrated into the crustacean collection and digital online database of Senckenberg (available online at http://sesam.senckenberg.de/).

Live observations on feeding behaviour

To observe feeding behaviour of the species of *Pinnotheres* inside bivalves, behavioural experiments were conducted with adult female specimens in manipulated

TABLE V

Nepinnotheres pinnotheres (Linnaeus, 1758) and *Pinnotheres pectunculi* Hesse, 1872 from the northeastern Atlantic coast (Brittany, France)

Date	*Nepinnotheres pinnotheres*	*Pinnotheres pectunculi*			
	Ascidia mentula Müller, 1776	*Glycymeris glycymeris* (Linnaeus, 1758)	*Venus verrucosa* Linnaeus, 1758	*Venus casina* Linnaeus, 1758	*Clausinella fasciata* (Da Costa, 1778)
May 1990, April 1991	–	>200♀, 1♂♀ (unknown)	11♀, 3♂ (unknown)	9♀, 2♂♀, 2♂ (28)	2♀, 2♂ (24)
April 2006	1♀, 1♂ (3)	1♀ (12)		–	–
March 2007	3♀, 1♀♂ (49)	–		–	–
April 2008	–	64♀ (89)		–	–
Infestation rate (%)	**12**	**64**	**?**	**53**	**17**

Numbers of examined hosts in parentheses.

TABLE VI

Nepinnotheres pinnotheres (Linnaeus, 1758) from ascidians and *Pinnotheres pisum* (Linnaeus, 1767) from bivalves, northern Adriatic Sea

Date	*Nepinnotheres pinnotheres*				*Pinnotheres pisum*	
	Ascidia mentula Müller, 1776	*Halocynthia papillosa* (Linnaeus, 1767)	*Phallusia mammillata* (Cuvier, 1815)	*Microcosmus* spp.	*Ostrea edulis* Linnaeus, 1758	*Mytilus gallo-provincialis* Lamarck, 1819
August 2003	3♀, 3♂, 1♂♀ (11)	– (10)	– (31)	2♀, 3♂ (253)	5♀, 2♂♀ (31)	(9)
March 2005	1♂ (6)	–	–	1♀, 2♂ (113)	–	–
August 2005	–	3♀ (21)	– (2)	10♀ (350)	14♀ (75)	1♀ (12)
August 2007	– (2)	–	2♀ (7)	2♀, 1♂ (145)	– (29)	(7)
August 2009	1♀ (1)	–	1♀ (19)	8♀, 4♂, 1♂♀ (228)	–	–
Total n/n	**9/20**	**3/31**	**3/59**	**34/1089**	**21/135**	**1/28**
Infestation rate (%)	**45**	**7**	**5**	**3**	**16**	**4**

Numbers of examined hosts in parentheses. n/n, number of findings/number of investigated hosts.

TABLE VII

Nepinnotheres pinnotheres (Linnaeus, 1758) and *Pinnotheres pisum* (Linnaeus, 1767) from the Mediterranean pen shell, *Pinna nobilis* Linnaeus, 1758, northern Adriatic Sea

Date	*Pinna nobilis*	*Nepinnotheres pinnotheres*	*Pinnotheres pisum*
August 2003	6/6	2♀, 1♂♀	1♀, 2♂♀
December 2003	2/2	1♀	1♂♀
March 2005	2/2	1♀	1♀
August 2005	0/1	–	–
August 2007	0/1	–	–
August 2009	1/1	–	1♂♀
Total n/n	**11/13**	**5**	**6**
Infestation rate (%)	Total **85**	**39**	**46**

n/n, number of findings/number of investigated hosts.

hosts in aquaria. *Pinnotheres pisum* was observed in *Ostrea edulis* Linnaeus, 1758 at Ruđer Bošković in 2005. *Pinnotheres pectunculi* was observed in *G. glycymeris* at Senckenberg in 2008. The specimens were extracted from their hosts and transferred into a bivalve belonging to the same species the crab was obtained from, with one shell removed to allow for observation. Pea crabs were transferred to fresh bivalves every 24 hours during the experiment. Three specimens of *P. pisum* and *P. pectunculi* each were observed for at least three days. Feeding behaviour was observed repeatedly during the experiment and images were taken with a Nikon Coolpix 4500 camera.

Fine structures of chelipeds (SEM)

The SEM study was conducted at Senckenberg. Six specimens of *P. pisum*, two of *P. pectunculi*, and four of *N. pinnotheres* were used. Specimens were preserved in 96% ethanol and cleaned in an ultrasonic bath for 30 s to 2 min. Samples were dried through a Balzor's CPD 030 critical point dryer and sputter coated with a gold/palladium-composite in a Edwards S 150B sputter coater for 3 min (equivalent to a coating of 20 nm thickness). Samples were subsequently examined with scanning electron microscope CamScan (Elektronenoptik) and photographs were taken with Orion® software. The description of setae types follows the nomenclature established by Garm (2004).

RESULTS

Host specificity

Data on hosts infested with *Pinnotheres pisum*, *P. pectunculi* and *Nepinnotheres pinnotheres* from different geographical regions are presented in tables I-VII

TABLE VIII

Summary of host species recorded for *Nepinnotheres pinnotheres* (Linnaeus, 1758), *Pinnotheres pisum* (Linnaeus, 1767) and *Pinnotheres pectunculi* Hesse, 1872 in the North Sea, Northeast Atlantic and Mediterranean

	Nepinnotheres pinnotheres	*Pinnotheres pisum*	*Pinnotheres pectunculi*
North Sea	–	*Modiolus modiolus* (Linnaeus, 1758) *Mytilus edulis* Linnaeus, 1758 *Mactra stultorum* (Linnaeus, 1758) *Spisula solida* (Linnaeus, 1758) *Spisula elliptica* (Brown, 1827) *Gari fervensis* (Gmelin, 1791) *Donax vittatus* (Da Costa, 1778)	–
NE Atlantic	*Ascidia mentula* Müller, 1776	*Mytilus edulis* Linnaeus, 1758 *M. galloprovincialis* Lamarck, 1819 *Spisula solida* (Linnaeus, 1758)	*Glycymeris glycymeris* (Linnaeus, 1758) *Venus verrucosa* Linnaeus, 1758 *Venus casina* (Linnaeus, 1758) *Clausinella fasciata* (Da Costa, 1778)
Mediterranean	*Pinna nobilis* Linnaeus, 1758 *Ascidia mentula* Müller, 1776 *Halocynthia papillosa* (Linnaeus, 1767) *Microcosmus* spp. Heller, 1877	*Pinna nobilis* Linnaeus, 1758 *Ostrea edulis* Linnaeus, 1758 *M. galloprovincialis* Lamarck, 1819	
	In ascidians and in the bivalve *Pinna nobilis*	**In many species of bivalves**	**In *Glycymeris glycymeris* bivalves and venerids**

Fig. 1. Selection of bivalves investigated from the North Sea (original sizes). Names of hosts of *Pinnotheres pisum* (Linnaeus, 1767) in white letters, names of non-infested species in grey letters. Photographs: Sven Tränkner, Senckenberg.

showing infestation rates and sex of pinnotherid specimens. An overview on the recorded hosts for each species is presented in table VIII for North Sea, northeastern Atlantic and Mediterranean, respectively. Fig. 1 shows bivalve species examined in the North Sea indicating hosts of *P. pisum* and bivalves that were not infested during our study. Fig. 2A shows *Ostrea edulis* and *Mytilus* spp., regular hosts of *P. pisum* but not of *P. pectunculi. Pinna nobilis* is a host of *P. pisum* and *N. pinnotheres* in the Mediterranean (fig. 2B). The bivalve host species of *P. pectunculi* are shown in fig. 2C.

Feeding behaviour

Adult female specimens of *P. pisum* and *P. pectunculi* were both observed in their manipulated bivalve hosts with one shell removed (fig. 3A-E). Crabs were sometimes hiding between the bivalve gills for a certain period of time but were repeatedly observed feeding during the experiments. When feeding, specimens were positioned close to the mouth of the bivalve and fed by brushing the bivalve gills with the setose inner ventral side of the cheliped (fig. 3C, E). Mucus strings from the gills stuck to setae on the chelipeds and were then conveyed towards the mouthparts where the setose third maxillipeds took over the mucus strings and conveyed them further into the mouth (fig. 3C). Specimens of *N. pinnotheres* were not observed for extended periods in their dissected host and therefore feeding behaviour was not observed (fig. 4).

Fine structure of chelipeds

The chelipeds of *P. pisum* and *P. pectunculi* both possess a distinct setal comb at the inner ventral face of the chelipeds (fig. 5A-F). Fig. 5A shows the cheliped of an adult female *P. pisum*, but males and juvenile female stages have the same types of setae with a similar distribution, also possessing a setal comb. The remaining surface of the cheliped is rather smooth except for the area around the cutting edges of the claw. Rows of denticles are formed on both the ventral and dorsal cutting edges, which are surrounded by scattered simple setae (fig. 5B, C). The setal comb is formed by long pappo-serrate setae, oriented regularly, directing towards the tip of the cheliped. Longitudinal rows of fine setules are running along setal shafts (fig. 5D, E). The distal tips are feathered increasing their surface (fig. 5F). *Pinnotheres pectunculi* differs from *P. pisum* only in having an additional small tooth at the cutting edge of the fixed finger. The chelipeds of male and female *N. pinnotheres* lack a distinct setal comb. The whole surface is setose instead (fig. 6A-F). Fields of long plumose setae and simple setae of different lengths are situated close to the cutting edges of the claw (fig. 6B, C). The remaining surface of the cheliped is completely covered by short plumose setae (fig. 6D-F), which are not only present on the chelipeds but cover the whole pilose body in *N. pinnotheres*.

Fig. 2. Bivalve hosts from the northeastern Atlantic and the Mediterranean (original sizes, except for B: see scale bar). A, Oyster and mussel hosts of *Pinnotheres pisum* (Linnaeus, 1767) in the Mediterranean; B, the Mediterranean pen shell *Pinna nobilis* Linnaeus, 1758 infested by *Nepinnotheres pinnotheres* (Linnaeus, 1758) and *P. pisum*, respectively; C, hosts of *Pinnotheres pectunculi* Hesse, 1872: the dog cockle *Glycymeris glycymeris* (Linnaeus, 1758) and venerid species. Photographs: A, B, Carola Becker; C, Sven Tränkner and Carola Becker; *Venus verrucosa* Linnaeus, 1758, Hans Hillewart, VLIZ, Belgium.

Fig. 3. Feeding of *Pinnotheres* species in bivalves. A, Adult female of *Pinnotheres pisum* (Linnaeus, 1767) in oyster, *Ostrea edulis* Linnaeus, 1758, with right valve removed; B, chelipeds are oriented ventrally by a distortion of the carpus; C, mucus strings are picked up by setal comb of the claw (black arrow on mucus string); D, E, *Pinnotheres pectunculi* Hesse, 1872 feeding in *Glycymeris glycymeris* (Linnaeus, 1758).

DISCUSSION

Host specificity of the European species

Our data on the host specificity of *Nepinnotheres pinnotheres*, *Pinnotheres pisum* and *P. pectunculi* is heterogeneous in terms of numbers of potential

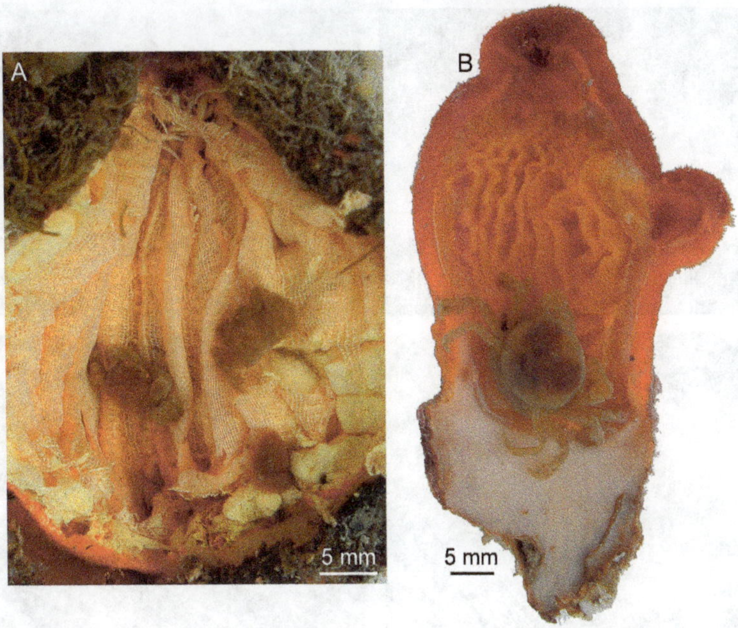

Fig. 4. *Nepinnotheres pinnotheres* (Linnaeus, 1758) in dissected ascidians. A, Pair of male and hard stage female in *Microcosmus* sp.; B, adult female in *Halocynthia papillosa* (Linnaeus, 1767).

hosts examined and in the coverage of sample sites from different geographic regions. Therefore, we may not have captured the complete host range of each species throughout their distribution but several points can be drawn. The host range observed for each species in different geographic regions is summarised in table VIII. As indicated by Castro (2015), pinnotherids are not host specific at a species level but often use certain groups of invertebrates as hosts. For example, *N. pinnotheres* infests several species of solitaire ascidians as well as the noble pen shell, *Pinna nobilis*, in the Mediterranean. Pea crabs from ascidians distributed in European waters were earlier described as separate species (*Pinnotheres ascidicola* Hesse, 1872 and *Pinnotheres marioni* Gourret, 1888), but both were later synonymized with *N. pinnotheres* (see Becker & Türkay, 2010).

For *P. pisum*, records of ascidian hosts are found in the literature (Lévi, 1951; Schmitt et al., 1973) but our results suggest that the species is restricted to bivalves. We suggest records of ascidian hosts to be based on misidentifications of *P. pisum* and confusion with *N. pinnotheres*. While males of both species can be distinguished by their first gonopods, females look very similar and can be easily mistaken. Ascidian hosts of *N. pinnotheres* are *Halocynthia papillosa*, *Ascidia mentula* Müller, 1776 and *Microcosmus* spp.; *Phallusia mammillata* however, high numbers of which were examined, is not a host. *Pinnotheres pisum* has a broad range of bivalve hosts, even including small-size species such as *Donax vittatus*

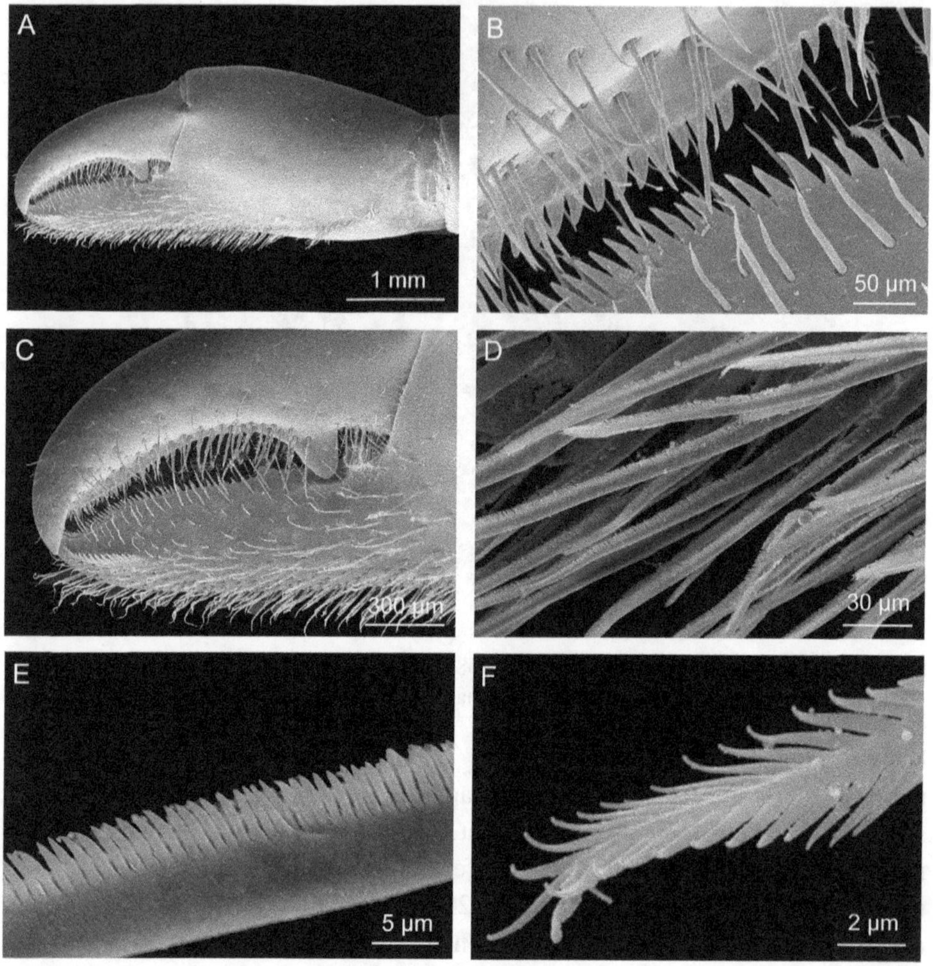

Fig. 5. Cheliped of adult female *Pinnotheres pisum* (Linnaeus, 1767) (SEM-micrographs). A, Palm of right cheliped showing setal comb; B, soft denticles and simple setae on cutting edge of claw; C, fixed (propodus) and movable finger (dactylus) of the claw showing setation; D, setal comb consisting of long regularly orientated pappo-serrate setae; E, higher magnification on setulation of setae shaft; F, distal tip of pappo-serrate setae.

(see fig. 1). *Pinnotheres pectunculi* is also restricted to bivalves. *Pinnotheres pectunculi* was originally described from *G. glycymeris*. Our study reveals that *P. pectunculi* is also found in several species of venerid bivalves (see table V, fig. 2C). There is no overlap in the host range of *P. pisum* and *P. pectunculi*, even not in specimens from the same geographical region (see table VIII). The only host which is used by more than one of the European pinnotherid species is *P. nobilis* in the Mediterranean, which can be infested by either *P. pisum* or *N. pinnotheres* and is the most frequented host out of all with an infestation rate

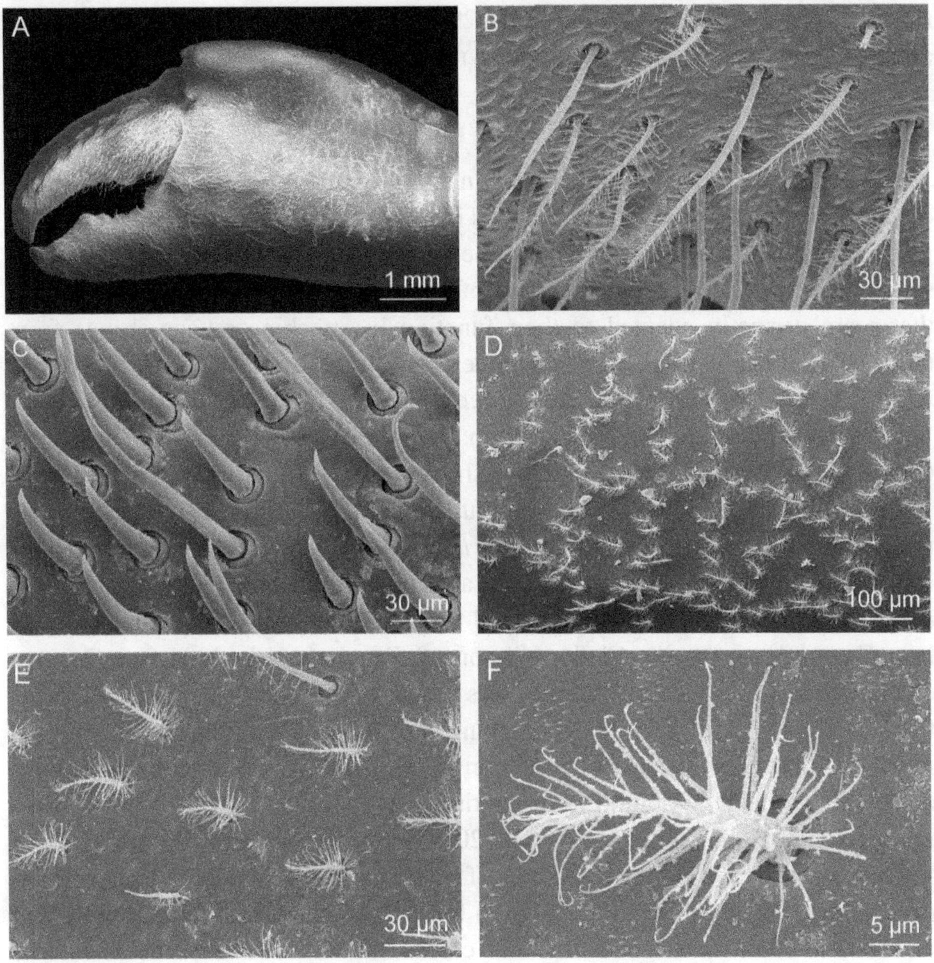

Fig. 6. Cheliped of adult female *Nepinnotheres pinnotheres* (Linnaeus, 1758) (SEM-micrographs). A, Claw with fixed and movable finger; B, field of long plumose setae around the cutting edges of the claw; C, simple setae of different lengths along the cutting edges of the claw; D, complete surface covered by different types of setae; E, short plumose setae that cover the whole body; F, higher magnification of plumose setae with fine soft setules.

of 85% ($n = 13$). Despite this, we never found more than one pair of the same species within its mantle cavity. The incidence of *P. pisum* inside a host seems to exclude the entry of *N. pinnotheres* and vice versa. Furthermore, Navarte & Saiz (2004) demonstrated in *Tumidotheres maculatus* (Say, 1818) that the presence of one berried female forecloses the intrusion of other females of the same species. Nevertheless, one host can hold several conspecific males together with one female (Silas & Agarswami, 1967). In *Arcotheres sinensis* (Shen, 1932), infesting *Mytilus galloprovincialis* Lamarck, 1819 in the Yellow Sea in China, one to six males were recorded from one mussel (Sun et al., 2005). This phenomenon, however, goes along with an extremely high infestation rate (Silas & Agarswami, 1967).

Feeding and structures used in food uptake

Observations on feeding behaviour in both species of *Pinnotheres* show the crucial function of their setose chelipeds in brushing the bivalve gills and thereby gather the gills mucus they feed on. The SEM study reveals fine structures of a setal comb in the chelipeds composed of pappo-serrate setae with feathered tips that seem perfectly adapted to their function in attaching mucus strings from the bivalve gills. Several other pinnotherid species that infest bivalves possess a similar setal comb, e.g., *Zaops ostreus* (Say, 1817) (as *Zaops ostreum* (Say, 1817)) (see Manning, 1993), *Fabia* spp. (Campos, 1996b) and *Arcotheres* spp. (Campos, 2001; Ahyong & Ng, 2007) shown in the line drawings of taxonomic studies, which might have a similar function in feeding. Such a setal comb is also present in *Calyptraeotheres garthi* (Fenucci, 1975), whose ontogenetic changes were described in Ocampo et al. (2017). A detailed study on feeding behaviour of *C. garthi* has demonstrated the role of chelipeds in feeding from its host *Crepidula cachimilla* Cledón, Simone & Penchaszadek, 2004 (see Ocampo et al., 2014). The feeding process was observed to be stereotyped involving the setose chelipeds of crabs in grasping mucus strings enriched with filtered phytoplankton. The successful ingestion of food in crabs was confirmed by a change in stomach coloration from whitish or translucent to greenish due to the ingested phytoplankton (Ocampo et al., 2014).

Kruczynski (1975) studied feeding of adult female *T. maculatus* (Say, 1818) (as *Pinnotheres maculatus* Say, 1818) in bivalves by the use of radioactive tracer to estimate food uptake showing that crabs feed on organic particles filtered and accumulated by their hosts. The experiments showed no food uptake in clawless crabs; intact crabs, however, fed also on phytoplankton from Petri dishes by picking planktonic organisms with their claws and cleaning themselves continuously (Kruczynski, 1975). In *T. maculatus* the elongated setose dactyli of the last pair of pereiopods grab mucus strings from the gills (Caine, 1975). Similarly elongated pereiopods are also present in members of the genera *Fabia* (see Campos, 1996b) and *Arcotheres* (see Ahyong & Ng, 2007). Some species have a single elongated leg and are therefore asymmetrical (Gordon, 1936; Griffin & Campbell, 1969; Campos & Manning, 2001; Campos, 2001). The body side of leg elongation in *Arcotheres* is related to the settlement of the bivalve host *Barbatia virescens* (Reeve, 1844) on either the left or right shell which results in a different feeding positon for the pea crab in relation to the mouth of the bivalve (Watanabe & Henmi, unpubl. data). Not only chelipeds but also pereiopods can thus have a crucial function in feeding of pinnotherids from their bivalve hosts.

Feeding behaviour of *N. pinnotheres* in ascidians was not observed during the present study. Infestation rates were very low and ascidians were not successfully maintained in aquaria for more than a few days. When ascidians were dissected to search for pea crabs, feeding was not observed in the short period of time before ascidians died. We can therefore only speculate on feeding in *N. pinnotheres*. The species lacks a specific setal comb on the claws. The whole chelipeds as well as the remaining body surface are pilose instead, being covered by short plumose setae. Additionally, *N. pinnotheres* has elongated dactyli in the last pair of pereiopods (Becker & Türkay, 2010), which could be involved in feeding, similar to the species mentioned above.

Bivalve and ascidian hosts are both suspension feeders by filtering organic matter from the seawater and accumulating food particles in a mucous secretion, which is ingested by the bivalve-infesting pinnotherids and probably also in species living in ascidians. The anatomy of the filtering structures in bivalves and ascidians is however different and therefore supposably also the feeding position of the crab. In bivalves, filtering is through gills consisting of several overlying valves, on which the *Pinnotheres* species roam. In ascidians, *N. pinnotheres* is situated inside the sac-like pharynx that filters sea water through pharyngeal slits. The overall pilosity in *N. pinnotheres* might play a role in accumulating mucus from the surrounding ascidian pharynx. During the preparation of samples for SEM observations, specimens from ascidians needed to be intensively cleaned because they were completely covered with debris. In the similarly pilose species *T. maculatus* feeding was observed to involve constant cleaning of the body to obtain the gills mucus (see above, Kruczynski, 1975). A similar feeding technique is conceivable for *N. pinnotheres*, but also grasping of mucus with the elongated last pair of pereiopods is possible. Whether feeding in *N. pinnotheres* is through the whole setose body surface, the elongated pereiopods, or both, can only be resolved through behavioural observations, for example by using an endoscopic camera introduced through the branchial siphon in future studies.

Factors of host choice

Next to morphological adaptations to feeding inside the host, several other factors determine the host specificity of pinnotherids. The first and foremost factor of host choice is the ability of the crab to find and recognize a suitable host in its habitat and to successfully enter this host.

Host recognition and host preferences.— Whether host recognition has a genetic basis or is a learned phenomenon was studied in New Zealand pea crabs (Stevens, 1990b). *Nepinnotheres atrinicola* (Page, 1983) (as *Pinnotheres atrinicola* Page, 1983) is host-specific, only infesting the fan mussel *Atrina zelandica* (Gray,

1835), whereas *Nepinnotheres novaezelandiae* (Filhol, 1885) (as *Pinnotheres novaezelandiae* Filhol, 1885) is a host-generalist. In behavioural experiments, it was not possible to induce a change in host recognition by conditioning crabs to novel hosts. Specimens of *N. novaezelandiae* showed a preference for its host *M. edulis* over its other host *Perna canaliculus* (Gmelin, 1791). This finding suggests that populations of *N. novaezelandiae* from different hosts represent biologically discrete units with distinct host recognition systems (Stevens, 1990b). This is also supported by a genetic differentiation between host races (Stevens, 1990a). A study on host odour attraction in host-generalist *T. maculatus* rather supports a plastic "chemical search image" and response specificity for certain species of hosts was interpreted as evidential for olfactory induction to hosts (Derby & Atema, 1980). Another experimental study on host choice in *T. maculatus* extracted from *Argopecten irradians concentricus* (Say, 1822), however, showed no preference for its original host over *Atrina rigida* (see Sastry & Menzel, 1962). Host recognition is supposed to be based on chemotactic stimuli (Sastry & Menzel, 1962) and antennules were identified as the principal structures of chemoreception used in host location in experiments with ablated antennules (Derby & Atema, 1980).

In a study on the ascidian-infesting species *Tunicotheres moseri* (Rathbun, 1918), crabs showed a preference for certain ascidian host species. Male crabs responded to conspecific non-berried females, but not to berried females or other males (Ambrosio & Brooks, 2011). The recognition of conspecifics and host choice were also studied in *Pinnixa chaetopterana* Stimpson, 1860 symbiotic with polychaetes *Chaetopterus variopedatus* (Renier, 1804) and *Amphitrite ornata* (Leidy, 1855) (see Grove & Woodin, 1996). Male and female crabs showed no attraction to empty hosts alone. They were attracted to isolated conspecifics instead. Ocampo et al. (2012) argued that host choice in host-generalists should be driven by reproductive benefits one host offers over another. The pinnotherid *C. garthi* inhabiting the limpets *C. cachimilla* and *Bostrycapulus odites* Collin, 2005, respectively, attains larger body sizes and therefore higher fecundities and brood weights in *C. cachimilla* than in *B. odites*. In behavioural experiments however, *C. garthi* preferred the host it was obtained from instead of choosing the reproductively more beneficial host (Ocampo et al., 2012).

Host entry.— The behaviour involved in the entry of bivalves has been studied in males of *T. maculatus* (as *P. maculatus*), entering the bay scallop *A. i. concentricus* (see Eidemiller, 1969) and in males of *N. novaezelandiae* entering the mussel *P. canaliculus* (see Trottier & Jeffs, 2015). In both studies, the male was observed to touch the mantle of the bivalve with chelipeds or legs attempting to gain entrance, which stimulated the valves to gape and allowed the pea crab to enter. Host entry was also observed in *Pinnixa tumida* Stimpson, 1858 symbiotic with the holothurian *Paracaudina chilensis* (Müller, 1850) (see Takeda et al., 1997).

The crab started the entry by touching the sea cucumber's posterior region with chelipeds and walking legs, which stimulated widening of the holothurian anus, so that the crab could crawl in. Specimens were observed to fight over a host if two crabs arrived at the holothurian posterior at the same time (Takeda et al., 1997).

In all observed cases of host entry, pinnotherids induced a reaction of the host by tactile stimuli to facilitate intrusion. The entry of a host still seems to be a critical event as males of *P. pisum* (which enter hosts repeatedly in search for females) often lack distal articles of pereiopods (pers. obs.), which probably got squashed within the bivalve shells while trying to enter. Ascidian hosts may require a slightly different entry strategy than bivalves or holothurians, but it is very likely that tactile stimuli are involved as well. One remarkable difference in ascidian hosts is the position of the branchial siphon, where pinnotherids supposedly enter, which is distant from the substrate in many species. Entry in such hosts should therefore only be possible in the hard stages of males and juvenile females, which are capable of swimming by paddling with their setose walking legs (Hartnoll, 1972).

Host size.— The size of hosts is an obvious criterion for host choice because some hosts offer larger shelter than others. Higher infestation rates in large-size hosts are present within one host species (Haines et al., 1994; Hsueh, 2003) but also among different hosts (see tables III-V and VII). The significance of shelter space is apparent in female pinnotherids reaching greater body sizes in large-size hosts (Houghton, 1963; Pearce, 1964; Seed, 1969; Pregenzer Jr., 1978; Palmer, 1995). Female *Z. ostreus* (Say, 1817) show direct correlation between body size and host dimension (as *Pinnotheres ostreum* Say, 1817; see McDermott, 1962). *Tumidotheres maculatus* also has several hosts of different sizes and specimens from the large-size pen shell, *Atrina rigida* (Lightfoot, 1786), are significantly larger than those from the small-size bay scallop, *Argopecten irradians* (Lamarck, 1819). This correlation is nevertheless only present in females but not in the partially free-living males (Kane & Farley, 2006). Large female body sizes lead to larger broods and are therefore reproductively beneficial to the species (Ocampo et al., 2012). A correlation between host and the size of crabs in the European species becomes obvious when comparing specimens of *N. pinnotheres* from ascidians and *P. nobilis* or specimens of *P. pisum* from *M. modiolus* and smaller bivalve hosts (Becker & Türkay, 2010). Larger hosts are also likely to offer greater food resources. It is therefore not surprising that *P. nobilis* is such a highly infested host. Damage to hosts is probably also related to host size and it is very likely that the infestation with pinnotherids impacts small-size hosts more severely than large-size hosts.

Ecological factors of host choice.— A number of other ecological factors play a role in host choice as in the case of the abundance and distribution of hosts. Water depth was studied for *P. pisum* in *M. edulis* demonstrating a significant increase of infestation rates from intertidal to subtidal (Houghton, 1963; Haines, 1994). The same is the case in *T. maculatus* (as *P. maculatus*; see Kruczynski, 1974) and *N. novaezelandiae* (as *P. novaezelandiae*; see Jones, 1977). In contrast to dense host aggregations as mussel beds, the distribution of hosts and the resulting abundance of pinnotherids can be patchy in some habitats, as in *P. nobilis* and in some other species of bivalves and ascidians. Few potential hosts were found in the samples from the Dogger Bank of the North Sea and the infestation with *P. pisum* was so rare that it is hard to figure how these pinnotherids find their conspecifics in such habitats.

Conclusions and outlook

Due to their small size and their cryptic way of life, pinnotherids are notoriously difficult to study and many issues in their taxonomy remain unresolved to date (Becker, 2010; Palacios Theil et al., 2016). Misidentifications of pinnotherid species have consequently led to unreliable data on host relations as the herein revised host range of *P. pisum*, *P. pectunculi* and *N. pinnotheres* demonstrates in comparison to previous data from the literature (Schmitt et al., 1973). Our observation on feeding behaviour and the identification of feeding structures reveal noteworthy adaptations. The setal comb used in feeding from gills in the bivalve-inhabiting species *P. pisum* and *P. pectunculi* is a character also present in many other species. A re-examination of the distribution of setal combs among pinnotherids, together with studies on host specificity, will potentially reveal a phylogenetic value of this character in future studies. The identification of structure involved in feeding is an important basis to better understand adaptations of pinnotherids to the anatomy of their hosts and may help to explain how the specific range of hosts is determined.

ACKNOWLEDGEMENTS

We thank all individual collectors for contributing specimens to this study: Cédric d'Udekem d'Acoz (Royal Belgium Institute of Natural Sciences, Brussels, Belgium), Axel Magdeburg and Martin Wagner (Goethe-University, Frankfurt, Germany), Moritz Sonnewald (Senckenberg) and Jutta Klein (formerly Senckenberg). We also thank Sven Tränkner (Senckenberg) for providing photographs and Marie-Louise Tritz and Dieter Fiege for their help with SEM-techniques. We are also grateful to Peter Castro for English corrections and comments to the manuscript and a second, anonymous reviewer for his helpful comments and suggestions.

PERSONAL NOTE

The data published in the present study was part of my Ph.D. thesis advised by Michael Türkay. He was fascinated by pea crabs for their great diversity, symbiotic lifestyle and complex life history, and has been collecting pinnotherids intensively since the 1980s. When I first walked into Senckenberg in 2002 to ask him for an internship on crabs, he suggested that I take on his pea crab collection. The material deposited at Senckenberg actually provided enough subject matter to follow up with a Diploma and a Ph.D. project, and altogether I stayed with Michael Türkay for over 10 years. He was the best mentor imaginable and I have wonderful memories of the times we spent together. I greatly appreciate his long lasting support and he is sadly missed.

REFERENCES

AHYONG, S. T. & P. K. L. NG, 2007. The pinnotherid type material of Semper (1880), Nauck (1880) and Bürger (1895) (Crustacea: Decapoda: Brachyura). Raffl. Bull. Zool., **16**: 191-226.

AMBROSIO, L. J. & W. R. BROOKS, 2011. Recognition and use of ascidian hosts, and mate acquisition by the symbiontic pea crab *Tunicotheres moseri* (Rathbun, 1918). Symbiosis, **53**: 53-61.

BECKER, C., 2010. European pea crabs — taxonomy, morphology, and host-ecology (Crustacea: Brachyura: Pinnotheridae). (Dissertation, Goethe-Universität, Frankfurt am Main).

BECKER, C. & M. TÜRKAY, 2010. Taxonomy and morphology of European pea crabs (Crustacea: Brachyura: Pinnotheridae). J. Nat. Hist., **44**: 1555-1575.

BELL, J. L., 1988. Distribution and abundance of *Dissodactylus mellitae* Rathbun (Pinnotheridae) on *Mellita quinquiesperforata* (Leske) (Echinodermata). J. Exp. Mar. Biol. Ecol., **117**: 93-114.

BERNER, L., 1952. Biologie de *Pinnotheres pisum* Penn. (Décapode Brachyoure). Extr. Bull. Soc. Zool. Fr., **77**: 344-349.

BEZERRA, L. E. A., A. O. DE ALMEIDA & P. A. COELHO, 2006. Occurrence of the family Pinnotheridae De Haan (Crustacea, Decapoda, Brachyura) on the coast of Ceará State, Brazil. Rev. bras. Zool., **23**: 1038-1043.

BIERBAUM, R. & S. FERSON, 1986. Do symbiotic pea crabs decrease growth rate in mussels? Biol. Bull., **170**: 51-61.

BIERBAUM, R. & S. E. SHUMWAY, 1988. Filtration and oxygen consumption in mussels, *Mytilus edulis*, with and without pea crabs, *Pinnotheres maculatus*. Estuaries, **11**: 264-271.

BOLOGNA, P. A. X. & K. L. HECK JR., 2000. Relationship between pea crab (*Pinnotheres maculatus*) parasitism and gonad mass of the bay scallop (*Argopecten irradians*). Gulf Carib. Res., **12**: 43-46.

CAINE, E. A., 1975. Feeding of *Pinnotheres maculatus* Say (Brachyura: Pinnotheridae). Forma e functio – Int. J. Funct. Biol., **8**: 395-404.

CAMPOS, E., 1990. *Calypotraeotheres*, a new genus for the limpet crab *Fabia granti* Glassell, 1933 (Crustacea, Brachyura). Proc. Biol. Soc. Wash., **103**: 364-371.

CAMPOS, E., 1996a. Partial revision of pinnotherid crab genera with a two-segmented palp on the third maxilliped (Decapoda: Brachyura). J. Crust. Biol., **16**: 556-563.

CAMPOS, E., 1996b. Partial revision of the genus *Fabia* Dana, 1851 (Crustacea: Brachyura: Pinnotheridae). J. Nat. Hist., **30**: 1157-1178.

CAMPOS, E., 2001. A new crab species of the genus *Arcotheres* Manning, 1993, from Thailand (Crustacea, Brachyura, Pinnotheridae). Zoosystema, **23**: 493-497.

CAMPOS, E., 2002. Two new genera of pinnotherid crabs from the Tropical Eastern Pacific (Decapoda: Brachyura: Pinnotheridae). J. Crust. Biol., **22**: 328-336.

CAMPOS, E. & R. B. MANNING, 2001. Authorship and diagnosis of the genus *Arcotheres* Manning, 1993 (Crustacea: Brachyura: Pinnotheridae). Raffl. Bull. Zool., **49**: 167-170.

CASTRO, P., 2015. Symbiotic Brachyura. In: P. CASTRO, P. DAVIE, D. GUINOT, F. R. SCHRAM & J. C. VON VAUPEL KLEIN (eds.), Decapoda: Brachyura. Treatise on Zoology — Anatomy, Taxonomy, Biology, **9C-I**: 543-581. (Brill, Leiden).

CHRISTENSEN, A. M. & J. J. MCDERMOTT, 1959. Life history and biology of *Pinnotheres ostreum* Say. Biol. Bull., **144**: 146-179.

DERBY, C. D. & J. ATEMA, 1980. Induced host odor attraction in the pea crab *Pinnotheres maculatus*. Biol. Bull., **158**: 26-33.

DULCE SUBIDA, M., A. M. ARIAS, P. DRAKE, J. E. GARCÍA-RASO, A. RODRÍGUEZ & J. A. CUESTA, 2011. On the occurrence of *Afropinnotheres monodi* Manning, 1993 (Decapoda: Pinnotheridae) in European waters. J. Crust. Biol., **31**: 367-369.

EIDEMILLER, A., 1969. Entry behavior of *Pinnotheres maculatus* Say. Q. J. Florida Acad. Sci., **32**: 266-274.

FELDMANN, R. M., D. I. MACKINNON, K. ENDO & L. CHIRINO-GALVEZ, 1996. *Pinnotheres laquei* Sakai (Decapoda: Pinnotheridae), a tiny crab commensal within the brachiopod *Laqueus rubellus* (Sowerby) (Terebratulida: Laqueidae). J. Paleontol., **70**: 303-311.

GARM, A., 2004. Revising the definition of the crustacean seta and setal classification systems based on examinations of the mouthpart setae of seven species of decapods. Zool. J. Linn. Soc., **142**: 233-252.

GEIGER, D. L. & J. W. MARTIN, 1999. The pea crab *Orthotheres haliotidis* new species (Decapoda: Brachyura: Pinnotheridae) in the Australian abalone *Haliotis asinina* Linnaeus, 1758 and *Haliotis squamata* Reeve, 1846 (Gastropoda: Vestigastropoda: Haliotidae). Bull. Mar. Sci., **64**: 269-280.

GEORGE, S. B. & S. BOONE, 2003. The ectosymbiont crab *Dissodactylus mellitae*–sand dollar *Mellita isometra* relationship. J. Exp. Mar. Biol. Ecol., **294**: 235-255.

GORDON, I., 1936. On a few Indopacific species of *Pinnotheres*, with special reference to asymmetry of walking legs. Linn. Soc. Lond., **60**: 163-180.

GRIFFIN, D. J. G. & B. M. CAMPBELL, 1969. The sub-littoral Goneplacidae and Pinnotheridae (Crustacea: Brachyura) of Moreton Bay. Mem. Qld. Mus., **15**: 141-163.

GROVE, M. W. & S. A. WOODIN, 1996. Conspecific recognition and host choice in a pea crab, *Pinnixa chaetopterana* (Brachyura: Pinnotheridae). Biol. Bull., **190**: 359-366.

HAINES, C. M. C., M. EDMUNDS & A. R. PEWSEY, 1994. The pea crab, *Pinnotheres pisum* (Linnaeus 1767), and its association with the common mussel, *Mytilus edulis* (Linnaeus, 1758), in the Solent (U.K.). J. Shellf. Res., **13**: 5-10.

HARTNOLL, R. G., 1972. Swimming in the hard stage of the pea crab, *Pinnotheres pisum* (L.). J. Nat. Hist., **6**: 475-480.

HOUGHTON, D. R., 1963. The relationship between tidal level and the occurrence of *Pinnotheres pisum* (Pennant) in *Mytilus edulis*. L. J. Anim. Ecol., **32**: 253-257.

HSUEH, P. W., 2003. Responses of the pea crab *Pinnotheres taichungae* to the life history patterns of its primary bivalve host *Laternula marilina*. J. Nat. Hist., **37**: 1453-1462.

JONES, J. B., 1977. Natural history of the pea crab in Wellington Harbour, New Zealand. New Zeal. J. Mar. Freshw. Res., **11**: 667-676.

KANE, K. & G. S. FARLEY, 2006. Body size of the endosymbiotic pea crab *Tumidotheres maculatus*: larger hosts hold larger crabs. Gulf Carib. Res., **18**: 27-33.

KRUCZYNSKI, W. L., 1972. The effect of the pea crab *Pinnotheres maculates* Say on growth of the bay scallop, *Argopecten irradians concentricus* (Say). Chesapeake Sci., **13**: 218-220.

KRUCZYNSKI, W. L., 1974. Relationship between depth and occurrence of pea crabs, *Pinnotheres maculatus*, in blue mussels, *Mytilus edulis*, in the vicinity of Woods Hole, Massachusetts. Chesapeake Sci., **15**: 167-169.

KRUCZYNSKI, W. L., 1975. A radioactive tracer study of food uptake by *Pinnotheres maculatus* in molluscan hosts. Biol. Bull., **148**: 60-67.

LÉVI, C., 1951. Inventaire de la faune marine de Roscoff. Tuniciers. Trav. Stat. Biol. Roscoff, **7**: 1-19.

MANNING, R. B., 1993. Three genera removed from the synonymy of *Pinnotheres* Bosc, 1802 (Brachyura: Pinnotheridae). Proc. Biol. Soc. Wash., **106**: 523-531.

MCDERMOTT, J. J., 1962. The incidence and host-parasite relations of pinnotherid crabs (Decapoda, Pinnotheridae). In: D. S. GORSLINE (ed.), Proceedings of the first coastal and shallow water research conference, October 1961, Baltimore (Maryland), Los Angeles (California), Tallahassee (Florida). Tallahassee (FL): 162-164.

MERCADO-SILVA, N., 2005. Condition index of the Eastern oyster, *Crassostrea virginica* (Gmelin, 1791) in Sapelo Island Georgia — effects of site, position on bed and pea crab parasitism. J. Shellf. Res., **24**: 121-126.

NAVARTE, M. A. & M. N. SAIZ, 2004. Effects of the pinnotherid crab *Tumidotheres maculatus* on the Tehuelche scallop *Aequipecten tehuelchus* in the San Matías Gulf, Argentina. Fish. Res., **67**: 207-214.

NG, P. K. L. & R. B. MANNING, 2003. On two new genera of pea crabs parasitic in holothurians (Crustacea: Decapoda: Brachyura: Pinnotheridae) from the Indo-West Pacific, with notes on allied genera. Proc. Biol. Soc. Wash., **116**: 901-919.

O'BEIRN, F. X. & R. L. WALKER, 1999. Pea crab, *Pinnotheres ostreum* Say, 1817, in the eastern oyster, *Crassostrea virginica* (Gmelin, 1791): Prevalence and apparent adverse effects on oyster gonad development. Veliger, **42**: 17-20.

OCAMPO, E. H., J. D. NUÑEZ, M. CLEDÓN & J. A. BAEZA, 2012. Host-specific reproductive benefits, host selection behavior and host use pattern of the pinnotherid crab *Calyptraeotheres garthi*. J. Exp. Mar. Biol. Ecol., **429**: 36-46.

OCAMPO, E. H., J. D. NUÑEZ, M. CLEDÓN & J. A. BAEZA, 2014. Parasitic castration in slipper limpets infested by the symbiotic crab *Calyptraeotheres garthi*. Mar. Biol., **161**: 2107-2120.

OCAMPO, E. H., E. D. SPIVAK, J. A. BAEZA & T. A. LUPPI, 2017. Ontogenetic changes in the external anatomy of the parasitic castrator crab *Calyptraeotheres garthi*: implications for the timing of host colonization and sexual behaviour. Biol. J. Linn. Soc., in press.

PALACIOS THEIL, E., J. A. CUESTA & D. L. FELDER, 2016. Molecular evidence for non-monophyly of the pinnotheroid crabs (Crustacea: Brachyura: Pinnotheroidea), warranting taxonomic reappraisal. Invert. Syst., **30**: 1-27.

PALMER, P., 1995. Occurence of a New Zealand pea crab, *Pinnotheres novaezelandiae*, in five species of surf clam. Mar. Freshw. Res., **46**: 1071-1075.

PEARCE, J. B., 1964. On reproduction in *Pinnotheres maculatus* (Decapoda: Pinnotheridae). Biol. Bull., **127**: 384.

PREGENZER JR., C., 1978. *Pinnotheres hickmani* (Guiler) in wild and cultured *Mytilus edulis* in Port Phillip Bay, Victoria. Aust. J. Mar. Freshw. Res., **29**: 127-139.

SASTRY, A. N. & R. W. MENZEL, 1962. Influence of hosts on the behavior of the commensal crab *Pinnotheres maculatus* Say. Biol. Bull., **123**: 388-395.

SCHMITT, W. L., J. C. MCCAIN & E. S. DAVIDSON, 1973. 3. Decapoda I Brachyura I: Pinnotheridae. In: H. E. GRUNER & L. B. HOLTHUIS (eds.), Crustaceorum Catalogus, p. 160. (W. Junk, Den Haag).

SEED, R., 1969. The incidence of the pea crab, *Pinnotheres pisum* in two types of *Mytilus* (Mollusca: Bivalvia) from Padstow, south-west England. J. Zool., **158**: 413-420.

SILAS, A. & A. ALAGARSWAMI, 1967. On an instance of parasitisation by the pea-crab (*Pinnotheres* sp.) on the backwater clam [*Meretrix casta* (Chemnitz)] from India, with a review of the work on the systematics, ecology, biology and ethology of pea crabs of the genus *Pinnotheres* Latreille. In: Proc. Symp. Crust., 1965, Jan 12-15; Ernakulam (India). Part III. Symp. Ser. 2. Mar. Biol. Ass. India: 1161-1227.

STEVENS, P. M., 1990a. A genetic analysis of the pea crabs (Decapoda: Pinnotheridae) of New Zealand. I. Patterns of spatial and host-associated genetic structuring in *Pinnotheres novaezelandidae* Filhol. J. Exp. Mar. Biol. Ecol., **141**: 195-212.

STEVENS, P. M., 1990b. Specifity of host recognition of individuals from different host races of symbiotic pea crabs (Decapoda, Pinnotheridae). J. Exp. Mar. Biol. Ecol., **143**: 193-207.

SUGIURA, Y., A. SUGITA & M. KIHARA, 1960. The ecology of pinnotherid crab as pest in culture of *Tapes japonica*.1. *Pinnotheres sinensis* living in *Tapes japonica* and the influence of the crab on the weight of the host's flesh. Bull. Jap. Soc. Sci. Fish., **26**: 89-94.

SUN, W., S. SUN, W. YUQI, Y. BAOWEN & S. WEIBO, 2005. The prevalence of the pea crab, *Pinnotheres sinensis*, and its impact on the cultured mussel, *Mytilus galloprovincialis*, in Jiaonan waters (Shandong Province, China). Aquaculture, **253**: 57-63.

TAKEDA, S., S. TAMURA & M. WASHIO, 1997. Relationship between the pea crab *Pinnixa tumdia* and its holothurian host *Paracaudina chilensis*. Mar. Ecol. Prog. Ser., **149**: 143-154.

TROTTIER, O. & A. G. JEFFS, 2015. Mate location and access behaviour of the parasitic pea crab, *Nepinnotheres novaezelandiae*, an important parasite of the mussel *Perna canaliculus*. Parasite, **22**(13): 1-11.

TROTTIER, O., D. WALKER & A. G. JEFFS, 2012. Impact of the parasitic pea crab *Pinnotheres novaezelandiae* on aquacultured New Zealand green-lipped mussels, *Perna canaliculus*. Aquaculture, **344-346**: 23-28.

First received 14 November 2016.
Final version accepted 14 February 2017.

[When citing this volume, please refer to Crustaceana 90 (2017) 771-1288]

Michael Türkay Memorial Issue

BROAD-RANGING LOW GENETIC DIVERSITY AMONG POPULATIONS OF THE YELLOW FINGER MARSH CRAB *SESARMA RECTUM* RANDALL, 1840 (SESARMIDAE) REVEALED BY DNA BARCODE

BY

RAQUEL C. BURANELLI[1]) and FERNANDO L. MANTELATTO[2])

Laboratory of Bioecology and Systematics of Crustaceans (LBSC), Department of Biology, Faculty of Philosophy, Sciences and Letters at Ribeirão Preto (FFCLRP), University of São Paulo (USP), Av. Bandeirantes 3900, 14040-901 Ribeirão Preto (SP), Brazil

ABSTRACT

Population genetic studies on marine taxa, specifically in the field of phylogeography, have revealed distinct levels of genetic differentiation in widely distributed species, even though they present long planktonic larval development. A set of factors have been identified as acting on gene flow between marine populations, including physical or physiological barriers, isolation by distance, larval behaviour, and geological and demographic events. In this way, the aim of this study was to analyse the genetic variability among populations of the crab species *Sesarma rectum* Randall, 1840 along the western Atlantic in order to check the levels of genetic diversity and differentiation among populations. To achieve this purpose, mtDNA cytochrome-*c* oxidase subunit I (COI) (DNA-barcode marker) data were used to compute a haplotype network and a Bayesian analysis for genetic differentiation, to calculate an Analysis of Molecular Variance (AMOVA), and haplotype and nucleotide diversities. Neutrality tests (Tajima's D and Fu's F_S) were accessed, as well as pairwise mismatch distribution under the sudden expansion model. We found sharing of haplotypes among populations of *S. rectum* along its range of distribution and no significant indication for restricted gene flow between populations separately over 6000 km, supporting the hypothesis of a high dispersive capacity, and/or the absence of strong selective gradients along the distribution. Nevertheless, some results indicated population structure suggesting the presence of two genetic sources (i.e., groups or lineages), probably interpreted as a result of a very recent bottleneck effect due to habitat losses, followed by the beginning of a population expansion.

Key words. — Bottleneck effect, Brachyura, genetic variability, mangrove, mtDNA cytochrome-*c* oxidase subunit I

ZUSAMMENFASSUNG

Populationsgenetische Studien an marinen Taxa, spezifisch im Feld der Phylogeographie, haben deutliche, genetisch basierte Abgrenzungsebenen weit verbreiteter Arten aufgezeigt, obwohl diese

[1]) e-mail: raquel.buranelli@gmail.com

[2]) Corresponding author; e-mail: flmantel@usp.br

eine lange, planktonische Larvalentwicklung haben. Es wurde eine Reihe von Einflussfaktoren auf den Genfluss zwischen marinen Populationen identifiziert, welche physische oder physiologische Barrieren, Isolation durch Distanz, Larvalverhalten, sowie geologische und demographische Ereignisse einbeziehen. Somit hatte diese Studie zum Ziel, die genetische Variabilität innerhalb von Populationen der Krabbenart *Sesarma rectum* Randall, 1840 entlang des Westatlantiks zu analysieren, um die Ebenen genetischer Diversität und Abgrenzung innerhalb von Populationen zu überprüfen. Zu diesem Zweck wurden Daten eines Fragments der mitochondrialen Cytochrom-*c*-Oxidase Untereinheit (DNA-Barcodemarker) benutzt, um ein Haplotypen-Netzwerk und eine Bayesische Analyse für genetische Abgrenzung zu berechnen, eine Analyse der molekularen Varianz (AMOVA) durchzuführen, und die Haplotyp- und Nukleotiddiversität zu ermitteln.

Weiterhin wurden Neutralitätstests (Tajima's D und Fu's F_S) durchgeführt und „pairwise mismatch distribution under the sudden expansion model" errechnet. Wir entdeckten gemeinsame Haplotypen der Populationen von *S. rectum* entlang ihres Verbreitungsgebietes und keinen signifikanten Hinweis auf eingeschränkten Genfluss zwischen Populationen, welche über 6000 km voneinander entfernt existieren. Dies unterstützt die Hypothese einer hohen Verteilungskapazität und/oder das Fehlen stark selektiver Gradienten entlang der Verbreitung. Jedoch liefern einige Ergebnisse Hinweise auf Populationsstrukturen, die auf die Anwesenheit zweier genetischer Gruppen (d.h., „lineages") hindeuten und möglicherweise als das Ergebnis eines sehr zeitnahen Flaschenhals-Effektes aufgrund von Habitatverlusten, gefolgt von einer beginnenden Populationsausdehnung, interpretiert werden können.

INTRODUCTION

Among marine invertebrates the planktonic larval development can potentially interconnect distant populations through ocean currents, resulting in a high dispersal capacity (Palumbi, 1994; Taylor & Hellberg, 2003; Silva et al., 2010; Ituarte et al., 2012). Thus, species with great dispersive capacity generally exhibit low genetic differentiation and lower levels of population structuring, even among distant populations, showing low potential to respond to local selection pressures (Hedgecock, 1986; Palumbi, 1994; Avise, 2004; Silva et al., 2010). Many previous population genetic studies, specifically in the field of phylogeography, revealed a positive correlation between a great dispersive capacity and the levels of gene flow among marine invertebrate species (e.g., Williams & Benzie, 1993; Oliveira-Neto et al., 2008; Kelly & Palumbi, 2010; Silva et al., 2010; Laurenzano et al., 2012). Other studies on marine taxa have revealed high levels of genetic differentiation in species with wide distribution, even though they present pelagic larval stages during their life cycle (e.g., Barber et al., 2000; Kelly & Palumbi, 2010; Ituarte et al., 2012; Terossi & Mantelatto, 2012; Laurenzano et al., 2013).

The detection of this variety of marine taxa with significant genetic and geographical differentiation indicates that a set of features is interfering with gene flow between populations, promoting either genetic divergence or homogeneity. Aspects acting on gene flow vary depending on the species, habitat, local oceanic conditions, and recent history (Palumbi, 1994). Besides, despite the fact that these limits to gene flow rarely create effective and absolute barriers, they can limit gene flow in some directions or temporarily (Palumbi, 1994).

Generally, the interruption of gene flow can be associated with the existence of physical or physiological barriers, favouring genetic differentiation among populations (Kimura & Weiss, 1964). Drastic salinity and temperature changes, for example, can impose barriers to larval dispersal (Burton & Feldman, 1982). Also, ocean currents may act as physical barriers, since the maximum distance over which larvae can be dispersed depends on the speed and direction of the currents that carry them (Scheltema, 1986), and drastic changes such as temperature, bifurcations, or gyres may also act as barriers to gene flow.

On the other hand, ocean current hydrodynamics may not necessarily act as a potential barrier to gene flow, but also could promote genetic homogeneity, especially considering that the currents perform a passive dispersion of planktonic larvae, and many marine organisms can potentially connect with distant conspecific populations (Burton, 1983; Taylor & Hellberg, 2003; Hamasaki et al., 2015). Besides, we may also have to bear in mind that planktonic larvae show behavioural features of active migration in the water column, which control the possibilities of either being retained or dispersed (Koehn, 1969; McMillen-Jackson et al., 1994; Warner & Cowen, 2002; Palumbi, 2003; Shanks, 2009).

Moreover, even when physical or physiological barriers do not exist, the size of the area inhabited may prevent the species from forming a single panmictic unit, characteristic of isolation by distance (Wright, 1943). Isolation by distance can significantly interfere in the genetic differentiation, since differentiation patterns are based on the magnitude of the migration rates between populations (Hellberg et al., 2002; Ituarte et al., 2012). Consequently, the reduction of gene flow becomes significantly likely between distant areas, especially considering species that do not exhibit planktonic phases (Chust et al., 2016) or exhibit larval stages of shorter duration.

Additionally, there are some other factors that may assist in identifying determinants of population structuring or homogenisation agents. Biotic factors that influence the dispersion may include food availability, predation, and larval behavioural responses to light, salinity, temperature and hydrostatic pressure (Burton, 1983). Also, species with high potential for dispersion may show genetic structure if they exhibit behavioural area fidelity, larval retention and/or self-recruitment (Barber et al., 2000; Ituarte et al., 2012). Furthermore, many non-apparent factors can also influence gene flow in marine organisms, such as past geological events and demographic history (Barber et al., 2000; Hellberg et al., 2002; Taylor & Hellberg, 2003; Avise, 2009). Sea level changes in the past, during glacial cycles for example, are likely to act as temporary barriers to dispersal to many species of marine invertebrates (e.g., Felder & Staton, 1994; Laurenzano et al., 2016).

Species widely distributed and having a long planktonic larval phase, as many marine decapod crustaceans, constitute great models to study genetic differentiation and population structure. The sesarmid species *Sesarma rectum* Randall, 1840 is a common crab inhabiting mangrove forests along the western Atlantic Neotropical coast and is recorded from Trinidad and Tobago (West Indies), the Lesser Antillean island of Grenada, Venezuela, the Guianas and Brazil (from the state of Amapá to Santa Catarina) (Abele, 1992; Melo, 1996; Schubart et al., 1999). This species constitutes an important member of the benthic fauna of mangroves due to its feeding behaviour and habit of digging burrows (Tavares & Albuquerque, 1989; Lee, 1998). So, based on the scenario previously described, *S. rectum* constitutes a great model for genetic variability analyses. The aim of this study was to check the levels of genetic diversity and differentiation among *S. rectum* populations by means of intraspecific variability analyses of mitochondrial data, in order to provide information on the degree of gene flow and connectivity among widely distributed fauna.

MATERIAL AND METHODS

Sample collection

Fresh specimens of *Sesarma rectum* were sampled following applicable state and federal laws of Brazil (DIFAP/IBAMA 126/05; permanent license to FLM for collection of Zoological Material No. 11777-1 MMA/IBAMA/SISBIO; temporary license to RCB for collection of Zoological Material No. 42887-1 MMA/IBAMA/SISBIO). Sampled specimens were preserved in 70-90% alcohol and deposited in the Crustacean Collection of the Biology Department, Faculty of Philosophy, Sciences and Letters at Ribeirão Preto, University of São Paulo (CCDB/FFCLRP/USP), São Paulo, SP, Brazil (permanent license for Crustacean Collection No. 071/2012/SECEX/CGEN). Complementary specimens were obtained on loan from the following museums: Museu de Zoologia da Universidade de São Paulo, São Paulo, Brazil (MZUSP); United States National Museum, Smithsonian National Museum of Natural History, Washington, DC, U.S.A. (USNM); and the Florida Museum of Natural History, University of Florida, Gainesville, FL, U.S.A. (UF) (table I). In our analysis we included a collection of specimens that encompass the entire geographic distribution of this species (fig. 1, table I).

DNA extraction, amplification and sequencing

Total genomic DNA was obtained using the salt extraction method (Miller et al., 1988), with changes aiming suitability to the material, as follows (Mantelatto et al.,

TABLE I

Data for the specimens of *Sesarma rectum* Randall, 1840 used in the analysis, with sampling locality, catalogue number (collection) and GenBank accession number

Locality	Collection	GenBank
Trinidad, Trinidad and Tobago	USNM 1188936	KY964587
Trinidad, Trinidad and Tobago	UF 8817	KU313336-KU313339
Trinidad, Trinidad and Tobago	UF 8827	KU313340
Brazil		
Bragança, PA	CCDB 4469	KU313342, KU313343
Bragança, PA	CCDB 4613	KU313344, KU313345
Salinópolis, PA	CCDB 4440	KU313341
Fortaleza, CE	CCDB 4369	KU313346-KU313349
Estuário Rio das Conchas, RN	MZUSP 22926	KU313354
Estuário Rio das Conchas, RN	MZUSP 29924	KU313352
Estuário Rio das Conchas, RN	MZUSP 29925	KU313353
Natal, RN	CCDB 3477	KU313350, KU313351
Passo do Camaragibe, AL	MZUSP 24878	KU313356
Passo do Camaragibe, AL	MZUSP 29914	KU313355
Passo do Camaragibe, AL	MZUSP 29927	KU313357
Porto Seguro, BA	CCDB 700	KU31335-KU313362
Anchieta, ES	MZUSP 29911	KU313364
Anchieta, ES	MZUSP 29912	KU313365
Marataízes, ES	CCDB 4005	KU313363
Ubatuba, SP	CCDB 158	KU313369
Ubatuba, SP	CCDB 2539	KU313368
Ubatuba, SP	CCDB 3914	KU313366, KU313367
Bertioga, SP	CCDB 4970	KU313370-KU313374
Ilha Comprida, SP	CCDB 3669	KU31337-KU313379

CCDB, Crustacean Collection of the Department of Biology (FFCLRP), São Paulo, SP, Brazil; MZUSP, Museu de Zoologia da Universidade de São Paulo, São Paulo, Brazil; USNM, United States National Museum, Smithsonian National Museum of Natural History, Washington, DC, U.S.A.; UF, Florida Museum of Natural History, Gainesville, FL, U.S.A.; PA, Pará; CE, Ceará; RN, Rio Grande do Norte; AL, Alagoas; BA, Bahia; ES, Espírito Santo; SP, São Paulo.

2006). Muscle tissue was incubated for 24 h in 600 μl of lysis buffer (10 mM TRIS, 100 mM EDTA, 1% sodium dodecyl sulphate, pH 7.5) and 7 μl of proteinase K (20 mg/ml) at 55°C. Prior to centrifugation (18°C, 10 min, 14 000 rpm), proteinase K was inactivated on ice for 10 min, followed by the addition of 200 μl of NH_4OAc (7.5 M). Thereafter 600 μl of cold isopropanol was added and samples were incubated for 48 h at −20°C, followed by centrifugation (18°C, 10 min, 14 000 rpm). The resulting pellet was washed with 20 μl of 70% EtOH, centrifuged (18°C, 10 min, 14 000 rpm), freeze-dried in an Eppendorf Concentrator 5301® and resuspended in 20 μl of TE buffer (10 mM TRIS, 1 mM EDTA, pH 8.0). Extracted DNA final concentration was measured using a Nanodrop spectrophotometer 2000®.

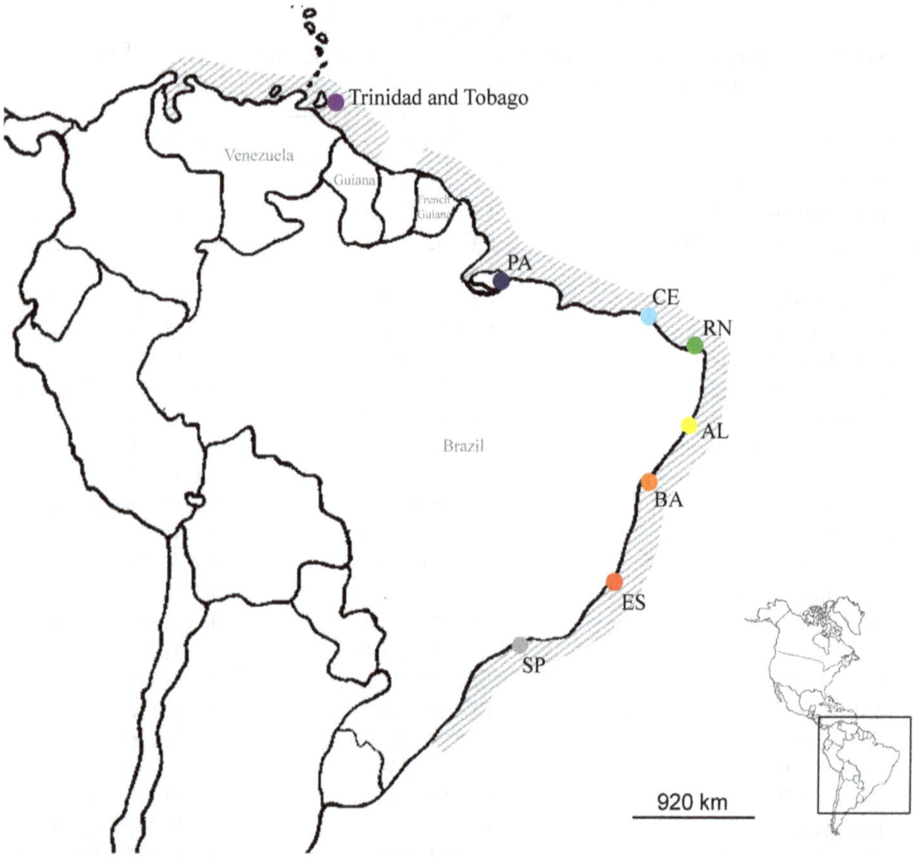

Fig. 1. Sample sites of *Sesarma rectum* Randall, 1840 along the western Atlantic. Coloured dots indicate the locations of the populations used in the analyses. Shaded areas indicate the reported distribution of *S. rectum* (countries indicated). PA, Pará; CE, Ceará; RN, Rio Grande do Norte; AL, Alagoas; BA, Bahia; ES, Espírito Santo; SP, São Paulo.

An approx. 600-bp region of the mitochondrial gene (mtDNA) cytochrome-*c* oxidase subunit I (COI) (DNA-barcode marker) was amplified by PCR (Sambrook et al., 1989) in a 96-well Thermal Cycler (Applied Biosystems®) with the primers COL6b (5′-ACAAATCATAAAGATATYGG-3′)/COH6 (5′-TADACTTCDGGRTGDCCAAARAAYCA-3′) (Schubart & Huber, 2006), performed with the following thermal cycle: initial denaturing for 2 min at 94°C; annealing for 35 cycles: 30 s at 94°C, 30 s at 50°C, 1 min at 72°C, final extension 4 min at 72°C. Each reaction was performed in a 25 μl total volume, containing: ultrapure water, betaine (5 M), DNTP (200 μM of each), PCR buffer (10×), $MgCl_2$ (25 mM), primers (10 μM each), 5 U/μl of *Taq* polymerase, and previously extracted DNA (50 ng/μl). PCR products were electrophoresed in a 1.4% agarose 1× TBE gel and photographed with an Olympus C-7070® digital camera on a UVP® UV transilluminator.

Positive PCR products were purified using the SureClean Plus® kit, following the manufacturer's protocol, and submitted to a sequencing reaction, containing: ultrapure water, 2.5× sequencing buffer (Save Money®), BigDye Terminator Cycle Sequencing v3.1 (Applied Biosystems®, Carlsbad, CA, U.S.A.), primers (10 pM each), and previously purified PCR product (50 ng/μl) (thermal cycle: initial denaturing for 1 min at 96°C; annealing for 39 cycles: 15 s at 96°C,15 s at 50°C, 4 min at 60°C). Products were purified and sequenced with the ABI Big-Dye Terminator Mix (Applied Biosystems®) in an ABI 3730xl DNA Analyzer (Applied Biosystems® automated sequencer), following the manufacturer's protocol.

Both strands (forward and reverse directions) of DNA were sequenced and a consensus sequence was obtained using BIOEDIT 7.0.5 (Hall, 1999). Primer regions and non-readable regions were omitted. Sequences were aligned using Clustal W (Thompson et al., 1994) with interface to BIOEDIT 7.0.5 (Hall, 1999) with default parameters. All sequences obtained were submitted to GenBank (table I).

Genetic variability analyses

The haplotype number was calculated in DnaSP 4.10.9 (Rozas & Rozas, 1999) and a haplotype network was constructed by statistical parsimony in TCS 1.21 (Clement et al., 2000). Ambiguous connection in the network was solved according to the criteria proposed by Excoffier & Langaney (1989). A Bayesian analysis for genetic differentiation among populations was conducted in BAPS 6.0, using a Bayesian clustering approach to probabilistically assign individuals to populations according to their haplotypes, based on 200 simulations from posterior haplotype frequencies (Corander et al., 2003, 2004, 2007). Analyses of Molecular Variance (AMOVA) were carried out using Arlequin 3.5.2.2 software (Excoffier et al., 1992), considering the variation at each nucleotide separately using a matrix of Euclidean squared distances for genetic distance calculations. AMOVA was applied with no hierarchical subdivision (all populations in a single group). Test of significance used a non-parametric permutation procedure (Excoffier et al., 1992), incorporating 10 000 permutations. Haplotype and nucleotide diversities were calculated using Arlequin 3.5.2.2 (Excoffier et al., 2005).

Neutrality tests and demographic analyses

The historical demography of *S. rectum* was inferred using Tajima's D (Tajima, 1989) and Fu's F_s (Fu, 1997) tests with 10 000 simulated samples in order to evaluate the neutrality of the sequence variation. Additionally, pairwise mismatch distributions were obtained and compared statistically to those expected under the sudden expansion model of demographic and spatial expansion (Rogers &

Harpending, 1992; Rogers, 1995), using the goodness-of-fit test based on the sum of squared deviations (SSD) (Schneider & Excoffier, 1999) and Harpending's raggedness index (HRI) (Harpending, 1994). The parameters and confidence intervals were estimated using the parametric bootstrap approach implemented in Arlequin 3.5.2.2, with the corresponding p-values with 10 000 bootstrap replicates. The neutrality analyses were run with no hierarchical structure (all populations in a single group).

RESULTS

Based on the 637-bp fragment of the mtDNA COI from 45 specimens of *Sesarma rectum* from eight populations that cover the total range of its distribution (fig. 1), 13 haplotypes were defined. Eleven sites were polymorphic and the average nucleotide composition was 38.65% A, 29.27% T, 16.43% C, and 15.63% G. Among the 13 haplotypes identified, 10 were unique to a single population and three were shared among populations (H1, H9 and H11) (fig. 2, table II). The haplotype network revealed that populations are distributed homogeneously and there are no discernible groups (fig. 2). All unique and two shared haplotypes radiate from a central one (H1), shared by 20 individuals (fig. 2, table II). This haplotype network revealed the absence of genetic structure (fig. 2). Although the population CE does not share haplotypes with none of the others, all haplotypes are arranged in close vicinity to the central one (fig. 2). Overall haplotype diversity found was 0.7283 and haplotype diversity within populations ranged from 0-1, while nucleotide diversity ranged from 0-0.00314 (table III).

The Bayesian analysis of genetic differentiation grouped all individuals in two clusters, supported by a posterior probability value of 1 (fig. 3). Cluster 1 included all individuals belonging to haplotypes H3, H11, H12, and H13, with specimens from Trinidad and Tobago, BA, and SP (fig. 3). The cluster 2 included the rest of the individuals and haplotypes (fig. 3), suggesting that this structure is not related to any geographic pattern (fig. 3).

AMOVA revealed that the within-population (71.76%) variance component exceeded the among-populations variance component (28.23%), which suggests limited population structuring, as observed in the haplotype network. Nevertheless, the obtained significant ($p = 0.00000$) average F_{ST}-value (0.2823) suggests an overall population structure with very high level of genetic differentiation ($F_{ST} > 0.25$) (Wright, 1978).

Average Tajima's D and Fu's F_s tests revealed significantly negative values ($D = -1.80710$, $p < 0.05$; $F_s = -10.63055$, $p < 0.05$), indicating a significant deviation from neutrality in overall sample and, probably the interference of

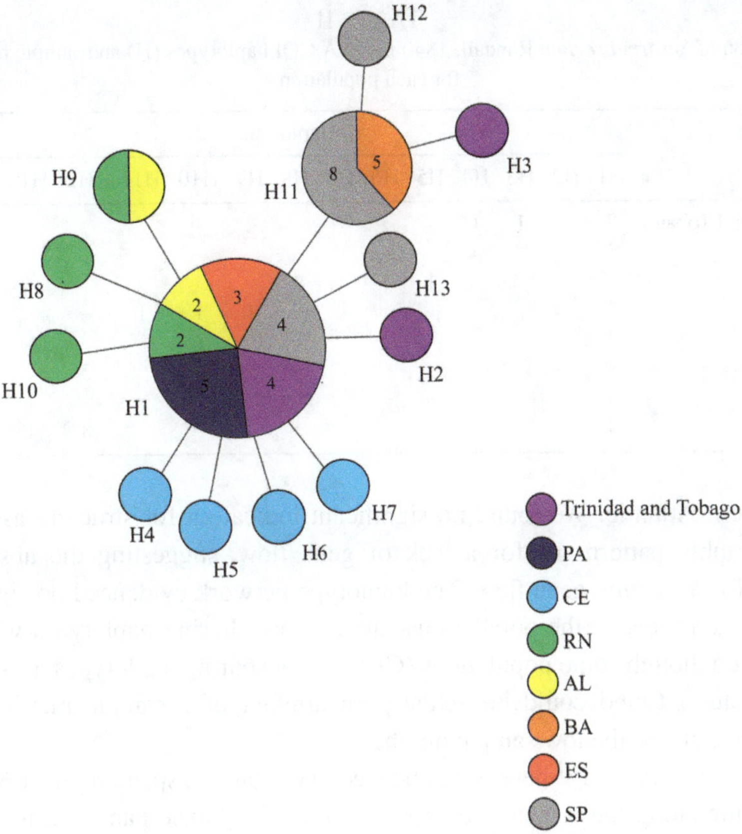

Fig. 2. Parsimony haplotype network for mtDNA COI from the populations of *Sesarma rectum* Randall, 1840, with the distribution of the 13 haplotypes (H) identified. The size of the circles is proportional to the haplotype frequency. Numbers inside haplotypes indicate the number of individuals within the haplotype if present more than once. Black circles indicate the median vector. Different colours represent different populations. PA, Pará; CE, Ceará; RN, Rio Grande do Norte; AL, Alagoas; BA, Bahia; ES, Espírito Santo; SP, São Paulo.

processes such as recent demographic expansion in the population, after a demographic reduction (Aris-Brosou & Excoffier, 1996). The mismatch distributions were not compatible with models of demographic and spatial expansion [p (SSD) and p (HRI) < 0.05]. Nevertheless, the skewed unimodal distribution found (fig. 4) is generally related to a recent bottleneck effect or sudden population expansion (Patarnello et al., 2007).

DISCUSSION

Based on our molecular data, we found a limited population structure and no significant geographic differentiation among populations of *Sesarma rectum* along

TABLE II
Distribution of *Sesarma rectum* Randall, 1840 mtDNA COI haplotypes (H) and sample number (n) for each population

Population	Haplotype														n
	H1	H2	H3	H4	H5	H6	H7	H8	H9	H10	H11	H12	H13	H14	
Trinidad and Tobago	3	1	1	1											6
PA	5														5
CE					1	1	1	1							4
RN	2								1	1	1				5
AL	2									1					3
BA												5			5
ES	3														3
SP	4											8	1	1	14

the western Atlantic. We found no significant indication for structure associated to geographic patterns or for a lack of gene flow, suggesting the absence of barriers for *S. rectum* gene flow. The haplotype network evidenced no significant structure and most of the populations sampled are sharing haplotypes with each other. Even though some populations (CE) are not sharing haplotypes, this genetic differentiation found could be actually an artefact of a sample missing some haplotypes, due to the low sample number.

This genetic exchange here found between populations spanning over 6000 km of coastline along the western Atlantic is impressive, since panmixia is rare and most species exhibit some level of differentiation among geographical localities (Avise, 2004; Addison et al., 2008). Nevertheless, marine species constitute a challenge to population structure and genetic analyses of species showing a slight or low geographical variation are common, especially in species with high dispersal potential (Palumbi, 1994, 2003; Gopurenko & Hughes, 2002; Silva et al., 2010),

TABLE III
Sampling effort (n), number of haplotypes (nh), haplotype diversity (h) and nucleotide diversity (π) for mtDNA COI data among *Sesarma rectum* Randall, 1840 populations

Population	n	nh	h	π
Trinidad and Tobago	6	4	0.6000 ± 0.2152	0.00157 ± 0.00140
PA	5	1	0.0000 ± 0.0000	0.00000 ± 0.00000
CE	4	4	1.0000 ± 0.1768	0.00314 ± 0.00263
RN	5	4	0.9000 ± 0.1610	0.00188 ± 0.00166
AL	3	2	0.6667 ± 0.3143	0.00104 ± 0.00130
BA	5	1	0.0000 ± 0.0000	0.00000 ± 0.00000
ES	3	1	0.0000 ± 0.0000	0.00000 ± 0.00000
SP	14	4	0.6264 ± 0.1098	0.00122 ± 0.00106

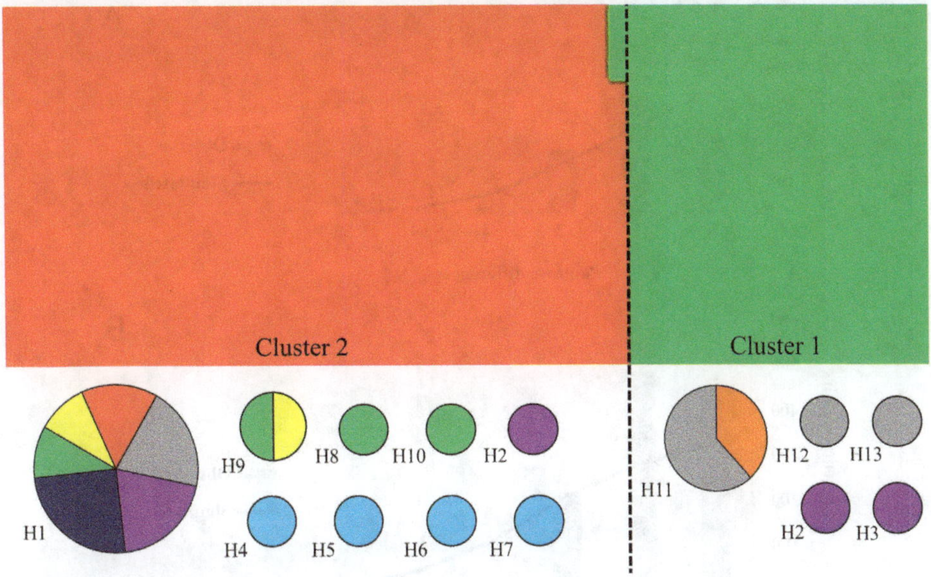

Fig. 3. Graphic generated by Bayesian analysis for genetic differentiation for mtDNA COI among populations of *Sesarma rectum* Randall, 1840, indicating the grouping of two clusters supported by a posterior probability value of 1. Haplotypes (H) belonging to each cluster are indicated. Different colours represent different populations. See fig. 2 for haplotypes relationships and colour of populations.

as observed for other marine/estuarine crabs, such as *Callinectes sapidus* Rathbun, 1896 (cf. McMillen-Jackson & Bert, 2004), *Cardisoma guanhumi* Latreille, 1828 (cf. Oliveira-Neto et al., 2008), *Uca annulipes* (H. Milne Edwards, 1837) (cf. Silva et al., 2010), *Uca uruguayensis* Nobili, 1901 (cf. Laurenzano et al., 2012) and *Uca maracoani* (Latreille, 1802) (cf. Wieman et al., 2014).

Moreover, considering that larval stages may be passively transported by marine currents, species with planktonic larval development can be carried over long distances (Shanks, 2009). Therefore, gene flow and genetic homogeneity found along a wide geographical area can be associated with a high dispersive capacity and the absence of a strong geographical or physiological barrier to gene flow, favouring panmixia (Waples, 1987; Laurenzano et al., 2012). Although the approx. 20-day larval development of *S. rectum* is considered short when compared to other sesarmids, it can be considered as extended compared to congeners which have a reduced larval phase (Anger & Moreira, 2004). Bearing in mind that our results indicated gene flow occurring along an extensive area, this 20-day period can also be considered extended and adequate to reach long distances. Consequently, the genetic structure of *S. rectum* can be considered panmictic, due to the amount of gene flow along a wide geographic area during planktonic larval stages (Silva et al., 2010).

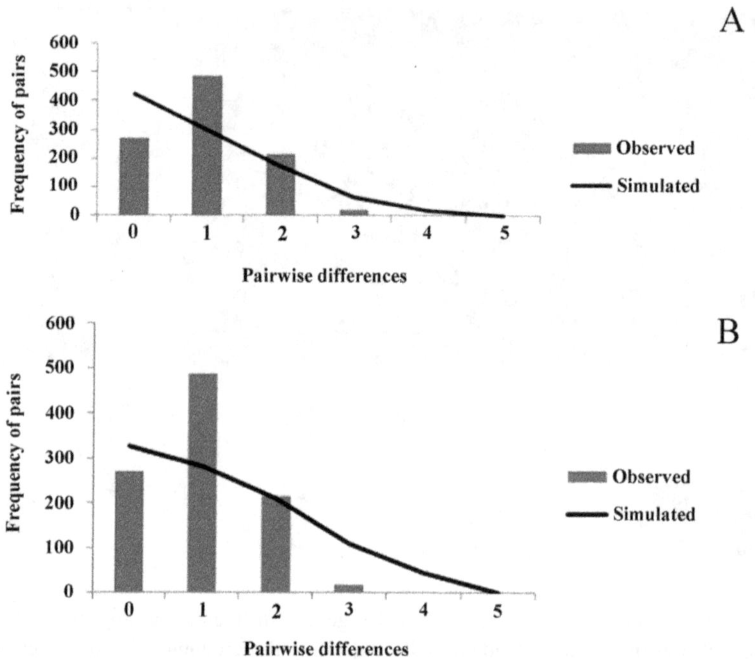

Fig. 4. Mismatch distribution for mtDNA COI of *Sesarma rectum* Randall, 1840 under the models of demographic (A) and spatial (B) expansion, showing the frequency distribution of the number of pairwise nucleotide differences among all individuals.

In this way, current systems associated with larval development duration can greatly influence the larval transport of marine species. Nevertheless, even if the larval carriage is mainly performed by passive transport, dispersion could be affected by larval behaviour and by larval interactions with ecological and physical processes (Burton, 1983; Barber et al., 2000; Hellberg et al., 2002; Palumbi, 2003; Shanks et al., 2003). In the South Atlantic, for example, the current systems highly influence the larval dispersion and, consequently, the population structure of many marine organisms (Rodríguez-Rey et al., 2014). The South-Equatorial Current crosses the Atlantic toward the west coast and suffers a bifurcation around 9-15°S (Cirano et al., 2006), acting as a potential barrier to gene flow between populations of either side of this current (Rodríguez-Rey et al., 2014). However, populations of *S. rectum* from both north and south sides of this current bifurcation are sharing haplotypes, indicating that for this species larval behaviour may also influence larval transportation and population structure. Therefore, larvae should not necessarily be considered as passive particles, but the integration of behavioural and physiological data as well as data on the duration of larval stages, in combination with hydrographic oceanographic models, makes it possible to increase the explanatory power of life histories driving genetic differentiation (Kelly & Palumbi, 2010).

Apart from this previous hypothesis, our data may also suggest that the connectivity here found for *S. rectum* populations may not be exclusively an outcome of high gene flow and high larval dispersal. In addition to this hypothesis, considering that the extent of gene flow will generally determine the genetic variation in the absence of localized selection (Altukhov, 1981), the low genetic differentiation here found can be an outcome of the absence of selective gradients that are strong enough to create small-scale local adaptation and maintain distinct traits among populations across the geographical range (Addison et al., 2008).

Some larval behaviour features for *S. rectum* can corroborate this last hypothesis. It was found that the larval development of *S. rectum* is partially independent of the sea and planktonic food sources due to (1) initial energy reserves; and (2) distinct degrees of larval salinity tolerance and better survival at lower salinities (Anger & Moreira, 2004). So, although the length of larval development can be considered characteristic of a high dispersive capacity due to the low genetic diversity here found, those larval behavioural characteristics may indicate the potential for adaptation to estuarine environments and limited but present retention of the larvae within the parental environment (Anger & Moreira, 2004).

If the larvae are retained next to parental populations by behaviour (Burton & Feldman, 1982) or physical oceanographic mechanisms (Cowen et al., 2000), then these populations have many chances of undergoing genetic differentiation and high levels of population structure (Taylor & Hellberg, 2003). Considering that we found low levels of genetic differentiation among *S. rectum*, even though this species shows larval behaviour retention, then the absence of a strong selective gradient along the distribution of this species is quite likely. Nevertheless, even in the absence of these selective gradients, persistent larval retention could allow the evolution of mate-recognition characters and reproductive isolation (Taylor & Hellberg, 2003), so further reproductive studies comparing distinct populations are necessary.

Although we found no associated geographic pattern and shared haplotypes among distant populations, the limited population structure evidenced by the AMOVA results and the significant high level of genetic differentiation found for the F_{ST} index could be associated with the presence of two clusters revealed by the Bayesian analysis of genetic differentiation and could be helpful to understand the past and current demographic historic of *S. rectum* populations. The results reported here may indicate that *S. rectum* populations could possibly be characterized by a stock currently growing from few genetic sources, probably represented by the two clusters found. It is important to notice that an increase of the dataset may change the number of haplotypes, clusters and the conformation found here.

Populations becoming established from few genetic sources may be indicative of a bottleneck effect in a recent past. The cost of a reduction on demographic size as a consequence of a bottleneck experience may influence the distribution of the genetic variation among and within populations (Wright, 1931; Nei et al., 1975). Populations known to have experienced a reduction in demographic size often show reduced genetic diversity (Wright, 1931; Nei et al., 1975; Ryman et al., 1995; Spencer et al., 2000), as we found here for *S. rectum*, represented by the two genetic clusters. Analyses of genetic variability have been widely used to identify species exhibiting low genetic differentiation among populations as a consequence of a population bottleneck effect (Spencer et al., 2000), such as the marine shrimp species *Artemesia longinaris* Spence Bate, 1888 (cf. Carvalho-Batista et al., 2014) and the commercial *Farfantepenaeus paulensis* (Pérez-Farfante, 1967) (cf. Teodoro et al., 2015). Especially for consumed species, the bottleneck effect can be associated with anthropogenic actions, including habitat degradation or overfishing (Ryman et al., 1995). Nevertheless, drastic reductions on demographic size can also be explained by and be associated with natural disasters or geological events, such as glacial cycles.

The conformation of a genetic stock composed from few genetic sources, i.e., groups or lineages, characterized by low genetic variability for *S. rectum* could be associated with a recent population retraction event followed by current, recently started expansion. Our suppositions can be corroborated by the diversity indices obtained for most of the populations. Low values of nucleotide diversity (π) (<0.005) associated to high levels (>0.5) of haplotype diversity (h) are indicative of an accumulation of mutations after a bottleneck effect, followed by a rapid population growth (Grant & Bowen, 1998). Our significantly negative results for Tajima's D and Fu's F_s also indicate a demographic expansion after a retraction event (Aris-Brosou & Excoffier, 1996). Additionally, star-shaped haplotype networks, like the one we found here for *S. rectum*, are expected in species that may have suffered demographic expansion from only a few sources, after a bottleneck event (Avise, 2009). Besides, the graphic for mismatch distributions presented a negative binomial curve, which provides evidence for either a very recent bottleneck or expansion (Patarnello et al., 2007).

These recent retraction events of *S. rectum* populations may mainly be associated with habitat losses, considering the degree of destruction of mangrove areas along South America in the past decades. This phenomenon involves large areas under the pressures of economic and human development (Ferreira & Lacerda, 2016). Geographical constraints and anthropogenic activities and occupation located upwards from the high tide mark and upstream along watersheds have limited mangrove expansion landwards and also increased proportionally higher losses in ecosystem and biodiversity in these regions (Ferreira & Lacerda, 2016). Considering that *S. rectum* inhabits mangrove fringe areas close to the high tide mark, specifically being found in drier grassy zones (Tavares & Albuquerque, 1989), the destruction of mangrove forests in these areas implies affecting an extremely sensitive biodiversity inhabiting mangrove areas up to the high tide mark. In this way, a bottleneck effect suffered by *S. rectum* populations may be associated with the loss of these mangrove areas in the past decades.

The current state of Brazilian mangroves still indicates a need for policies that entail mangrove protection, but indeed some restoration efforts have already offset the losses (Ferreira & Lacerda, 2016). Although still few, some areas with destroyed ecosystems have naturally recuperated during the last two decades (Lacerda et al., 2007) or were restored by means of planting mangrove stands in the last years, promoting an early return of the functional fauna group constituted by brachyuran crabs (Ferreira & Lacerda, 2016). Thus, along with mangrove restorations, the *S. rectum* populations at present may be self-recovering and, although at a low rate, start expanding, mainly due to the capacity of larval dispersion, as already discussed.

Overall, our mitochondrial data for *S. rectum* populations revealed low genetic differentiation and haplotype sharing along a wide geographical range. These results support the hypothesis of a high dispersive capacity, since we found high levels of gene flow among populations spanning over 6000 km. Also, our results and larval behavioural features support a second hypothesis, according to which the absence of strong selective gradients along the species' distribution might be a key element underlying low genetic differentiation. Finally, although we found no geographic structure, some results indicated population structure, thus suggesting the presence of two genetic sources, i.e., lineages, probably interpreted as a result of a very recent bottleneck effect as a consequence of habitat losses, followed by the beginning of a population expansion.

ACKNOWLEDGEMENTS

This article was part of a PhD thesis by RCB supported by scientific fellowship from São Paulo Research Foundation — FAPESP (DD 2012/06299-5). Financial support was provided by research grants from FAPESP (2002/08178-9, Temático Biota 2010/50188-8; Coleções Científicas 2009/54931-0), Conselho Nacional de Desenvolvimento Científico e Técnológico — CNPq (472746/2004-9, 491490/2004-6, 473050/2007-2, 471011/2011-8, 504322/2012-5; Research Scholarships PQ 302748/2010-5, 304968/2014-5), and Coordenação de Aperfeiçoamento de Pessoal de Nível Superior — CAPES (Ciências do Mar II Proc.

2005/2014 — 23038.004308/201414) to FLM. RCB is grateful to CNPQ (PROTAX 150462/2016-6) for Post-Doctoral scholarship. We are grateful to the Postgraduate Program in Comparative Biology of the FFCLRP/USP for partial financial support via PROAP/CAPES. We are grateful to many colleagues and friends (Abner Carvalho, Darryl Felder, Edvanda Souza Carvalho, Fabrício Carvalho, Fernando Alvarez, Fernando Abrunhosa, Gustav Paulay, Jose Luis Villalobos, Luigi Pavoni Cerantola, Luis Ernesto Bezerra, Mariana Negri, Mariana Terossi, Mateus Lopes, Rafael Lemaitre and Rafael Robles) for helping in samplings, loans, and critical discussions for this manuscript. We thank the suggestions of the anonymous reviewers that improved the quality of the manuscript. Finally, the authors, especially FLM, were saddened and shocked by the sudden loss of our colleague Michael Türkay. His long years of intense and unconditional dedication to the study of decapod crustaceans have left significant marks for the knowledge of this taxon. Michael has co-authored two new sesarmid crabs from the New World and following his enthusiasm we are honoured to dedicate this contribution on genetic variability of *Sesarma rectum* for this commemorative edition to him.

REFERENCES

ABELE, L. G., 1992. A review of the grapsid crab genus *Sesarma* (Crustacea: Decapoda: Grapsidae) in America, with the description of a new genus. Smithsonian Contribution to Zoology, **527**: 1-60.

ADDISON, J. A., B. S. ORT, K. A. MESA & G. H. POGSON, 2008. Range-wide genetic homogeneity in the California sea mussel (*Mytilus californianus*): a comparison of allozymes, nuclear DNA markers, and mitochondrial DNA sequences. Molecular Ecology, **17**: 4222-4232.

ALTUKHOV, Y. P., 1981. The stock concept from the viewpoint of population genetics. Canadian Journal of Fisheries and Aquatic Sciences, **38**: 1523-1538.

ANGER, K. & G. S. MOREIRA, 2004. Biomass and elemental composition of eggs and larvae of a magrove crab, *Sesarma rectum* Randall (Decapoda: Sesarmidae) and comparison to a related species with abbreviated larval development. Scientia Marina, **68**: 117-126.

ARIS-BROSOU, S. & L. EXCOFFIER, 1996. The impact of population expansion and mutation rate heterogeneity on DNA sequence polymorphism. Molecular Biology and Evolution, **13**: 494-504.

AVISE, J. C., 2004. Molecular markers, natural history and evolution (2nd ed.): 1-532. (Sinauer & Associates, Sunderland, MA).

AVISE, J. C., 2009. Phylogeography: retrospect and prospect. Journal of Biogeography, **36**: 3-15.

BARBER, P. H., S. R. PALUMBI, M. V. ERDMANN & M. K. MOOSA, 2000. A marine Wallace's line? Nature, **406**: 692-693.

BURTON, R. S., 1983. Protein polymorphisms and genetic differentiation of marine invertebrate populations. Marine Biology Letters, **4**: 193-206.

BURTON, R. S. & M. W. FELDMAN, 1982. Population genetics of coastal and estuarine invertebrates: does larval behavior influence population structure? In: V. S. KENNEDY (ed.), Estuarine comparisons: 537-551. (Academic Press, San Diego, CA).

CARVALHO-BATISTA, A., M. NEGRI, L. G. PILEGGI, A. L. CASTILHO, R. C. COSTA & F. L. MANTELATTO, 2014. Inferring population connectivity across the range of distribution of the stiletto shrimp *Artemesia longinaris* Spence Bate, 1888 (Decapoda, Penaeidae) from DNA barcoding: implications for fishery management. ZooKeys, **457**: 271-288.

CHUST, G., E. VILLARINO, A. CHENUIL, X. IRIGOIEN, N. BIZSEL, A. BODE, C. BROMS, S. CLAUS, M. L. F. PUELLES, S. FONDA-UMANI, G. HOARAU, M. G. MAZZOCCHI, P. MOZETIČ, L. VANDEPITTE, H. VERÍSSIMO, S. ZERVOUDAKI & A. BORJA, 2016. Dispersal similarly shapes both population genetics and community patterns in the marine realm. Scientific Reports, **6**: 28730.

CIRANO, M., M. M. MATA, E. J. D. CAMPOS & N. F. R. DEIR, 2006. A circulação oceânica de larga-escala na região oeste do Atlântico Sul com base no modelo de circulação global OCCAM. Revista Brasileira de Geofísica, **24**: 209-230.

CLEMENT, M., D. POSADA & K. A. CRANDALL, 2000. TCS: a computer program to estimate gene genealogies. Molecular Ecology, **9**: 1657-1660.

CORANDER, J., M. GYLLENBERG & T. KOSKI, 2007. Random partition models and exchangeability for Bayesian identification of population structure. Bulletin of Mathematical Biology, **69**: 797-815.

CORANDER, J., P. WALDMANN, P. MARTTINEN & M. J. SILLANPÄÄ, 2004. BAPS 2: enhanced possibilities for the analysis of genetic population structure. Bioinformatics, **20**: 2363-2369.

CORANDER, J., P. WALDMANN & M. J. SILLANPÄÄ, 2003. Bayesian analysis of genetic differentiation between populations. Genetics, **163**: 367-374.

COWEN, R. K., M. M. L. KAMAZIMA, S. SPONAUGLE, C. B. PARIS & D. B. OLSON, 2000. Connectivity or marine populations: open or closed? Science, **287**: 857-859.

EXCOFFIER, L. & A. LANGANEY, 1989. Origin and differentiation of human mitochondrial DNA. American Journal of Human Genetics, **44**: 73-85.

EXCOFFIER, L., G. LAVAL & S. SCHNEIDER, 2005. Arlequin (version 3.0): an integrated software package for population genetics data analysis. Evolutionary Bioinformatics Online, **1**: 47-50.

EXCOFFIER, L., P. E. SMOUSE & J. M. QUATTRO, 1992. Analysis of molecular variance inferred from metric distances among DNA haplotypes: application to human mitochondrial DNA restriction data. Genetics, **131**: 479-491.

FELDER, D. L. & J. L. STATON, 1994. Genetic differentiation in trans-Floridian species complexes of *Sesarma* and *Uca* (Decapoda: Brachyura). Journal of Crustacean Biology, **14**: 191-209.

FERREIRA, A. L. & L. D. LACERDA, 2016. Degradation and conservation of Brazilian mangroves, status and perspectives. Ocean & Coastal Management, **125**: 38-46.

FU, Y. X., 1997. Statistical tests of neutrality of mutations against population growth, hitchhiking and background selection. Genetics, **146**: 915-925.

GOPURENKO, D. & J. M. HUGHES, 2002. Regional patterns of genetic structure among Australian populations of the mud crab, *Scylla serrata* (Crustacea: Decapoda): evidence from mitochondrial DNA. Marine and Freshwater Research, **53**: 849-857.

GRANT, W. S. & B. W. BOWEN, 1998. Shallow population histories in deep evolutionary lineages of marine fishes: insights from sardines and anchovies and lessons for conservation. Journal of Heredity, **89**: 415-426.

HALL, T. A., 1999. BioEdit: a user-friendly biological sequence alignment editor and analysis program for Windows 95/98/NT. Nucleic Acids Symposium Series, **41**: 95-98.

HAMASAKI, K., C. IIZUKA, A. OJIMA, M. SUGIZAKI, M. SUGIMOTO, D. SHIGEKI & S. KITADA, 2015. Genetic diversity and demographic history of the terrestrial hermit crabs *Birgus latro* and *Coenobita brevimanus* in the north-western Pacific region. Journal of Crustacean Biology, **35**(6): 793-803.

HARPENDING, H., 1994. Signature of ancient population growth in a low-resolution mitochondrial DNA mismatch distribution. Human Biology, **66**: 591-600.

HEDGECOCK, D., 1986. Is gene flow from pelagic larval dispersal important in the adaptation and evolution of marine invertebrates? Bulletin of Marine Science, **39**: 550-564.

HELLBERG, M. E., R. S. BURTON, J. E. NEIGEL & S. R. PALUMBI, 2002. Genetic assessment of connectivity among marine populations. Bulletin of Marine Science, **70**: 273-290.

ITUARTE, R. B., A. D'ANATRO, T. A. LUPPI, P. D. RIBEIRO, E. D. SPIVAK, O. O. IRIBARNE & E. P. LESSA, 2012. Population structure of the SW Atlantic estuarine crab *Neohelice granulata* throughout its range: a genetic and morphometric study. Estuarine and Coasts, **35**: 1249-1260.

KELLY, R. P. & S. R. PALUMBI, 2010. Genetic structure among 50 species of the northeastern Pacific rocky intertidal community. PLoS ONE, **5**: e8594.

KIMURA, M. & G. H. WEISS, 1964. The stepping stone model of population structure and the decrease of genetic correlation with distance. Genetics, **49**: 561-576.

KOEHN, R. K., 1969. Esterase heterogeneity: dynamics of a polymorphism. Science, **163**: 943-944.

LACERDA, L. D., M. O. T. MENEZES & M. M. MOLISANI, 2007. Changes in mangrove extension at the Pacoti River estuary, CE, NE Brazil due to regional environmental changes between 1958 and 2004. Biota Neotropica, **7**: 67-72.

LAURENZANO, C., T. M. COSTA & C. D. SCHUBART, 2016. Contrasting patterns of clinal genetic diversity and potential colonization pathways in two species of western Atlantic fiddler crabs. PLoS ONE, **11**: e0166518.

LAURENZANO, C., N. E. FARÍAS & C. D. SCHUBART, 2012. Mitochondrial genetic structure of two populations of *Uca uruguayensis* fails to reveal an impact of the Rio de la Plata on gene flow. Nauplius, **20**: 15-25.

LAURENZANO, C., F. L. MANTELATTO & C. D. SCHUBART, 2013. South American homogeneity versus Caribbean heterogeneity: population genetic structure of the western Atlantic fiddler crab *Uca rapax* (Brachyura, Ocypodidae). Journal of Experimental Marine Biology and Ecology, **449**: 22-27.

LEE, S. Y., 1998. Ecological role of grapsid crabs in mangrove ecosystems: a review. Marine and Freshwater Research, **49**: 335-343.

MANTELATTO, F. L., R. ROBLES, R. BIAGI & D. L. FELDER, 2006. Molecular analysis of the taxonomic and distributional status for the hermit crab genera *Loxopagurus* Forest, 1964 and *Isocheles* Stimpson, 1858 (Decapoda, Anomura, Diogenidae). Zoosystema, **28**: 495-506.

MCMILLEN-JACKSON, A. L. & T. M. BERT, 2004. Mitochondrial DNA variation and population genetic structure of the blue crab *Callinectes sapidus* in the eastern United States. Marine Biology, **145**: 769-777.

MCMILLEN-JACKSON, A. L., T. M. BERT & P. STEELE, 1994. Population genetics of the blue crab *Callinectes sapidus*: modest population structuring in a background of high gene flow. Marine Biology, **118**: 53-65.

MELO, G. A. S., 1996. Manual de identificação dos Brachyura (caranguejos e siris) do litoral brasileiro: 1-604. (Plêiade, São Paulo).

MILLER, S. A., D. D. DYKES & H. F. POLESKY, 1988. A simple salting out procedure for extracting DNA from human nucleated cells. Nucleic Acids Research, **16**(3): 1215.

NEI, M., T. MARAYUMA & R. CHAKRABORTY, 1975. The bottleneck effect and genetic variability in populations. Evolution, **29**: 1-10.

OLIVEIRA-NETO, J. F., M. R. PIE, M. A. CHAMMAS, A. OSTRENSKY & W. A. BOEGER, 2008. Phylogeography of the blue land crab, *Cardisoma guanhumi* (Decapoda: Gecarcinidae) along the Brazilian coast. Journal of the Marine Biological Association of the United Kingdom, **88**: 1417-1423.

PALUMBI, S. R., 1994. Genetic divergence, reproductive isolation, and marine speciation. Annual Review of Ecology and Systematics, **25**: 547-572.

PALUMBI, S. R., 2003. Population genetics, demographic connectivity, and the design of marine reserves. Ecological Applications, **13**: s146-s158.

PATARNELLO, T., F. A. M. VOLCKAERT & R. CASTILHO, 2007. Pillars of Hercules: is the Atlantic–Mediterranean transition a phylogeographical break? Molecular Ecology, **16**: 4426-4444.

RODRÍGUEZ-REY, G. T., A. M. SOLÉ-CAVA & C. LAZOSKI, 2014. Genetic homogeneity and historical expansions of the slipper lobster, *Scyllarides brasiliensis*, in the south-west Atlantic. Marine and Freshwater Research, **65**: 59-69.

ROGERS, A. R., 1995. Genetic evidence for a Pleistocene population explosion. Evolution, **49**: 608-615.

ROGERS, A. R. & H. HARPENDING, 1992. Population growth makes waves in the distribution of pairwise genetic differences. Molecular Biology and Evolution, **9**: 552-569.

ROZAS, J. & R. ROZAS, 1999. DnaSP version 3.0: an integrated program for molecular population genetics and molecular evolution analysis. Bioinformatics, **15**: 174-175.

RYMAN, N., F. UTTER & L. LAIKRE, 1995. Protection of intraspecific biodiversity of exploited fishes. Reviews in Fish Biology and Fisheries, **5**: 417-446.

SAMBROOK, J., E. F. FRITSCH & T. MANIATIS, 1989. In vitro amplification of DNA by the polymerase chain reaction. In: J. SAMBROOK, E. F. FRITSCH & T. MANIATIS (eds.), Molecular cloning: a laboratory manual: 14.1-14.35. (Cold Spring Harbor Laboratory Press, New York, NY).

SCHELTEMA, R. S., 1986. On dispersal and planktonic larvae of benthic invertebrates: an eclectic overview. Bulletin of Marine Science, **39**: 290-322.

SCHNEIDER, S. & L. EXCOFFIER, 1999. Estimation of past demographic parameters from the distribution of pairwise differences when the mutation rates vary among sites: application to human mitochondrial DNA. Genetics, **152**: 1079-1089.

SCHUBART, C. D., D. HORST & R. DIESEL, 1999. First record of *Sesarma rectum* Randall (Brachyura, Grapsidae, Sesarminae) from the Lesser Antilles. Crustaceana, **72**: 537-538.

SCHUBART, C. D. & M. G. J. HUBER, 2006. Genetic comparisons of German populations of the stone crayfish, *Austropotamobius torrentium* (Crustacea: Astacidae). Bulletin Français de la Pêche et de la Pisciculture, **380-381**: 1019-1028.

SHANKS, A. L., 2009. Pelagic larval duration and dispersal distance revisited. Biological Bulletin, **216**: 373-385.

SHANKS, A. L., B. A. GRANTHAM & M. H. CARR, 2003. Propagule dispersal distance and the size and spacing of marine reserves. Ecological Applications, **13**: S159-S169.

SILVA, I. C., N. MESQUITA & J. PAULA, 2010. Lack of population structure in the fiddler crab *Uca annulipes* along an East African latitudinal gradient: genetic and morphometric evidence. Marine Biology, **157**: 1113-1126.

SPENCER, C. C., J. E. NEIGEL & P. L. LEBERG, 2000. Experimental evaluation of the usefulness of microsatellite DNA for detecting demographic bottlenecks. Molecular Ecology, **9**: 1517-1528.

TAJIMA, F., 1989. Statistical method for testing the neutral mutation hypothesis by DNA polymorphism. Genetics, **123**: 585-595.

TAVARES, M. S. & E. F. ALBUQUERQUE, 1989. Levantamento preliminar dos Brachyura (Crustacea: Decapoda) da lagoa de Itaipu, Rio de Janeiro, Brasil. Atlântica, **11**: 101-108.

TAYLOR, M. S. & M. E. HELLBERG, 2003. Genetic evidence for local retention of pelagic larvae in a Caribbean reef fish. Science, **299**: 107-109.

TEODORO, S. S. A., M. TEROSSI, R. C. COSTA & F. L. MANTELATTO, 2015. Genetic homogeneity in the commercial pink shrimp *Farfantepenaeus paulensis* revealed by COI barcoding gene. Estuarine, Coastal and Shelf Science, **166**: 1-7.

TEROSSI, M. & F. L. MANTELATTO, 2012. Morphological and genetic variability in *Hippolyte obliquimanus* Dana, 1852 (Decapoda, Caridea, Hippolytidae) from Brazil and the Caribbean Sea. Crustaceana, **85**: 685-712.

THOMPSON, J. D., D. G. HIGGING & T. J. GIBSON, 1994. CLUSTALW: improving the sensitivity of progressive multiple sequence alignment through sequence weighting specific gap penalties and weight matrix choice. Nucleic Acids Research, **22**: 4673-4680.

WAPLES, R. S., 1987. A multispecies approach to the analysis of gene flow in marine shore fishes. Evolution, **41**: 385-400.

WARNER, R. R. & R. K. COWEN, 2002. Local retention of production in marine populations: evidence, mechanisms, and consequences. Bulletin of Marine Science, **70**(1): 245-249.

WIEMAN, A. C., P. B. BERENDZEN, K. R. HAMPTON, J. JANG, M. J. HOPKINS, J. JURGENSON, J. C. MCNAMARA & C. L. THURMAN, 2014. A panmictic fiddler crab from the coast of Brazil? Impact of divergent ocean currents and larval dispersal potential on genetic and morphological variation in *Uca maracoani*. Marine Biology, **161**: 173-185.

WILLIAMS, S. T. & J. A. H. BENZIE, 1993. Genetic consequences of long larval life in the starfish *Linckia laevigata* (Echinodermata: Asteroidea) on the Great Barrier Reef. Marine Biology, **117**: 71-77.

WRIGHT, S., 1931. Evolution in Mendelian populations. Genetics, **16**: 97-159.

WRIGHT, S., 1943. Isolation by distance. Genetics, **38**: 114-138.

WRIGHT, S., 1978. Evolution and the genetics of populations, **4**, Variability within and among natural populations: 1-590. (University of Chicago Press, Chicago, IL).

First received 21 September 2016.
Final version accepted 7 April 2017.

[When citing this volume, please refer to Crustaceana 90 (2017) 771-1288]

Michael Türkay Memorial Issue

FEEDING ECOLOGY OF THE SHRIMP *CRANGON ALLMANNI* KINAHAN, 1860 (DECAPODA, CRANGONIDAE) IN THE NORTH AND WHITE SEAS

BY

R. N. BURUKOVSKY[1])
Kaliningrad State Technical University, Kaliningrad 236022, Russia

ABSTRACT

The food composition of the shrimp *Crangon allmanni* from the Helgoland Trench (North Sea) and Onega Bay (White Sea) is described. The main food items (>60%) include detritus, representatives of about 30 benthic species, dominated by polychaetes, malacostracans, ophiuroids, bivalves and ophistobranchs (*Cylichna*, *Diaphana* spp.) as well as plant remains. In the North Sea, *C. allmanni* is closer to an attacking predator than to a predator-gatherer like in the White Sea. This difference may be caused by diverging habitat- and community characteristics and dissimilar size composition of the two studied shrimp populations. Moreover, *C. allmanni* changes its foraging mode (grazing, gathering, attacking) during ontogenesis. A comparison of the obtained food composition data of *C. allmanni* with literature data on six other species of the same genus showed all to be benthos feeders, predator-gatherers with elements of detrito- and necrophagia, using grains of sand as millstones in their gastric mill.

ZUSAMMENFASSUNG

Hier wird die Nahrungszusammensetzung der Garnele *Crangon allmanni* aus der Helgoländer Tiefen Rinne (Nordsee) und der Onega-Bucht (Weißes Meer) beschrieben. Die Nahrungs-Hauptbestandteile (>60%) sind Detritus und Pflanzenreste, sowie Vertreter von ungefähr 30 benthischen Arten, dominiert von den Tiergruppen Polychaeta, Malacostraca, Ophiuroidea, Bivalvia und Ophistobranchia (*Cylichna*, *Diaphana* spp.). In der Nordsee ist *C. allmanni* eher ein angreifender Jäger gegenüber einem Jäger und Sammler im Weißen Meer. Dieser Unterschied könnte durch eine Divergenz der Habitate und Lebensgemeinschaften, sowie der Größenklassen der beiden Garnelenpopulationen verursacht werden. Weiterhin ändert *C. allmanni* die Art seiner Nahrungssuche (grasen, sammeln, attackieren) im Laufe seiner Entwicklung. Ein Vergleich der erhaltenen Nahrungszusammensetzungsdaten von *C. allmanni* mit Literaturdaten von sechs weiteren Arten derselben Gattung zeigt, dass diese durchweg Benthosfresser und Jäger/Sammler mit Zügen von Reste- und Aasfressern sind, welche Sandkörner als "Mühlsteine" in ihren Magenmühlen verwenden.

[1]) e-mail: burukovsky@klgtu.ru

INTRODUCTION

The genus *Crangon* Fabricius, 1798 includes 20 species (De Grave & Fransen, 2011); however, the eastern North Atlantic is inhabited by only two of them: *Crangon crangon* (Linnaeus, 1758) and *C. allmanni* Kinahan, 1860. The first one is the target species for fishery in the North Sea (Schwinn et al., 2014).

Crangon allmanni is a mobile epibenthic species that regularly occurs in bottom trawl catches but it does not form dense aggregations. Data on the biology of *C. allmanni* are very scarce, and information on its food composition is limited to three paragraphs in a fundamental paper by Allen (1960). This shrimp is very sensitive to effects of stressful environmental factors, due to which it can serve as an environmental indicator (Blahudka & Türkay, 2002). Because of this, *C. allmanni* was chosen for a study in the framework of project INTAS 51-5458 "Population and trophic biology of *Crangon allmanni*" with project coordinator Dr. M. Türkay (Crustacean Section, Senckenberg Society for Nature Research, Frankfurt am Main, Germany), the initiator of sampling for this study. The study of the food composition of *C. allmanni* in different parts of its range was my task in this project. The results of this investigation with comparison on other *Crangon* species are presented in this paper.

MATERIAL AND METHODS

For this study, four samples from the Helgoland Trench (North Sea) and one from the Onega Bay (White Sea) were analysed.

The Helgoland Trench samples were collected on 11 August 2004, 8 August 2005, 22 July 2006 and 27 July 2006 at 54°08′N 7°50′-7°53′E at a depth range of 55-57 m. In total 639 stomachs of shrimps as well as their carapace length (from the end of the rostrum to the posterior edge of the carapace) were analysed. It was found that 351 stomachs contained food remains and 92 stomachs were full with food.

The second sampling was performed in July 2006 in the southeastern part of Onega Bay at 64°03.676′-64°37.147′N 36°53.107′-37°54.998′E at a depth of 6.9-31.8 m. In total 169 stomachs were analysed and the carapace length of the respective males and females was determined. We found that 103 stomachs contained food remains and 27 stomachs were full.

The shrimp carapace length was measured to the nearest 0.1 mm, using the micrometer of a binocular microscope (MBS-9), as the shortest distance between the anterior end of the rostrum to the posterior edge of the carapace in the median line of its dorsal side.

For this feeding study the methodology of Burukovsky (Burukovsky & Trunova, 2007; Burukovsky, 2009) was used. After stomach opening, the degree of its filling was determined on a 4-point scale: 0, the stomach is empty; 1, the food occupied less than half the stomach volume; 2, food occupied about half (1/3-2/3) of the volume of the stomach; 3, the stomach is full.

The food lump from each stomach was studied in a drop of water in a Petri dish. Identification of the taxonomic status of the food remains usually was made on the level of class or order (for example, Gastropoda or Bivalvia, Mysidacea, Euphausiacea or Isopoda), trying to determine the taxonomic affiliation of the prey as accurately as possible, ideally to the species level. Special attention was paid to find out to which particular lifestyle, including the specific habitat (pelagic, benthic, sessile, burrows etc.), the prey belonged to.

The food organism remains were counted and measured with the ruler of the ocular micrometer of the microscope. Measuring the prey length entirely was rarely possible because the remains were relatively crushed. Therefore, the parts of the body that could be measured (primarily skeletal elements, scales, eye lenses, otoliths or vertebrae of fish, the chaetae of Chaetognatha and Polychaeta, statoliths of mysids etc.) were used for size determination.

Regardless of stomach fullness, the composition of food items was identified in all stomachs with food. In full stomachs, the components of the food lump were visually estimated by volume with an accuracy of 10%. Full stomachs were chosen for visual estimation to avoid the impact of different degrees of digestion of the food remains in the stomachs on the obtained result. The food and inedible items that constituted less than 10% of the total volume were just noted. According to Burukovsky (2009), the results of that study were calculated using:

(1) The frequency of occurrence (FO, percent of occurrence of given food component from the total number of examined stomachs with food).
(2) The Froerman Index (mean number of prey items in the full stomach excepting grains of sand, detritus and plant remains). The Froerman Index was calculated as the sum of all food group's FO in percent, divided by 100.
(3) The frequency of prevalence (FP, frequency of occurrence of full stomachs (in %) where one of the food items consists of 60% and more of the food lump volume, i.e., prevailed).
(4) The reconstructed averaged (virtual) food lump volume (VFLV, mean share of each food component in the total food lump volume in %).

VFLV and FP were calculated using only data on full stomachs. The term "food components" means both alive and non-living remains (grains of sand, spicules of sponges) that were detected in the stomachs, in contrast to the "food items", i.e., those components that are directly used by shrimps as food. The terminology for describing the methods of shrimp hunting (feeding behaviour) of Burukovsky (2009) was used.

These feeding characteristics should be used together because they complement each other. Separately they give an one-sided image of the diet of the studied animals. The role of foraminifera in the diet of some shrimp species has been demonstrated to be misinterpreted (Burukovsky, 2009). Their FO values can reach up to 60-70%, so they fall into the category of the most common and important food group. But their share in the VFLV was less than 0.1-0.2%. Hence, studying the ontogenetic variability of food composition using both parameters can be especially effective. In some species during ontogenesis the FO of given food item gradually decreases, but its VFLV increases. Therefore, adult shrimps feed on a given prey less frequently but in larger quantities (Burukovsky, 2009). Thus, the use of these two parameters (FO and VFLV) together gives a realistic idea on the role of given food groups in the diet of a shrimp. This double approach also provides an effective "instrument" for reconstruction of feeding behaviour (Burukovsky, 2009).

A stepwise characterization of the food composition was used:

(1) Description of food remains. This is very important, as it allows to estimate the type (pattern) of feeding. This is the first step towards descriptive results against a mere listing of food composition.
(2) Description of the frequency of occurrence of food items in all stomachs with food, regardless of its amount in the stomach.
(3) Description of the volume ratios of food components in full stomachs. It is also very important, as it allows to reconstruct the virtual food lump.

RESULTS

The North Sea

The carapace length of *Crangon allmanni* from the North Sea ranged from 3.2 to 14.0 mm. The food components obtained here include sand grains, detritus, plant remains, non-identified animal remains and remains that taxonomic status could be identified at least to class or order levels (table I). The sand grains were presented in nearly all stomachs. Their numbers varied from 3-10 to hundreds of grains with sizes from 0.05 to 0.7 mm (modal 0.3-0.5 mm), i.e., it was the fraction of medium sand (Petelin, 1967). The sand FO was 81.8%. In approximately half full stomachs the share of sand in their total food lump volume was less than 10%, varying in limits of 60-80% in 14.3% of full stomachs. The sand content in the VFLV was 16.3%.

TABLE I

Food composition in stomachs of the shrimp *Crangon allmanni* Kinahan, 1860 in the North Sea and the White Sea

Component	Frequency of occurrence (FO, %)		Share of virtual food lump volume (VFLV, %)		Frequency of prevalence (FP, %)	
	North Sea	White Sea	North Sea	White Sea	North Sea	White Sea
Detritus	72.4	86.8	16.3	31.6	6.5	15.8
Polychaeta	29.1	54.4	21.3	5.8	21.7	5.2
Mysidacea	15.1	–	13.7	–	14.1	–
Ophiuroidea	14.5	–	4.1	–	4.3	–
Amphipoda	12.8	5.9	5.6	1.1	3.3	–
Echinoidea	11.4	–	0.7	–	–	–
Foraminifera	9.1	16.2	–	4.2	–	–
Bivalvia	8.8	48.5	1.4	12.1	–	–
Gastropoda	8.0	2.9	2.2	–	2.2	–
Copepoda and Cladocera	8.0	22.1	–	11.0	–	10.5
Cumacea	–	4.4	–	6.8	–	5.2
Pisces	4.3	–	1.7	–	2.2	–
Euphausiacea	3.7	–	0.6	–	–	–
Indetermined eggs	3.1	19.1	–	–	–	–
Paguroidea	2.6	–	–	–	3.3	–
Shrimps	2.3	–	4.1	–	4.3	–
Ostracoda	2.0	14.7	–	4.7	–	–
Gastropoda eggs	1.7	–	–	–	–	–
Brachyura and other "Reptantia"	1.4	–	0.2	–	–	–
Isopoda	1.4	–	0.2	–	–	–
Cnidaria (Hydrozoa)	1.4	–	1.0	–	1.1	–
Chaetognatha	1.4	1.5	0.2	–	–	–
Nematoda	1.4	4.4	–	–	–	–
Plant debris	1.4	1.5	0.7	–	1.1	–
Diatomea	1.1	–	–	–	–	–
Tintinoidea	0.8	–	–	–	–	–
Acari and their nymphs	0.6	–	–	–	–	–
Insecta	0.6	–	1.7	–	3.3	–
Oligochaeta	0.6	–	0.5	–	–	–
Brachiopoda	0.3	–	–	–	–	–
Holothurioidea	0.3	–	–	–	–	–
Priapulida	0.3	–	0.4	–	0.4	–
Tanaidacea	–	2.9	–	1.6	–	–
Undetermined remains	20.5	10.9	6.7	–	5.4	–
Sand-grains	81.8	79.4	14.3	21.0	8.7	15.8
Spicules and other debris	3.1	1.5	–	–	–	–
Total stomachs	351	103	92	27	92	27
Froerman Index	2.20	2.93	–	–	–	–
Frequency of prevalence (%)	–	–	–	–	73.3	36.7

The obtained detritus was a loose mass of grey or greyish-beige, rarely almost black, colour. It is more or less uniform and does not contain inclusions. The detritus FO was 72.4%, i.e., it also occurs in the majority of studied stomachs, at least in trace amounts. The detritus in full stomachs usually does not exceed half the volume of the food lump, however, in 6.5% of the stomachs, its share exceeded 60%, and two of the stomachs were filled solely by detritus. Its share in the VFLV was 16.3%.

Plant remains occurred very rarely — only in 5 stomachs from 351 studied (1.4%), and their share in the VFLV was 0.7%. It was not possible to identify these remains. In one full stomach their share was 60% of the total food volume. Plant remains should be considered as an accidental food component.

Unidentified food components were divided into two groups. The first one included pieces of chitin in different stages of degradation, that were impossible to identify. Probably, they are crustacean remains that were eaten during the previous feeding event, or maybe remains of dead, decomposed prey. Their FO was 6.5% and as a rule they occurred in stomachs that were hardly full, but in three full stomachs they occupied 10-20% of the volume of food lump. Their share in the VFLV was very small (0.6%).

The other group of non-identified food components includes tissue fragments with fibrous or gelatinous consistency, wrinkled and covered with detritus. Their FO was 14.0%. Probably this were the half-decayed remains of dead animals that were ingested. Most likely *C. allmanni* feeds on dead animals, as evidenced by the occurrence of adults and nymphs of mites and imagines of insects in its stomach that only could drop to a depth of about 50 m, as they were already dead. One insect specimen had a length of 2.5 mm. In two stomachs the share of insects in the volume of the food lump was 60 and 100%, respectively. By preliminary estimation necrophagy in *C. allmanni* occurred at least in 20%, and the share of these dead animals was 8.4% of the VFLV. The last three groups of food components in total accounted for 39% of the VFLV.

Food items eaten alive can be divided into three groups, depending on their FO values. The first group includes polychaetes, which are found almost 2.5 times less frequently than detritus (FO 29.1%), being completely dominated by other food groups. Polychaetes presented at least 4-5 species that were mainly the representatives of errant life forms of the families Polinoidae and Glyceridae (*Glycera* sp.). They occupied 80-100% of the lump food volume in 15 of 92 full stomachs. Their body lengths were about 3.0-35.0 mm, commonly 3.0-4.0 mm. These pieces were the fragments of single specimens. Only once more than 10 fragments of polychaetes of 3 or 4 species were found, each having a length of 1-2 mm. The share of polychaetes (21.3%) in the VFLV is almost 1.5 times that of detritus, which, in contrast, occurs 2.5 times more often (table I). In the first place, polychaetes had a FP value of 21.7%.

Secondary food items include mysids, amphipods and ophiuroids, the largest fraction (FO 15.1%) of which were mysids. In the stomachs they occurred either as whole specimens or, more often, as statoliths. Mysids constitute 13.7% of the VFLV and are the dominating group in full stomachs (FP 14.1%). In 10 stomachs they occupy 90-100% of the volume. Usually, in the stomachs we found 1-3 specimens of adult mysids with a length of 13-14 mm. Female marsupiums were found to contain embryos or larvae. Conclusively, polychaetes and mysids were found to be the main prey of *C. allmanni*.

Ophiuroidea remains occurred in 14.5% of all studied stomachs. Their body fragments were easily identified by skeleton elements of arms that most frequently occurred in stomach contents. Their share in the VFLV was 4.1%, probably due to dominating small fragments. Whole specimens with a disc diameter of nearly 2 mm were found in only two instances.

The amphipod FO (12.8%) and the value of share in the VFLV (5.6%) were nearly the same as those for Ophiuroidea. In two cases more than half of the full stomach volume consisted solely of amphipod remains. Among the amphipods, the representatives of Gammarida dominated the diet. As a rule, they were represented by remains of single specimens (maximum of 4) with a length of 0.8-11.0 mm, mainly 2.0-3.5 mm. Also fragments of Caprellidae occurred relatively often, and once there were found specimens of Hyperiidae with 5.0-6.0 mm length.

All other food items were insignificant or rare. The insignificant food is considered to be the components that had a relatively high frequency of occurrence but VFLV of full stomachs never reached or exceeded 10%. Among others, sea urchins and foraminifera were found (table I).

Rare food items with low FO can be found in relatively large numbers, occupying more than 60% of the food lump volume. For example, there were bivalves (*Macoma* sp. with shell length 1.2 mm) and gastropods. The opisthobranch *Diaphana* sp. and its egg masses were dominating among the gastropods. In one stomach, up to 5 specimens of *Diaphana* sp. were found, having shell sizes of 2.0-3.5 mm. In one stomach, *Natica* sp. with nearly the same shell size as *Diaphana* sp. was detected.

The rare food group also includes various crustaceans: juveniles of hermit crabs, crabs, juveniles of *Munida* sp. and the small crustacean family Axiidae with a length of 3-4 mm. Interesting data were obtained for shrimps. Their FO was only 2.3%, but the share in the VFLV had a relatively high value (4.1%), being a clear indication for cannibalism: most of the shrimp remains were represented exclusively by juveniles of *C. allmanni*. Shrimps only occurred in 8 stomachs with

4 full stomachs being occupied up to 90-100% of their food lump volume. The total lengths of shrimps were 12-13 mm, probably being molting specimens.

A large number of dominant food items (12), high total the FP value (79.4%) and a relatively low Froerman Index value (2.20) suggested that the feeding behavior of *C. allmanni* in the North Sea can be characterized as an attacking predator in combination with detritophagy and necrophagy, with active feeding on living animals dominating (about 60% of the VFLV).

The White Sea

At Onega Bay, the carapace length of males studied was 4.2-6.9 mm, while females reached 3.5-12.3 mm. Sand grains occurred in most of the stomachs studied with an FO of 79.4%. Their number varied from 2-3 up to hundreds, on average 10-20. Grain size was 0.05-0.8 mm (modal 0.15-0.3 mm) (Petelin, 1967). The share of sand in the VFLV was 21%. In 16% of the stomachs studied sand occupied more than 60% of the food lump volumes.

Detritus in the stomachs of *C. allmanni* from the Onega Bay (table I) is represented by two types. The first type of detritus was grey or brownish in colour, disintegrating into flakes in a drop of water. As a rule, this type occurred in stomachs with small food remains, but in one instance the share of detritus was 90% of the food lump volume. The second type of detritus was in the form of a pulp consisting of very fine pieces of chitin. All transitions of these chitin pieces were found during the maceration process. In several stomachs, copepod and cladoceran parts, as well as other skeleton elements, were found among these pieces of chitin, being a mixture of body parts with inclusions of particles of detritus. Thirty specimens of whole harpacticoid copepods with length 0.4-0.5 and 0.7-1.5 mm only occurred in two stomachs. The whole cladocerean *Podon leuckartii* (Sars, G. O., 1862) was found once. In the remaining cases, cladocerean residues were identified by the morphology of their mandibles. That is why the cladoceran FO and the share in the VFLV are underestimated, while the situation for copepods is the opposite. Therefore, the data on cladocerans and copepods were combined (table I).

Most probably shrimps in Onega Bay primarily feed on dead copepods and cladocerans in their places of aggregation at the bottom (for example, in the hydrologic situations that promote the sedimentation and accumulation of these dead crustaceans at the bottom). In these places the dead bodies undergo all stages of transformation to detritus and serve as a food for *C. allmanni*. Detritus occured more often than sand (86.8%) and it occupied 31.6% of VFLV. Plant remains were presented by significantly macerative pieces of brown algae (Phaeophyta) in the only one stomach.

It is very difficult to estimate the role of dead animals in shrimp feeding ecology. Doubtless, dead (before ingestion) animals (fish larvae with an intact integument, insects, mites, etc.) were absent in the stomachs from Onega Bay. There, little worn pieces of chitin and different fibrous fragments with the traces of decomposition were found. They were classified as unidentified remains; these fragments occurred only rarely (FO 10.9%) and in very small amounts. In reality, according to our data the role of necrophagy in this shrimp is probably undervalued. Copepods and cladocerans are a relatively important food group (sometimes they amount to 80% of the food lump volume in some stomachs). However, it is impossible to determine which of them were eaten dead or alive. The mixture of these small crustaceans with detritus occurred in 22.1% of all observed stomachs and it amounted 11.0% of the VFLV. By the rank of their FP value (10.5%), copepods and cladocerans are in 3rd place after sand grains and detritus.

Detritus together with the mixture of copepods and cladocerans at different stages of decomposition occupied almost half of the VFLV (42.6%). They were the main food groups for *C. allmanni*.

Polychaetes and bivalve molluscs with FO values of, respectively, 54.4 and 48.5% occupy the first two places among animals eaten alive. Polychaetes were represented almost exclusively by errant forms. Only once the chaetae of the sedentary polychaete family Spionidae were found. In general, relatively small fragments of bodies and chaetae were found in the stomachs. In one stomach there was a piece of a worm belonging to the family Polinoidae of 1.4 mm length and almost the whole specimen of 2.7 mm length with few elytra on the dorsal side. The VFLV of polychaetes was at 5.8%.

In the Onega Bay stomachs of *C. allmanni* bivalves were presented by *Macoma* sp. (shell length 0.7 mm), *Musculus* sp. (1.25 mm) occurring only once. More important in shrimp food were fragments of the bivalve family Cardiidae (*Cerastoderma* sp.?): 1-3 specimens with a shell size of 0.65-1.25 mm, mainly 0.7-0.8 mm. Although the mollusc FO is a little less than that of polychaetes, its share in the VFLV is twice as high (12.1%). Gastropods in the stomachs were very rare (FO 2.9%) and they were represented by single specimens, mostly by the young opisthobranch *Cylichne* sp. with a size of 0.9-1.7 mm.

Moreover, unidentified, spherical eggs of a diameter from 0.07-0.25 mm occurred relatively often (FO 19.1%), but they were found solitary and thus do not play a significant role in the diet of the shrimp species. On the other hand, foraminifera with the FO 16.2% had VFLV values of up to 4.2%. They were represented by the agglutinated species *Ammobaculites cassis* (Parker, 1870) [currently as: *Ammotium cassis* (Parker, 1870)]. The size of whole animals varied from 0.7 to 2.5 mm.

The remaining part of the VFLV consisted of crustaceans (table I). Among them, ostracods were the most important ones (FO 14.7% and VFLV 4.7%); they occurred singly or rarely with 2-3 specimens with a shell length of 0.1-0.9 mm. The FO value of cumaceans was 4.4% and an VFLV value of 6.8% (due to the much larger body size of 3-4 mm). Based on their body's degree of preservation it can be assumed that they were probably eaten alive. Fragments of other malacostracans were used for body length reconstruction and had roughly similar sizes: amphipods 1-2 mm and tanaidaceans 3 mm.

Thus, the virtual food lump of *C. allmanni* in the Onega Bay consisted of three main components: (1) the detritus together with inclusions of copepods and cladocerans (VFLV 42.6%), (2) crustaceans (14.2%) and (3) bivalves (12.1%). Polychaetes (VFLV 5.8%) and foraminiferans (4.2%) were a secondary food and they had an appreciable role in the shrimp diet.

In the studied sample the total FP value was the relatively low (table I). Primarily it includes the detritus together with inclusions of copepods and cladocerans. Judging by the value of the Froerman Index (2.93), *C. allmanni* combines the hunting types of grazing and gathering, but rather, it is a predator-gatherer. A large share of detritophagous brings this shrimp to a predator-opportunist by type of hunting.

DISCUSSION

Comparison of food composition in *Crangon allmanni* at different geographic areas

In this study the food composition of *Crangon allmanni* in the Northumberland area of the western part of the North Sea (Allen, 1960) is compared with the food composition at the Helgoland Trench (North Sea) and at Onega Bay (White Sea).

In the Northumberland area, shrimps fed on live crustaceans (FO 63.3%) and polychaetes (51.6%). Molluscs, foraminifers and Ophiuroidea occurred significantly less, as well as scales of the whiting (a fish, *Merlangius merlangus* (L., 1758)). The last was found only in stomachs of the shrimps that were caught at shallow depths. In food lumps silt was always present (personal remark: most likely it was detritus) together with sand grains. Among eaten polychaetes, *Nephthys* sp. had dominated (FO about 90%) and *Glycera* sp. occurred at about in 10% of stomachs. Crustaceans were represented by small cumaceans, amphipods, copepods, and juveniles of *C. allmanni*. Among the identified molluscs, the bivalves *Dosinia lupinus* (L., 1758), *Venus striatula* Da Costa, 1778 [currently as: *Chamelea striatula* (Da Costa, 1778)] and the opisthobranch gastropod *Cylichna cylindrica* (Bruguière, 1792) [currently as: *Cylichna cylindracea* (Pennant, 1777)] were the most common ones. The food composition of shrimps that were caught at inshore and offshore stations and in males and females insignificantly differed (Allen, 1960).

These data have both a marked difference and a definite similarity in comparison with our results. There are important details that indicate a similarity in the diet of the shrimp at different parts of its distributional range: in all three studied areas, *C. allmanni* prefer to eat opisthobranchs among the gastropods — *Cylichne cylindrica* in Northumberland waters (Allen, 1960), *Diaphana* sp. in the Helgoland Trench and *Cylichne* sp. in the Onega Bay.

Unfortunately, it is impossible to carry out a more complete comparison of Allen's (1960) data with ours because her description of food composition was presented in a much more generalized form. Comparison of the food composition of *C. allmanni* in the Helgoland Trench and Onega Bay finds both general similarities and many particular differences.

In the Helgoland Trench area (excluding diatoms and tintinnidean ciliats, which were probably a transit food: Nigmatullin & Toporova, 1982) food remains belong to 29 taxa and in Onega Bay 15 species were found in shrimp stomachs. In the last area, 7 taxa of malacostracans and all echinoderms and fish found in the Helgoland trench were absent, but here cladocerans were found. There were significant differences in the FO, VFLV and FP values for different food groups (table I). This is reflected in the values of the Froerman Index and total values of the FP. The Froerman Index in shrimp from the Helgoland Trench and from Onega Bay was 2.20 and 2.93, respectively, while the total FP were 73.3 and 36.7%, respectively. *C. allmanni* inhabiting the North Sea is closer to an attacking predator, while in the White Sea it can be considered as a predator-gatherer. This difference, probably, may be explained by differences in habitat and community characteristics and the size composition of studied shrimps in these parts of the range.

The *C. allmanni* habitat depth geographically varies between the different parts of its range. In the western part of the North Sea it occurs at depths from 20-160 m, mainly 40-100 m (Walker, 1892; Allen, 1960), but it was also caught together with *C. crangon* at depths less than 15 m (Allen, 1960). In the Helgoland Trench, it was found at depths of 30-50 m on a relatively hard bottom, mainly on shell limestone (Blahudka & Türkay, 2002). In the Denmark Strait and Faeroe plateau slopes its maximum depth of occurrence is 900 m (Spiridonov et al., 2008). In the Barents Sea this species was caught at depths of 85-104 m, in the northern part of the Throat of the White Sea it was found at depths of 35.6-48.3 m in June 2004, and in Onega Bay soft bottoms at depths of 6.9-31.8 m in early August 2006.

In Onega Bay, shrimps live at depths less than the optimum specified by Allen (1960) and in the Helgoland Trench it occurs almost within the most common

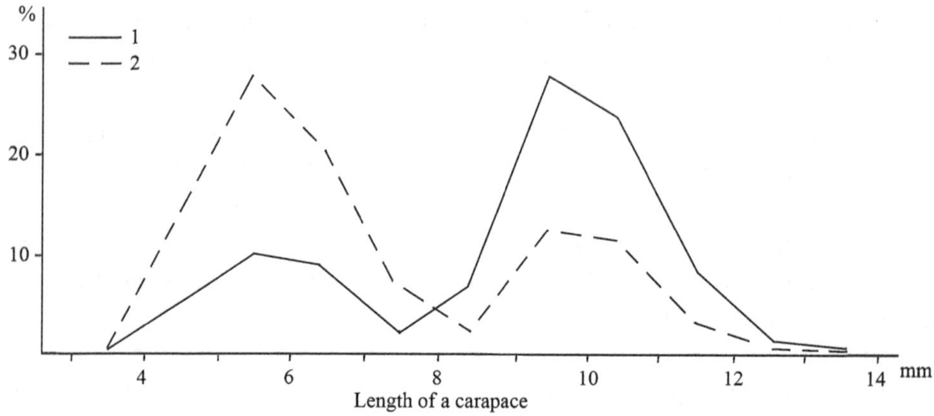

Fig. 1. Size composition of the studied shrimp, *Crangon allmanni* Kinahan, 1860 from (1) the Helgoland Trench area (North Sea) and (2) Onega Bay (White Sea).

depths for this species. Salinity at these two locations is different: it is 24-30‰ in Onega Bay and 32-34‰ in the Helgoland Trench, respectively (Spiridonov et al., 2008). Despite such differences in habitat conditions, the range of the shrimp sizes and their bimodal (5.5 and 9.5 mm) structure in both areas is almost identical (fig. 1). However, in Onega Bay about 60% of the studied shrimps had a size of less than 8 mm, and in the Helgoland Trench, in contrast, about 60% of shrimps had a size of more than 8 mm (fig. 1).

Crangon crangon and *C. allmanni* are close sister species. The composition of food and types of hunting of *C. crangon* in the North and White Sea (Burukovsky & Trunova, 2007; Burukowsky, 2009) changes during ontogenesis. Juvenile shrimps mainly feed on copepods (Harpacticoida) and behave as grazing predators. Sub-adult middle-sized shrimps become predator-gatherers and adult mature specimens with carapace lengths of more than 9 mm change their hunting method to the one of an attacking predator. There are also very tentative data on the same ontogenetic shifts in feeding behaviour of *C. allmanni* (cf. Burukovsky, unpublished data). Probably these ontogenetic distinctions explain the difference in diet composition of the shrimp *C. allmanni* in the Helgoland Trench area of the North Sea and Onega Bay (White Sea).

Comparative characteristics of food composition in some shrimps of the genus *Crangon*

The genus *Crangon* actually comprises 20 species (De Grave & Fransen, 2011). All species of this genus are the inhabitants of the shelf, mostly its upper part, often from the water's edge. The most common deep-water species, *Crangon dalli* Rathbun, 1902 from the Far East, inhabits depths from 3 to

630 m (Hayashi & Kim, 1999), as does *C. allmanni* (see above). All species are morphologically very similar. They show the characteristics of a buried shrimp lifestyle: body dorsoventrally flattened and rostrum in form of small ledge. Interspecific morphological differences are mainly the presence or absence of single or paired keels on the posterior segments of abdomen; the rounded, flattened or sulcated dorsal side of these segments, and other relatively poorly expressed traits (Hayashi & Kim, 1999).

These facts suggest that their morpho-functional trophological complex allows to use the quasi-same food organism spectrums for nutrition. Therefore, these shrimps occupy similar habitats and use comparable food resources within the limits of their ranges, i.e., members of identical life forms. Consequently *Crangon* shrimps probably occupy a similar position in the food webs of their communities. There is evidence that, on the one hand, shrimps of the genus *Crangon* are a food item of the adult flatfishes and, on the other, they act as predators against the newly settled flatfish larvae (Hayashy & Kim, 1999; Hanamura & Matsuoki, 2003; Taylor, 2005).

The composition of food in varying details was studied in 7 species: *C. allmanni*, *C. crangon*, *C. septemspinosa* Say, 1818, *C. franciscorum* Stimpson, 1856, *C. nigricauda* Stimpson, 1856, *C. uritai* Hayashi & J. N. Kim, 1999, and *C. affinis* De Haan, 1849 (cf. Plagman, 1940; Price, 1961; Wilcox & Jeffries, 1974; Sitts & Knight, 1979; Siegfried, 1982; Whale, 1985; Hanamura & Matsuoka, 2003; Burukovsky & Trunova, 2007).

Unfortunately, the methods of quantitative estimation of different food items and their rations in the food lumps used by different authors are not fully comparable. It does not allow a detailed analysis of the similarities and differences in the food composition of all these studied species. Authors often paid attention to quite specific aspects of feeding, while ignoring what they considered to be unimportant details. Therefore, further comparison of the food of *Crangon*-shrimps based on data of *C. crangon* and *C. allmanni*, i.e., two species that I studied using one methodology (see above and Burukovsky & Trunova, 2007), is not possible. What could be done in a future study is to extrapolate the results of this comparison to other species for which the data on food are available.

Crangon crangon is distributed from the White Sea to Morocco coast and the Mediterranean, Black and Baltic Seas. In the North Sea it is subject to the fishing industry. The range of *C. allmanni* is limited by the Northeastern Atlantic from Iceland and the Bay of Biscay to the White Sea. *Crangon crangon* inhabits sandy and silt-sandy bottoms of the upper sublittoral (mainly from the water's edge down to 50 m), and it is rare in waters of the outer shelf. *Crangon allmanni* is distributed on the inner shelf at depths, mainly 40-100 m.

Consequently, these very closely related species have ranges that overlap spatially and somewhat differ bathymetrically and can be considered as vicarious species that divided the upper part of the North-East Atlantic shelf. Because bathymetric parameters of their ecological niches are distinguishable, we should expect a divergence of the food spectra of these species according to the features of the bathymetric distribution of their food organisms.

The similarity is determined by morpho-functional features of both species, and the differences by the peculiarities of specific ecological niches. Indeed, both species are characterized by the presence of sand in almost every stomach. Grains of sand probably play the role of millstones in their gastric mill: in *C. crangon*, the FO value was 91.9% and in *C. allmanni* 78.6%, while the VFLV was 22.0 and 21-14.3%, respectively. However, in the North Sea the sand VFLV value in the stomachs of *C. allmanni* never exceeded 50%, which is probably connected with the peculiarities of the bottom in its habitat (Burukovsky & Trunova, 2007).

The FO and VFLV of detritus in both species are comparable to those values of sand. Detritus and sand occurred in almost every stomach and together they make up almost half of the VFLV, 41.7% in *C. crangon* and 28.4-52.6% in *C. allmanni*. The third common food item for both species were the remains of dead animals. They were represented in the stomachs by indefinable fragments of tissue of aquatic organisms with evident signs of post-mortem maceration, and in the presence of terrestrial invertebrates (mites, adult insects) and also entire specimens of dead, young fish with a body size disabling the shrimp to ingest these in a living state.

Both species are benthos feeders with a preference for relatively sedentary prey (although *C. allmanni* to a lesser degree). Both species feed on relatively mobile animals (amphipods in *C. crangon*, mysids in *C. allmanni*) only in the later stages of ontogeny, i.e., a large-sized shrimp.

Differences in food composition clearly originate in different food availability being habitat-dependent. In the diet of more deep-water *C. allmanni*, almost no plant remains, as well as no larvae of chironomids and other aquatic insect larvae were detected. In stomachs of *C. crangon*, the remains of echinoderms, crabs and hermit crabs have never been encountered.

According to the method of food procurment, both species can be classified as predator-gatherers with elements of detrito- and necrophagia. In the early stages of ontogenesis, *C. crangon* predominantly feed on meiobenthos (harpacticoids, nematodes). Young *C. allmanni* predominantly show attacking predator behavior, especially in the North Sea. Both species (mentioned above) demonstrate such modes of foraging as grazing, gathering and attacking, changing them in the process of ontogenesis.

The application of this approach to other studied species for comparison is very difficult because different researchers are studying the very same species had described food items in different ways. For example, Siegfried (1982) notes that *C. franciscorum* the FO of inorganic material (i.e., sand) reached 26%, plant remains 42% and the non-identified (usually crustaceans: "unidentified fragments of crustacean exoskeletons"; Siegfried, 1982: p. 133), probably dead animals, 44%. In contrast, Whale (1985) does not mention the presence of sand and unidentified remains in food lumps of the same species, and the FO of plant remains was only 4%. Comparison of the data of Siegfried (1982) on the food composition of *C. franciscorum* gives a substantial similarity with *C. crangon* and *C. allmanni*, but the opposite when compared to the data and description of Whale (1985).

Wilcox & Jeffries (1974) directly compared the food composition of *C. affinis* and *C. septemspinosa* with *C. crangon* and *C. allmanni*. They classified all these species as predators that are capable of swallowing anything (i.e., silt, sand, algae and detritus). However, they are hesitant to identify some of the components of the food lumps as the remains of dead animals: in their opinion it is part of detritus or sometimes it is result of predation. Hanamura & Matsuoka (2003) described almost the same for *C. uritai*. However, they specifically stress that plant remains were absent in stomachs, but they perfectly demonstrated the decrease of the detritus role and the increasing role of amphipods and mysids in the diet of *C. uritai* in parallel with body size growth.

We can conclude that all these members of the genus *Crangon* are benthos feeders, predators-gatherers, feeding on relatively sedentary preys. The most mobile of them are amphipods and mysids, they start to play a significant role in feeding of the largest shrimps. Probably, shrimps of the genus *Crangon* all use grains of sand as millstones in their gastric mill, collecting detritus and plant remains (the latter at the depths where they are presented), and they are also all necrophagous to some degree.

ACKNOWLEDGEMENTS

I cordially thank the project coordinator, the late Michael Türkay, for the transfer of the shrimp stomach samples and their biological data from the Helgoland Trench; E. Yu. Soljanko and V. A. Spiridonov for collection shrimp samples in the White Sea; A. V. Trunova and S. Yu. Grigorenko for their help in cameral study of stomach contents; and Ch. M. Nigmatullin for the MS reading, its translation and useful comments, as well as M. Sonnewald for editing the scientific English of the manuscript.

REFERENCES

ALLEN, J. A., 1960. On the biology of *Crangon allmani* Kinahan in Northumberland waters. J. Mar. Biol. Ass. U.K., **39**: 481-508.

BLAHUDKA, S. & M. TÜRKAY, 2002. A population study of the shrimp *Crangon allmanni* in the German Bight. Helgol. Mar. Res., **56**: 190-197.

BURUKOVSKY, R. N., 2009. Feeding and feeding relationships of shrimp: 1-408. (FGOU VPO KGTU Publishing, Kaliningrad). [In Russian with English abstract.]

BURUKOVSKY, R. N. & A. V. TRUNOVA, 2007. On the feeding of shrimp *Crangon crangon* (Decapoda, Crangonidae) in Kandalaksha Gulf (the White Sea) in July and September 2004. In: Collected papers on memory of famous Russian hydrobiologist B. G. Ivanov "Marine commercial invertebrates and algae (biology and fishery)". Trudy VNIRO, **147**: 181-203. (VNIRO Publishing, Moscow). [In Russian with English abstract.]

DE GRAVE, S. & C. H. J. M. FRANSEN, 2011. Carideorum catalogus. The recent species of the dendrobranchiate, stenopodidean, procarididean and caridean shrimps (Crustacea, Decapoda). Zool. Meded. Leiden, **85**: 195-588.

HANAMURA, Y. & M. MATSUOKA, 2003. Feeding habits of the sand shrimp, *Crangon uritai* Hayashi & Kim, 1999, in the central Seto Inland Sea, Japan. Crustaceana, **76**: 1017-1024.

HAYASHI, K.-I. & J. N. KIM, 1999. Revision of the East Asian species of *Crangon* (Decapoda: Caridea: Crangonidae). Crust. Res., **28**: 62-103.

NIGMATULLIN, CH. M. & N. M. TOPOROVA, 1982. Food spectrum of squid *Sthenoteuthis pteropus* (Steenstrup, 1855) in the epipelagic zone of the Tropical Atlantic. Collected papers "Feeding and food relations of fishes and invertebrates in the Atlantic Ocean": 3-8. (AtlantNIRO Publishing, Kaliningrad). [In Russian with English abstract.]

PETELIN, V. P., 1967. Granulometric analysis of sea bottom deposits. (Nauka, Moscow). [In Russian.]

PLAGMANN, J., 1940. Ernährungsbiologie der Garnele (*Crangon vulgaris* Fabr.). Helgol. Wiss. Meeresunters., **2**: 113-162.

PRICE, JR., K. S., 1961. Biology of the sand shrimp, *Crangon septemspinosa*, in the shore zone of the Delaware Bay region. Univ. Delaware Mar. Lab. Contrib., **29**: 244-255.

SCHWINN, M., M. TÜRKAY & M. SONNEWALD, 2014. Decapod fauna of the Helgoland Trench (Crustacea) a long-term study in a biodiversity hotspot. Mar. Biodiv., **44**: 491-517.

SIEGFRIED, C. A., 1982. Trophic relation of *Crangon franciscorum* Stimpson and *Palaemon macrodactylus* Rathbun: predation on the opossum shrimp, *Neomysis mercedis* Holmes. Hydrobiologia, **89**: 129-139.

SITTS, R. M. & A. W. KNIGHT, 1979. Predation by the estuarine shrimps *Crangon franciscorum* Stimpson and *Palaemon macrodactylus* Rathbun. Biol. Bull., **156**: 356-368.

SPIRIDONOV, V. A., V. I. SOKOLOV, I. KRÖNKE & M. TÜRKAY, 2008. How is the fragmented distribution range of the reddish sand shrimp *Crangon allmanni* Kinahan formed? In: Presentations of the Scientific Conference dedicated to the 70 years anniversary of the White Sea Biological Station of the Moscow University: 121-123. (Grif & K Publishing, Moscow).

TAYLOR, D. L., 2005. Predation on post-settlement winter flounder *Pseudopleuronectes americanus* by sand shrimp *Crangon septemspinosa* in NW Atlantic estuaries. Mar. Ecol. Prog. Ser., **289**: 245-262.

WAHLE, R. A., 1985. The feeding ecology of *Crangon franciscorum* and *Crangon nigricauda* in San Francisco Bay, California. J. Crust. Biol., **5**: 311-326.

WALKER, A. O., 1892. Revision of the Podophtalmata and Cumacea of Liverpool Bay to May, 1892. Proc. Trans. Liverpool Biol. Soc., **6** (Session 1891-1892): 96-104.

WILCOX, J. R. & H. P. JEFFRIES, 1974. Feeding habits of the sand shrimp *Crangon septemspinosa*. Biol. Bull., **146**: 424-434.

First received 19 April 2016.
Final version accepted 23 June 2016.

[When citing this volume, please refer to Crustaceana 90 (2017) 771-1288]

Michael Türkay Memorial Issue

SPECIES DIVERSITY OF FRESHWATER DECAPOD CRUSTACEANS (CRABS AND SHRIMPS) FROM COLOMBIA

BY

MARTHA R. CAMPOS[1,3]) and DIÓGENES CAMPOS[1,2])

[1]) Universidad Nacional de Colombia, Instituto de Ciencias Naturales, Ciudad Universitaria, Cra 30, calle 45, Bogotá D.C., Colombia

[2]) Universidad La Gran Colombia, Carrera 6 No. 12B-12, Bogotá, Colombia

ABSTRACT

The aim of this paper is to characterize the species diversity of freshwater decapod crustaceans (crabs and shrimps) from Colombia by using a dataset containing 964 digitized records and 13 881 specimens collected between 1910 and 2016, information that has been assembled with data from 21 museums and other institutions worldwide. The characterization of species diversity is based on the estimation of relative abundance of species and it is calculated in three separate analyses in which the data were partitioned as follows: (a) by decapod families (six in Colombia) (b) the unpartitioned dataset of all 139 Colombian species and (c) by the five biogeographic regions in the country. In each case, Campos & Isaza's species diversity index and the Shannon entropy index were calculated. The calculations performed here also allowed the identification of predominant species and those that are least represented in collections and might therefore be at greater risk of extinction.

Key words. — Decapoda, diversity index, Simpson index, South America, species richness

ZUSAMMENFASSUNG

Ziel dieser Publikation ist es, die Artenvielfalt der dekapoden Crustaceen (Krabben und Garnelen) aus dem Süßwasser von Kolumbien zu charakterisieren. Dies geschieht unter Zuhilfenahme eines Datensatzes mit 964 digitalisierten Serien, welche 13.881 Individuen enthalten, die zwischen 1910 und 2016 gesammelt wurden. Der Datensatz wurde durch Daten von 21 Museen und anderen weltweit verteilten Instituten erstellt. Die Charakterisierung der Artenvielfalt basiert auf der Schätzung der relativen Abundanzen der Arten und wird in drei eigenständigen Analysen berechnet, für welche die Daten wie folgt aufgeteilt wurden: (a) nach Dekapodenfamilien (sechs in Kolumbien) (b) der ungeteilte Datensatz aller 139 Kolumbianischen Arten (c) nach den fünf biogeografischen Regionen im Land. In jedem Fall wurde der Arten-Diversitätsindex nach Campos & Isaza, sowie der Shannon-Entropie-Index berechnet. Die hier durchgeführten Berechungen ermöglichten zudem die Identifizierung der vorherrschenden Arten und derer, welche zumindest in Sammlungen vorhanden sind, aber von einem größeren Risiko des Aussterbens betroffen sein könnten.

Schlüsselwörter. — Decapoda, Arten-Diversitätsindex, Simpson-Index, Südamerika, Artenvielfalt

[3]) Corresponding author; e-mail: mhrochad@unal.edu.co

INTRODUCTION

This paper focuses on the use of diversity indices to quantify the species diversity of freshwater decapod crustaceans from Colombia by using information on material of Colombian shrimps and crabs deposited in museums and other institutions worldwide. This study covers records spanning the period 1910 to 2016 and includes 139 species that belong to two families of freshwater crabs (Pseudothelphusidae and Trichodactylidae) and four families of freshwater shrimps (Atyidae, Palaemonidae, Euryrhynchidae and Sergestidae). According to Cumberlidge et al. (2009), Colombia is the most species-rich country for freshwater crabs in South America, and the second worldwide after China (Cumberlidge et al., 2011). Globally, Colombia has the highest degree of species level endemism in freshwater crabs (82%).

The Andean orogeny is one of the most relevant factors in order to explain species distribution patterns in Colombia, because the Andes are characterized by a multiplicity of environments in which allopatric speciation has occurred, in particular for decapod crustaceans. According to its orography, Colombia can be divided into five natural regions: the Andean region is made up of three major divisions formed by chains of high mountains (cordilleras), that are known as the Western, Central and Eastern Cordilleras. The Caribbean region comprises the area adjacent to the Caribbean Sea, which includes desert areas in La Guajira, the Sierra Nevada of Santa Marta and rainforests in the Gulf of Uraba. The Pacific region has varied climates and is considered one of the wettest places on the planet. The Orinoco region includes the foothills of the Eastern Cordillera and the vast plains of the Orinoco River Basin. The Amazonian region comprises part of the rainforest and the Guyana Shield (Rangel-Ch., 2015).

The aim of this study is to illustrate the importance of collecting specimens for scientific studies and the formulation of proposals for species conservation. In fact, we aimed to quantify the species diversity of freshwater decapod crustaceans for Colombia and determine, in particular, the species diversity for each one of the families and the natural regions of the country. We hypothesize that data from collection-preserved specimens provide information on the diversity of species in Colombia, and that the species diversity index proposed by Campos & Isaza (2009) allows quantifying the degree of diversity and graphically representing the information in terms of the six families of freshwater decapods, of the five natural regions of Colombia, and of the 139 species currently recorded from the whole country.

We are aware that, in the natural world, the species diversity of freshwater decapod crustaceans from Colombia has been affected over time by natural events and by the impact of anthropogenic actions, e.g., deforestation, environmental pollution, habitat fragmentation and climate change, among many others. Therefore, the results of the calculations represent only an approximation to the degree of diversity of species that currently exists in the country. Notwithstanding, the results provide baseline information that can be used for future observations and estimation of the species diversity, and to assess whether diversity is been preserved, declined or increased.

MATERIAL AND METHODS

Database

From Colombia 106 species of freshwater crabs belonging to the families Pseudothelphusidae and Trichodactylidae, and 33 species of freshwater shrimps of the families Atyidae, Palaemonidae, Euryrhynchidae and Sergestidae (Campos, 2014) have been recorded. It is important to note that four subspecies of the complex *Hypolobocera bouvieri* (Rathbun, 1898) (Pseudothelphusidae) were included separately, because they are distributed in different regions of Colombia.

As a starting point, we consolidated a database with available information on the species of freshwater decapod crustaceans from Colombia deposited in 21 museums and institutions worldwide, which included records from 1910 to 2016 (hereafter, "the database"). The database consists of a matrix containing nine columns associated with the information for each record and 964 rows representing the number of records. A record is referred to a set that corresponds to one or more specimens for which the following information is available: The family (column 1); the genus (column 2); the scientific name (column 5); the Colombian natural region where the material was collected (column 6); the number of specimens (column 7); the year in which the specimens were gathered (column 8); and the collection where the specimens are deposited (column 9). Within a family, each species is labelled with a number code 1 (column 3) that consecutively lists the species included into the family and, similarly, the code 2 (column 4) corresponds to the number assigned by considering as a whole the 139 species existing in Colombia. Likewise, the genera within a family are successively coded, using the symbol G followed by the number to identify the genus (column 2). The full database is included as supplementary material (table AI in the Appendix in the online edition of this journal, which can be accessed via http://booksandjournals.brillonline.com/content/journals/15685403) as well as the list of codes assigned to the six families and 139 species of freshwater decapods existing in Colombia (table AII in the Appendix in the online edition of this journal, which can be accessed via http://booksandjournals.brillonline.com/content/journals/15685403).

A total of 88.48% of the specimens included in the database (74.48% of the 964 records) corresponds to specimens deposited in the collection of the Instituto de Ciencias Naturales, Museo de Historia Natural, Universidad Nacional de

Colombia, Bogotá (ICN-MHN). The remaining specimens (11.52%) come from other museums or institutions, as follows: British Museum (Natural History), London (BM), Museo de Biología Marina, Universidad del Valle, Cali (CRBMUV), the Field Museum of Natural History, Chicago (FMNH), Instituto Colombiano de Desarrollo Rural, Cartagena (INCODER), Instituto Nacional de Pesquisas da Amazônia, Manaus (INPA), Instituto Venezolano de Investigaciones Científicas, Caracas (IVIC), Museo de la Sociedad de Ciencias Naturales La Salle, Caracas (LS), Biology Institute, University of Ljubljana (LU), Museo de Biología, Universidad Central de Venezuela, Caracas (MB-UCV), Museo de La Salle, Bogotá (MLS), Muséum National d'Histoire Naturelle, Paris (MP), Museo Universidad de Antioquia, Medellín (MUUA), Naturhistorisches Museum Basel (NHMB), Natural History Museum, London (NHML), Naturhistorisches Museum Wien (NHMW), Musée de Strasbourg (SM), Research Institute and Natural History Museum Senckenberg, Frankfurt am Main (SMF), Museum of Natural History of Tulane University (TU), the National Museum of Natural History, Smithsonian Institution, Washington, D.C. (USNM), Universidad de los Andes, Bogota (UAND), Universidad de Antioquia, Medellín (UANT), Zoologische Staatssammlungen, Munich (ZSM).

Relative weights of the species

The criteria used for organizing the data at species level were the following:

(A) Six subsets constituted by the families, i.e., two families of freshwater crabs (PS, Pseudothelphusidae; TR, Trichodactylidae) and four families of freshwater shrimps (AT, Atyidae; PA, Palaemonidae; EU, Euryrhynchidae; SE, Sergestidae).

(B) A single set in which Colombia was considered as a whole with 139 species (COL).

(C) Five subsets conformed by the regions, in which only the species that occur in each region were considered (AN, Andean; CA, Caribbean; AM, Amazonian; OR, Orinoquian; PAC, Pacific region).

We took into account three ways to extract information from the database. In each case, the total number of specimens ($N_0 = 13\,881$) was divided according to the criterion (A, B or C) under consideration. Next, the specimen numbers were translated into relative weights of the species (or relative abundances) by assigning the value $P_n = N_n/N_0$ to the nth species, where N_n is the total number of specimens for the nth species. For each subset of A, B or C, the number of species was identified with the letter S.

Data analysis

For analysing the data we applied the species diversity index proposed by Campos & Isaza (2009) that has a geometric origin, involves the Simpson index and allows a biological interpretation of the data. In addition to the valuation based on this index, calculations with Shannon index were also included.

Shannon index.— In this section, we want to answer to the following question: what other information can be obtained from the database (table AI in the Appendix in the online edition of this journal, which can be accessed via http://booksandjournals.brillonline.com/content/journals/15685403). Note that, with the information of the previous sections, the following vectors of species relative abundances (also called probability distributions) are now available: (A) The vectors P^{PS}, P^{TR}, P^{AT}, P^{PA}, P^{EU}, and P^{SE} that are associated with the families and identified by the respective superscripts; (B) the vector P^{COL} that corresponds to the country as a whole with 139 species, and (C) the vectors P^{AN}, P^{CA}, P^{AM}, P^{OR}, and P^{PAC} related to the natural regions of Colombia and identified by the respective superscripts.

In general, omitting the superscripts notation above, for a given S-species community (species of a family, species in the whole country or species in a natural region of Colombia), the number of specimens in the different collections are translated into relative weights of the species $P = (P_1, P_2, \ldots, P_S)$, where $P_n = N_n/N_0$ is the relative abundance of the nth species, N_n is the number of specimens for the nth species, N_0 is the total number of specimens in the S-species community, and the sum of all the relative weights gives one, $\sum_{n=1}^{S} P_n = 1$.

The so-called abundance vector $P = (P_1, P_2, \ldots, P_S)$ is used for the calculation of traditional diversity indices, such as the Shannon index (or Shannon entropy)

$$H = -\sum_{n=1}^{S} P_n \ln(P_n),$$

where H is measured in nats, since the natural logarithm $\ln(\ldots)$ is used in the definition of H, together with the convention $0\ln(0) = 0$. Since $I_n := -\ln(P_n)$ is known as the surprise associated with the the nth species, whose weight or relative abundance is P_n, then H is the mean value of the surprises associated with the set of S species. After recalling that $0 \leqslant P_n \leqslant 1$, for $n = 1, 2, \ldots, S$, note that: (a) if P_n is small and the species is found in a sampling, then we will have a big surprise, and on the contrary (b) if P_n approaches 1, then the surprise goes to zero.

Species Diversity Index.— In this section, the index for measuring biological diversity proposed by Campos & Isaza (2009) is used in order to characterize the species diversity of freshwater decapod crustaceans from Colombia. Different from the Shannon index, a major practical advantage of this index is that it is sensitive to species richness and relative abundance of species. It acts as an ordering system that allows a mathematical and visual representation of the biodiversity value in a two-dimensional plane and the comparison between various systems (e.g., the families or the natural regions).

Briefly, the species diversity index $B_1(S, r)$ has a geometric origin (Campos & Isaza, 2009), and it is given by the relation between a monotonous increasing function $\alpha(S)$ of the number S of species, and a radius r, that is related to the Simpson index D and to the probability distribution $P = (P_1, P_2, \ldots, P_S)$. Explicitly, $B_1(S, r)$ is calculated by the relationships

$$B_1(S, r) = \frac{\alpha(S)}{r}, \qquad r = \sqrt{D} = \sqrt{P_1^2 + P_2^2 + \cdots + P_S^2},$$

where

$$\alpha(S) = \frac{1}{\sqrt{\pi}} \frac{\Gamma\left(\frac{S+3}{2}\right)}{\Gamma\left(\frac{S+2}{2}\right)}.$$

Here, D is the Simpson index and $\Gamma(x)$ is the well-known Gamma function for real arguments, x. Fig. 1 displays the behaviour of $\alpha(S)$ as a function of the number of species, S. It is important to note that the diversity index is sensitive to the number of different species and to the relative abundance of them, due to the presence of the function $\alpha(S)$ and the radius r, respectively. Because, for any

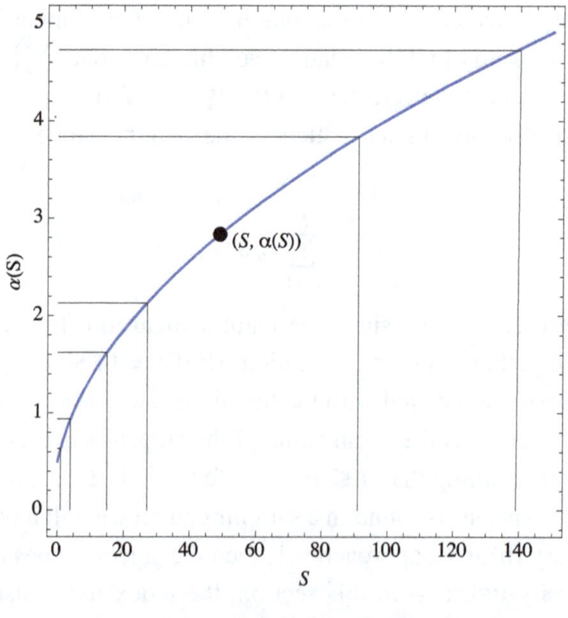

Fig. 1. Behaviour of $\alpha(S)$ as function of the species richness, S.

TABLE I
Number of records and specimens for each of the collections with holdings of the freshwater decapod fauna from Colombia (see text for list of the collections)

Collection	No. of records	No. of specimens
ICN	718	12 282
IVIC	14	62
MB-UCV	2	19
SMF	22	286
USNM	18	172
TU	20	68
MLS	40	306
CRBMUV	70	407
BM	5	26
MUAICR	24	131
INPA	3	9
FMNH	2	3
MP	1	4
UL	2	5
GM	1	4
NHMB	1	1
MNHN	1	1
UAND	6	16
UANT	2	13
INCODER	3	10
INV	9	56

The total number of records is 964, and the number of specimens is 13 881.

finite probability distribution $P = (P_1, P_2, \ldots, P_S)$, the Simpson index D is a measure of homogeneity or concentration, and the species diversity index $B_1(S, r)$ is defined in terms of the reciprocal of r, then the diversity index increases as r decreases. The conditions $0 \leqslant P_n \leqslant 1$ and $P_1 + P_2 + \cdots + P_S = 1$ imply that $0 < r \leqslant 1$.

RESULTS

Primary information included in the database

Table I gives an overview of the material contained in the database that has been organized for this study taking into account information from different Colombian crustaceans collections or cited in the carcinological literature. The total number of records in the database is 964 and the number of specimens reaches 13 881. Fig. 2 illustrates the level of sampling performed over the years, showing the number of records obtained per decade between 1910 and 2016. This figure makes it clear that the period of intensive research began in the 1970s.

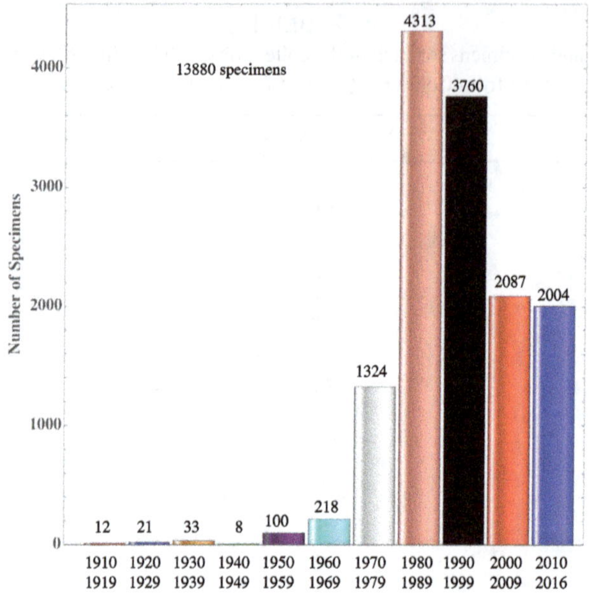

Fig. 2. Number of specimens recorded decennially between 1910 and 2016 of Colombian freshwater decapod crustaceans for the different collection databases (see text for list of the collections).

Table II summarizes the information contained in the database. The Pseudothelphusidae is predominant in number of species and specimens, followed by Palaemonidae, Trichodactylidae and Atyidae. In contrast, Euryrhynchidae and Sergestidae have low representation in the database, particularly the latest, which is represented by a single species and only two records.

TABLE II

Overview of the material contained in the database of freshwater decapod crustaceans from Colombia

Family	Genera	Species	Records	Specimens	Years
PS	15	91	499	5991	61
TR	9	15	186	1443	53
AT	2	4	18	260	13
PA	3	27	251	5901	53
EU	1	1	8	188	7
SE	1	1	2	98	2
COL	31	139	964	13 881	72

Abbreviations: PS, Pseudothelphusidae; TR, Trichodactylidae; AT, Atyidae; PA, Palaemonidae; EU. Euryrhynchidae; SE, Sergestidae; COL, Colombia.

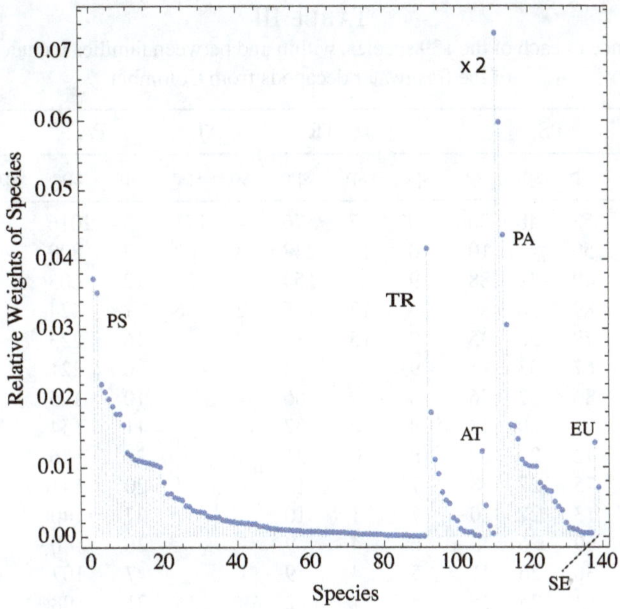

Fig. 3. Relative weights for each family: PS, Pseudothelphusidae; TR, Trichodactylidae; AT, Atyidae; PA, Palaemonidae; EU, Euryrhynchidae; SE, Sergestidae. The sum of weights associated with the 139 species is normalized to one.

Relative weights of the species

(A) Subsets constituted by the families.— Fig. 3 shows graphically the relative importance of each species within its own family and in respect to the other families. In this figure, the species occur from left to right in the same order in which they appear in table III, where the "sp" columns refer to the number that identifies the species (1 to 139, according to Code 2 in tables AI and AII in the Appendix in the online edition of this journal, which can be accessed via http://booksandjournals.brillonline.com/content/journals/15685403), and the columns "spe" contain the number of specimens. At the end of the table, the total numbers of species and specimens are indicated for each family, so that the relative weight of each species within the family can be calculated by dividing the number of specimens of each species, and the total number of specimens of the respective family.

The palaemonid species *Macrobrachium brasiliense* (Heller, 1862) (number 111) is the preponderant when the relative weights of each among all the 139 species is taken into account (fig. 3), and its relative weight was halved in the graph. The two most and the two less representative species of each family according to their relative weights are listed in table IV.

TABLE III

Relative importance of each of the 139 species, within and between families included in the database of the freshwater decapods from Colombia

	PS						TR		AT		PA		EU		SE	
	sp	spe	sp	spe	sp	spe	sp	spe	sp	spe	sp	spe	sp	spe	sp	spe
	53	517	52	48	30	10	7	576	4	171	5	2016	1	188	1	98
	16	489	50	41	10	10	12	248	3	62	2	829				
	17	305	40	39	88	9	5	154	1	21	12	603				
	54	290	89	38	39	9	13	125	2	6	1	424				
	51	276	77	38	35	9	15	88			16	223				
	49	261	67	33	11	9	8	72			6	221				
	70	246	86	32	76	8	9	66			10	195				
	14	246	23	30	43	8	2	37			11	154				
	72	224	42	29	29	8	10	31			24	146				
	59	168	75	27	78	7	3	15			20	141				
	62	163	13	27	20	7	1	10			27	140				
	66	155	69	26	32	6	14	9			26	140				
	9	153	36	26	41	5	4	9			7	107				
	48	151	84	25	25	5	11	2			21	98				
	44	149	37	25	6	5	6	1			17	92				
	65	148	33	20	4	5					18	90				
	21	145	12	20	90	4					3	69				
	60	143	57	18	47	4					15	59				
	56	140	64	16	38	4					19	52				
	87	109	19	15	31	3					8	27				
	28	86	81	14	91	2					13	17				
	5	86	45	13	71	2					22	15				
	83	78	24	13	61	2					14	13				
	85	75	18	13	22	2					23	9				
	15	73	34	12	3	2					25	7				
	46	61	1	12	2	2					9	7				
	26	60	80	11	82	1					4	7				
	63	52	74	10	68	1										
	79	50	73	10	58	1										
	7	49	55	10	27	1										
					8	1										
Total					91	5991	15	1443	4	260	27	5901	1	188	1	98

Abbreviations: PS, Pseudothelphusidae; TR, Trichodactylidae; AT, Atyidae; PA, Palaemonidae; EU, Euryrhynchidae; SE, Sergestidae; sp, species; spe, specimens. For the respective species codes, see tables AI and AII in the Appendix.

(B) Colombia as a whole with 139 species.— Fig. 4 and table V contain information about the 139 species that are recorded from Colombia, when the relative weights assigned to them are sorted using a decreasing order: the "sp" columns identify the species (from 1 to 139, according to Code 2 in tables AI and AII in the Appendix in the online edition of this journal, which can be accessed via

TABLE IV

Overview of the two most and the two least representative species of each family of freshwater decapods from Colombia, according to the assigned weights

Family	Species	No. of specimens	Relative weights	
			Families	COL
PS	*Neostrengeria guenteri* (Pretzmann, 1965)	517	0.08630	0.03724
PS	*Hypolobocera bouvieri bouvieri* (Rathbun, 1898)	489	0.08162	0.03522
PS	*H. malaguena* Prahl, 1988	1	0.00017	0.00007
PS	*Eidocamptophallus chacei* (Pretzmann, 1967)	1	0.00017	0.00007
TR	*Moreirocarcinus emarginatus* (H. Milne Edwards, 1853)	576	0.3992	0.04150
TR	*Sylviocarcinus piriformis* (Pretzmann, 1968)	248	0.1719	0.01787
TR	*S. pictus* (H. Milne Edwards, 1853)	2	0.00139	0.00014
TR	*Fredilocarcinus raddai* (Pretzmann, 1978)	1	0.00069	0.00007
AT	*Potimirim glabra* (Kingsley, 1878)	171	0.6577	0.01232
AT	*Atya scabra* (Leach, 1816)	62	0.2385	0.00447
AT	*A. crassa* (Smith, 1871)	21	0.0808	0.00151
AT	*A. innocous* (Herbst, 1792)	6	0.0231	0.00043
PA	*Macrobrachium brasiliense* (Heller, 1862)	2016	0.3416	0.14523
PA	*M. amazonicum* (Heller, 1862)	829	0.1405	0.05972
PA	*M. digueti* (Bouvier, 1895)	7	0.0012	0.00050
PA	*M. atabapense* Pereira, 1986	7	0.0012	0.00050
EU	*Euryrhynchus amazoniensis* Tiefenbacher, 1978	188	1	0.01354
SE	*Acetes paraguayensis* Hansen, 1919	98	1	0.00706

Abbreviations: PS, Pseudothelphusidae; TR, Trichodactylidae; AT, Atyidae; PA, Palaemonidae; EU, Eryrhinchidae; SE, Sergestidae. The fourth and the fifth column represent the weights within the family and within the set of 139 species, respectively.

http://booksandjournals.brillonline.com/content/journals/15685403) and the "spe" columns contain the number of specimens. Table VI shows the information of the three most, and the three less representative species in Colombia. The three highest relative weight values, 0.14523, 0.05972 and 0.04344, correspond to the palaemonid species *Macrobrachium brasiliense*, *M. amazonicum* (Heller, 1862) and *M. ferreirai* Kensley & Walker, 1982, respectively.

(C) Subsets conformed by the regions.— Fig. 5 displays the relative weight of the species when it is calculated taking into account the number of specimens in each natural region: Andean (AN), Caribbean (CA), Amazonian (AM), Orinoquian (OR), and Pacific (PAC). The table VII is associated with the fig. 3, in which the species are classified in the natural region where they occur, and ordered according

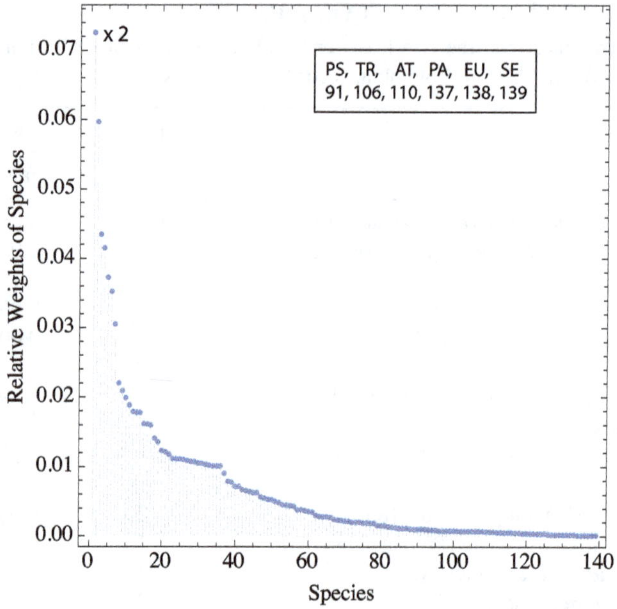

Fig. 4. Relative weights of the 139 species included in the decapod crustaceans collection database, with the set normalized to one. The symbol ×2 indicates that the weight of *Macrobrachium brasiliense* (Heller, 1862) is twice the value indicated in the graph, i.e., it is 0.14523.

to the number of specimens in the database. Table VIII shows the two most and the two less relevant species and their relative weights for each natural region.

Table IX summarizes the distribution of the 139 species and the 13 381 specimens according to the families of decapod crustaceans and the natural regions of Colombia. It is important to note that there are 16 species that share two regions, coded in table AII in the Appendix (available in the online edition of this journal, which can be accessed via http://booksandjournals.brillonline.com/content/journals/15685403) (Code 2) as follows: Trichodactylidae (TR): 98, 100, 103, 104, 106; Palaemonidae (PA): 112, 115, 122, 126, 124, 129, 131, 134, 136, 137; and Euryrhynchidae (EU): 138. Thus, when Colombia is considered as a whole, the apparent number of the species is equivalent to 155 species.

Shannon index

Table X shows the results of calculating the Shannon entropies for the 12 abundance vectors annotated above. We observe that the Shannon entropy combines in a single number the total number of species in the community (species richness *S*) and the evenness that expresses how evenly the specimens in the *S*-species community are distributed over the different species (relative weights of the *S* species).

TABLE V
Relative importance of the 139 species included in the database of the freshwater decapods from Colombia when they are considered as a whole in the country (COL), i.e., avoiding any families or region partitions

COL							
sp	spe	sp	spe	sp	spe	sp	spe
115	2016	56	140	39	9	7	49
112	829	104	125	35	9	52	48
122	603	87	109	11	9	50	41
98	576	117	107	76	8	40	39
53	517	139	98	43	8	89	38
16	489	131	98	29	8	77	38
111	424	127	92	135	7	93	37
17	305	128	90	119	7	67	33
54	290	106	88	114	7	86	32
51	276	28	86	78	7	101	31
49	261	5	86	20	7	23	30
103	248	57	18	108	6	42	29
70	246	123	17	32	6	118	27
14	246	64	16	41	5	75	27
72	224	132	15	25	5	13	27
126	223	94	15	6	5	69	26
116	221	19	15	4	5	36	26
120	195	81	14	90	4	84	25
138	188	124	13	47	4	37	25
110	171	45	13	38	4	107	21
59	168	24	13	31	3	33	20
62	163	18	13	102	2	12	20
66	155	34	12	83	78	91	2
121	154	1	12	85	75	71	2
96	154	80	11	15	73	61	2
9	153	92	10	99	72	22	2
48	151	74	10	113	69	3	2
44	149	73	10	100	66	2	2
65	148	55	10	109	62	97	1
134	146	30	10	46	61	82	1
21	145	10	10	26	60	68	1
60	143	133	9	125	59	58	1
130	141	105	9	129	52	27	1
137	140	95	9	63	52	8	1
136	140	88	9	79	50		
Total						139	13 881

Abbreviation: sp, species; spe, specimens. For the respective species codes, see tables AI and AII in the Appendix.

TABLE VI

Overview of the three most, and the three less representative species in Colombia included in the database of the freshwater decapods from Colombia, according to the relative weights of the 139 species registered for Colombia

Country	Species	Specimens COL	Relative weights
COL	*Macrobrachium brasiliense* (Heller, 1862)	2016	0.14523
COL	*M. amazonicum* (Heller, 1862)	829	0.05972
COL	*M. ferreirai* Kensley & Walker, 1982	603	0.04344
COL	*Neostrengeria macarenae* Campos, 1992	1	0.00007
COL	*Hypolobocera malaguena* Prahl, 1988	1	0.00007
COL	*Eidocamptophallus chacei* (Pretzmann, 1967)	1	0.00007

Thus, as remarked by Kolasa & Biesladka (1984), the combination of S and P in the Shannon index leads to a value of unknown ecological meaning.

Species diversity index

Tables XI and XII give the values of some quantities related with the calculation of the index for measuring species diversity, when the partition of the 13 881 specimens of the database is done according to the decapod crustaceans families

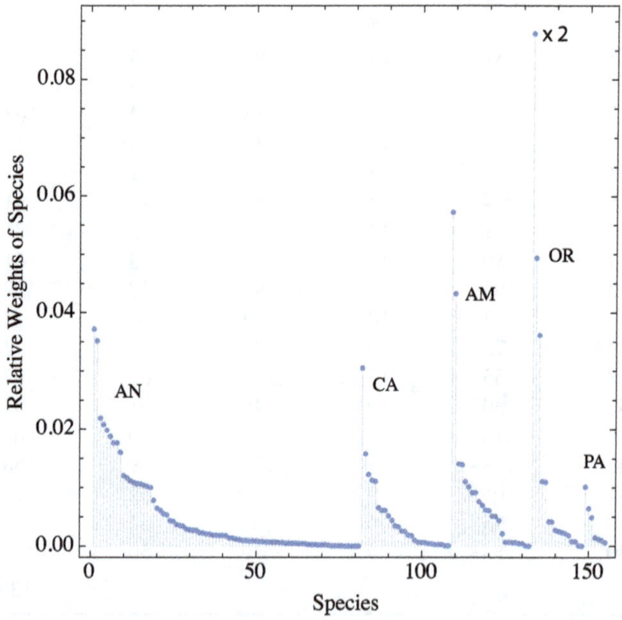

Fig. 5. Relative weights of the species of freshwater decapod crustaceans for each natural region of Colombia: AN, Andean region; CA, Caribbean region; AM, Amazonian region; OR, Orinoquian region; PAC, Pacific region.

TABLE VII
Species present in the different natural regions of Colombia, decreasingly ordered according to the number of specimens

AN						CA		AM		OR		PA	
sp	spe	sp	spe	sp	spe	sp	spe	sp	spe	sp	spe	sp	spe
53	517	52	48	55	10	111	424	115	795	115	1221	130	141
16	489	50	41	30	10	116	221	122	602	112	687	128	90
17	305	40	39	35	9	110	171	126	197	98	503	113	69
54	290	89	38	11	9	103	157	120	195	96	154	107	21
51	276	77	38	124	8	121	154	138	154	9	153	123	17
49	261	67	33	76	8	127	92	112	142	134	59	132	15
70	246	86	32	43	8	106	86	137	128	125	59	133	9
14	246	23	30	29	8	5	86	136	128	104	39		
72	224	42	29	78	7	15	73	117	107	131	36		
59	168	75	27	20	7	109	62	139	98	138	34		
62	163	13	27	32	6	7	49	134	87	100	31		
66	155	69	26	129	5	129	47	104	86	126	26		
48	151	36	26	41	5	93	37	98	73	137	12		
44	149	84	25	25	5	100	35	99	72	136	12		
65	148	33	20	90	4	118	27	131	62	106	2		
21	145	12	20	47	4	37	25	101	31	122	1		
60	143	57	18	38	4	94	15	74	10				
56	140	64	16	31	3	92	10	10	10				
87	109	19	15	91	2	88	9	105	9				
103	91	81	14	71	2	39	9	95	9				
28	86	45	13	61	2	119	7	135	7				
83	78	24	13	22	2	108	6	114	7				
85	75	18	13	82	1	124	5	102	2				
46	61	34	12	68	1	6	5	97	1				
26	60	1	12	58	1	4	5						
63	52	80	11	27	1	3	2						
79	50	73	10	8	1	2	2						
Total				81	5657	27	1821	24	3012	16	3029	7	362

For each region the number of species and the total number of specimens included in the database of the freshwater decapods from Colombia is given. Abbreviations: AN, Andean region; CA, Caribbean region; AM, Amazonian region; OR, Orinoquian region; PAC, Pacific region; sp, species; spe, specimens. For respective species codes, see tables AI and AII in the Appendix.

(table XI) or according to natural regions of the country (table XII). The information is also represented in figs. 6 and 7 by using a two-dimensional Cartesian plane (biodiversity plane), that has radius r as x-axis and diversity index $B_1(S, r)$ as y-axis.

From tables XI, XII, and figs. 6, 7 one can observe that:

TABLE VIII

The two most and the two least relevant species for each natural region of Colombia, classified according the relative weights respect to the region and the whole country

Region	Species	No. of specimens	Relative weight	
			Regions	COL
AN	*Neostrengeria guenteri* (Pretzmann, 1965)	517	0.09139	0.03725
AN	*Hypolobocera bouvieri bouvieri* (Rathbun, 1898)	489	0.08644	0.03523
AN	*H. malaguena* Prahl, 1988	1	0.00018	0.00007
AN	*Eidocamptophallus chacei* (Pretzmann, 1967)	1	0.00018	0.00007
CA	*Macrobrachium acanthurus* (Wiegmann, 1836)	424	0.2328	0.03055
CA	*M. carcinus* (Linnaeus, 1758)	221	0.1214	0.01592
CA	*Chaceus curumanensis* Campos & Valencia, 2004	2	0.0011	0.00014
CA	*Ch. cesarensis* Rodríguez & Viloria, 1992	2	0.0011	0.00014
AM	*Macrobrachium brasiliense* (Heller, 1862)	795	0.2639	0.05727
AM	*M. ferreirai* Kensley & Walker, 1982	602	0.1999	0.04337
AM	*Sylviocarcinus pictus* (H. Milne Edwards, 1853)	2	0.0007	0.00014
AM	*Fredilocarcinus raddai* (Pretzmann, 1978)	1	0.0003	0.00007
OR	*Macrobrachium brasiliense* (Heller, 1862)	1221	0.4031	0.08796
OR	*M. amazonicum* (Heller, 1862)	687	0.2268	0.04949
OR	*Trichodactylus quinquedentatus* Rathbun, 1893	2	0.0007	0.00014
OR	*Macrobrachium ferreirai* Kensley & Walker, 1982	1	0.0003	0.00007
PAC	*Macrobrachium rathbunae* Holthuis, 1950	141	0.3895	0.01016
PAC	*M. panamense* Rathbun, 1912	90	0.2486	0.00648
PAC	*M. tenellum* (Smith, 1871)	15	0.0414	0.00108
PAC	*M. transandicum* Holthuis, 1950	9	0.0249	0.00065

Abbreviaiotns: AN, Andean region; CA, Caribbean region; AM, Amazonian region; OR, Orinoquian region; PAC, Pacific region; COL, Colombia.

(a) The diversity index $B_1(S, r)$ describes changes in the species diversity as the combination of two contributions, species richness and abundance distribution (Campos & Isaza, 2009).

(b) For each subset (e.g., families, natural regions of Colombia), the diversity index is always located on the curve $\alpha(S)/r$ corresponding to the number of species S of the subset under consideration. The function $\alpha(S)/r$ is monotonically decreasing function and it is defined in the range $0 < r \leqslant 1$.

(c) When the information of the database is organized in terms of the families, the set of probability distributions P^{PS}, P^{TR}, P^{AT}, P^{PA}, P^{EU}, and P^{SE}, and the quantities given in table X, allow to calculate the weighted mean value of the radii $\bar{r}$ and theirs uncertainties Δr, the weighted species diversity indices and theirs uncertainties ΔB. The resulting values are represented in fig. 6 by a big black point in the middle of the rectangle and by the same rectangle. The species diversity index of each family is represented by a point located on the curve $\alpha(S)/r$ that corresponds to the number S of species in the family. The weight assigned to each

TABLE IX
Distribution of species (sp) and specimens (spe) between the families and the natural regions of Colombia

Region	PS		TR		AT		PA		EU		SE		Total	
	sp	spe	sp	spe	sp	spe	sp	spe	sp	spe	sp	spe	sp	spe
AN	78	5553	1	91	0	0	2	13	0	0	0	0	81	5657
CA	10	265	6	340	3	239	8	977	0	0	0	0	27	1821
AM	2	20	8	283	0	0	12	2457	1	154	1	98	24	3012
OR	1	153	5	729	0	0	9	2113	1	34	0	0	16	3029
PAC	0	0	0	0	1	21	6	341	0	0	0	0	7	362
Total COL	91	5991	15	1443	4	260	28	5901	1	188	1	98	139	13 881
Λ			5				10		1				16	

In the last row the number of species for each family that is shared with other region is indicated. Families: PS, Pseudothelphusidae; TR, Trichodactylidae; AT, Atyidae; PA, Palaemonidae; EU, Euryrhynchidae; SE, Sergestidae. Regions: AN, Andean region; CA, Caribbean region; AM, Amazonian region; OR, Orinoquian region; PAC, Pacific region; COL, Colombia.

TABLE X
Shannon entropies calculated for the probability distributions associated with communities of S species: (A) the families, (B) Colombia as a whole and (C) the natural regions of Colombia, in each case using the relative species abundance vector

	S	Entropy (nat)
Families		
PS	91	3.684
TR	15	1.917
AT	4	0.908
PA	27	2.391
EU	1	0
SE	1	0
Total COL	139	3.953
Regions		
AN	81	3.580
CA	27	2.613
AM	24	2.442
OR	16	1.755
PAC	7	1.562
Total COL	139	3.953

Mean entropy families = 2.823, mean entropy regions = 2.755.
Families: PS, Pseudothelphusidae; TR, Trichodactylidae; AT, Atyidae; PA, Palaemonidae; EU, Euryrhynchidae; SE, Sergestidae. Regions: AN, Andean region; CA, Caribbean region; AM, Amazonian region; OR, Orinoquian region; PAC, Pacific region; COL, Colombia.

TABLE XI

Values of some quantities involved in the calculation of the species diversity index, when the families of freshwater decapod crustaceans of Colombia are considered

	Family							$\overline{x}$	Δx
	PS	TR	AT	PA	EU	SE	COL		
Specimens	5991	1443	260	5901	188	98	13 881		
S	91	15	4	27	1	1	139		
$\alpha(S)$	3.84	1.62	0.938	2.13	0.637	0.637	4.73		
$r \approx$	0.191	0.466	0.705	0.40	1.0	1.0	0.196	0.335	0.156
$B_1(S,r) \approx$	20.097	3.476	1.331	5.318	0.637	0.637	24.089	11.334	7.693

The last two columns give the mean values ($\overline{x}$) and uncertainties (Δx): either of the radius r or the diversity index. PS, Pseudothelphusidae; TR, Trichodactylidae; AT, Atyidae; PA, Palaemonidae; EU, Euryrhynchidae; SE, Sergestidae; COL, Colombia.

family is given by the relation between the number of specimens of the family and the total number of specimens in the collections (13 881).

(d) A similar procedure to the above is followed when the analysis is done in terms of the regions, represented by the set of probability distributions P^{AN}, P^{CA}, P^{AM}, P^{OR}, and P^{PAC}. In this case, the analysis is supported by table XII and fig. 7.

(e) A comparison of the results when the specimens are partitioned into families and regions is shown in fig. 8. It is important to note that the mean diversity index for the families is slightly larger than the value for the regions, whereas the average value of the radii is slightly lower (compare data in tables XI and XII).

(f) As a result of an additional calculation, it is interesting to note that 89 and 68 are the effective numbers of species, for families and regions, that give rise to curves $\alpha(S)/r$ passing through the points located in the centres of the rectangles, respectively.

TABLE XII

Values of some quantities involved in the calculation of the species diversity index, when the natural regions of Colombia are considered

	Region						$\overline{x}$	Δx
	AN	CA	AM	OR	PAC	COL		
Specimens	5657	1821	3012	3029	362	13 881		
S	81	27	24	16	7	139		
$\alpha(S)$	3.624	2.130	2.015	1.669	1.164	4.729		
$r \approx$	0.199	0.324	0.364	0.498	0.508	0.196	0.325	0.119
$B_1(S,r) \approx$	18.140	6.576	5.540	3.352	2.293	24.089	10.249	6.632

The last two columns give the mean values ($\overline{x}$) and uncertainties (Δx): either of the radius r or the diversity index. PS, Pseudothelphusidae; TR, Trichodactylidae; AT, Atyidae; PA, Palaemonidae; EU, Euryrhynchidae; SE, Sergestidae; COL, Colombia.

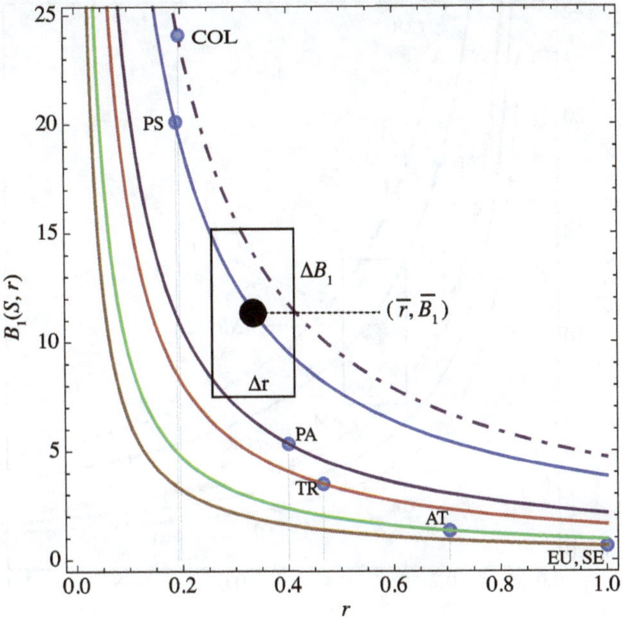

Fig. 6. Species diversity index ($B_1(S, r)$) as function of radius r, calculated according to a partition in terms of families. The weighted mean value of the diversity index is represented by the big black point, and the rectangle is constructed with the uncertainties of the radii and the diversity indices. PS, Pseudothelphusidae; TR, Trichodactylidae; AT, Atyidae; PA, Palaemonidae; EU, Euryrhynchidae; SE, Sergestidae.

DISCUSSION

Using a database assembled with quantitative data on Colombian freshwater decapod crustaceans obtained from natural history collections, we were able to identify the predominant as well as the less representative species occurring in Colombia in terms of the families, the natural regions and the country as a whole.

Relative weights of the species

(A) Subsets constituted by the families (fig. 3, table III).— Among the 91 species of Pseudothelphusidae, the most representative is *Neostrengeria guenteri* (Pretzmann, 1965) with a relative weight of 0.03724. This species is relevant because it is widespread in the foothills of the Eastern Cordillera from Colombia, has large population densities and lives in preserved environments, all factors that guarantee the conservation of this species. The subspecies *Hypolobocera bouvieri bouvieri* (Rathbun, 1898) is next in importance (0.03522), characterized by having an extensive distribution in the Central and Eastern Cordilleras. The adults reach a large body size, which facilitates their capture and explains the large number of specimens found in collections. On the other hand, *Hypolobocera malaguena*

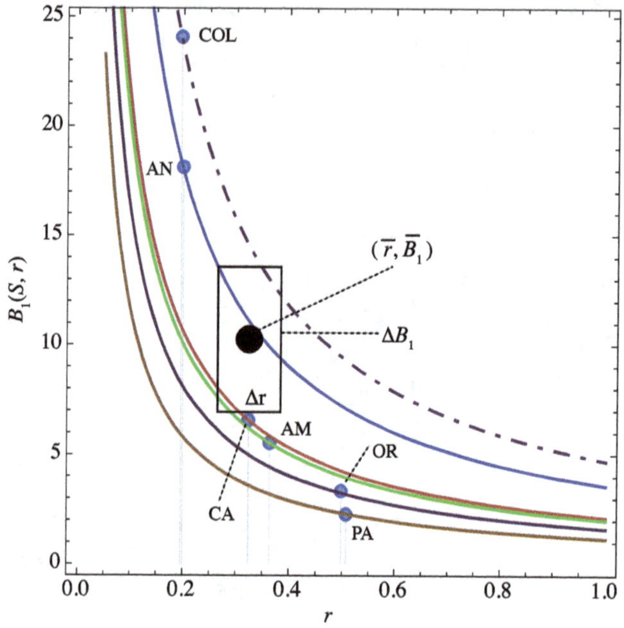

Fig. 7. Biodiversity index ($B_1(S, r)$) as function of radius r, calculated by using a partition of specimens in terms of the natural regions from Colombia. The black point represents the mean value calculated with the weighted contributions of the regions, and the rectangle is constructed with the uncertainties of the radii and the diversity indices of the regions. AN, Andean region; CA, Caribbean region; AM, Amazonian region; OR, Orinoquian region; PAC, Pacific region; COL, Colombia.

Prahl, 1988, and *Eidocamptophallus chacei* (Pretzmann, 1967) share the lower relative weight of 0.00007. This value reflects their scarcity of records, with only one for each of them. *H. malaguena* was described by Prahl (1988), and since then no more specimens could be captured, which allows us to assume that this species has a very reduced distributional range. Campos & Lasso (2015), applying the IUCN red list protocols, have categorized this species as Data Deficient (DD). The first documented location of *E. chacei* has only recently been recorded from the Andean region of Colombia (Campos & Magalhães, 2014), being known only from one locality.

Within the Trichodactylidae, with 15 species recorded from Colombia, the more representative species is *Moreirocarcinus emarginatus* (H. Milne Edwards, 1853), with a relative weight of 0.0415. This species has an extensive distributional range in the Orinoco and the Amazon basins of Colombia, and its distribution extends to Venezuela, and southward to Brazil, Ecuador and Peru. The widespread distribution of *M. emarginatus* explains the high number of specimens deposited in the different collections. The second most representative trichodactylid is *Sylviocarcinus piriformis* (Pretzmann, 1968) (0.01787), which only occurs in Colombia and Venezuela, and is distributed in the upper and median course

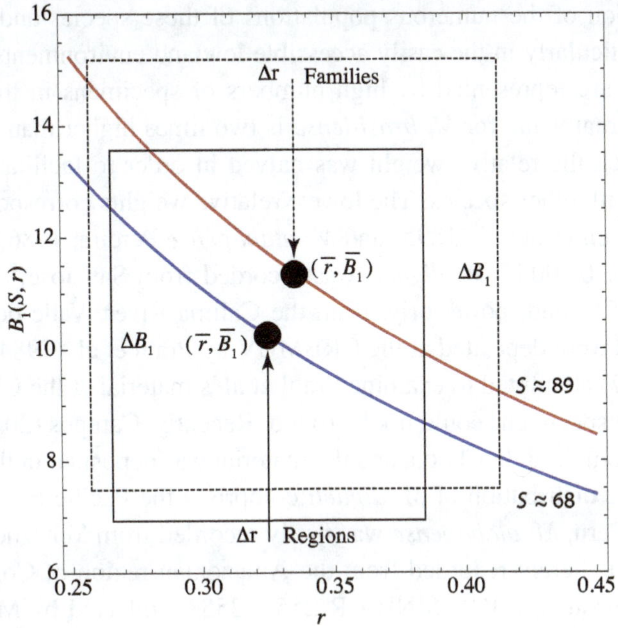

Fig. 8. Comparison of mean radii and diversity indices when the collected specimens are partitioned into freshwater decapod Crustaceans families or in terms of the natural regions of Colombia. The uncertainty rectangles are also shown.

of the Magdalena River and in the Maracaibo basin. Intensive fishing in the Magdalena River explains its large number of specimens included in the database. *Sylviocarcinus pictus* (H. Milne Edwards, 1853) and *Fredilocarcinus raddai* (Pretzmann, 1978) have the lowest relative weight values into the family. *S. pictus* (0.00014) is restricted to the Amazonian region, and only two specimens have been recorded from Colombia. *F. raddai* (0.00007) was recorded for the first time in 2015, and is known only from one locality in the Amazonian region of Colombia.

Among the only four shrimp species of Atyidae from Colombia, *Potimirim glabra* (Kingsley, 1878) has the highest relative weight (0.01232), which is associated with its widespread distributional range along the Caribbean and Pacific regions. In contrast, *Atya innocous* (Herbst, 1792) has the lowest relative weight of 0.00043, with a single record from the Caribbean region. Nevertheless, it is important to note that this species extends its distribution to the Caribbean and Pacific regions.

Palaemonidae is represented in Colombia by 27 species, of which *Macrobrachium brasiliense* and *M. amazonicum* have the highest relative weights, 0.14523 and 0.05972, respectively. These values can be explained because both species have wide distributional ranges along the Orinoco and Amazon basins of Colombia, and are associated with low altitude in costal habitats. The commercial exploitation of the numerous populations of these species and the intensive sampling, particularly in the easily accessible lowland environments, explain that these species are represented by high numbers of specimens in the collections. Note that the real value for *M. brasiliense* is two times higher than that shown in fig. 3, in which the relative weight was halved in order to facilitate the graphic comparison with other species. The lowest relative weights correspond to *Macrobrachium digueti* (Bouvier, 1895) and *M. atabapense* Pereira, 1986, both of them with weight of 0.00050. *M. digueti* was recorded from San José, SW Colombia (Holthuis, 1952), and, posteriorly, from the Calima River, Valle del Cauca, and the specimens were deposited at the CRBMUV by Prahl et al. (1984). Valencia & Campos (2007) attempted to examine Prahl et al.'s material at the CRBMUV collection but the specimens could not be found. Recently, Campos (2014) registered this species from Acandí, Chocó, and the material was deposited in the ICN-MHN collection. The distribution of *M. digueti* comprises the Pacific region from Baja California to Peru. *M. atabapense* was firstly recorded from Venezuela by Pereira (1986), but it is herein recorded from the Amazonian region of Colombia based on species deposited at ICN-MNH-CR 2552, 2553, collected by M. R. Campos on 23-24 March 1998, in Guainía Department, municipality Inírida, Yucuta and Agujón creeks, altitude 70 m.

Finally, Euryrhynchidae and Sergestidae are represented with one species each in Colombia: *Euryrhynchus amazoniensis* Tiefenbacher, 1978 and *Acetes paraguayensis* Hansen, 1919, with relative weights of 0.01354 and 0.00706, respectively.

(B) Colombia as a whole with 139 species.— The highest values shown by the three palaemonid shrimps when Colombia is considered as a whole (fig. 4, table VI) can be explained by the widespread distributions of *Macrobrachium brasiliense* and *M. amazonicum* in the Orinoco and Amazon basins, while *M. ferreirai* is restricted to the Amazon basin of Colombia. On the other hand, the pseudothelphusid crabs *Neostrengeria macarenae* Campos, 1992, *Hypolobocera malaguena* and *Eidocamptophallus chacei* share the lowest relative weights, with the value of 0.00007. Recently, in Red Book of the Freshwater Crabs from Colombia, Campos & Lasso (2015) categorized *N. macarenae* as Endangered (EN) because it has a reduced distributional range and it lives in disturbed habitats associated with a high rate of deforestation and the increasing agricultural and livestock frontier. *H. malaguena* and *E. chacei* are known only from the type locality, which implies that they have a very restricted geographic distribution (Campos, 2014).

(C) Subsets conformed by the regions.— For the Andean region, the pseudothelphusid crabs appears as the predominant group (table VIII). According to Rodriguez (1986), the great diversity of this family is associated with their widespread distribution in the Andes Mountains and highland massifs of northern and central South America and its disjunct isolation in freshwater systems, resulting from a series of geologic events. The pseudothelphusids are considered to be montane fauna due that most of its species are distributed between altitudes of 300 and 3000 m asl (Rodriguez, 1981).

Neostrengeria guenteri has the highest relative weight (0.03725) followed by *Hypolobocera bouvieri bouvieri*, with a weight of 0.03523 (table VIII); the reasons for such values are discussed above, under (A) Subsets constituted by the families. The species *Hypolobocera malaguena* and *Eidocamptophallus chacei* share the lowest relative weight of 0.00007. Similarly, the value of both species is discussed above.

For the Caribbean region, the palaemonid shrimps *Macrobrachium acanthurus* (Wiegmann, 1836) and *M. carcinus* (Linnaeus, 1758) rank with the highest relative weights of 0.03055 and 0.01592 (table VIII). These species have widespread distributions in the Caribbean region of Colombia, and they are species commercially exploited due to their large body size; these factors and intensive surveys in the coastal area made it possible to gather a large number of specimens in collections. The lowest relative weight of 0.00014 is shared by the pseudothelphusid crabs *Chaceus curumanensis* Campos & Valencia, 2004, and *C. cesarensis* Rodríguez & Viloria, 1992. No new record of *C. curumanensis* was published after its description in 2004. Campos & Lasso (2015) categorized *C. curumanensis* as Endanger (EN) because it has a very reduced distributional area and its habitat is highly disturbed. Likewise, no new records of *C. cesarensis* are known other than its type locality.

In the Amazonian region, the highest relative weights correspond to the shrimp species *Macrobrachium brasiliense* and *M. ferreirai*, with 0.05727 and 0.04337, respectively (table VIII). The former is distributed along an extensive area of the Orinoquian and Amazonian regions, whereas the latter occurs in the Amazonian region and has only one record from Casanare, in the Orinoquian region. The lowest relative weights correspond to the trichodactylid crabs *Sylviocarcinus pictus* and *Fredilocarcinus raddai* with the values of 0.00014 and 0.0007, respectively. These species are restricted to the Amazonian region. The low number of specimens in the database is due to the scarceness of explorations in that region.

In the Orinoquian region, the shrimp species *Macrobrachium brasiliense* ranks first with a relative weight of 0.08796 (table VIII). This high value is consistent with the fact that the species has an extensive distribution in the Orinoco and Amazon rivers and their tributaries, and that intensive surveys in the region have contributed with high number of specimens deposited in collections. Ranked second, *M. amazonicum* (0.04949) has a very wide distributional range in the Orinoco and Amazon basins and is commercially exploited in this region, similarly to *M. brasiliense*. The lowest relative weights of 0.000014 and 0.00007 correspond, respectively, to the trichodactylid crab *Trichodactylus quinquedentatus* Rathbun, 1893 and the palaemonid shrimp *Macrobrachium ferreirai*, each of which only having one record from this region.

With regards to the Pacific region, the shrimp species *Macrobrachium rathbunae* Holthuis, 1950 and *M. panamense* Rathbun, 1912, are considered the most representative species, with relative weights of 0.01016 and 0.00648 (table VIII). These species have extensive distributions in the region and are economically relevant in this area. On the opposite side, the shrimp species *Macrobrachium tenellum* (Smith, 1871) and *M. transandicum* Holthuis, 1950, recorded the lowest relative weights. The explanation for such lower values is related to the fact that the region is poorly surveyed for decapods.

Diversity indices

Notwithstanding that the Shannon index is a traditional measure of biological diversity, and that this index is effective in quantify the uncertainty of probability distribution, it has an unknown ecological meaning (Kolasa & Biesladka, 1984). On the contrary, the Campos & Isasa (2009) species diversity index used in this study is able to handle simultaneously and independently the levels of species richness (S) and the species relative abundances (probability distributions P).

The index for measuring diversity of species (Campos & Isaza, 2009) has been successfully applied to characterize and quantify the diversity of freshwater decapod crustacean species. It has allowed representing the results of the calculations in a biodiversity plane, as shown in figs. 6, 7 and 8. The diversity index allows interpreting the species diversity as a combination of the species richness (S) and the abundance distribution of them, and the biodiversity plane is organized in layers formed by curves with a defined number of species (isonumber S-curves). The Shannon index has no analogous properties.

In summary, in this paper we were able to show that the database with information from Colombian freshwater decapod crustaceans is valuable for scientific purposes, because the information there included has allowed getting a general view on the freshwater decapod crustacean diversity of Colombia. It also shows critical aspects on some species as a consequence of deficient information, which is a very important element for future research.

ACKNOWLEDGEMENTS

The authors would like to thank for the opportunity for participating in the special issue of the journal Crustaceana in honour of the late Dr. Michael Türkay, carcinologist of the Research Institute and Natural History Museum Senckenberg, Frankfurt am Main, Germany, who became a prominent biologist through their research on marine and the freshwater crustaceans. One of us (MRC) had the opportunity to meet him in 1991 during a visit to review the crustacean collection of the Senckenberg Institute and with the expectation of learning from a specialist of great international recognition. The result was the publication of a new species of freshwater crab for Colombia, *Chaceus curumanensis*, based on the specimens collected during his visit to this country in 1978. The authors are grateful to Dr. Célio Magalhães for the careful reading of the manuscript and valuable recommendations, and to Dr. David Hudson for helpful comments. We also thank the large number of collectors who have contributed to the knowledge of the diversity of freshwater decapod crustaceans from Colombia, and the Universidad Nacional de Colombia for providing the research support.

REFERENCES

CAMPOS, D. & J. F. ISAZA, 2009. A geometrical index for measuring species diversity. Ecological Indicators, **9**: 651-658.

CAMPOS, M. R., 2014. New records of *Macrobrachium digueti* (Bouvier, 1895) for Colombia (Crustacea: Decapoda: Palaemonidae). Revista Academia Colombiana de Ciencias Exactas, Físicas y Naturales, **38**: 191-194.

CAMPOS, M. R. & C. A. LASSO, 2015. Libro rojo de cangrejos dulceacuícolas de Colombia: 1-168. (Instituto de Investigación de Recursos Biológicos Alexander von Humboldt, Instituto de Ciencias Naturales de la Universidad Nacional de Colombia, Bogotá, D.C., Colombia).

CAMPOS, M. R. & C. MAGALHÃES, 2014. *Colombiathelphusa*, a new genus of freshwater crab from Colombia, and the first location record of *Eidocamptophallus chacei* (Pretzmann, 1967) (Crustacea: Decapoda: Pseudothelphusidae). Zootaxa, **3860**: 571-579.

CUMBERLIDGE, N., P. K. L. NG, D. C. J. YEO, C. MAGALHÃES, M. R. CAMPOS, F. ÁLVAREZ, T. NARUSE, S. R. DANIELS, L. J. ESSER & F. Y. K. ATTIOPE, 2009. Freshwater crabs and the biodiversity crisis: importance, threats, status, and conservation challenges. Biological Conservation, **142**: 1665-1673.

CUMBERLIDGE, N., P. K. L. NG, D. C. J. YEO, T. NARUSE, K. S. MEYER & L. J. ESSER, 2011. Diversity, endemism and conservation of the freshwater crab of China (Brachyura, Potamidae and Gecarcinucidae). Integrative Zoology, **6**: 45-55.

HOLTHUIS, L. B., 1952. A general revision of the Palaemonidae (Crustacea: Decapoda: Natantia) of the Americas II. The subfamily Palaemoninae. Allan Hancock Foundation Publications, Occasional Papers, **12**: 1-396.

KOLASA, J. & B. BIESLADKA, 1984. Diversity in ecology. Acta Biotheoretica, **33**: 145.

PEREIRA, G., 1986. Freshwater shrimps from Venezuela I: seven species of Palaemonidae (Crustacea: Decapoda: Palaemonidae). Proceedings of the Biological Society of Washington, **99**: 198-210.

PRAHL, H., 1988. Fresh-water crabs (Crustacea: Decapoda: Pseudothelphusidae) of the Pacific drainage of Colombia. Zoologische Jahrbücher für Systematik, **115**: 171-186.

PRAHL, H., C. CAICEDO & R. RÍOS, 1984. Camarones Palaemonidos (Crustacea, Caridea, Palaemonidae) de agua dulce y salobre del Departamento del Valle del Cauca. Cespedesia, **13**: 45-58.

RANGEL-CH., J. O., 2015. La biodiversidad de Colombia: significado y distribución regional. Revista de la Academia Colombiana de Ciencias Exactas, Físicas y Naturales, **39**: 176-200.

RODRÍGUEZ, G., 1981. Decapoda. In: S. H. HURLBERT, G. RODRÍGUEZ & N. D. DOS SANTOS (eds.), Aquatic Biota of Tropical South America, part 1. Arthropoda: 1-323. (San Diego State University, San Diego, CA).

RODRÍGUEZ, G., 1986. Centers of radiation of fresh-water crabs in the neotropics. In: R. H. GORE & K. L. HECK (eds.), Biogeography of the Crustacea. Crustacean Issues, **4**: 51-67.

VALENCIA, D. M. & M. R. CAMPOS, 2007. Freshwater prawns of the genus *Macrobrachium* Bate, 1868 (Crustacea: Decapoda: Palaemonidae) of Colombia. Zootaxa, **1456**: 1-44.

First received 23 March 2016.
Final version accepted 8 April 2017.

[When citing this volume, please refer to Crustaceana 90 (2017) 771-1288]

Michael Türkay Memorial Issue

WESTERN PACIFIC EURYPLACIDAE STIMPSON, 1871 AND GONEPLACIDAE MACLEAY, 1838 (DECAPODA, BRACHYURA, GONEPLACOIDEA) IN THE SENCKENBERG NATURMUSEUM, FRANKFURT

BY

P. CASTRO[1])
Biological Sciences Department, California State Polytechnic University, Pomona, CA 91768, U.S.A.

ABSTRACT

Examination of the western Pacific material of goneplacoid crabs (families Euryplacidae and Goneplacidae) in the Senckenberg Research Institute and Natural History Museum Frankfurt a. M. (Germany) particularly uncatalogued material examined by Tune Sakai, allowed the discovery of the holotype of *Eucrate formosensis* Sakai, 1974 (Euryplacidae), as well as the identification of some rarely collected species.

Key words. — Holotype *Eucrate formosensis*, Japan, Taiwan, taxonomy, Tune Sakai

ZUSAMMENFASSUNG

Die Untersuchung westpazifischer Krabben der Überfamilie Goneplacoidea (Familien Euryplacidae und Goneplacidae) im Senckenberg Forschungsinstitut und Naturmuseum Frankfurt a. M. (Deutschland), darunter insbesondere unkatalogisiertes Material aus Untersuchungen von Tune Sakai, ermöglichte die Entdeckung des Holotyps von *Eucrate formosensis* Sakai, 1974 (Euryplacidae), sowie die Identifikation einiger nur wenig gesammelter Arten.

INTRODUCTION

Thanks to the efforts of the late Michael Türkay, the Senckenberg Naturmuseum, Frankfurt (SMF) holds a rich collection of brachyuran crabs from Japan and nearby waters, notably those collected and identified by Tune Sakai (1903-1986). The goneplacoid crabs (superfamily Goneplacoidea MacLeay, 1838) belonging to this collection, many of which were unidentified and uncatalogued, were examined during a 2010 visit to the museum. It included material belonging to several

[1]) e-mail: pcastro@cpp.edu

species that are rarely collected as well as the holotype of *Eucrate formosensis* Sakai, 1974 (Euryplacidae), which was thought to be lost (Castro & Ng, 2010). Material part of the important T. Sakai collection now at Senckenberg Naturmuseum had been studied previously for members of Palicidae and Crossotonotidae (see Castro, 2000), Ethusidae (see Castro, 2005), and Trapeziidae (unpublished).

MATERIAL AND METHODS

Carapace length (cl) was measured across the middle of the carapace from the middle of the front to the middle portion of the posterior border of the carapace; carapace width (cw) across the widest breadth of the carapace between the largest anterolateral teeth. All material examined is deposited in the Senckenberg Naturmuseum (SMF).

SYSTEMATICS

Order DECAPODA Latreille, 1802
Infaorder BRACHYURA Latreille, 1802
Superfamily GONEPLACOIDEA MacLeay, 1838
Family EURYPLACIDAE Stimpson, 1871
Genus *Eucrate* De Haan, 1835

Eucrate crenata (De Haan, 1835)

Cancer (*Eucrate*) *crenatus* De Haan, 1835: 51, pl. 15, fig. 1.
Eucrate crenata — Sakai, 1976: 535, pl. 192, fig. 1 — Castro & Ng, 2010: 21, figs. 2A, B, 3A-G, 14D-F [synonymy and references].

Material examined.— One male, Seto Inland Sea, Bizan Seto near Sakaide, Kagawa-ken, Japan, 34°20′N 133°52′E, October 1978, trawl, 10-45 m (SMF 37784); one female, Seto Inland Sea, near Ohshima, Imabari, Ehime-ken, Japan, June 1979, T. Sakai leg. (SMF 37781); two males, Tsuyazaki, Fukuoka-ken, Japan, 33°36.747′N 130°22.362′E, 10 m, 5 November 1979 (SMF 37782); one female, Seto Inland Sea, Bizan Seto, near Sakaide, Kawaga-ken, Japan, 34°20′N 133°52′E, stn. HB BP, 150 m, June 1980 (SMF 37780); three males, Tanabe Bay, near Tanabe, Wakayama-ken, Japan, 33°43′N 135°19′E, 26-27 October 1988, M. Türkay leg. (SMF 3779); one male, Naruto, Tokushima-ken, Japan, 34°11′N 134°37′E, 2 September 1995 (SMF 37778).

Remarks.— *Eucrate crenata* was described from Japan (De Haan, 1835). It is widely distributed in shallow water from the Mediterranean Sea (introduced) and across the Indo West Pacific region from the Red Sea to the western Pacific.

Eucrate formosensis Sakai, 1974
(fig. 1)

Eucrate formosensis Sakai, 1974: 94 — Castro & Ng 2010: 30, figs. 2E, F, 8A-F, 14J-L [synonymy and references].

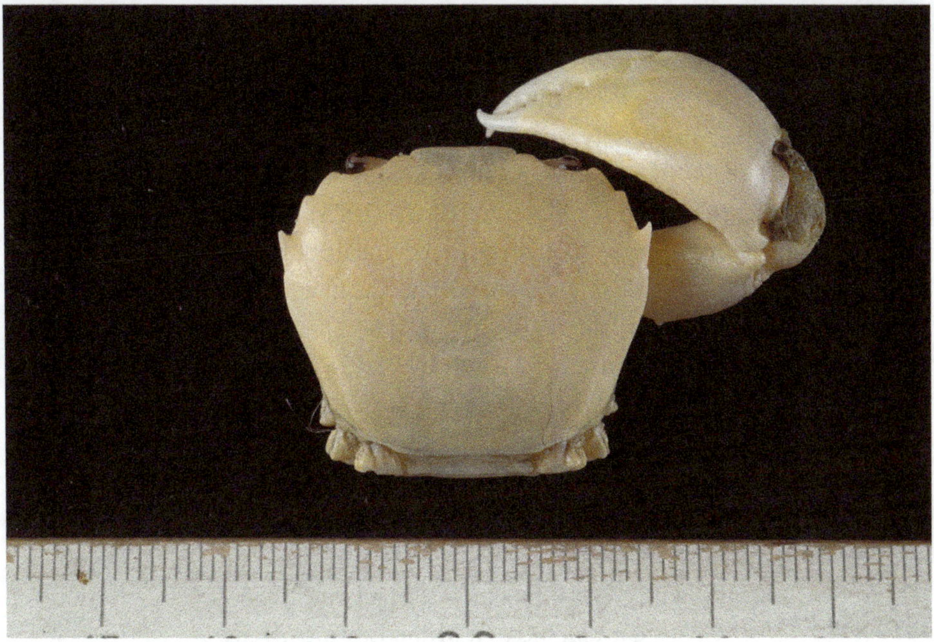

Fig. 1. *Eucrate formosensis* Sakai, 1974, male holotype (SMF 37525), cl 23.2 mm, cw 29.0 mm, Kaohsiung, Taiwan, dorsal view of carapace.

Eucrate alcocki Sakai, 1976: 536, pl. 192, fig. 2 (part) (not *E. alcocki* Serène in Serène & Lohavanijaya, 1973).

Material examined.— Male holotype, cl 23.2 mm, cw 29.0 mm, Kaohsiung, Taiwan (SMF 37525).

Remarks.— Sakai (1974: 94) described *Eucrate formosensis* as a new species, with "Kao-hsung, Formosa [Taiwan]" as type locality, and the holotype collected by T. Watabe (Manazuru Marine Laboratory) on December 1971. Sakai (1976) subsequently synonymized his species with *E. alcocki* Serène in Serène & Lohavanijaya, 1973 (type locality Vietnam). Castro & Ng (2010) resurrected Sakai's species, notwithstanding their close morphological similarities, noting that the two species can be distinguished mainly by their respective colour patterns. In *E. formosensis* the anterior half or most of the dorsal surface of the carapace and chelipeds have small, red-brown dots (Castro & Ng, 2010, fig. 2E), larger dots throughout the carapace in small individuals (Castro & Ng, 2010, fig. 2F). In contrast, diagnostic for *E. alcocki* is the presence of one large, round spot on the median portion of the dorsal surface of the carapace and numerous, smaller red spots laterally and anteriorly to the large spot (Serène & Lohavanijaya, 1973, pl. 16, figs. C, D; Castro & Ng, 2010, figs. 2C, 4A, B). *Eucrate formosensis* is known only from Taiwan, whereas *T. alcocki* has been recorded from southern China to

Singapore (Castro & Ng, 2010). Both species are known from the shallow subtidal, as deep as 15 m in *E. formosensis* and 37 m in *E. alcocki* (Castro & Ng, 2010).

Castro & Ng (2010: 30) listed the type material of *E. formosensis* as "unknown" and concluded that the holotype was probably not extant because it could not be located in Japanese or European collections. A dry specimen with detached pereiopods at the Senckenberg Museum (SMF 37525) (fig. 1), unknown because it had remained uncatalogued, proved to be the lost holotype. The specimen showed the characteristic carapace shape and the diagnostic small dots on the anterior portion of the carapace still visible. The specimen was indicated with the label "Type" together with three additional labels: "TS 00231, Box 7-1" and "*Eucrate sexdentata* Haswell 1881 [1882] (see Campbell 1969)" and "*Eucrate alcocki* Serène". Its measurements (cl 23.2 mm, cw 29.0 mm), are very close to the measurements given by Sakai (1974: 94): cl 24 mm, cw 29.5. The distinctive colour pattern of the species is still discernible on the specimen (fig. 1).

Family GONEPLACIDAE MacLeay, 1838
Genus *Carcinoplax* H. Milne Edwards, 1852
Carcinoplax inaequalis (Yokoya, 1933)

Pilumnoplax inaequalis Yokoya, 1933: 194, 217, 220, fig. 63.
Homoioplax haswelli — Sakai, 1976: 540, fig. 287.
Carcinoplax inaequalis — Castro, 2007: 633 [synonymy and references].

Material examined.— One male, cl 20.0 mm, cw 28.5 mm, Kiinagashima, Mie-ken, Japan. T. Sakai leg. ["TS 00243, Box 7-1"] (SMF 37547).

Remarks.— The male from Kiinagashima, Japan (SMF 37547) is most probably the specimen from the same locality listed and photographed by Sakai (1976: 540, fig. 287, as *Homoioplax haswelli*). The specimen was re-measured and its size is similar to that given by Sakai (1976: 540): cl 20.5, cw 29 mm. The species is known only from Japan at depths of 35-384 m (Castro, 2007).

Carcinoplax tomentosa Sakai, 1969

Carcinoplax tomentosa Sakai, 1969: 270 [in list], 271, figs. 16a, 17c, 18a — Castro, 2007: 648 [synonymy and references].

Material examined.— One male, Tosa Bay, Kōchi-ken, Japan, T. Sakai leg. [TS 00236, Box 7-1] (SMF 37541).

Remarks.— The male specimen was included in a lot identified as "*Carcinoplax surugensis*" (= *Pycnoplax surugensis* (Rathbun, 1932), see below), so it is possible that it is one of the three males from Tosa Bay erroneously identified as such by Sakai (1976: 525). It had a label, "hime-enkogani", the Japanese name used by

Sakai (1976: 525, as *Carcinoplax surugensis*) for *P. surugensis*, which was also included in one correctly identified lot of *P. surugensis* (see below).

Carcinoplax tomentosa is known only from Japan (type locality) and Taiwan at depths of 150-300 m (Castro, 2007).

Genus *Entricoplax* Castro, 2007

Entricoplax vestita (De Haan, 1833)

Cancer (*Curtonotus*) *vestitus* De Haan, 1833: 51, pl. 5, fig. 3.
Carcinoplax vestita — Sakai, 1976: 525, pl. 190, fig. 3.
Entricoplax vestita — Castro, 2007: 656, fig. 11 [synonymy and references].

Material examined.— One male, cl 15.5 mm, cw 22.0 mm, 3 females, largest cw 22.5 mm, cw 32.7 mm, unknown location, T. Sakai leg. (SMF 37549); three males, unknown location, T. Sakai leg. (SMF 37548).

Remarks.— The material examined was most probably that listed by Sakai (1976: 525), who included specimens from numerous Japanese locations. *Entricoplax vestita* is known only from Japan and China at depths of 15-110 m (Castro, 2007; unpublished).

Genus *Exopheticus* Castro, 2007

Exopheticus insignis (Alcock, 1900)

Psopheticus insignis Alcock, 1900: 310 — Sakai, 1976: 531, pl. 193, fig. 2.
Exopheticus insignis — Castro, 2007: 747, figs. 48, 49 [synonymy and references].

Material examined.— One male, one female, East China Sea, S. Ohishi, T. Sakai leg. (SMF 37527); one female, dry [T-1], Tainan, Taiwan, T. Sakai leg. (SMF 37521).

Remarks.— *Exopheticus insignis* is thus far unknown from Japan, the specimen recorded by Sakai (1976) being from Taiwan. It is known from the Andaman Sea (type locality) and the western Pacific from Taiwan to Fiji at depths of 110-300 m (Castro, 2007). The SMF specimen from the East China Sea (SMF 37527) represents the northernmost known limit of the species.

Exopheticus hughi (Rathbun, 1914)

Psopheticus hughi Rathbun, 1914: 144 — Sakai, 1976: 530, pl. 193, fig. 1.
Exopheticus hughi — Castro, 2007: 749, fig. 50 [synonymy and references].

Material examined.— Five males, one female, Mimase, Tosa Bay, Kōchi-ken, Japan, 33°51′N 133°56′E, trawl, December 1961, T. Sakai leg ("T.S.") (SMF 37534).

Remarks.— The species is known from Japan to the Philippines (type locality) at depths of 208-418 m.

Genus *Psopheticus* Wood-Mason, 1892
Psopheticus musicus Guinot, 1990

Psopheticus stridulans — Sakai, 1976: 530, pl. 193, fig. 3.
Psopheticus musicus Guinot, 1990: 355, figs. 35-39, 52-54, 56 — Castro, 2007: 742 [synonymy and references].

Material examined.— One pre-adult female, Mimase, Tosa Bay, Kōchi-ken, trawl, December 1961, T. Sakai leg. ("T.S.") (SMF 37535).

Remarks.— The species is known from Japan to the Philippines (type locality) at depths of 75-150 m (Sakai, 1976; as *Psopheticus stridulans*).

Genus *Pycnoplax* Castro, 2007
Pycnoplax surugensis (Rathbun, 1932)

Carcinoplax surugensis Rathbun, 1932: 34 — Sakai, 1976: 525, pl. 188, fig. 3.
Pycnoplax surugensis — Castro, 2007: 663, figs. 13A, 14 [synonymy and references].

Material examined.— One male, Mimase, Tosa Bay, Kōchi-ken, Japan, ca 1961, T. Sakai leg. (SMF 37537); one male, Tosa Bay, Kōchi-ken, Japan, T. Sakai leg. ["hime-enkogani", "TS 00236, Box 7-1"] (SMF 37542); one female, Mimase, Tosa Bay, Kōchi-ken, Japan, T. Sakai leg. ["TS 00242, Box 7-1"] (SMF 37536); two males, dry, Tosa Bay, Kōchi-ken, Japan, T. Sakai leg. (SMF 37522).

Remarks.— Some of the material examined was included among the material collected by K. Sakai and identified by Sakai (1976). The species is known from Japan (type locality) to New Caledonia at depths of 65-496 m (Castro, 2007).

ACKNOWLEDGEMENTS

I dedicate this article to the memory of Michael Türkay, an old friend who left us far too early with far too many projects left unfinished. My appreciation to Kristin Pietratus (SMF) for help with the collections, Sven Tränkner (SMF) for taking the photograph, and Peter K. L. Ng (National University of Singapore) and Tohru Naruse (University of the Ryukyus, Japan) for their comments to the manuscript.

REFERENCES

ALCOCK, A., 1900. The Brachyura Catometopa or Grapsoidea. Materials for a carcinological fauna of India, no. 6. Journal of the Asiatic Society of Bengal, **69**: 279-456.

CAMPBELL, B. M., 1969. The genus *Eucrate* (Crustacea: Goneplacidae) in eastern Australia and the Indo-West Pacific. Memoirs of the Queensland Museum, **15**: 117-140.

CASTRO, P., 2000. Crustacea Decapoda: a revision of the Indo-west Pacific species of palicid crabs (Brachyura Palicidae). In: A. CROSNIER (ed.), Résultats des Campagnes MUSORSTOM, **21**. Mémoires du Muséum national d'Histoire naturelle (Paris), **184**: 437-610.

CASTRO, P., 2005. Crabs of the subfamily Ethusinae Guinot, 1977 (Crustacea, Decapoda, Brachyura, Dorippidae) of the Indo-West Pacific region. Zoosystema, **27**: 499-600.

CASTRO, P., 2007. A reappraisal of the family Goneplacidae MacLeay, 1838 (Crustacea, Decapoda, Brachyura) and revision of the subfamily Goneplacinae, with the description of 10 new genera and 18 new species. Zoosystema, **29**: 609-774.

CASTRO, P. & P. K. L. NG, 2010. Revision of the family Euryplacidae Stimpson, 1871 (Crustacea: Decapoda: Brachyura: Goneplacoidea). Zootaxa, **2375**: 1-130.

DE HAAN, W., 1833-1850. Crustacea. In: P. F. VON SIEBOLD, Fauna Japonica sive Descriptio Animalium, quae in Itinere per Japoniam, Jussu et Auspiciis Superiorum, qui Summum in India Batava Imperium Tenent, Suscepto, Annis 1823-1830 Collegit, Notis, Observationibus et Adumbrationibus Illustravit. Leiden, i-xvii, i-xxxi, ix-xvi, 1-243, pls. A-J, L-Q, 1-55, circ. tab. 2.

GUINOT, D., 1990. Crustacea Decapoda: le genre *Psopheticus* Wood-Mason, 1892 (Goneplacidae). In: A. CROSNIER (ed.), Résultats des Campagnes MUSORSTOM, **6**. Mémoires du Muséum national d'Histoire naturelle (Paris), sér. A, **145**: 331-367.

HASWELL, W. A., 1882. On some new Australian Brachyura. Proceedings of the Linnean Society of New South Wales, **6**: 540-551.

MILNE EDWARDS, H., 1852. Observations sur les affinités zoologiques et la classification naturelle des crustacés. Annales des Sciences naturelles, Zoologie (Paris), sér. 3, **18**: 109-166, pls. 3, 4.

RATHBUN, M. J., 1914. A new genus and some new species of crabs of the family Goneplacidae. Scientific Results of the Philippine cruise of the Fisheries Steamer "Albatross", 1907-1910 — No. 32. Proceedings of the United States National Museum, **48**(2067): 137-154.

RATHBUN, M. J., 1932. Preliminary descriptions of new species of Japanese crabs. Proceedings of the Biological Society of Washington, **45**: 29-38.

SAKAI, T., 1969. Two new genera and twenty-two new species of crabs from Japan. Proceedings of the Biological Society of Washington, **82**: 243-280.

SAKAI, T., 1974. Notes from the carcinological fauna of Japan (V). Researches on Crustacea (Tokyo), no. **6**: 86-102, 1 frontispiece.

SAKAI, T., 1976. Crabs of Japan and the adjacent seas. Kodansha, Tokyo, vol. 1 [English text], xxix + 773 pp., figs. 1-379, maps 1-3; vol. 2 [Japanese text], 461 pp., figs. 1, 2; vol. 3 [plates], 16 pp. + pls. 1-251.

SERÈNE, R. & P. LOHAVANIJAYA, 1973. The Brachyura (Crustacea: Decapoda) collected by the Naga Expedition, including a review of the Homolidae. In: E. BRINTON, W. A. NEWMAN & W. S. WOOSTER (eds.), Scientific Results of Marine Investigations of the South China Sea and the Gulf of Thailand, 1959-1961. Naga Report **4** (4): 1-187.

WOOD-MASON, J., 1892. Illustrations of the Zoology of the Royal Indian Marine Surveying Steamer Investigator, Under the Command of Commander A. Carpenter R.N., D.S.O. and Commander R. F. Hoskyn, R.N. Crustacea, Part 1, Office of the Superintendent of Government Printing, Calcutta [= Kolkata], pls. 1-5.

YOKOYA, Y., 1933. On the distribution of decapod crustaceans inhabiting the continental shelf around Japan, chiefly based upon the materials collected by S.S. Sôyô-Maru, during the years 1923-1930. Journal of the College of Agriculture, Tokyo Imperial University, **12**: 1-226.

First received 1 March 2016.
Final version accepted 28 June 2016.

[When citing this volume, please refer to Crustaceana 90 (2017) 771-1288]

Michael Türkay Memorial Issue

REDESCRIPTION OF *POTAMONAUTES WALDERI* (COLOSI, 1924) FROM THE LOWER CONGO RIVER BASIN IN CENTRAL AFRICA (BRACHYURA, POTAMOIDEA, POTAMONAUTIDAE)

BY

NEIL CUMBERLIDGE[1])
Department of Biology, Northern Michigan University, Marquette, MI, U.S.A.

ABSTRACT

Potamonautes walderi (Colosi, 1924), is redescribed from the lectotype and the male gonopods, abdomen, mouthparts, chelipeds, and sternum are illustrated. Photographs of the lectotype are provided and its distribution and conservation status is discussed. *Potamonautes walderi* is also compared to the other species of this genus that are found in the Congo River basin in Central Africa.

Key words. — Freshwater crab, Democratic Republic of the Congo, Republic of the Congo, taxonomy, conservation

ZUSAMMENFASSUNG

Potamonautes walderi (Colosi, 1924) wird anhand des Lectotyps, der männlichen Gonopoden, des Abdomens, der Mundwerkzeuge, der Chelipeden und des Sternums neu beschrieben und illustriert. Fotografien des Lectotyps werden hier gezeigt und dessen Verbreitung und Schutzstatus werden diskutiert. *Potamonautes walderi* wird auch mit den anderen Arten dieser Gattung verglichen, welche im Congobecken in Zentralafrika zu finden sind.

INTRODUCTION

The freshwater crabs reported on here belong to the African freshwater crab family Potamonautidae Bott, 1970, and are redescribed here based on the lectotype of *Potamon* (*Potamonautes*) *walderi* Colosi, 1924, from Kingoyo in the Lower Congo River basin in the Democratic Republic of the Congo (D. R. Congo). This work also includes more recent material from the Lower Congo River basin in the Republic of the Congo and the D. R. Congo. A redescrip-

[1]) e-mail: ncumberl@nmu.edu

tion of this species is necessary because it is difficult for the non-specialist to distinguish between the many species of *Potamonautes* from Central Africa, not least because the only available identification keys (Rathbun, 1921; Chace, 1942; Bott, 1955) are now incomplete and the classifications used in those works are out of date (Cumberlidge et al., 2008). This means that the only reliable way to identify freshwater crabs from this part of Africa is to refer to the original type material of all relevant taxa, a task made more difficult by the fact that the types are deposited in a number of different museums. Important taxonomic characters of the gonopods, mouthparts, pereiopods, and thoracic sternum of the lectotype of *P. walderi* are illustrated, its distributional range revised based on all known specimens (Balss, 1936; Bott, 1955; Cumberlidge, 1997, 1998), and its conservation status discussed (Cumberlidge, 2008, 2011).

METHODS

Carapace width (CW) is the distance across the carapace at the widest point; carapace length (CL) is measured along the median line, from the anterior to the posterior margin; carapace height (CH) is the maximum height of the cephalothorax; front width (FW) is measured along the anterior frontal margin between the orbits. These measurements were made with digital calipers. The following abbreviations are used: e, thoracic episternite; s4/e4, s5/e5, s6/e6, s7/e7, episternal sulci between adjacent thoracic sternites and episternites; G1, first gonopod; G2, second gonopod; s, thoracic sternite; s1/s2, s2/s3, s3/s4, s4/s5, s5/s6, s6/s7, s7/s8, sternal sulci between adjacent thoracic sternites; p1-p5, pereiopods 1-5; NHMW, Naturhistorisches Museum, Vienna, Austria; MRAC, Museum Royal d'Afrique Centrale, Tervuren, Belgium; SMF, Senckenberg Museum, Frankfurt, Germany; SMNH, Swedish Museum of Natural History, Stockholm, Sweden; and ZSM, Zoologische Staatssammlung München, Munich, Germany. All measurements are given in mm. The terminology is adapted from Cumberlidge (1999), and the higher classification used here follows that of Ng et al. (2008). Line drawings were prepared using a Leica MZ 16 binocular microscope. The habitus photographs were taken with a digital camera in combination with a Leitz MZ 95 adapter. Post processing was done using Adobe Photoshop CS5.

SYSTEMATIC ACCOUNT

Family POTAMONAUTIDAE Bott, 1970
Subfamily POTAMONAUTINAE Bott, 1970
Genus *Potamonautes* MacLeay, 1838
Potamonautes walderi (Colosi, 1924)
Walder's freshwater crab
(figs. 1-5, table I)

Potamon (*Potamonautes*) *Walderi* Colosi, 1924: 8-9, fig. 5, 5a, b.
Potamonautes Walder — Balss, 1936: 107-108, fig 2, 3a, b.
Potamon walderi — Chace, 1942: 223.
Potamonautes (*Tripotamonautes*) *walderi* — Bott, 1955: 264, fig. 32, 33, pl. XIII fig. 2a-d.
Potamonautes walderi — Cumberlidge, 1997: 578; Cumberlidge, 1998: 204; Ng et al., 2008: 171; Cumberlidge et al., 2009: appendix item 942.

Type material examined.— Lectotype, adult male (CW 36, CL 25, CH 13, FW 11) SMNH 11817 (formerly 6388), Kingoyo (5.1500S, 13.9167E) in the Lower Congo River basin, D. R. Congo, J. A. Wahlberg, no date.

Other material.— DEMOCRATIC REPUBLIC OF THE CONGO. Adult male, CW 29.9, ZSM 1208/l, Kai Ndunga, Mayumbe District, between Boma and Tshela, Congo River mouth, coll. H. Schouteden, 10 Oct. 1920. Adult male, CW 28.4, subadult male, CW 22.5, ZSM 1208/2, Kai Ndunga, Mayumbe District, between Boma and Tshela, Congo River mouth, coll. H. Schouteden, Oct. 1920. Five males, 2 females, MRAC 33022-33028, Kindumba-N'Goma, Shiloango River, no date, 6 males, 1 female, MRAC 33030-33036, Kindumba-N'Goma, Shiloango River, no date. One male, 3 juvenile females, MRAC 32921-32923, 4 juvenile males, 3 juvenile females, MRAC 33013-33021, 2 juvenile males, 3 juvenile females, MRAC 33455-33459, 2 males (1 adult, 1 juvenile), 2 females (1 adult, 1 juvenile), MRAC 38373-38376, Gombe Matadi, Luazi River, no date. Two juvenile males, 2 juvenile females, MRAC 31348, Ganda Sundi, no date. One juvenile female, MRAC 31349, Tshela, Lubuzi River. No date. REPUBLIC OF THE CONGO: Adult, CW 29, subadult, CW 25, NHMW 13293, NE de Pointe Noire, Route de Sounda, coll. A. Crosnier, 18 Jun. 1964. Juvenile male, CW 18.5, NHMW 13294, northeast of Pointe Noire, Route de Sounda, coll. A. Crosnier, 26 Aug. 1964. Subadult female, CW 23, 2 juvenile males, CWs 15.5, 17.5, NHMW 13295, northeast of Pointe Noire, coll. A. Crosnier. Two subadult males, CWs 20, 25.5, NHMW 13296, northeast of Pointe Noire, Route de Sounda, coll. A. Crosnier, no date. One male, MRAC 22196, Kaye, km 101 du Chemin de fer Congo-Océan (CFCO), Mayumbe District, coll. E. Dartevelle, Jun. 1938.

Diagnosis.— Postfrontal crest sharp-edged, complete, lateral ends meeting anterolateral margins, epibranchial tooth lacking (figs. 1A, 3A), anterolateral margin smooth posterior to epibranchial tooth (figs. 1A, 3A), distal part of G1 terminal article strongly upcurved, ending in pointed tip, terminal article conspicuously widened in middle by raised medial, lateral lobes (figs. 2E, 4D-F), ischium of third maxilliped smooth, lacking visible sulcus (fig. 4A).

Redescription.— Postfrontal crest complete, epigastric lobes continuous with postorbital crests, lateral ends of postorbital crests meeting anterolateral margins. Exorbital tooth small, low; epibranchial tooth lacking; anterolateral margin between exorbital and epibranchial teeth short, smooth, lacking intermediate tooth;

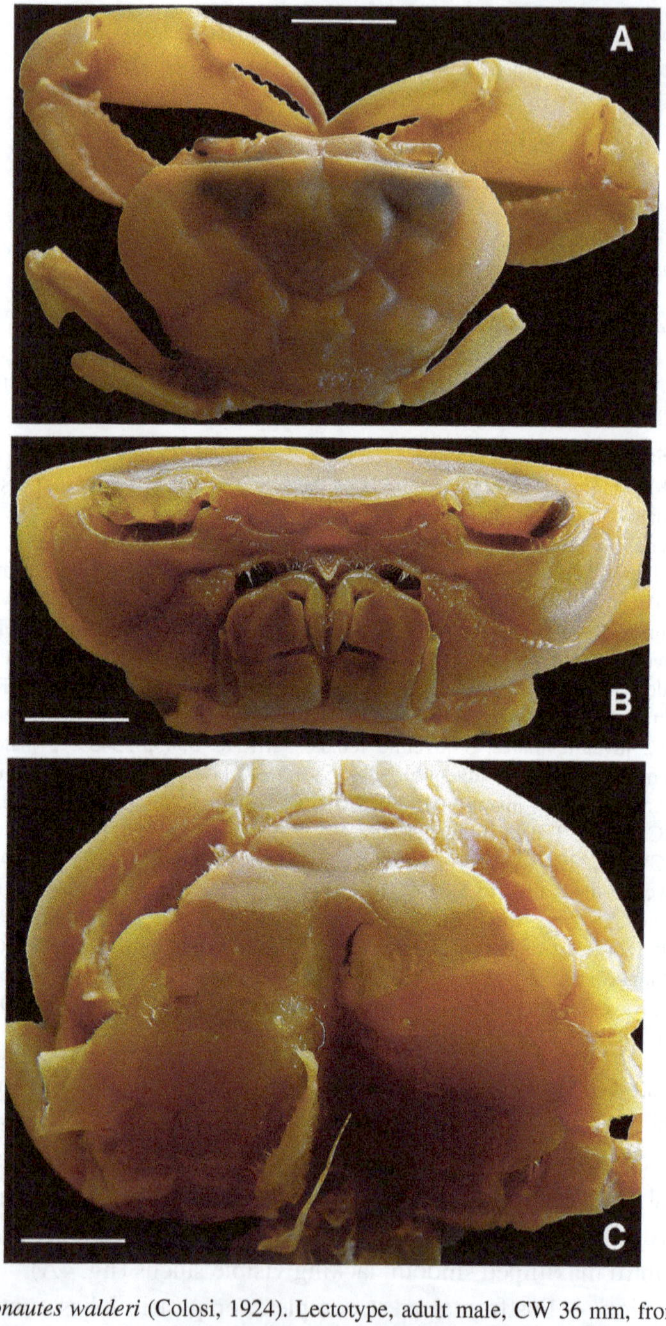

Fig. 1. *Potamonautes walderi* (Colosi, 1924). Lectotype, adult male, CW 36 mm, from the Congo River at Kingoyo, D. R. Congo (SMNH 6386), whole animal. A, dorsal view; B, frontal view; C, ventral view. Scale bar: 12 mm (A-C).

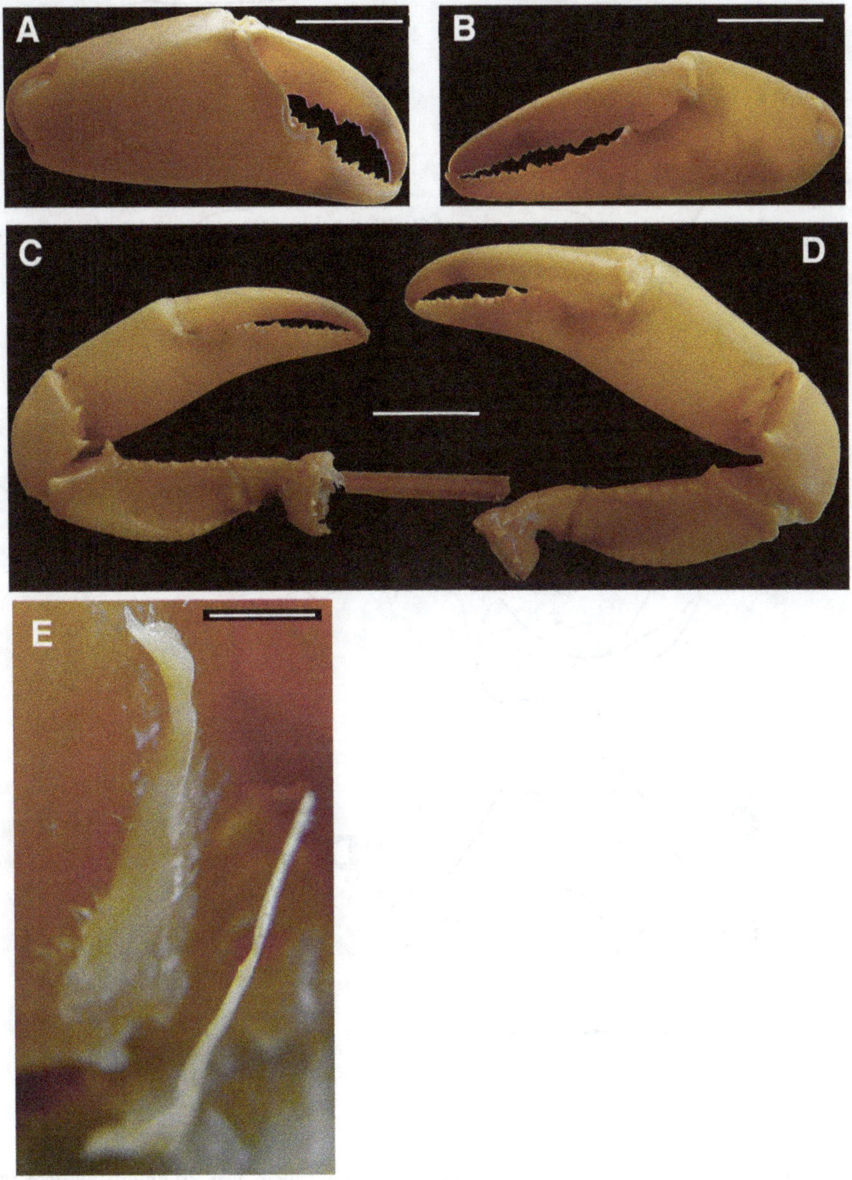

Fig. 2. *Potamonautes walderi* (Colosi, 1924). Lectotype, adult male, CW 36 mm, from the Congo River at Kingoyo, D. R. Congo (SMNH 6386). A, right cheliped, frontal view; B, left cheliped, frontal view; C, merus, carpus, fingers of left cheliped, dorsal view; D, merus, carpus and fingers of right cheliped, dorsal view; E, ventral view with abdomen pulled back showing right first gonopod (right), and right second gonopod (left). Scale bar: 10.2 (C-D), 7.8 (A-B), 8.48 mm (E).

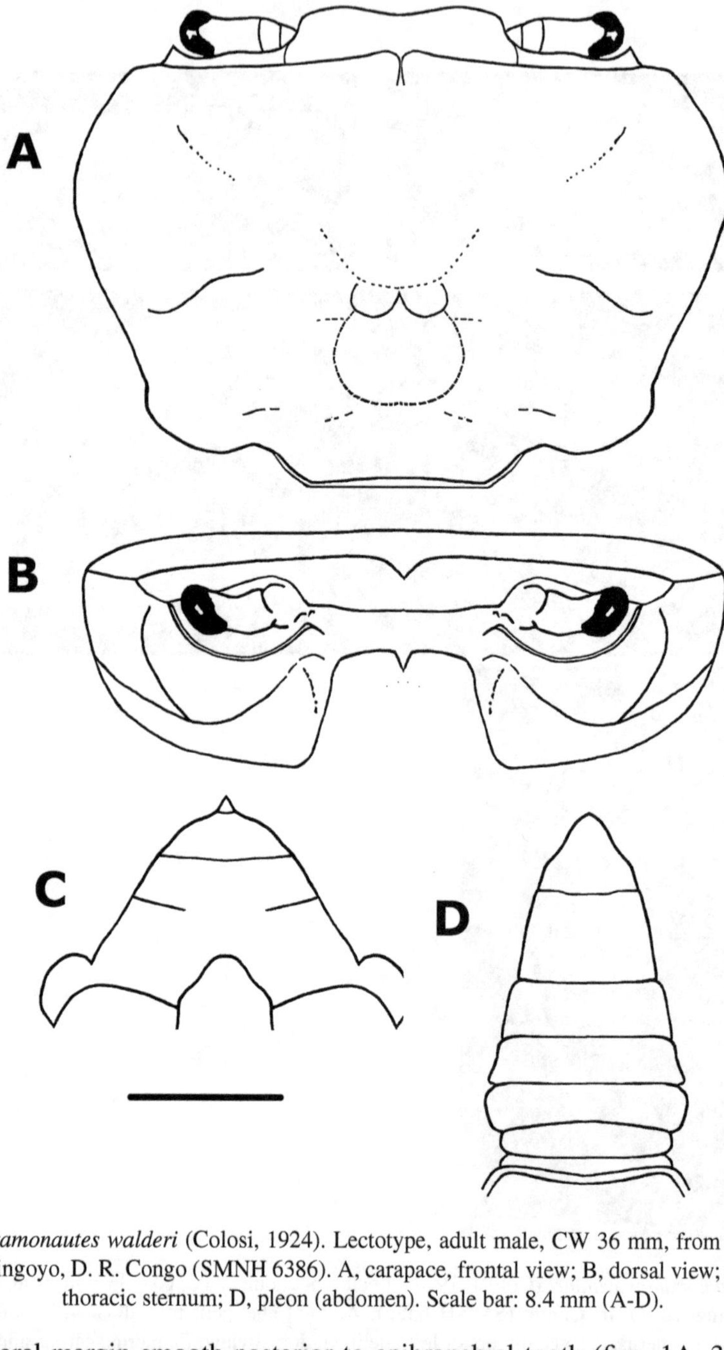

Fig. 3. *Potamonautes walderi* (Colosi, 1924). Lectotype, adult male, CW 36 mm, from the Congo River at Kingoyo, D. R. Congo (SMNH 6386). A, carapace, frontal view; B, dorsal view; C, anterior thoracic sternum; D, pleon (abdomen). Scale bar: 8.4 mm (A-D).

anterolateral margin smooth posterior to epibranchial tooth (figs. 1A, 3A). Suborbital margin smooth. Carapace medium height (CH/FW 1.18), front almost one-third width of carapace (FW/CW 0.31) (figs. 1B, 3B). Semi-circular, cervical carapace grooves faint, urogastric, cardiac, posterior carapace grooves all dis-

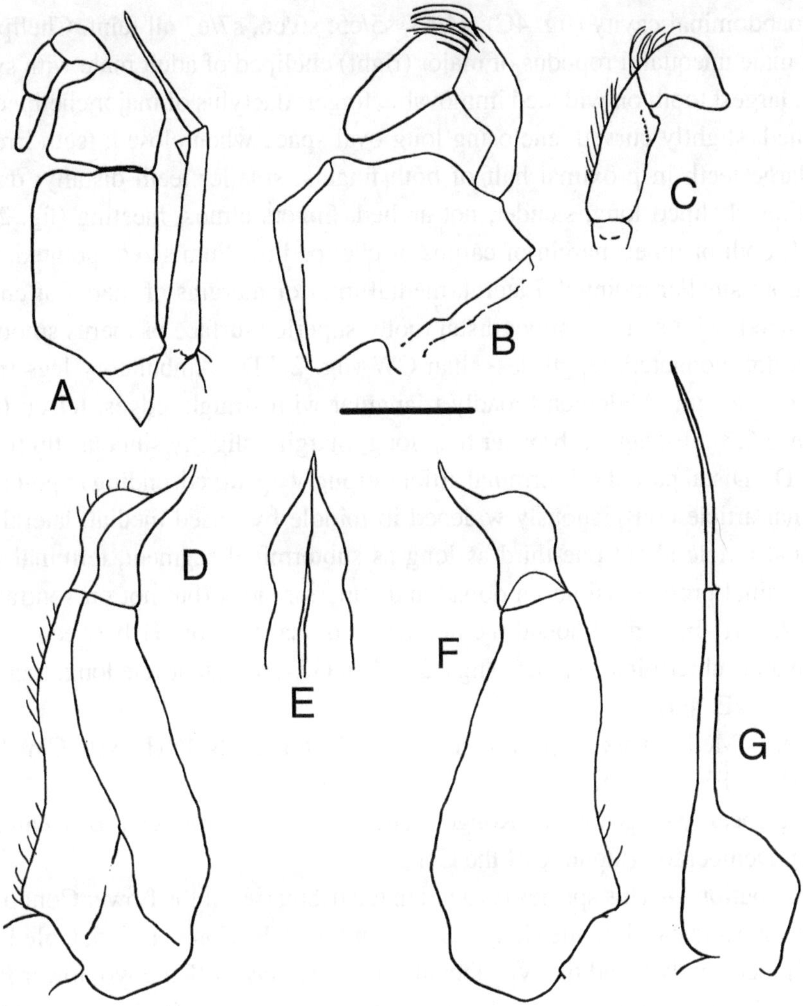

Fig. 4. *Potamonautes walderi* (Colosi, 1924). Lectotype, adult male, CW 36 mm, from the Congo River at Kingoyo, D. R. Congo (SMNH 6386). A, third maxilliped frontal view; B, mandible frontal view; C, mandibular palp superior view; D, G1 ventral view; E, superior view of terminal article; F, G1 dorsal view; G, G2 ventral view. Scale bar: 8.4 mm (A), 2.1 mm (B-G).

tinct. Epigastric crests clear, median sulcus between crests short, forked posteriorly (figs. 1A, 3A). Sidewall of carapace smooth, with distinct vertical sulcus, meeting longitudinal sulcus, dividing sidewall into three parts (subhepatic, subbranchial, pterygostomial regions). Exopod of third maxilliped with long flagellum, ischium smooth, lacking visible sulcus (fig. 4A). Epistomial tooth large, triangular (figs. 1B, 3B). Mandibular palp two-segmented, terminal segment simple (fig. 4B-C). S2/s3 deep, running horizontally across sternum; s3/s4 incomplete, deep at sides, absent in middle, sides slanted inward toward anterior margin of sternoabdominal cavity (fig. 4C); s4/e4, s5/e5, s6/e6, s7/e7 all faint. Chelipeds of adult male unequal. Propodus of major (right) cheliped of adult male with swollen palm, largest tooth on widened immovable finger; dactylus of major cheliped long, widened, slightly curved, enclosing long oval space when closed; teeth irregular, two large teeth in proximal half of both fingers, smaller teeth distally; dactylus of minor cheliped long, slender, not arched, fingers almost meeting (fig. 2A-D). Distal tooth of inner margin of carpus of cheliped medium sized, pointed, proximal tooth smaller, pointed. Lateral, medial inferior margins of merus of cheliped granulated, with single distinct distal tooth; superior surface of merus smooth; p1 merus not elongated, length less than CW (fig. 2C-D). Ambulatory legs (p2-p5) moderate length. Abdomen broadly triangular with straight edges; telson (a7) as long as a6, subtriangular, broader than long, margins slightly sinuous, tip rounded (fig. 3D). Distal part of G1 terminal article strongly upcurved, ending in pointed tip, terminal article conspicuously widened in middle by raised medial, lateral lobes terminal article about one-third as long as subterminal segment, terminal article longitudinal groove visible on dorsal and superior sides (but not on ventral side) (figs. 2E, 4D-F). Broad dorsal membrane on dorsal face of G1 between terminal article and subterminal segment (figs. 2E, 4F). G2 terminal article long, flagellum-like (figs. 2E, 4G).

Size.— Medium sized species, adult male lectotype, SMNH 6386, CW 36, CL 25, CH 13, FW 11 mm.

Type locality.— Kingoyo, Kongo Central Province (formerly Bas-Zaire Province), Democratic Republic of the Congo.

Distribution.— This species is found in the tributaries of the Lower Congo River at Kingoyo and east into the Congo district of the D. R. Congo (fig. 5, table I). Bott (1955) incorrectly listed the type locality of *P. walderi* as 'Kingoyo, lower French Congo': Kingoyo is actually in the Bas-Zaire district of the D. R. Congo. Most of Bott's (1955) material listed under *P. walderi* has been included here: Kindumba-N'Goma, Shiloango River (MRAC 33022-33028, 33030-33036), Gombe Matadi, Luazi River (MRAC 32921-32923, 33013-33021, 33455-33459, 38373-38376), Ganda Sundi (MRAC 31348), Tshela, Lubuzi River (MRAC 31349), and Kaye, Mayumbe (MRAC 22196). The three localities for *P. walderi* in the Upper Congo River in the D. R. Congo reported by Cumberlidge (2008) have not been included here due to doubts over the validity of the identification of these specimens.

Habitat.— Rivers and streams in the Lower Congo River basin.

Conservation status.— *Potamonautes walderi* was listed as Least Concern (IUCN 2003; Cumberlidge et al., 2009; Cumberlidge, 2011) because it was then thought to occur widely throughout the Congo River basin in the D. R. Congo, and because there were no known long-term threats to its habitat from disturbance

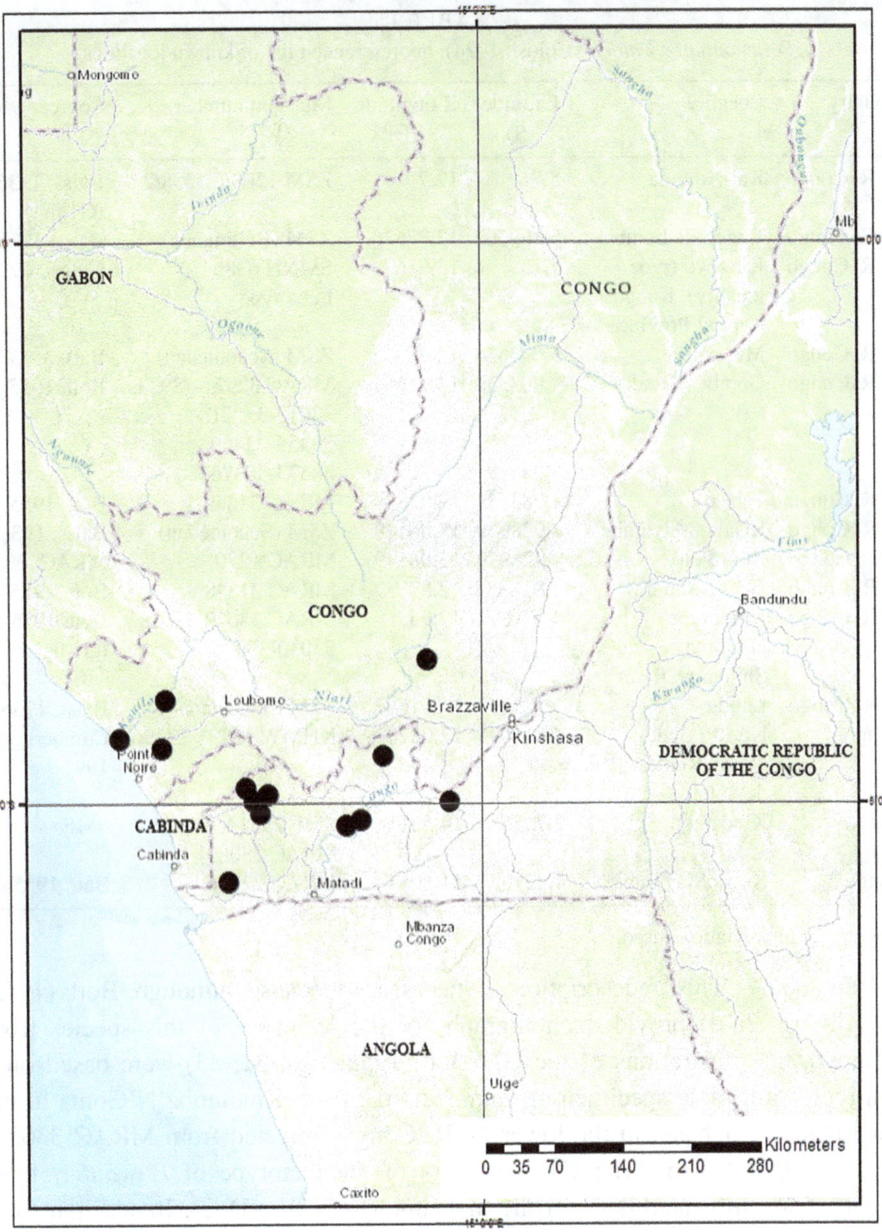

Fig. 5. Map of the Congo Republic and D. R. Congo in Central Africa showing the known localities of *Potamonautes walderi* (Colosi, 1924) (black circles).

and pollution (Cumberlidge, 2008). The present study provides a revised (smaller) Extent of Occurrence (EOO) for this species (47 180 km^2, http://geocat.kew.org) and adds new sites from the Republic of the Congo (fig. 5).

TABLE I

Potamonautes walderi (Colosi, 1924): georeferenced list of known localities

Country	Locality	Latitude (S)	Longitude (E)	Museum number	Source
D. R. Congo	Kai-Ndunda	5.7	12.7	ZSM 1208/1, 1208/2	Balss, 1936 (G1 fig.)
D. R. Congo	Congo da Lemba	5.199002	13.77526	ZSM (Bellinghen)	Balss, 1936
D. R. Congo	Kingoyo (type locality), Kongo Central Province	5.15	13.9167	SMNH 6386 Lectotype	Colosi, 1924
D. R. Congo	Mayumbe	5.093634	12.98962	ZSM (Schouteden)	Balss, 1936
D. R. Congo	Gombe Matadi	4.983333	14.71667	MRAC 32921-32923, 33013-33021, 33455-33459, 38373-38376	Bott, 1955
D. R. Congo	Tshela	4.98333	12.9333	MRAC 31349	Bott, 1955
D. R. Congo	Kisala, Mayumbe	4.928979	13.06149	ZSM (Schouteden)	Balss, 1936
D. R. Congo	Butu Polo	4.928979	13.06149	MRAC 1270	MRAC
D. R. Congo	Ganda Sundi	4.87	12.87	MRAC 31348	Bott, 1955
D. R. Congo	Kindumba-N'Goma, Shiloango River	4.77	12.93	MRAC 33022-33028, 33030-33036	Bott, 1955 (G1 figs.)
D. R. Congo	Lundu	4.583333	14.11667	ZSM (Schouteden)	Balss, 1936
Congo	NE of Pointe Noire, Route de Sounda	4.515095	12.08702	NHMW 13293-13296	Cumberlidge, 1997
Congo	Kindamba	3.7275	14.5211	SMF 2381A (ex. MRAC 33080)	SMF
Congo	Kaye, Mayumbe	4.43167	11.6961	MRAC 22196	Bott, 1955

See text for abbreviations used.

Remarks.— This redescription is necessary because although Bott (1955, pl. XIII fig. 2a-d) provided photographs of the lectotype of this species from Kingoyo, his illustrations of the G1 (Bott, 1955, figs. 32, 33) were based on a non-type adult male specimen of unknown size from Kindumba N'Goma in the Shiloango River basin in the lower D. R. Congo (selected from MRAC 33022-33028, 33030-33036). The G1 characters of the lectotype of *P. walderi* from Kingoyo are illustrated here for the first time (figs. 2E, 4D-G). In addition, Bott (1955) described the ischium of the third maxilliped of *P. walderi* to have a distinct vertical sulcus, whereas that of the lectotype is smooth and lacks a visible sulcus (figs. 1B, 4A). Bott (1955) also described the p1 merus of this species as being very elongated (longer than the CW) but this was not found to be the case here. It is possible that Bott's (1955) description of elongated chelipeds is an artifact arising from the fact that both chelipeds of the lectotype are detached and had been carefully positioned and temporarily reattached for the photo (Bott, 1955, pl. Xlll, fig. 2a-d) making the cheliped meri appear to be longer than they are. Other differences between Bott's (1955) description and the present work include the carapace sidewall which is divided onto three parts by two sulci (rather than two parts divided by one sulcus, according to Bott (1955)). Moreover, Bott's (1955) classification of the large African genus *Potamonautes* is also questionable, because he placed *P. walderi* in the subgenus *P.* (*Tripotamonautes*) Bott, 1955, as the type species of a subgenus that also included *P. loveridgei* (Rathbun, 1933) from Tanzania. However, my examination of the holotype of *P.* (*P.*) *loveridgei* Rathbun, 1933, from the Uzungwe Mountains, Tanzania (MCZ 7676) for this study casts doubt on a close relationship between these two species given the significant morphological differences in the gonopods and somatic characters of these two taxa, and the large geographical distance separating river drainages in Tanzania from the Lower Congo River basin (Reed & Cumberlidge, 2006). Further, Cumberlidge (1997, 1998), Ng et al. (2008) and Cumberlidge et al. (2009) all recognized *P. walderi* as a valid species of *Potamonautes* but they did not accept the subgenus assignment.

Comparisons.— *Potamonautes walderi* can be recognized by the characters listed above in the diagnosis and by the illustrations and photographs of the lectotype from Kingoyo provided here (figs. 1-4) and by Bott (1955, pl. Xlll fig. 2a-d). Both Balss (1936, fig. 3a-b) and Bott (1955, figs. 32, 33) provided a sketch of the G1 of different non-type specimens of *P. walderi* from the Lower Congo River basin in the D. R. Congo (from Kai-Ndunda and Kindumba-N'Goma respectively) that resemble those of the lectotype in most respects.

Potamonautes walderi is superficially similar to six other species of *Potamonautes* that occur in the Congo River basin in Central Africa: *P. congoensis* (Rathbun, 1921), *P. stanleyensis* (Rathbun, 1921), *P. lueboensis* (Rathbun, 1904), *P. dybowskii* (Rathbun, 1905), *P. emini* (Hilgendorf, 1892) and *P. perparvus* (Rathbun, 1921). All of these species have a distinct postfrontal crest across the carapace that reaches the anterolateral margins, and smooth and untoothed anterolateral carapace margins (Bott, 1955; Cumberlidge, 2011, 2015). *Potamonautes walderi* can be distinguished from these species by its G1 (which has extremely widened rounded lobes on the terminal article), and by the ischium of the third maxilliped (which lacks a vertical sulcus). *Potamonautes walderi* can also be distinguished on geographical grounds because it is restricted to the Lower Congo River basin in the D. R. Congo and Congo, while these other species are only found either to the east in the Upper Congo River basin, or in the Middle Congo River basin in the southwest of the D. R. Congo (Meyer & Cumberlidge, 2011).

Potamonautes walderi is closest morphologically to *P. lueboensis* in that in addition to sharing a complete postfrontal crest and smooth anterolateral margins of the carapace, both species both have a third maxilliped ischium that is smooth and lacking a vertical sulcus. Unfortunately, additional characters of the gonopod, thoracic sternum, and major cheliped are not available for *P. lueboensis* because the holotype of *Potamon* (*Potamonautes*) *lueboensis* Rathbun, 1904 (CW 40.6, CL 30, FW 10.9 mm) is a female. Nevertheless, these two species can be separated on geographical grounds — *P. walderi* is restricted to the Lower Congo River basin, while *P. lueboensis* is restricted to Luebo in the Kasai Province of southern-central D. R. Congo, is not known to occur elsewhere in the Congo River basin (Bott, 1955) and was last collected in 1934.

Potamonautes walderi can be distinguished from *P. stanleyensis* and *P. congoensis* from the Upper Congo River basin by the lack of a vertical sulcus on the third maxilliped ischium (which is deep in *P. stanleyensis* and *P. congoensis*) (Cumberlidge, 2015). *Potamonautes walderi* can be distinguished from *P. dybowskii* from Bangui, Central African Republic, in the Central Congo River basin by s3/s4, which is deep at both sides and shallow in the middle in *P. walderi* (figs. 1C, 3C), but deep and completely crosses the sternum in *P. dybowskii* (Rathbun, 1921; Capart, 1954; Bott, 1955; Cumberlidge, 1998). *Potamonautes walderi* can be distinguished from *P. emini* from the Upper Congo by the G1 terminal article which is distinctly widened in *P. walderi* (figs. 1E, 4D-E), but slim and not significantly widened in *P. emini* (Reed & Cumberlidge, 2006). *Potamonautes walderi* can be distinguished from *P. perparvus* from the Upper Congo by s3/s4, which is deep at the sides but absent in the middle in *P. walderi* (figs. 1C, 3C), but completely crossing the sternum in *P. perparvus* (Meyer & Cumberlidge, 2011).

Molecular data from a broad sampling of Afrotropical freshwater crabs provided by Daniels et al. (2015) indicated that the large genus *Potamonautes* comprises at least four separate evolutionary lineages. Interestingly, species from the Lower Congo River basin (such as *P. ballayi* (A. Milne-Edwards, 1886)) belong to a separate lineage to species from the Upper Congo River basin (such as *P. langi* (Rathbun, 1921) and *P. stanleyensis* (Rathbun, 1921)) (Daniels et al., 2015). Unfortunately, DNA sequence data are still deficient for a large number of species from the Congo River basin (including *P. walderi*) and further sampling from this part of Africa is clearly needed.

ACKNOWLEDGEMENTS

This work is dedicated to the memory of Dr. Michael Türkay for his numerous contributions to the taxonomy of decapod crustaceans that includes many important publications on freshwater crabs. Michael Türkay was also a valued friend and colleague who will be missed by the whole scientific community. K. Sindemark (SMNH), A. Baldinger (MCZ), R. Jocqué (MRAC) and L. Tiefenbacher (ZSM) are each thanked for hosting a visit to their museum by the author and/or for loaning specimens that were used in this work.

REFERENCES

BALSS, H., 1936. Beitrage zur Kenntnis der Potamonidae (Süßwasserkrabben) des Kongogebietes. Rev. Zool. Bot. d'Afrique, **28**: 165-204, figs. 1-29.

BOTT, R., 1955. Die Süßwasserkrabben von Afrika (Crust., Decap.) und ihre Stammesgeschichte. Ann. Mus. Congo belge, (Tervuren, Belgique) C-Zoologie, (3,3) **3**(1): 209-352.

BOTT, R., 1970. Betrachtungen uber die Entwicklungsgeschichte der Süßwasserkrabben nach der Sammlung des Naturhistorischen Museums in Genf/Schweiz. Rev. Suisse Zool., **77**: 327-344, pls. 1, 2.

CAPART, A., 1954. Révision des types des espèces de Potamonidae de l'Afrique Tropicale conserves au Muséum d'Histoire Naturelle de Paris. Volume Jubilaire de Victor Van Strallen, Director de l'Institut royal des Sciences naturelles de Belgique, 1925–1934, **II**: 819-847.

CHACE, F. A., 1942. Scientific results of a fourth expedition to forested areas in eastern Africa, m. Decapod Crustacea. Bull. Mus. Comp. Zool., Harvard College, **91**: 185-233.

COLOSI, G., 1924. Potamonides africains du Muséum de Stockholm. Arkiv für Zoologie, **16**: 1-24.

CUMBERLIDGE, N., 1997. The African and Madagascan freshwater crabs in the Museum of Natural History, Vienna (Crustacea: Decapoda: Brachyura: Potamoidea). Ann. Naturhist. Mus. Wien, **99B**: 571-589.

CUMBERLIDGE, N., 1998. The African and Madagascan freshwater crabs in the Zoologische Staatssammlung, Munich (Crustacea: Decapoda: Brachyura: Potamoidea). Spixiana, **21**: 193-214.

CUMBERLIDGE, N., 1999. The freshwater crabs of west Africa. Family Potamonautidae. Faune et Flore Tropicales **35**: 1-382. (Institut de recherche pour le développement IRD (ex-ORSTOM), Paris).

CUMBERLIDGE, N., 2008. *Potamonautes walderi*. The IUCN Red List of Threatened Species 2008: e.T134247A3925382. Available online at http://dx.doi.org/10.2305/IUCN.UK.2008.RLTS.T134247A3925382.en (accessed 30 August 2016).

CUMBERLIDGE, N., 2011. The status and distribution of freshwater crabs, pp. 71-78. Chapter 6. In: E. G. E. BROOKS, D. J. ALLEN & W. R. T. DARWALL (Compilers). The Status and Distribution of Freshwater Biodiversity in Central Africa. (IUCN, Gland).

CUMBERLIDGE, N., P. K. L. NG, D. C. J. YEO, C. MAGALHÃES, M. R. CAMPOS, F. ALVAREZ, T. NARUSE, S. R. DANIELS, L. J. ESSER, F. Y. K. ATTIPOE, F.-L. CLOTILDE-BA, W. R. T. DARWALL, A. MCIVOR, J. E. M. BAILLIE, B. COLLEN & M. RAM, 2009. Freshwater crabs and the biodiversity crisis: importance, threats, status, and conservation challenges. Biol. Cons., **142**: 1665-1673.

CUMBERLIDGE, N., R. V. STERNBERG & S. R. DANIELS, 2008. A revision of the higher taxonomy of the Afrotropical freshwater crabs (Decapoda: Brachyura) with a discussion of their biogeography. Biol. J. Linn. Soc., **93**: 399-413.

DANIELS, S. R., E. PHIRI, S. KLAUS, C. ALBRECHT & N. CUMBERLIDGE, 2015. Multi-locus phylogeny of the Afrotropical freshwater crab fauna reveals an Eocene cladogenesis and the impact of paleodrainage rearrangements. Syst. Biol., **64**(6): 549-567.

HILGENDORF, F., 1869. Ober eine neue Art der kurzschwanzigen Krebse aus den Sammlungen des Baron von der Decken, *Deckenia imitatrix*. Sitz.-Bericht Gesells. nat. Freunde zu Berlin, **1868**(1): 2.

HILGENDORF, F., 1892. Über eine neue ostafrikanische Süßwasserkrabbe (*Telphusa emini*). Sitz.-Bericht Gesells. nat. Freunde zu Berlin, **1892**(1): 11-13.

IUCN, 2003. Guidelines for application of IUCN Red List criteria at regional levels: version 3.0. IUCN Species Survival Commission. Gland, Switzerland and Cambridge, UK: IUCN.

MACLEAY, W. S., 1838. Brachyurous Decapod Crustacea, illustrations of the zoology of South Africa 5; being a portion of the objects of natural history chiefly collected during an expedition into the interior of South Africa, under the direction of Dr. Andrew Smith, in the years 1834,

1835, and 1836; fitted out by "The cape of Good Hope Association for exploring Central Africa." In: A. SMITH, Illustrations of the zoology of South Africa; consisting chiefly of figures and descriptions of the objects of natural history collected during an expedition into the interior of South Africa, in the years 1834, 1835, and 1836; fitted out by "The cape of Good Hope Association for exploring Central Africa", 5, Invertebrata, **3**: 53-71.

MEYER, K. S. & N. CUMBERLIDGE, 2011. A revision of the freshwater crabs (Crustacea: Decapoda: Brachyura: Potamonautidae) of the Lake Kivu drainage basin in Central and East Africa. Zootaxa, **3011**: 45-58.

NG, P. K. L., D. GUINOT & P. J. F. DAVIE, 2008. Systema Brachyurorum: Part 1. An annotated checklist of extant brachyuran crabs of the world. Raffles Bull. Zool., (Suppl.) **17**: 1-286.

RATHBUN, M. J., 1904. Les crabes d'eau douce (Potamonidae). Nouv. Arch. Mus. Hist. nat. Paris, **6**(4): 255-312.

RATHBUN, M. J., 1905. Les crabes d'eau douce (Potamonidae). Nouv. Arch. Mus. Hist. nat. Paris, **7**(4): 159-322.

RATHBUN, M. J., 1921. Brachyuran crabs of the Belgian Congo. Bull. Am. Mus. Nat. Hist., **43**: 379-468, pls. 15-64, figs. 1-33.

RATHBUN, M. J., 1933. Reports on the scientific results of an expedition to the southwestern highlands of Tanganyika territory. V. Crabs. Bull. Mus. Comp. Zool. Harvard College, **75**(5): 250-262, pls. 7.

REED, S. K. & N. CUMBERLIDGE, 2006. Taxonomy and biogeography of the freshwater crabs of Tanzania, east Africa (Brachyura: Potamoidea: Potamonautidae, Platythelphusidae, Deckeniidae). Zootaxa, **1262**: 1-139.

First received 2 September 2016.
Final version accepted 18 November 2016.

[When citing this volume, please refer to Crustaceana 90 (2017) 771-1288]

Michael Türkay Memorial Issue

POLYMORPHISM IN THE CHELAE OF MATURE MALES OF THE LAND CRABS *JOHNGARTHIA LAGOSTOMA* AND *EPIGRAPSUS* SPP.

BY

RICHARD G. HARTNOLL[1,5]), NICOLA WEBER[2,3]), SAM B. WEBER[2,3]) and HUNG-CHANG LIU[4])

[1]) School of Environmental Sciences, University of Liverpool, Liverpool L69 3BX, U.K.
[2]) Centre for Ecology and Conservation, University of Exeter, Penryn, Cornwall TR10 9EZ, U.K.
[3]) Ascension Island Government Conservation Department, Georgetown, Ascension Island
[4]) No. 53, Chengong 11th Street, Jhubei City, Hsinchu County, Taiwan 302

ABSTRACT

The growth of the chelae of mature males was examined in three gecarcinid land crabs — *Johngarthia lagostoma* on Ascension Island, *Epigrapsus notatus* on Taiwan, and *E. politus* on Moorea. Chelar dimorphism was found in each species, with a mixture of homochelous and heterochelous males. In *J. lagostoma* there was progressive polymorphism, with the heterochelous condition appearing only in a proportion of the larger mature males. The situation in *Epigrapsus* was less clear. The proportion of heterochelous males increased in the larger mature size classes, but progressive polymorphism is yet to be confirmed.

ZUSAMMENFASSUNG

Das Wachstum der Scheren adulter Männchen wurde bei drei gecarciniden Landkrabben untersucht — *Johngarthia lagostoma* von Ascension, *Epigrapsus notatus* von Taiwan und *E. politus* von Moorea. Scherendimorphismus wurde mit einer Mischung aus Homo- und Heterochelie bei den Männchen jeder Art gefunden. Bei *J. lagostoma* trat fortgeschrittener Polymorphismus auf, wobei lediglich ein Teil der größeren, geschlechtsreifen Männchen Heterochelie aufwies. Der Sachverhalt bei *Epigrapsus* war weniger klar. Der Anteil heterocheler Männchen nahm mit zunehmenden Größenklassen geschlechtsreifer Tiere zu, aber der Nachweis von progressivem Polymorphismus steht noch aus.

INTRODUCTION

Johngarthia lagostoma (H. Milne Edwards, 1837) occurs only on four small Atlantic Islands: Ascension Island in the Central South Atlantic, and the offshore West Atlantic Brazilian islands of Trindade, Fernando de Noronha, and Atol das

[5]) Corresponding author; e-mail: hartnoll@manx.net

Rocas (Hartnoll et al., 2006b). The biology of the species in Ascension Island has been described in Hartnoll et al. (2006a, b, 2009, 2010, 2014). *Epigrapsus notatus* (Heller, 1865) and *E. politus* Heller, 1862 are broadly distributed on islands in the Indo-West Pacific (Türkay, 1974). The only previous study is of reproduction in *E. notatus* in Taiwan (Liu & Jeng, 2005). The growth of the chelae has not been previously investigated in any of the above.

This study investigates chelar growth in these species. Early in a study of chelar growth in *J. lagostoma* it became apparent that the chelae of mature males might show progressive polymorphism, a situation whereby different chelar morphologies succeed each other within the mature phase. Such a condition has not been previously described for the Brachyura, though known for some macrurans (Hartnoll, 2012). Subsequent examination of data on the two species of *Epigrapsus* also indicated polymorphism in mature males. The aim of this paper is to describe chelar growth in the three species, to confirm or refute progressive polymorphism in mature males of *J. lagostoma*, and to clarify the situation in the two species of *Epigrapsus*.

MATERIAL AND METHODS

All measurements were made on living specimens in the field, mostly at night: they were released unharmed at the site of capture. This constrained the measurements which could be accurately taken. For the small species of *Epigrapsus* the only readily determined chelar measure was propodus height.

Johngarthia lagostoma

Crabs were measured on Ascension Island, South Atlantic Ocean, in March and April 2014. They were measured alive at night in the field, and released unharmed at the same location. Females (32) were measured in the main breeding area of North East Bay. Males (60) were also measured there, and an additional 31 in the permanent residential area along the NASA Road (see Hartnoll et al., 2010 for site locations). The larger number of males does not reflect the population sex ratio, which is near unity (Hartnoll et al., 2006, 2009): the greater complexity of male chelar growth required a larger data set. The crabs were wrapped tightly in a plastic bag whilst being measured, and the chelae exposed one at a time. Measurements were made with dial callipers, and repeat measures agreed to 0.1 mm.

The sex and colour type (see Hartnoll et al., 2006) were recorded for each crab, and the following measurements were made as specified (the abbreviations will be used in the text and tables where appropriate). Unless otherwise stated, 'chela' implies chelar propodus.

Maximum carapace width (CW).

Major and minor chelar propodus length, from the ventral articulation with the carpus to the tip of the fixed finger (MaChL; MiChL).

Major and minor chelar propodus height just proximal to the insertion of the dactylus (MaChH; MiChH).

Major and minor maximum chelar propodus width, located about half way along the palm (MaChW; MiChW).

Major and minor chelar propodus gape, as the gap between opposing teeth half way along the dactylus (MaChG; MiChG).

It should be noted that the terminology relating to chelar measurement varies in the literature. The terms used here are correctly related to the topography of the appendage. Chela length is unambiguous. Chela height is measured in the vertical plane, so that the propodus and dactylus are seen in profile. Chela width is measured in the antero-posterior plane (the dactylus being seen from above). Some other studies are vague. Thus 'chela breadth' is not explained in Oyenekan (1995). Others define measurements differently: thus 'propodus width' in Turner et al. (2011) is what we define as 'propodus height'. If comparisons are made, it is necessary to ensure that like is being compared to like.

On females only the CW, MaChL, MiChL, and chela gape were measured. On Type A males (morphotypes are explained in the results) all the above measurements were made on 51 crabs, but only CW and morphotype were recorded for the remainder. All 24 Type B males were fully measured. Several ratios were calculated from these measurements.

Heterochely ratio (HR) as the size of the major chela dimension divided by that of the minor chela.

The ratio of ChH/ChL.

The ratio of ChW/ChL.

Epigrapsus spp.

Epigrapsus notatus was sampled from May 1998 to July 2001 and in September 2014 at Banana Bay (= Hsiangchiaowan), Kenting, Taiwan. Thirty three females, CW 16.9 to 34.8 mm, and 41 males, CW 22.0 to 35.0 mm, were sampled. The smallest mature female found was 16.9 mm CW (Liu & Jeng, 2005), and this was taken as the size of maturity in both sexes.

E. politus was sampled in Moorea in October 2001. Fifty-seven males, CW 6.5 to 18.9 mm, were sampled. The smallest mature female found was 11.1 mm CW, and this was taken as the size of maturity in both sexes.

Additional material of *Epigrapsus* was examined in the Senckenberg Forschungsinstitut und Naturmuseum, Frankfurt (SMF) to compare morphology. Only CW and the height of each chelar propodus were measured for *Epigrapsus* spp.

For both genera regressions of one measurement on another were calculated using $\log_{10}$ transformed data. The slope of such regressions indicates the level of allometry. Regressions of ratios on CW and on chelar gape were calculated using untransformed data.

RESULTS

Details of the linear regressions referred to in the following text are presented in table I.

Johngarthia lagostoma

Female chelae.— A total of 32 females were measured, ranging in size from 63.9 to 101.3 mm CW. This covered most of the recorded size range of mature females, from approx. 60 to 110 mm CW (Hartnoll et al., 2009). There was mild preferential right-handedness ($p<0.05$): 17 were right-handed, 13 left-handed, and 2 equal.

The relative growth of the female chelae is straightforward. There was low positive allometry of 1.16 for the major chela and 1.14 for the minor chela lengths. There was a very low level of heterochely, with ChLHR ranging from 1.00 to 1.04, with a mean value of 1.01. ChLHR did not change with CW — the slope of the regression is 0.0001. With one exception, there was no gape: the largest female had a gape of 0.5 mm for the major chela.

Male chelae.— A total of 91 males were measured, 67 Type A males, ranging from 68.5 to 113.2 mm CW, and 24 Type B, ranging from 95.3 to 116.0 mm CW (see below for definition of 'types'). As for females, sampling covered most of the mature size range, from approx. 60 to 120 mm CW (Hartnoll et al., 2009). As in females there was preferential right-handedness ($p<0.01$): 44 were right-handed, 31 left-handed.

Early in sampling it became clear that growth of male chelae was more complex than that of females. Up to about 100 mm CW all males had chelae of very similar size and shape (fig. 1C, D). These are designated Type A morphs. Above about 100 mm CW around half of the males retained this morphology. However, the others transformed to Type B morphs. These had strong heterochely in length, and marked differences in chelar morphology (fig. 1A, B). The major chela had a pronounced gape, and reduced dentition on both the dactylus and fixed finger. The minor chela had no or very little gape, and the teeth were not reduced: apart from size (see below), it was little different from the chelae of a Type A male of the same CW. Both morphs occurred in both colour forms, and with either handedness. The two morphs were readily discriminated before measurement, and

TABLE I

Linear regressions of chelar dimensions

Y-variable	X-variable	$\log_{10}$	Sex	n	Intercept	Slope	r^2	p	Allometry	
Johngarthia lagostoma (H. Milne Edwards, 1837)										
MaChL	CW	Yes	F	32	−0.541	1.155	0.919	<0.001	+*	($F_{1,31} = 6.1$, $p = 0.02$)
MiChL	CW	Yes	F	32	−0.526	1.144	0.917	<0.001	+*	($F_{1,31} = 5.3$, $p = 0.03$)
ChLHR	CW	No	F	32	0.0001	1.002	0.014	0.91	n.a.	
MaChL	CW	Yes	Ma	51	−1.3820	1.6392	0.951	<0.001	+*	($F_{1,49} = 145.1$, $p<0.001$)
MiChL	CW	Yes	Ma	51	−1.3468	1.6155	0.927	<0.001	+*	($F_{1,49} = 107.7$, $p<0.001$)
ChLHR	CW	No	Ma	51	1.0044	0.0002	0.0102	0.45	n.a.	
MaChH	CW	Yes	Ma	50	−0.6966	1.1064	0.852	<0.001	0	($F_{1,48} = 2.5$, $p = 0.12$)
MiChH	CW	Yes	Ma	50	−0.7426	1.1221	0.755	<0.001	0	($F_{1,48} = 1.9$, $p = 0.18$)
ChHHR	CW	No	Ma	50	1.0668	0.0003	0.0025	0.77	n.a.	
MaChH/ChL	CW	No	Ma	50	0.6661	−0.0025	0.6724	<0.001	n.a.	
MiChH/ChL	CW	No	Ma	50	0.6463	−0.0023	0.615	<0.001	n.a.	
MaChW/ChL	CW	No	Ma	50	0.4501	−0.0019	0.6729	<0.001	n.a.	
MiChW/ChL	CW	No	Ma	50	0.3671	−0.0014	0.5474	<0.001	n.a.	
MaChL	CW	Yes	Mb	23	−1.4672	1.7041	0.4988	<0.001	+*	($F_{1,22} = 3.6$, $p = 0.07$)
MiChL	CW	Yes	Mb	24	−1.4248	1.6321	0.1703	0.045	+?	($F_{1,23} = 0.68$, $p = 0.42$)
ChLHR	CW	No	Mb	23	1.0198	0.0002	0.0037	0.78	n.a.	
MaChH	CW	Yes	Mb	24	−0.3901	0.9656	0.272	0.009	0	($F_{1,23} = 0.01$, $p = 0.92$)
MiChH	CW	Yes	Mb	24	−1.4012	1.3975	0.245	0.014	+?	($F_{1,23} = 0.58$, $p = 0.45$)
ChHHR	CW	No	Mb	24	1.9621	−0.0056	0.0407	0.34	n.a.	
MaChH/ChL	CW	No	Mb	23	0.7157	−0.0031	0.1533	0.064	n.a.	
MiChH/ChL	CW	No	Mb	23	0.4457	−0.0009	0.0103	0.66	n.a.	
MaChW/ChL	CW	No	Mb	23	0.4512	−0.002	0.1442	0.11	n.a.	
MiChW/ChL	CW	No	Mb	23	0.3624	−0.0015	0.0597	0.25	n.a.	
ChLHR	MaChG	No	Mb	23	1.3213	−0.0121	0.0686	0.23	n.a.	

TABLE I
(Continued)

Y-variable	X-variable	$\log_{10}$	Sex	n	Intercept	Slope	r^2	p	Allometry	
Epigrapsus notatus (Heller, 1865)										
Taiwan										
MaChH	CW	Yes	F	33	−0.8851	1.1727	0.942	<0.001	+*	($F_{1,32} = 10.9$, $p = 0.002$)
MiChH	CW	Yes	F	33	−0.8986	1.1739	0.928	<0.001	+*	($F_{1,32} = 8.8$, $p = 0.006$)
MaChH	CW	Yes	Ma	25	−1.3979	1.7177	0.5651	<0.001	+*	($F_{1,24} = 5.2$, $p = 0.032$)
MiChH	CW	Yes	Ma	25	−1.3645	1.6837	0.5429	<0.001	+*	($F_{1,24} = 4.5$, $p = 0.044$)
MaChH	CW	Yes	Mb	16	−1.2832	1.6445	0.4987	<0.001	+*	($F_{1,10} = 5.2$, $p = 0.048$)
MiChH	CW	Yes	Mb	16	−0.9145	1.2714	0.2907	0.69	+?	($F_{1,10} = 1.0$, $p = 0.35$)
Epigrapsus politus Heller, 1862 Moorea										
MaChH	CW	Yes	Ma	32	−1.3133	1.7659	0.9004	<0.001	+*	($F_{1,30} = 51.0$, $p<0.001$)
MiChH	CW	Yes	Ma	32	−1.3107	1.7566	0.8933	<0.001	+*	($F_{1,30} = 46.6$, $p<0.001$)
MaChH	CW	Yes	Mb	21	−1.7352	2.1506	0.9209	<0.001	+*	($F_{1,19} = 63.3$, $p<0.001$)
MiChH	CW	Yes	Mb	21	−1.2848	1.6118	0.8786	<0.001	+*	($F_{1,19} = 19.8$, $p<0.001$)

As indicated the regressions are, where appropriate, calculated from $\log_{10}$-transformed data. Ma, Type A males; Mb, Type B males. Significance of allometry level different from 1.0 indicated by an asterisk. +, positive allometry; 0, isometry; +?, putative positive allometry, but not significantly different from 1.0 (see Discussion).

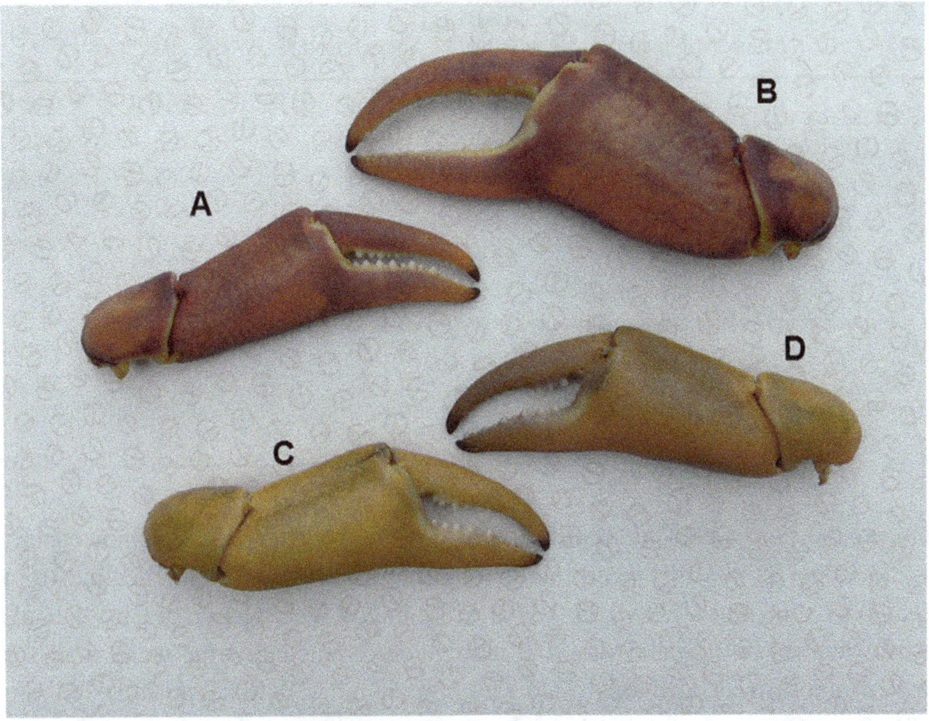

Fig. 1. *Johngarthia lagostoma* (H. Milne Edwards, 1837), Ascension Island. A, B, chelae of heterochelous (Type B) male, CW 110 mm, ChLHR 1.33; C, D, chelae of homochelous (Type A) male, CW 109 mm, ChLHR 1.02.

were analysed separately. The distribution of the two morphs by size classes is shown in table II. Of the 31 males sampled in the 80-99 mm CW range, only one of these, of 95 mm CW, was Type B. Conversely, in the 100-119 mm CW range, 30 were Type A and 23 were Type B. It is not known why the proportion of Type B males was lower in the largest size class (110-119 mm CW), but the sample size was small.

For Type A morphs growth of ChL was positively allometric, with slopes of 1.63 and 1.61 for major and minor chelae respectively. The ChLHR was low, with a mean of 1.03 and a range of 1.00-1.09 (one specimen had a ratio of 1.16 — its MiChL was unusually small, presumably regenerating). The ratio did not increase with CW, the regression slope being 0.0002. ChH was isometric, with slopes of 1.11 and 1.12 for major and minor chelae (slopes not significantly different from 1.0). The HR for ChH has a mean of 1.04, and a range of 0.96-1.26 (a few specimens have high values). Values of <1.0 are possible since major and minor chelae were categorised by ChL. The HR did not change with CW, the regression slope being 0.0003. As can be predicted from their respective regression

TABLE II
Numbers of the two male morphotypes in each mature CW size class

	CW (mm)	Type A	Type B	% Type B
Johngarthia lagostoma (H. Milne Edwards, 1837)	60-69	3	0	0
	70-79	3	0	0
	80-89	15	0	0
	90-99	16	1	6
	100-109	23	22	49
	110-119	7	2	22
Epigrapsus notatus (Heller, 1865) Taiwan	20.1-25	5	1	17
	25.1-30	11	3	21
	0.1-35	9	12	57
Epigrapsus politus Heller, 1862 Moorea	10.1-15	19	12	39
	15.1-20	7	9	56

slopes, the ratio ChH/ChL declines with CW for both major and minor chelae. Ratio values predicted by the regressions for major/minor chelae decline from 0.50/0.49 at 68 mm CW to 0.38/0.39 at 113 mm CW. The relation of ChW to ChL was examined in the same way, and ratio values predicted by the regressions for major/minor chelae declined from 0.32/0.27 at 68 mm CW to 0.24/0.21 at 113 mm CW. So as CW increased the chelae became proportionately longer, but the chelae themselves became relatively narrower on both axes. The gape also increased with CW, similarly in both chelae. Up to 105 mm CW, the gape seldom exceeded 1 mm, Above that size some, but not all, crabs had a larger gape of up to 4.5 mm (mean gape for crabs >95 mm CW was 1.7 mm).

For Type B morphs growth in ChL was putatively positively allometric, with slopes of 1.70 and 1.63 for major and minor chelae respectively (despite the high slope values, they were not significantly different from 1.0: see Discussion). The ChLHR was high, with a mean of 1.26. However, it was very variable, with a range of 1.06-1.62. The ratio did not increase with CW, with a slope of 0.0002. Growth in ChH was isometric for the major chela, and putatively positively allometric for the minor, with slopes of 0.97 and 1.40 (despite the high slope values, this was not significantly different from 1.0: see Discussion) respectively. The ChHHR had a mean of 1.37, and a range of 1.14-1.56. The ratio declined slightly with CW, with a slope of −0.0056. The ratio ChH/ChL declined with CW, with values predicted from the regression for major/minor chelae of 0.43/0.36 at 95 mm CW, and 0.36/0.34 at 116 mm CW. The relations of ChW to ChL were analysed in the same way. Predicted ratios for major/minor chelae were 0.26/0.22 at 95 mm CW, and 0.22/0.19 at 116 mm CW. So the general trend is as for the Type A chelae, with ChH and ChW both declining in relation to ChL as CW increases, producing relatively narrower chelae. There is a large difference in gape between the two chelae. In the minor chela there was generally no gape, though in 3 out of 24 crabs there was a gape of <2.0 mm. In the major chela there was always a clear gape, with a mean of 5.4 mm, and a range of 1.5-9.8 mm. The correlation between two indices of the 'strength' of the Type B condition, ChLHR and MaChG, was examined. None was found, with a slope of −0.0121, and $r^2 = 0.0686$.

Chelar comparisons.— The most relevant comparisons are between Type A and Type B males, and between males and females. The predictions of chela length for several CWs are calculated from the regressions in table I, and listed in table III. Several features stand out. There is minimal heterochely in females and Type A males throughout their size ranges ('subequal' in the accepted terminology). In contrast heterochely is marked in Type B males. Type A males have longer chelae than females, and the disparity increases with size. The major chela of Type B males is longer than that of Type A males, and morphologically different, whereas the minor chela is shorter, though still longer than that of females (fig. 2).

Epigrapsus notatus

Female chelae.— A total of 33 females of 16.9 to 34.8 mm CW were sampled. There was very little heterochely. The ChHHR ranged from 1.00 to 1.10, with a mean of 1.03. There was little evidence of handedness: 13 were right-handed, 11 left-handed, and 9 equal. The relative growth of the female chelae was straightforward. There was low positive allometry of 1.17 for both major and minor chela height. Both regressions had r^2 values > 0.9.

Male chelae.— Not all of the specimens measured were available for subsequent examination in this study. However, it seemed clear from material available that males exhibited polymorphism. Some males had morphologically similar chelae with only minor heterochely, others were clearly heterochelous with different morphologies in the major and minor chelae — Type A and Type B males respectively. The chelae of Type A males had a prominent clump of teeth midway along the fixed finger, and a very prominent proximal tooth on the dactylus. The chelae gaped markedly (fig. 3A). The major chela of Type B males was similar, but the minor chela had less prominent dentition, and did not gape (fig. 3B). Where specimens were not available males were assigned to morphotypes on the basis of ChHHR. If this was $\leqslant 1.1$ it was deemed Type A, if >1.1 then Type B. Both types occurred in each mature size class, but the proportion of Type B increased with CW from 17 to 57% (table II).

Type A males ($N = 25$) ranged from 22.0 to 35.0 mm CW. There was a tendency to left handedness, with 9 right handed, 15 left handed, and 1 equal. The ChHHR ranged from 1.00 to 1.10, with a mean of 1.03. Both chelae were

TABLE III

Predicted measures of chelar propodus at selected carapace widths, derived from the regressions in table I, plus the ratios of chela propodus measure for Male B/Male A, and Male A/female

	CW (mm)	Ch. measure (mm)						Chelar ratio		
		Male A		Male B		Female		Male B/Male A		Male A/Female
		Major	Minor	Major	Minor	Major	Minor	Major	Minor	Major
Johngarthia lagostoma (H. Milne Edwards, 1837)	70	43.9	43.0	–	–	38.9	38.8	–	–	1.13
	95	72.4	70.5	80.0	63.5	55.4	54.5	1.10	0.90	1.31
	110	92.1	89.3	102.7	80.7	65.6	64.5	1.12	0.90	1.40
Epigrapsus notatus (Heller, 1865)	25	10.1	9.8	10.4	7.3	5.68	5.53	1.03	0.74	1.78
	30	13.8	13.3	14.0	9.2	7.03	6.58	1.01	0.69	1.96
	35	18.0	17.2	18.0	11.2	8.43	8.20	1.00	0.65	2.14
Epigrapsus politus Heller, 1862	11	3.36	3.30	3.19	2.48	–	–	0.95	0.75	–
	15	5.80	5.69	6.23	4.80	–	–	1.07	0.72	–
	19	8.89	8.62	10.53	5.97	–	–	1.16	0.69	–

The measure is chelar propodus length for *Johngarthia lagostoma*, chelar propodus height for *Epigrapsus* species.

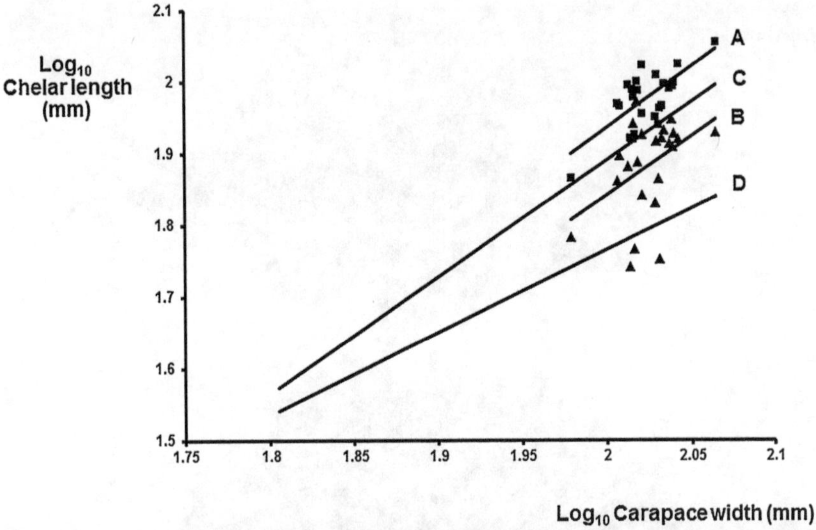

Fig. 2. *Johngarthia lagostoma* (H. Milne Edwards, 1837), Ascension Island. $\log_{10}$ (chela propodus length) plotted against $\log_{10}$ (carapace width). A, Type B males, major chelae (squares); B, Type B males, minor chelae (triangles); C, Type A males, common regression for major and minor chelae; D, Females, common regression for major and minor chelae. For clarity data points omitted for Type A males and females.

positively allometric, with levels of 1.72 and 1.69 for the major and minor chelae respectively. Both regressions had r^2 values > 0.5.

Type B males ($N = 16$) ranged from 23.5 to 34.7 mm CW. There was little tendency to handedness, with 7 right handed, and 9 left handed. The ChHHR ranged from 1.16 to 2.76, with a mean of 1.59. Both chelae were putatively positively allometric, with levels of 1.64 and 1.27 for the major and minor chelae respectively (though slopes were not significantly different from 1.0: see Discussion). Whereas the major chela regression had an r^2 value of 0.5, that for the minor chela was only 0.29.

Chelar comparisons.— The most relevant comparisons are between Type A and Type B males, and between males and females. The predictions of chela height for several CWs are calculated from the regressions in table I, and listed in table III. There is minimal heterochely in females and Type A males throughout their size ranges ('subequal' in the accepted terminology). In contrast, heterochely is very marked in Type B males. Type A males have higher chelae than females, and the disparity increases with size. The major chela of Type B males is similar in size to that of Type A males, and does not differ morphologically, whereas the minor chela is much lower (though still higher than that of females) and morphologically distinct (fig. 4).

Fig. 3. *Epigrapsus notatus* (Heller, 1865), Taiwan. A, Homochelous (Type A) male, CW 30.6 mm, ChHHR 1.01; B, Heterochelous (Type B) male, CW 27.5 mm, ChHHR 1.40.

Epigrapsus politus

Male chelae.— Not all specimens measured were available for subsequent examination. However, it appeared that males exhibited polymorphism. Some had morphologically similar chelae with minimal heterochely, others were clearly heterochelous with different morphologies in the major and minor chelae — Type A and Type B respectively. The chelar morphologies of the two types are shown in fig. 5: note however that the specimens figured are not from Moorea, the location of the sample analysed. Specimen from Moorea were not available, and males were assigned on the basis of ChHHR. If this was $\leqslant 1.1$ it was deemed Type A, if >1.1 then Type B. Type B occurred in both mature size classes, with the proportion increasing with CW from 39 to 56% (table II).

Type A males ($N = 32$) ranged from 6.5 to 18.5 mm CW. There was little tendency to handedness, with 9 right handed, 7 left handed, and 16 equal. The ChHHR ranged from 1.00 to 1.07, with a mean of 1.02. Both chelae were positively allometric, with levels of 1.77 and 1.76 for the major and minor chelae respectively. Both regressions had r^2 values of 0.9.

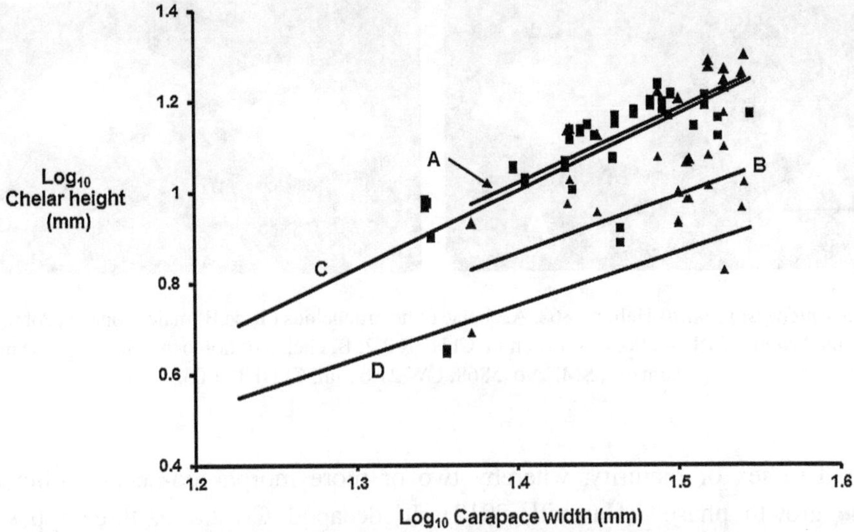

Fig. 4. *Epigrapsus notatus* (Heller, 1865), Taiwan. Log_{10} chela propodus height plotted against log_{10} (carapace width). A, Type B males, major chelae (squares); B, Type B males, minor chelae (triangles); C, Type A males, common regression for major and minor chelae; D, Females, common regression for major and minor chelae. For clarity data points omitted for Type A males and females.

Type B males ($N = 21$) ranged from 10.1 to 18.9 mm CW. There was some tendency to handedness, with 8 right handed, and 13 left handed. The ChHHR ranged from 1.12 to 1.92, with a mean of 1.48. Both chelae were positively allometric, with levels of 2.15 and 1.61 for the major and minor chelae respectively. The major chela regression had an r^2 value of 0.92, that for the minor chela was 0.88.

Chelar comparisons.— The relevant comparison is between Type A and Type B males. The predictions of chela height for several CWs are calculated from the regressions in table I, and listed in table III. There is minimal heterochely in Type A males throughout their size ranges. In contrast heterochely is very marked in Type B males. The major chela of Type B males is slightly higher than that of Type A males, and morphologically similar, whereas the minor chela is much lower, and morphologically distinct (see fig. 5).

DISCUSSION

The main focus of this study is polymorphism of the chelae of mature males — some related aspects of chelar growth are reviewed briefly for the Gecarcinidae. In this context, polymorphism "relates to differences in form which are not

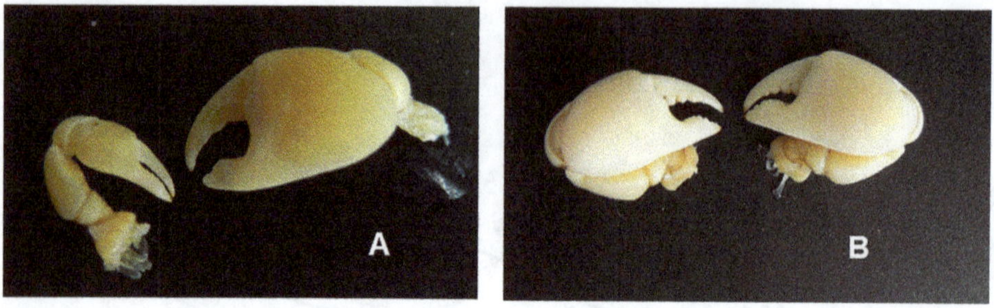

Fig. 5. *Epigrapsus politus* Heller, 1862. A, chelae of heterochelous (Type B) male, Ponape [Pohnpei], Caroline Islands, SMF #4290, CW 17.8 mm, ChHHR 2.2; B, chelae of homochelous (Type A) male, Sumatra, SMF No. 5868, CW 21.6 mm, ChHHR 1.03.

related to sex or maturity, whereby two or more morphs co-occur within the same growth phase" (Hartnoll, 2012). In decapod Crustacea, three types of such polymorphism have been described, all occurring within the mature phase. Cyclical polymorphism involves moulting on a seasonal basis from a reproductive instar to a non-reproductive instar, and back again. Definitive polymorphism is where, at the puberty moult, an individual moults into one or other of the mature morphs, in which it then remains for its adult life. Progressive polymorphism is where the morphs form an ontogenetic sequence within the mature phase. Examples of each are described in Hartnoll (2012). The first two patterns have been described from the Brachyura (see Hartnoll, 2012), but progressive polymorphism has not been previously found in that taxon.

Johngarthia lagostoma

An initial consideration was that the dimorphism might be an artefact of chela loss and regeneration. This was rejected on several grounds.

1. Evidence of regeneration in Type A morphs was rare. Of the 51 Type A males measured, only one potential regenerant had an anomalously high ChLHR of 1.16 (the others were 1.08 or less). Consequently there is no reason why males of 100 mm CW and over should be found regenerating in nearly 50% of specimens. No males were encountered with a missing chela.
2. In the regenerating specimen cited above, the minor chela was of normal form and proportions for a Type A chela of that ChL. The Type B minor chelae differed in proportions and morphology from Type A chelae.
3. Even if regeneration could account for the reduced length of Type B minor chelae, it would also have to account for the increased length and

distinctive morphology of Type B major chelae. There are no records, in an extensive literature, that the loss of one chela can result in the disproportionate growth of the other (except where the loss of a major chela results in the transformation of the minor chela into a major).

If regeneration is not the cause of chelar dimorphism, then polymorphism occurs. Definitive polymorphism is ruled out, since the polymorphism appears only late in the mature phase. Consequently progressive polymorphism is strongly indicated. The definitive confirmation would be observation of mature males moulting from Type A to Type B: unfortunately observations of moulting in gecarcinids are rare (Hartnoll et al., 2006a), and unknown for *J. lagostoma.* Assuming progressive polymorphism, what might mediate the switch from Type A to Type B in some individuals, but not in others? Environmental control is possible, as occurs in the prawn *Macrobrachium rosenbergii* (De Man, 1879). There, progression through the series of four mature male morphs is mediated by the social structure of the population (Kuris et al., 1987). In a high density of the dominant blue-claw morphs other males can be held indefinitely at earlier stages of the progression. However, this seems a poor model for *J. lagostoma*, which does not live in dense social groups. Nor does it explain the limitation of the switch to the larger mature size groups. An alternative is that the polymorphism is genetic, and persists in a balanced status. In large mature males perhaps alternative mating strategies allow both morphs to achieve equal reproductive success, and thereby persist. The heterochely in Type B may improve efficiency in both feeding dexterity and intraspecific combat.

In Type B males the major chela generally shows a good fit to the regressions, though with more variation than in the Type A males of the same size range. In contrast, the minor chela shows great variation in Type B (fig. 2), as does the ChLHR (this explains why regression slopes as high as 1.6 are not significantly different from 1.0). A possible explanation is that transformation to the Type B morphology is not accomplished at a single moult, but is only completed at subsequent ones. The ChLHR does not increase consistently with CW, as might be expected under that scenario. However, many crabs can become mature over a wide size range (Vernet-Cornubert, 1958; Hartnoll, 1963; Hines, 1989; Alunno-Bruscia & Sainte-Marie, 1998; McLay & Van den Brink, 2009), so possibly in *J. lagostoma* the morph switch may also occur over a range of sizes, confounding size-related patterns.

J. lagostoma occurs in four discrete island populations. The mature male polymorphism described here from Ascension Island appears to occur in Fernando de Noronha. A series of photos of males (taken by RGH, Oct 2006, but for a different purpose) clearly shows both Type A (CW 101 mm; ChLHR approx.

1.01) and Type B (CW 107 mm; ChLHR approx. 1.26) morphs. However, available images from Trindade and Atol das Rocas are insufficiently clear. Since they share the same series of colour morphs, it would not be surprising if they shared the same chelar growth patterns.

Epigrapsus spp.

The data for *Epigrapsus* spp. are less comprehensive, and conclusions remain provisional. In both species there were both homochelous and heterochelous mature males, but with several marked difference from *J. lagostoma*.

1. Both morphs occurred throughout the mature size range, though the proportion of heterochelous males increased with body size. In *J. lagostoma* heterochelic males are restricted to the upper third of the mature size range.
2. The major chelae of Type B males were morphologically similar to that of Type A males — in *J. lagostoma* the Type B major chelae are different from Type A chelae. The minor chelae of Type B males differs from that of Type A males — in *J. lagostoma* they are similar.
3. The major chelae of Type B males were similar in size to Type A males in *E. notatus*, but larger in *E. politus* — it was larger in *J. lagostoma*. The minor chelae of Type B males was smaller than Type A males in both genera, but more so in *Epigrapsus*.

Some of these results support regeneration as a putative causal factor of dimorphism in *Epigrapsus*. These are the presence of polymorphism over the whole mature size range, and the morphological similarity of Type A chelae and Type B major chelae. However, others do not — the increased incidence of Type B morphs with size, the morphological difference of Type B minor chelae from Type A chelae, and the very small size of Type B minor chelae. An additional fact is the lack of obvious signs of limb loss and regeneration in the samples. So regeneration is not a proven cause. Definitive polymorphism would be supported by much of the data, except for the increased incidence of Type B males with size: this could result from better survival of Type B males, but there is no evidence for this. Progressive polymorphism would entail the switch to Type B occurring in some small mature males, and continuing through the size range to add to the incidence of Type B: this is quite possible. No firm conclusion can be drawn at present.

Handedness, sexual dimorphism, and heterochely

Polymorphism is only one aspect of chelar growth. Three others have been considered above in the context of polymorphism — handedness, sexual chelar dimorphism, and heterochely. They will in turn be briefly reviewed in the Gecarcinidae.

Handedness is not often commented on, but there is generally only limited preference. "No prominent handedness" in *Cardisoma guanhumi* Latreille, 1828 (cf. Herrreid, 1967), though Gifford (1962) reports a 302:244 preference (NS) for right handedness. In *Cardisoma armatum* Herklots, 1851 Akin-Oriola et al. (2005) report a 64:56 preference (NS) for left handedness. In *Cardisoma carnifex* (Herbst, 1796), Alexander (1976) reported a 396:391 left handedness (NS). These data, plus those in the present study, support the limited preference model.

Sexual chelar dimorphism is universal in gecarcinids, with males having larger chelae, as is general in the Brachyura (see Hartnoll, 2012). The degree of sexual dimorphism varies (table IV), from modest in *Gecarcinus ruricola* (Linnaeus, 1758) and *Discoplax hirtipes* (Dana, 1851), substantial in *J. lagostoma* and *C. guanhumi*, to very marked in *E. notatus*: dimorphism invariably increases with body size.

Heterochely is variable. 'Subequal' is regularly used in this context, but is not clearly defined: we take it to imply that heterochely is not obvious on superficial inspection, and that a ChLHR of <1.10 could be considered subequal. Heterochely is examined in the three lineages of the Gecarcinidae — *Cardisoma* and *Discoplax*, *Epigrapsus*, and *Gecarcinus*, *Gecarcoidea* and *Johngarthia* (Ng et al., 2008).

The *Cardisoma*/*Discoplax* lineage shows apparent generic contrast (Ng & Davie, 2012), surprising given that the two genera appear to be very closely linked. In the literature heterochely is not recorded in *Discoplax*. In *D. celeste* "heterochely not noted" (Turner et al., 2011, as *D. hirtipes*); "chelipeds subequal" (Ng & Davie, 2012); photos show no heterochely (Orchard, 2012). *Discoplax rotunda* (Quoy & Gaimard, 1824) photos show no heterochely (Türkay, 1973, 1974; Orchard, 2012), nor does *D. hirtipes* (Türkay, 1974). *Discoplax gracilipes* Ng & Guinot, 2001 and *Discoplax longipes* A. Milne-Edwards, 1867, descriptions and photos indicate subequal chelae (Türkay, 1974; Ng & Guinot, 2001). However, material of *D. hirtipes* from Java in the Senckenberg Forschungsinstitut und Naturmuseum (SMF No. 4147) showed marked heterochely in larger males. The situation in this genus needs further examination. In contrast, all *Cardisoma* spp. show very marked heterochely in both sexes, most notably in males (see table IV). Formal (though limited) analyses are available for *C. guanhumi* (Herreid, 1967; see also fig. 1 in Gifford, 1962, and fig. 64 in Chace & Hobbs, 1969) and *C. armatum* (Oyenikan, 1995), but accounts and photos from a diversity of sources indicate equally pronounced heterochely in *Cardisoma crassum* Smith, 1870 (e.g., Türkay, 1970) and *C. carnifex* (e.g., Türkay, 1973, 1974).

There are few data for *Epigrapsus*. In *E. notatus* Liu & Jeng (2005) refer to the 'major chela', implying heterochely. In *E. politus* Orchard (2012) states "males sometimes have one claw larger than the other", a feature very obvious from

TABLE IV

Heterochely ratios (major/minor) for mature males and females, and sex ratios (male/female) for major chela of mature specimens

Species	Reference	Heterochely ratio				Major chela sex ratio	
		Male		Female		Small	Large
		Small	Large	Small	Large		
Johngarthia lagostoma (H. Milne Edwards, 1837)	Present study[a]	1.02	1.03	1.00	1.00	1.13	1.40
	Present study[b]		1.28				1.56
Gecarcinus ruricola (Linnaeus, 1758)	Hartnoll et al. (2006a)	1.10	1.09	1.04	1.02	1.15	1.11
Discoplax celeste Ng & Davie, 2012	Turner et al. (2011)[c]					0.99	1.12
Cardisoma guanhumi Latreille in Latreille, Le Peletier, Serville & Guérin, 1828	Herreid (1967)[d]	1.33	1.94	1.42	1.39	0.94	1.40
Cardisoma armatum Herklots, 1851	Oyenekan (1995)		1.68		1.40		
Epigrapsus notatus (Heller, 1865)	Liu & Jeng (2005)[e]					1.41	2.19
Epigrapsus notatus (Heller, 1865)	Liu (unpublished)[e]	1.01	1.38	1.03	1.03		

'Small' and 'large' refer to sizes near the lower and upper bounds of the mature range for the species concerned. Unless otherwise noted, ratios are based on chela propodus length. Not all ratios determinable for some species.

[a] Based on Type A males.

[b] Based on Type B males.

[c] As *D. hirtipes* "heterochely not noted". Measured only right chela.

[d] Based on cheliped length, not propodus length. Common regression for male and female minor chelae (difference ns).

[e] Based on chela propodus height.

his photos. The present study reports marked heterochely in some males of both species. In contrast, Ng (2002) describes the chelae of *Epigrapsus villosus* Ng, 2003 as "subequal": however, this is based on a single male of 9.4 mm CW, which could well be immature.

In the third lineage, for the otherwise intensively studied *Gecarcoidea natalis* (Pocock, 1889), there is little reference to the chelae: Orchard (2012) states "claws are usually of equal size", a view supported by a wide range of photos in various sources. Photos of *Gecarcoidea lalandii* H. Milne Edwards, 1837 (cf. Türkay, 1974; Orchard, 2012) and *Gecarcoidea humei* (Wood-Mason, 1874) (cf. Orchard, 2012) indicate a similar condition. In *Gecarcinus ruricola* (Linnaeus, 1758) the chelae are subequal (table 4: Hartnoll et al., 2006b). The substantial literature on *G. lateralis* (Guérin, 1832)/*quadratus* Saussure, 1853 does not mention chela growth, but none of the many photos indicate heterochely. In *Johngarthia*, the situation in *J. lagostoma* from Ascension Island is as described above: females and Type A males are subequal, Type B males are markedly heterochelous. The population on Fernando de Noronha could be similar. In *J. cocoensis*, Perger et al. (2011) state "specimens <CW 100 mm homochelous, males >CW 100 mm slightly heterochelous (left palm bigger, biggest difference in palm length 20%)": this suggests heterochely in large males, but it is based on a limited sample of 23 specimens. There are no data for *Johngarthia weileri* (Sendler, 1912) or *Johngarthia planata* (Stimpson, 1860).

Apparently heterochely has developed several times within the Gecarcinidae, at least once within each of the lineages. However, polymorphism appears restricted to the examples reported in this study.

ACKNOWLEDGEMENTS

Prior to RGH's visit to Ascension Island in 2014 Michael Türkay asked if I could collect some specimens of *Johngarthia lagostoma* for the Senckenberg Forschungsinstitut und Naturmuseum, which had none from that location. Early in my visit I collected a pair of large males, but then noted that one had a seemingly regenerating chela. On seeking a replacement, I found that the situation was not uncommon. This generated a programme of nocturnal measurement (and a number of sore fingers), from which this study resulted. So Michael was the unwitting cause of this work, which is perhaps a fitting tribute to offer. The work on Ascension Island was carried out under the auspices of the Flagship Species Fund of the Department of Environment, Food and Rural Affairs (DEFRA) and Flora & Fauna International (FFI) with funding from DEFRA and other donors. We are grateful to the late Michael Türkay for facilitating access to and photography of material in the Senckenberg Forschungsinstitut und Naturmuseum.

REFERENCES

AKIN-ORIOLA, G., M. ANETEKHAI & K. OLOWONIREJUARO, 2005. Morphometric and meristic studies in two crabs: *Cardisoma armatum* and *Callinectes pallidus*. Turkish J. Fish. Aquat. Sci., **5**: 85-89.

ALEXANDER, H. G. L., 1976. An ecological study of the terrestrial decapod Crustacea of Aldabra. Ph.D. thesis, University of London, London.

ALUNNO-BRUSCHIA, M. & B. SAINTE-MARIE, 1998. Abdomen allometry, ovary development, and growth of female snow crab, *Chionoecetes opilio* (Brachyura, Majidae), in the northwestern Gulf of St. Lawrence. Can. J. Fish. Aquat. Sci., **55**: 459-477.

CHACE, F. A. & H. H. HOBBS, 1969. The freshwater and terrestrial decapod crustaceans of the West Indies with special reference to Dominica. Bull. U.S. nat. Mus., **292**: 1-258.

GIFFORD, C. A., 1982. Some observations on the general biology of the land crab *Cardisoma guanhumi* (Latreille) in South Florida. Biol. Bull., **97**: 207-223.

HARTNOLL, R. G., 1963. The biology of Manx spider crabs. Proc. Zool. Soc. London, **141**: 423-496.

HARTNOLL, R. G., 2012. Relative growth: description and analysis. In: J. FOREST & J. C. VON VAUPEL KLEIN (eds.), Treatise on Zoology. Anatomy, taxonomy, biology. The Crustacea, revised and updated from the Traité de Zoologie, **3** (18): 365-401. (Brill, Leiden).

HARTNOLL, R. G., M. S. P. BAINE, Y. GRANDAS, J. JAMES & H. ATKIN, 2006a. Population biology of the black land crab, *Gecarcinus ruricola*, in the San Andres Archipelago, Western Caribbean. J. crust. Biol., **26**: 316-325.

HARTNOLL, R. G., A. C. BRODERICK, B. J. GODLEY, S. MUSICK, M. PEARSON, S. A. STROUD & K. E. SAUNDERS, 2010. Reproduction in the land crab *Johngarthia lagostoma* on Ascension Island. J. Crust. Biol., **30**: 83-92.

HARTNOLL, R. G., A. C. BRODERICK, B. J. GODLEY & K. E. SAUNDERS, 2009. Population structure of the land crab *Johngarthia lagostoma* on Ascension Island. J. crust. Biol., **29**: 57-61.

HARTNOLL, R. G., T. MACKINTOSH & T. J. PELEMBE, 2006b. *Johngarthia lagostoma* (H. Milne Edwards, 1837) on Ascension Island: a very isolated land crab population. Crustaceana, **79**: 197-215.

HARTNOLL, R. G., C. RÉGNIER-MCKELLAR, N. WEBER & S. B. WEBER, 2014. Return to the land; the stages of terrestrial recruitment in land crabs. Crustaceana, **87**: 531-539.

HERREID, C. F., 1967. Skeletal measurements and growth of the land crab, *Cardisoma guanhumi* Latreille. Crustaceana, **13**: 39-44.

HINES, A. S., 1989. Geographical variation in size at maturity in brachyuran crabs. Bull. mar. Sci., **45**: 356-368.

KURIS, A. M., Z. RA'ANAN, A. SAGI & D. COHEN, 1987. Morphotypic differentiation of male Malaysian giant prawns, *Macrobrachium rosenbergii*. J. crust. Biol., **7**: 219-237.

LIU, H.-C. & M.-S. JENG, 2005. Reproduction of *Epigrapsus notatus* (Brachyura: Gecarcinidae) in Taiwan. J. crust. Biol., **25**: 135-140.

MCLAY, C. L. & A. M. C. VAN DEN BRINK, 2009. Relative growth and size at sexual maturity in *Halicarcinus cookii* (Brachyura: Hymenosomatidae): why are some crabs precocious moulters? J. mar. biol. Ass. U. K., **89**: 743-752.

NG, P. K. L., 2002. On a new species of *Epigrapsus* (Decapoda, Brachyura, Gecarcinidae) from a cave in Guam. Crustaceana, **75**: 1219-1227.

NG, P. K. L. & P. J. F. DAVIE, 2012. The blue crab of Christmas Island, *Discoplax celeste*, new species (Crustacea: Decapoda: Brachyura: Gecarcinidae). Raffles Bull. Zool., **60**: 89-100.

NG, P. K. L. & D. GUINOT, 2001. On the land crabs of the genus *Discoplax* A. Milne Edwards, 1867 (Crustacea: Decapoda: Brachyura; Gecarcinidae), with description of a new cavernicolous species from the Philippines. Raffles Bull. Zool., **49**: 311-338.

NG, P. K. L., D. GUINOT & P. J. F. DAVIE, 2008. Systema brachyorum: part I. An annotated checklist of extant brachyuran crabs of the world. Raffles Bull. Zool. (Suppl.), **17**: 1-286.

ORCHARD, M., 2012. Crabs of Christmas Island. (Christmas Island Natural History Association, Christmas Island).

OYENEKAN, J. A., 1995. Growth patterns in three brachyuran crabs in Lagos, Nigeria. Arch. Hydrobiol., **134**: 533-546.

PERGER, R., R. VARGAS & A. WALL, 2011. *Johngarthia cocoensis*, a new species of Gecarcinidae MacLeay, 1838 (Crustacea, Decapoda, Brachyura) from Cocos Island, Costa Rica. Zootaxa, **2911**: 57-68.

TÜRKAY, M., 1970. Die Gecarcinidae Americas. Mit einem Anhang über *Ucides* Rathbun (Crustacea: Decapoda). Senckenbergiana biol., **51**: 333-345.

TÜRKAY, M., 1973. Die Gecarcinidae Afrikas (Crustacea: Decapoda). Senckenbergiana biol., **51**: 81-103.

TÜRKAY, M., 1974. Die Gecarcinidae Asiens und Ozeaniens (Crustacea: Decapoda). Senckenbergiana biol., **55**: 223-259.

TURNER, L. M., J. P. HALLAS & S. MORRIS, 2011. Population structure of the Christmas Island blue crab, *Discoplax hirtipes* (Decapoda: Brachyura: Gecarcinidae) on Christmas Island, Indian Ocean. J. crust. Biol., **31**: 450-457.

VERNET-CORNUBERT, G., 1958. Biologie générale de *Pisa tetraodon*. Bull. Inst. océanogr. Monaco, **1113**: 1-52.

First received 14 June 2016.
Final version accepted 3 September 2016.

[When citing this volume, please refer to Crustaceana 90 (2017) 771-1288]

Michael Türkay Memorial Issue

PALAEOGENE AND NEOGENE BRACHYURANS OF THE AMAZON BASIN: A REVISED FIRST APPEARANCE DATE FOR PRIMARY FRESHWATER CRABS (BRACHYURA, TRICHODACTYLIDAE)

BY

SEBASTIAN KLAUS[1,6]), CÉLIO MAGALHÃES[2]), RODOLFO SALAS-GISMONDI[3]), MARTIN GROSS[4]) and PIERRE-OLIVIER ANTOINE[5])

[1]) Department of Ecology and Evolution, J.W. Goethe Universität — Biologicum, Max-von-Laue-Strasse 13, D-60438 Frankfurt a. M., Germany

[2]) Instituto Nacional de Pesquisas da Amazônia, Av. André Araújo 2936, 69067-375 Manaus, AM, Brazil

[3]) Departamento de Paleontología de Vertebrados, Museo de Historia Natural, Universidad Nacional Mayor de San Marcos, Avenida Arenales 1256, Lima 14, Peru

[4]) Department for Geology and Palaeontology, Universalmuseum Joanneum, Weinzöttlstrasse 16, 8045 Graz, Austria

[5]) Institut des Sciences de l'Evolution, UMR 5554 CNRS, IRD, EPHE, Université de Montpellier, Place Eugène Bataillon, 34095 Montpellier cedex 5, France

ABSTRACT

We describe claw fragments of fossil primary freshwater crabs from three areas in the Amazon basin, Tarapoto (Early Oligocene) and Contamana (Middle Eocene to early Late Miocene) in Peru, and Eirunepé (Late Miocene) in Brazil. All these fragments most likely belong to the family Trichodactylidae. We show a continuous presence of primary freshwater crabs in proto-Amazonian lowlands from the Middle Eocene to the Late Miocene and can thus shift the earliest appearance date of freshwater-adapted brachyurans into the Eocene, at least in the Neotropics.

Key words. — Freshwater crabs, Trichodactylidae, Cenozoic, Proto-Amazonia, Neotropics

ZUSAMMENFASSUNG

Wir beschreiben fossile Scherenfragmente von primären Süßwasserkrabben, die von drei Lokalitäten innerhalb des Amazonasbeckens stammen: Aus Tarapoto (Frühes Oliogozän) und Contamana (Mittleres Eozän – frühes Spätes Miozän) in Peru, und aus Eirunepé (Spätes Miozän) in Brasilien. Alle Fragmente gehören vermutlich zur Familie der Trichodactylidae. Damit können wir vom Mittleren Eozän bis zum Späten Miozän eine kontinuierliche Präsenz von primären Süßwasserkrabben im proto-Amazonischen Tiefland nachweisen; und, zumindest für die Neotropen, das erste Auftreten von Süßwasser-angepassten Brachyuren bis in das Eozän verschieben.

[6]) Corresponding author; e-mail: klaus@bio.uni-frankfurt.de

INTRODUCTION

A long-standing opinion regarding freshwater crabs is that their fossil record is notoriously poor. Contrasting the extant diversity patterns, we currently have the best knowledge on fossil freshwater crabs from Europe and the Near East, with 35 documented sites ranging from the early Middle Miocene to the Pleistocene (Klaus & Gross, 2010; Klaus & Prieto, 2014). Much scarcer is our knowledge of the fossil record of Asian freshwater crabs that date from the Miocene/Pliocene (Klaus & Gross, 2010; Klaus et al., 2011, in press) and Pleistocene/Holocene (Naruse et al., 2003; Rabett et al., 2009; Ng & Cranbrook, 2014; Conrad et al., 2016), and from Africa (Morris, 1976; Carriol & Secrétan, 1992; Martin & Trautwein, 2003). The latter landmass yields the earliest representative of Old World freshwater crabs, the Oligocene *Tanzanonautes tuerkayi* Feldmann, O'Connor, Stevens, Gottfried, Roberts, Ngasala, Rasmusson & Kapilima, 2007 from the East African Rift Valley. The fossil freshwater crabs of the Neotropics, however, so far represent a terra incognita for both systematists and palaeontologists. Currently there is only one undisputed record of fossil claw fragments, originating from the Middle Miocene Honda group in the upper Río Magdalena River, at La Venta, Colombia (Rodríguez, 1997). These have been assigned to the extant species *Sylviocarcinus piriformis* (Pretzmann, 1968), of the family Trichodactylidae. Trace fossils from fluvial deposits of the Early Oligocene to Miocene Vinchina Formation, La Rioja, Argentina have been suggested to represent trichodactylid crab burrows (Melchor et al., 2010), but this evidence is indirect. Additionally, subfossil remains of *Eudaniela garmani* (Rathbun, 1898) (= *Rodriguezus garmani*) of the Pseudothelphusidae were found in an anthropogenic shell midden disposed between 2925 and 3485 years ago near the Guacharo cave, Monagas state, Venezuela (Rodríguez & Díaz, 1977).

In the Neotropics two families of freshwater crabs occur today, the Pseudothelphusidae (278 species) and the Trichodactylidae (51 species) (Yeo et al., 2008), that most likely represent independent colonization events of the limnic habitat as indicated by morphology (Sternberg & Cumberlidge, 2003) and molecular phylogenetics (Tsang et al., 2014). The centre of taxonomic diversity of the Pseudothelphusidae is northwestern South America and Central America (Rodríguez, 1982, 1986), while the range of the Trichodactylidae extends much more south, with a disjunctive occurrence of few species in Central America (Magalhães & Türkay, 2012).

Here, we describe freshwater crab remains from several localities of Palaeogene and Neogene age in the Amazon basin that push the first appearance of primary freshwater crabs back to the Eocene and point to their probable continuous presence in the area throughout the last 40 million years in Neotropical lowlands. For nomenclature used to describe the claw morphology, see fig. 1.

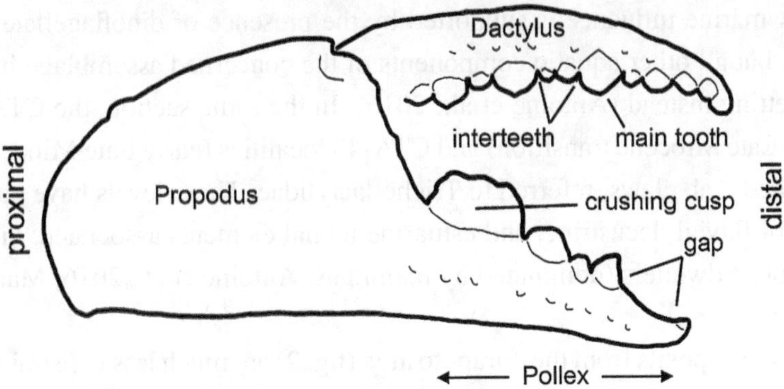

Fig. 1. Habitus of the brachyuran claw (right chela).

GEOLOGICAL SETTING

The claw fragments originate from three locations: Eirunepé area, Amazonas state, Brazil; Tarapoto area, San Martín province, Peru; and from several sites ranging from the Middle Eocene to the early Late Miocene in the Contamana area, Loreto province, Peru (fig. 2).

Sedimentological and palaeontological evidence (e.g., palynomorphs, charophytes, molluscs, fish, anurans, turtles and crocodiles) are primarily pointing to freshwater environments at Contamana for the crab-yielding levels, be them referred to the Pozo Formation (low-energy small streams of fluvial origin: CTA-47, CTA-27, CT-66, and CTA-29, Antoine et al., 2012, 2016) or to the overlying Chambira Formation (oxbow lake: CTA-32; Antoine et al., 2016a, b; Boivin et al.,

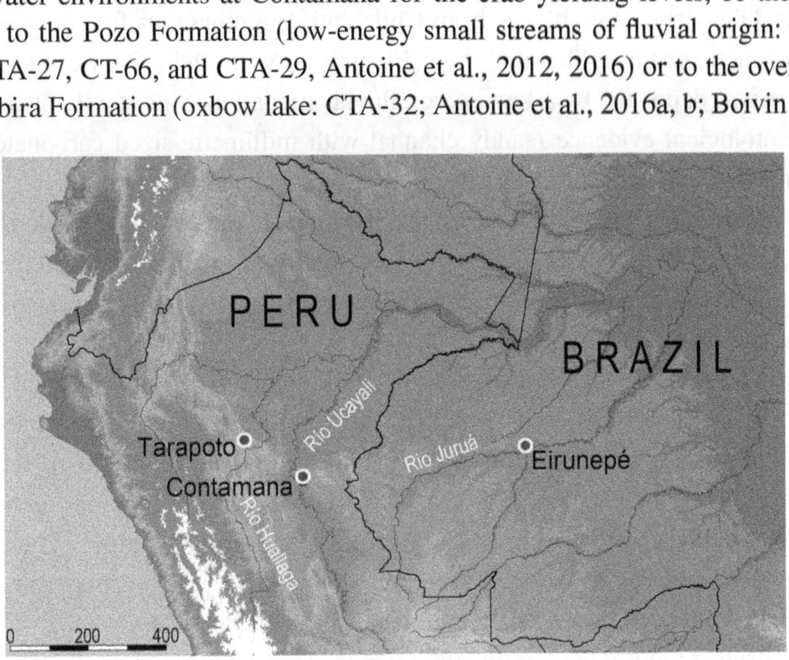

Fig. 2. Location of the sampling sites within the Amazon drainage in Peru and Brazil.

2017). A marine influence is supported by the presence of dinoflagellate cysts in CTA-47, but all other aquatic components of the concerned assemblage indicate a fluvial setting instead (Antoine et al., 2016). In the same section, the CTA-44 top (Middle-Late Miocene transition) and CTA-43 localities (early Late Miocene) have also yielded crab claws, referred to Trichodactylidae. These levels have gathered a mixture of fluvial, lacustrine, and estuarine faunal elements associated with tropical rainforest dwellers (dominated by mammals: Antoine et al., 2016; Marivaux et al., 2016).

Cenozoic deposits from the Tarapoto area (fig. 2) are much less constrained from a chronostratigraphical point of view (Hermoza et al., 2005; Eude et al., 2015). Continuing fieldwork by an international team of palaeontologists coordinated by POA and RSG led to the recent discovery of a rich vertebrate locality termed TAR-01, currently under study (Antoine et al., data not shown). The concerned level was considered to be part of the Oligocene-Miocene Chambira Formation (Hermoza et al., 2005), but it might instead be referred to the upper part of the underlying Pozo Formation (Late Eocene to Early Oligocene; Roddaz et al., 2010). Be it as it may, the preliminary faunal list of TAR-01 points to an early Oligocene age, based on the co-occurrence of undisputable prepidolopid/polydolopid and argyrolagoid polydolopimorphian metatherians. Such a co-occurrence was only attested in early Oligocene assemblages from Chile and Argentina thus far (Tinguiririca South American Land Mammal Age; Flynn & Wyss, 1999; Goin et al., 2010). TAR-01 was deposited by a low-energy fluvial stream, as consistently attested by sedimentological evidence (sandy channel with millimetre-sized carbonate nodules and blue clayish matrix) and its palaeontological content (charophytes, freshwater malacofauna and ichthyofauna, and putative gharial caimans). Nevertheless, the presence of myliobatids (eagle rays) and pristids (sawfishes) lower in the same stratigraphical section suggests a deltaic/marine influence for the corresponding member (unpubl. data).

Close to the city of Eirunepé three sections (Pau d'Alho, Morada Nova, Remanso) yielded fossil crab remains. The exposed strata belong to the upper part of the Solimões Formation and are biostratigraphically dated to the Late Miocene (Gross et al., 2011, 2013). Sedimentological observations indicate a fluvial, possibly anastomosing river system as depositional setting. The chelae originate from layers which were interpreted as partly crevasse splay influenced floodplain lake sediments. Palaeontological and geochemical ($\delta^{18}O$, $\delta^{13}C$) data derived from molluscs and ostracods refer to exclusively freshwater conditions (Gross et al., 2011, 2013).

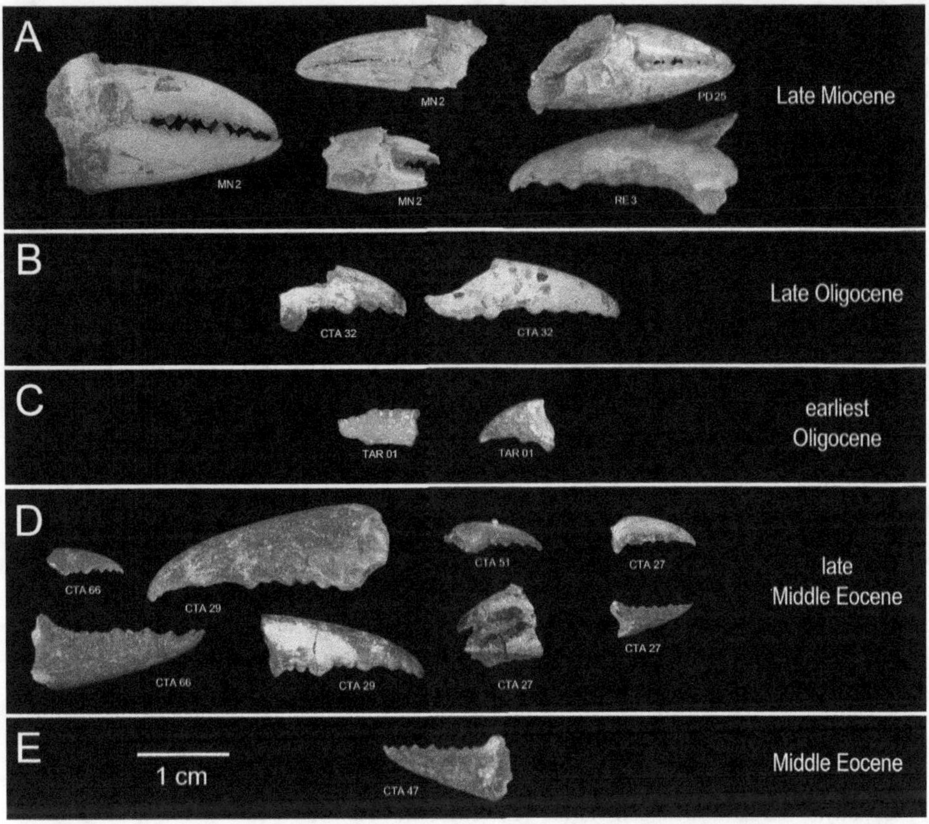

Fig. 3. Freshwater crab claw fragments from Peru and Brazil grouped according to their stratigraphical position: A, Eirunepé, Brazil; B, D, E, sites at Contamana, Peru; C, Tarapoto, Peru. Likely identity as pollex or dactylus is indicated by orientation of the claws' cutting edge (upwards vs. downwards). All claws are of the same scale.

SYSTEMATIC PALAEONTOLOGY

Infraorder BRACHYURA Latreille, 1802
Family TRICHODACTYLIDAE H. Milne Edwards, 1853

Trichodactylidae indet.

Material.— Four articulated claws (propodus and dactylus), 2 pollices (fig. 3A), housed in the Universalmuseum Joanneum, Graz under references UMJG&P 211469, 211470, 211471.

Locality.— Eirunepé municipality; left bank of Juruá River, downstream from Eirunepé, Amazonas State, Brazil (Gross et al., 2011, 2013); sites Pau d'Alho: layer PD 25 (UMJG&P 211470); Morada Nova: layer MN 2 (UMJG&P 211469); Remanso: layer: RE 3 (UMJG&P 211471).

Stratigraphy and age.— Solimões Formation; Late Miocene (Gross et al., 2011; Gross et al., 2013).

Palaeoenvironment.— Floodplain lake within a possibly anastomosing river system.

Description.— Well preserved fragments. On the cutting edge of the larger claw's pollex (fig. 3A, left) triangular main teeth and interteeth are alternating; on the dactylus the number of smaller interteeth between the main teeth varies between one and three. One more slender claw (fig. 3A, upper row middle) most likely presents a smaller claw of a left-handed crab with hardly visible teeth on the dactylus, and on the pollex several larger teeth in the distal third and equally sized smaller teeth in the proximal part with a gap in-between.

Trichodactylidae indet.

Material.— Sixty-five claw fragments, housed in the Museo de Historia Natural de la Universidad Nacional Mayor San Marcos de Lima, Peru (MUSM) (fig. 3B).

Locality.— Contamana area, Loreto, Peru (site CTA-32; Antoine et al., 2016).

Stratigraphy and age.— Base of Chambira Formation; Late Oligocene (Antoine et al., 2016a).

Palaeoenvironment.— Oxbow lake (see Boivin et al., 2017).

Description.— Most claws are highly fragmented. The better preserved specimens (dactyli) show distally up to six larger, triangular teeth followed by two interteeth and one large main tooth.

Trichodactylidae indet.

Material.— More than 340 smaller claw fragments, housed in the MUSM (fig. 3C).

Locality.— Tarapoto area, San Martín, Peru (site TAR-01).

Stratigraphy and age.— Pozo Formation, upper member; most probably earliest Oligocene (Tinguiririca n, approximately 33-30 Ma) based on mammalian biochronology (unpubl. data).

Palaeoenvironment.— Low-energy fluvial setting, with deltaic/estuarine influence (presence of myliobatid rays in the assemblage; unpubl. data).

Description.— The claws from site TAR-01 are highly fragmented and appear to be strongly eroded. Nevertheless, main and interteeth can be identified.

Trichodactylidae indet.

Material.— Fifty-one claw fragments (site CTA-27: 35; site CTA-29: 5; site CTA-51: 9; site CTA-66: 2), housed in the MUSM (fig. 3D).

Locality.— Contamana area, Loreto, Peru (sites CTA-27; CTA-29; CTA-51; CTA-66; Antoine et al., 2016a).

Stratigraphy and age.— Pozo Formation, lower member; Late Middle Eocene (early Barrancan, >41.6-40.94 Ma) based on mammalian biochronology; site CTA-29: 43.44 ± 2.5 Ma ($^{40}Ar/^{39}Ar$ age; Antoine et al., 2016a).

Palaeoenvironment.— Fluvial setting.

Description.— The claw fragments are of different size and include one fragment of an articulated claw. The dactylus' distal teeth are very small (abraded? gap?). The claw fragments from CTA-66 are of different size bearing triangular teeth, the smaller fragment showing more prominent teeth in the distal part. The larger pollex bears at its tip a small cup formed by two teeth that held the tip of the dactylus.

Trichodactylidae indet.

Material.— One claw fragment, housed in the DIPP-MUSM (fig. 3E).
Locality.— Contamana area, Loreto, Peru (site CTA-47; Antoine et al., 2016a).

Stratigraphy and age.— Pozo Formation, lower member; Middle Eocene (Barrancan) (Antoine et al., 2016a).

Palaeoenvironment.— Fluvial setting, with marine influence (presence of dinoflagellates in the palynomorph assemblage; Antoine et al., 2016a).

Description.— The fragment represents most likely a pollex with nine preserved triangular teeth, the distal five being larger than the following teeth.

DISCUSSION

Claw morphology is hardly used in brachyuran systematics as a diagnostic character, especially as its variability within populations and — ontogenetically — within an individual is not explored yet. As claw morphology appears to be highly susceptible to selection related to ecology (Marijnissen, 2007), its phylogenetic bearing is difficult to assess, especially in a palaeontological context when neither carapace and gonopodial morphology nor genetic information is available. For freshwater crabs there are few endeavours to assign isolated fossil claw fragments to extant species. Ng & Cranbrook (2014) used comparative morphology to assign claw fragments from historical cave deposits to extant species. Pretzmann (1972) and Bachmayer & Pretzmann (1971) investigated patterns of main and interteeth in extant and fossil *Potamon*, but failed to find consistent data within the fossil assemblages and in the extant species. More successful was a morphometric approach that could assign (sub)fossil Pleistocene claws to an extant species, and also point out evolutionary changes (Ashkenazi et al., 2005). For these approaches, however, large series of completely preserved claws of the same size are necessary, what is not given with the present material. Therefore, we have to rely on a mere qualitative approach that makes species identification impossible and also hampers family and genus assignment.

With a reasonable degree of certainty, the present findings belong to the freshwater crab family Trichodactylidae. Claws of Pseudothelphusidae usually have a well-marked row of rather similar sized teeth and longitudinal rows of granules along dactylus and pollex. The latter might be eroded in fossil remains, but especially the fragments from Eirunepé (fig. 3A) are quite well-preserved and lack any granules. Ecological differences, i.e., differential altitudinal distribution of both Neotropical freshwater crab families, would also corroborate the assignment of the present remains to the Trichodactylidae. The species of this family are widely distributed in the lowland areas of the main South American river basins, with the highest diversity in lentic and lotic environments of the Amazon drainage (Rodríguez, 1981; Magalhães, 2003). The pseudothelphusid crabs, on the other hand, are a mountainous group that supposedly evolved along the Andean mountains (Rodríguez, 1981, 1982) and are not distributed in the periodically flooded areas of the Amazon drainage (Magalhães, 2003). Considering that the fossil claw fragments originate from floodplain deposits of the palaeo-Amazon-Orinoco lowland system characterized by rivers, small streams and lakes, with recurrent marine incursions and with no palynological trace of high elevation habitats by the considered interval (Lundberg et al., 1998; Hoorn et al., 2010; Antoine et al., 2016a), it is very likely that the crabs inhabiting this region were indeed Trichodactylidae (given that the occurrence at higher altitudes represents the ancestral state for South American pseudothelphusids). All the specimens described here originate from oxbow lakes and low-energy streams, whereas higher-energy fluvial deposits from the Cenozoic of Western Amazonia (CTA-61, Late Oligocene, Contamana: Marivaux et al., 2016; Fitzcarrald local fauna, Middle Miocene: Tejada-Lara et al., 2015) and/or the Subandean Zone (MD-61, Early Miocene: Marivaux et al., 2012; MD-67, early Middle Miocene: Antoine et al., 2013) have yielded no crab remains thus far.

A putative Middle Miocene decapod claw fragment from a drilling core at the Ituí river, Amazonas state, Brazil (Linhares & Ramos, 2011) is most likely not of brachyuran origin, as its very small size (0.2 mm length) is far below the range even of juveniles' claws. Two claws from the Pebas section, Peru (Middle Miocene, 14-13 Ma; Mollusc Zone 7 of Wesselingh et al., 2006) are illustrated by Wesselingh & Ramos (2010: 309, fig. 18.6c, d). Based on the palaeoenvironment these fragments might be of freshwater crab origin, as they were found in the "type" locality of the Pebas Megawetland System, known to predate the major Miocene marine incursion. This record is slightly prior to the Colombian occurrence of trichodactylids at La Venta (Rodríguez, 1997), which would have biogeographical implications concerning potential dispersal pathways, as La Venta and Pebas belong to two distinct (palaeo)biogeographical areas within the Pebas System at that time (Salas-Gismondi et al., 2015; Salas-Gismondi et al., 2016). Morphologically, however, the claw fragments are unlikely to be of trichodactylid origin. One of the fragments (Wesselingh & Ramos, 2010:

309, fig. 18.6d) cannot be assigned unequivocally to the Trichodactylidae due to the extremely elongated distal tooth, while the other (Wesselingh & Ramos, 2010: 309, fig. 18.6c) is a strongly curved dactylus, a situation that very rarely (and less prominent) occurs in very large trichodactylid specimens only, e.g., in *Trichodactylus fluviatilis* Latreille, 1828 from southeast Brazil.

Most of the remains described here show affinities to the claws of the trichodactylid tribe Valdiviini (the genera *Sylviocarcinus*, *Valdivia* and *Zilchiopsis*), based on the presence of generally triangular teeth with alternating main and interteeth. This applies for fig. 3A left, 3D, 3E middle and 3F. However, several claws could as well belong to genera of the Dilocarcinini (*Dilocarcinus*, *Fredilocarcinus* and *Moreirocarcinus*), e.g., fig. 3A left, 3A upper row middle, 3E lower row middle, 3F. One claw from Eirunepé is extremely slender and resembles the smaller chela of species of the genus *Valdivia* (fig. 3A upper row middle). Table I gives an overview of the extant species occurring in the drainages of the three fossil sites. However, based on the preservation of most claws and the minor phylogenetic signal in freshwater crab claws in general, the present fragments might or might not be of conspecific origin.

The significance of these findings — despite the unsatisfying inability for a more precise taxonomic identification — is the Middle Eocene age of the earliest sites (CTA-47, 51, 27, 66 and 29), as this pushes the first appearance date of primary freshwater crabs more than ten million years deeper into the Palaeogene (compared to Early Oligocene *Tanzanonautes tuerkayi* from East

TABLE I

Occurrence of extant trichodactylid species within the drainages of the fossil sites of Tarapoto (Río Huallaga), Contamana (Río Ucayali) and Eirunepé (Rio Juruá). Based on Magalhães (2003) and Magalhães & Türkay (1996a, b, 2008a, b)

Species	Río Huallaga	Río Ucayali	Rio Juruá
Dilocarcinus pagei Stimpson, 1861	✓	✓	✓
Dilocarcinus septemdentatus (Herbst, 1783)			✓
Fredilocarcinus musmuschiae (Pretzmann & Mayta, 1980)		✓	
Fredilocarcinus raddai (Pretzmann, 1978)	✓		
Fredilocarcinus apyratii Magalhães & Türkay, 1996			✓
Goyazana rotundicauda Magalhães & Türkay, 1996		✓	
Rotundavaldivia latidens (A. Milne-Edwards, 1869)		✓	
Sylviocarcinus devillei H. Milne Edwards, 1853	✓	✓	✓
Sylviocarcinus maldonadoensis (Pretzmann, 1978)		✓	
Sylviocarcinus pictus (H. Milne Edwards, 1853)		✓	
Trichodactylus faxoni Rathbun, 1906	✓		
Valdivia serrata White, 1847		✓	
Zilchiopsis cryptodus (Ortmann, 1893)		✓	

Africa). This reduces the vast gap between the freshwater crab's first appearance date in the fossil record and the age of primary freshwater crabs as estimated by molecular clock techniques. The divergence of the monophyletic clade of Old World freshwater crabs (Potamidae, Gecarcinucidae and Potamonautidae) and Pseudothelphusidae from their marine relatives was estimated at 151-127 Ma; and that of the Trichodactylidae at about 102-73 Ma (Tsang et al., 2014). The temporal discrepancy between the hypothesised split and the available fossil record is thus still enormous, and is usually explained by the incompleteness of the latter for freshwater crabs (Klaus et al., 2011; Tsang et al., 2014). It has to be stressed, however, that also molecular clock calibration is subject to error that might not be reflected by credibility intervals of divergence time alone, and includes uncertainty of taxonomical assignment of fossils used to calibrate the clock, as well as the dating of these fossils (Klaus & Prieto, 2014).

The present study is only the second undisputed report of fossil primary freshwater crabs at the South American scale (since Rodríguez, 1997). Given that, it is likely that more intense future work — and a better awareness of crustacean remains by palaeontologists — will lead to a more detailed picture of the ?Cretaceous-Palaeogene freshwater crab fauna of South America. This might not only shed light on the biogeographical connection between the crab faunas of South and Middle America and the Caribbean, but can also contribute to our understanding of tropical palaeoecosystems, given the importance of freshwater crabs in extant tropical limnic food webs as main shredders (Dobson et al., 2002).

ACKNOWLEDGEMENTS

Fieldwork and post-field analyses in Peru were carried out thanks to the support from the CNRS 'Eclipse 2', CNRS 'Palaeo2', and Toulouse University 'SPAM' programs, from the Institut des Sciences de l'Evolution, the ANR PALASIAFRICA (ANR-08-JCJC-0017, ANR-ERC), the L.S.B. Leakey Foundation, the National Geographic Society (grant 9679-15), and the Campus France program of the French Ministry of Foreign Affairs. This work was further supported by an "Investissements d'Avenir" grant managed by the Agence Nationale de la Recherche (CEBA, ANR-10-LABX-25-01) and by the cooperative programs ECOS-FonCyT (A14-U01) and CNRS-CONICET, in the frame of the on-going cooperation agreement between the Museo de Historia Natural de la Universidad Nacional Mayor San Marcos (Lima, Peru) and the Institut des Sciences de l'Evolution de Montpellier-Université de Montpellier, France. CM thanks the Conselho Nacional de Desenvolvimento Científico e Tecnológico (CNPq - Brazil) for an ongoing research grant (Proc. PQ 304736/2015-5). SK is supported by a grant of the German Research Foundation (DFG; grant KL2378/2-1). The sampling campaign in the Eirunepé region was enabled by the Austrian Science Fund (project P21748-N21 to MG). We also thank Peter K. L. Ng and Javier Luque for their helpful comments on the initial manuscript.

REFERENCES

ANTOINE, P.-O., M. A. ABELLO, S. ADNET, A. J. ALTAMIRANO SIERRA, P. BABY, G. BILLET, M. BOIVIN, Y. CALDERÓN, A. CANDELA, J. CHABAIN, F. CORFU, D. A. CROFT, M. GANERØD, C. JARAMILLO, S. KLAUS, L. MARIVAUX, R. E. NAVARRETE, M. J. ORLIAC, F. PARRA, M. E. PÉREZ, F. PUJOS, J.-C. RAGE, A. RAVEL, C. ROBINET, M. RODDAZ, J. V. TEJADA-LARA, J. VÉLEZ-JUARBE, F. P. WESSELINGH & R. SALAS-GISMONDI, 2016 (cf. a). A 60-million-year Cenozoic history of western Amazonian ecosystems in Contamana, eastern Peru. Gondwana Research, **31**: 30-59.

ANTOINE, P.-O., L. MARIVAUX, D. A. CROFT, G. BILLET, M. GANERØD, C. JARAMILLO, T. MARTIN, M. J. ORLIAC, J. TEJADA, A. J. ALTAMIRANO, F. DURANTHON, G. FANJAT, S. ROUSSE & R. S. GISMONDI, 2012. Middle Eocene rodents from Peruvian Amazonia reveal the pattern and timing of caviomorph origins and biogeography. Proceedings of the Royal Society of London Series B: Biological Sciences, **279**: 1319-1326.

ANTOINE, P.-O., M. RODDAZ, S. BRICHAU, J. V. TEJADA-LARA, R. SALAS-GISMONDI, A. J. ALTAMIRANO, M. LOUTERBACH, L. LAMBS, T. OTTO & S. BRUSSET, 2013. Middle Miocene vertebrates from the Amazonian Madre de Dios Subandean Zone, Perú. Journal of South American Earth Sciences, **42**: 91-102.

ANTOINE, P.-O., R. SALAS-GISMONDI, F. PUJOS, M. GANERØD & L. MARIVAUX, 2016 (cf. b). Western Amazonia as a hotspot of mammalian biodiversity throughout the Cenozoic. Journal of Mammalian Evolution, **24**: 5-17.

ASHKENAZI, S., U. MOTRO, N. GOREN-INBAR, R. BITON & R. RABINOVICH, 2005. New morphometric parameters for assessment of body size in the fossil freshwater crab assemblage from the Acheulian site of Gesher Benot Ya'aqov, Israel. Journal of Archaeological Science, **32**: 675-689.

BACHMAYER, F. & G. PRETZMANN, 1971. Krebsreste aus den altpliozänen Süßwasserablagerungen des Eichkogels bei Mödling, Niederösterreich. Annalen des naturhistorischen Museums Wien, **75**: 283-291.

BOIVIN, M., L. MARIVAUX, A. M. CANDELA, M. J. ORLIAC, F. PUJOS, R. SALAS-GISMONDI, J. V. TEJADA-LARA & P.-O. ANTOINE, 2017. Late Oligocene caviomorph rodents from Contamana, Peruvian Amazonia. Papers in Palaeontology, **3**: 69-109.

CARRIOL, R.-P. & S. SECRÉTAN, 1992. Présence dans le Bassin du Lac Albert (Ouganda), dès le Miocène supérieur, de *Potamonautes* (*Acanthothelphusa*) *niloticus* (Crustacea, Brachyura). Comptes rendus de l'Académie des sciences. Série 2, Mécanique, Physique, Chimie, Sciences de l'Univers, Sciences de la Terre, **314**: 411-417.

CONRAD, C., C. HIGHAM, M. EDA & B. MARWICK, 2016. Palaeoecology and forager subsistence strategies during the Pleistocene – Holocene transition: A reinvestigation of the zooarchaeological assemblage from Spirit Cave, Mae Hong Son Province, Thailand. Asian Perspectives, **55**(1): 2-27.

DOBSON, M., A. MAGANA, J. M. MATHOOKO & F. K. NDEGWA, 2002. Detritivores in Kenyan highland streams: more evidence for the paucity of shredders in the tropics? Freshwater Biology, **47**: 909-919.

EUDE, A., M. RODDAZ, S. BRICHAU, S. BRUSSET, Y. CALDERÓN, P. BABY & J.-C. SOULA, 2015. Controls on timing of exhumation and deformation in the northern Peruvian eastern Andean wedge as inferred from low-temperature thermochronology and balanced cross section. Tectonics, **34**: 715-730.

FELDMANN, R. M., P. M. O'CONNOR, N. J. STEVENS, M. D. GOTTFRIED, E. M. ROBERTS, S. NGASALA, E. L. RASMUSSON & S. KAPILIMA, 2007. A new freshwater crab (Decapoda: Brachyura: Potamonautidae) from the Paleogene of Tanzania, Africa. Neues Jahrbuch für Geologie und Paläontologie — Abhandlungen, **244**: 71-78.

FLYNN, J. J. & A. R. WYSS, 1999. New marsupials from the Eocene-Oligocene transition of the Andean Main Range, Chile. Journal of Vertebrate Paleontology, **19**: 533-549.

GOIN, F. J., M. A. ABELLO & L. CHORNOGUBSKY, 2010. Middle Tertiary marsupials from central Patagonia (early Oligocene of Gran Barranca): understanding South America's Grande Coupure. In: R. H. MADDEN, A. A. CARLINI, M. G. VUCETICH & R. F. KAY (eds.), The paleontology of Gran Barranca: evolution and environmental change through the Middle Cenozoic of Patagonia: 71-107. (Cambridge University Press, Cambridge).

GROSS, M., W. E. PILLER, M. I. RAMOS & J. DOUGLAS DA SILVA PAZ, 2011. Late Miocene sedimentary environments in south-western Amazonia (Solimoes Formation; Brazil). Journal of South American Earth Sciences, **32**: 169-181.

GROSS, M., M. I. RAMOS, M. CAPORALETTI & W. E. PILLER, 2013. Ostracods (Crustacea) and their palaeoenvironmental implication for the Solimoes Formation (Late Miocene; Western Amazonia/Brazil). Journal of South American Earth Sciences, **42**: 216-241.

HERMOZA, W., S. BRUSSET, P. BABY, W. GIL, M. RODDAZ, N. GUERRERO & R. BOLAÑOS, 2005. The Huallaga foreland basin evolution: thrust propagation in a deltaic environment, northern Peruvian Andes. Journal of South American Earth Sciences, **19**: 21-34.

HOORN, C. M., F. P. WESSELINGH, H. TER STEEGE, M. A. BERMUDEZ, A. MORA, J. SEVINK, I. SANMARTIN, A. SANCHEZ-MESEGUER, C. L. ANDERSON, J. P. FIGUEIREDO, C. JARAMILLO, D. RIFF, F. R. NEGRI, H. HOOGHIEMSTRA, J. LUNDBERG, T. STADLER, T. SARKINEN & A. ANTONELLI, 2010. Amazonia through time: Andean uplift, climate change, landscape evolution, and biodiversity. Science, **330**: 927-931.

KLAUS, S., M. BÖHME, S. SCHNEIDER, J. PRIETO & B. PHETSOMPHOU, 2011. Evidence of the earliest freshwater decapod fossil from Southeast Asia (Crustacea: Decapoda: Brachyura). The Raffles Bulletin of Zoology, **59**: 47-51.

KLAUS, S. & M. GROSS, 2010. Synopsis of the fossil freshwater crabs of Europe (Crustacea: Brachyura: Potamidae). Neues Jahrbuch für Geologie und Paläontologie — Abhandlungen, **256**: 39-59.

KLAUS, S. & J. PRIETO, 2014. New occurrence of Miocene freshwater crabs (Brachyura, Potamidae) in the North Alpine Foreland Basin, Germany, with a note on fossil *Potamon* to calibrate molecular clocks. In: D. C. J. YEO, N. CUMBERLIDGE & S. KLAUS (eds.), Advances in freshwater decapod systematics and biology. Crustacean Monographs, **19**: 161-172. (Brill, Leiden).

KLAUS, S., B. SINGH, L. HARTMANN, K. KRISHAN, A. GHOSH & R. A. PATNAIK, in press. A fossil freshwater crab from the Pliocene Tatrot Formation (Siwalik group) in Northern India (Crustacea, Brachyura, Potamidae). Palaeoworld. http://dx.doi.org/10.1016/j.palwor.2016.08.00

KLAUS, S., D. C. J. YEO & S. T. AHYONG, 2011. Freshwater crab origins — laying Gondwana to rest. Zoologischer Anzeiger, **250**: 449-456.

LATREILLE, P. A., 1802. Histoire naturelle, générale et particulière des crustacés et des insectes, **3**. (F. Dufart, Paris).

LATREILLE, P. A., 1828. Trichodactyle, *Trichodactylus*. In: Encyclopédie Méthodique. Histoire Naturelle. Entomologie, ou Histoire Naturelle des Crustacés, des Arachnides & des Insectes, **10**: 705.

LINHARES, A. P. & M. I. F. RAMOS, 2011. Evidence for marine influx during the Miocene in southwestern Amazonia, Brazil — Evidencias de influencia marina durante el Mioceno en el Sudoeste de la Amazonia (Brasil). Geología Colombiana, **36**: 91-104.

LINNAEUS, C., 1758. Systema naturae per regna tria naturae, secundum classes, ordines, genera, species, cum characteribus, differentiis, synonymis, locis. (Lars Salvi, Stockholm).

LUNDBERG, J. G., L. G. MARSHALL, J. GUERRERO, B. HORTON, M. MALABARBA & F. P. WESSELINGH, 1998. The stage for Neotropical fish diversification: a history of tropical South American rivers. In: L. R. MALABARBA, R. E. REIS, R. P. VARI, Z. M. LUCENA & C. A. LUCENA (eds.), Phylogeny and classification of Neotropical fishes: 13-48. (EDIPUCRS, Porto Alegre, Brazil).

MAGALHÃES, C., 2003. Brachyura: Pseudothelphusidae e Trichodactylidae. In: G. A. MELO (ed.), Manual de Identificação dos Crustáceos decápodos de água doce brasileiros: 143-297. (Edições Loyola, São Paulo).

MAGALHÃES, C. & M. TÜRKAY, 1996a. Taxonomy of the Neotropical freshwater crab family Trichodactylidae II. The genera *Forsteria*, *Melocarcinus*, *Sylviocarcinus*, and *Zilchiopsis* (Crustacea: Decapoda: Brachyura). Senckenbergiana biologica, **75**: 97-130.

MAGALHÃES, C. & M. TÜRKAY, 1996b. Taxonomy of the Neotropical freshwater crab family Trichodactylidae III. The genera *Fredilocarcinus* and *Goyazana* (Crustacea: Decapoda: Brachyura). Senckenbergiana biologica, **75**(1-2): 131-142.

MAGALHÃES, C. & M. TÜRKAY, 2008a. Taxonomy of the Neotropical freshwater crab family Trichodactylidae. IV. The genera *Dilocarcinus* and *Poppiana*. Senckenbergiana biologica, **88**: 185-215.

MAGALHÃES, C. & M. TÜRKAY, 2008b. Taxonomy of the Neotropical freshwater crab family Trichodactylidae. V. The genera *Bottiella* and *Rotundovaldivia* (Crustacea, Decapoda, Trichodactylidae). Senckenbergiana biologica, **88**: 217-230.

MAGALHÃES, C. & M. TÜRKAY, 2012. Taxonomy of the Neotropical freshwater crab family Trichodactylidae: VI. The genera *Avotrichodactylus* and *Rodriguezia* (Decapoda: Brachyura: Trichodactylidae). Nauplius, **20**: 27-40.

MARIJNISSEN, S. A., 2007. Lake Tanganyika crabs: evolution, ecology, and implications for conservation. (PhD Thesis, University of Amsterdam, Amsterdam).

MARIVAUX, L., S. ADNET, A. J. ALTAMIRANO SIERRA, M. BOIVIN, F. PUJOS, A. RAMDARSHAN, R. SALAS-GISMONDI, J. V. TEJADA-LARA & P.-O. ANTOINE, 2016. Neotropics provide insights into the emergence of New World monkeys: new dental evidence from the late Oligocene of Peruvian Amazonia. Journal of Human Evolution, **97**: 159-175.

MARIVAUX, L., R. SALAS-GISMONDI, J. TEJADA, G. BILLET, M. LOUTERBACH, J. VINK, J. BAILLEUL, M. RODDAZ & P.-O. ANTOINE, 2012. A platyrrhine talus from the early Miocene of Peru (Amazonian Madre de Dios Sub-Andean Zone). Journal of Human Evolution, **63**: 696-703.

MARTIN, J. W. & S. TRAUTWEIN, 2003. Fossil crabs (Crustacea, Decapoda, Brachyura) from Lothagam. In: M. G. LEAKEY & J. HARRIS (eds.), Lothagam: the dawn of humanity in Eastern Africa: 67-73. (Columbia University Press, New York, NY).

MELCHOR, R. N., J. F. GENISE, J. L. FARINA, M. V. SÁNCHEZ, L. SARZETTI & G. VISCONTI, 2010. Large striated burrows from fluvial deposits of the Neogene Vinchina Formation, La Rioja, Argentina: a crab origin suggested by neoichnology and sedimentology. Palaeogeography, Palaeoclimatology, Palaeoecology, **291**: 400-418.

MILNE EDWARDS, H., 1853. Mémoire sur la famille des Ocypodiens. Annales des sciences naturelles, **20**: 163-226.

MORRIS, S. M., 1976. A new fossil freshwater crab from the Ngorora Formation (Miocene) of Kenia. Bulletin of the British Museum (Natural History). Geology, **27**: 295-300.

NARUSE, T., H. KARASAWA, S. SHOKITA, T. TANAKA & M. MORIGUCHI, 2003. A first fossil record of the terrestrial crab *Geothelphusa tenuimanus* (Miyake & Minei, 1965) (Decapoda, Brachyura, Potamidae) from Okinawa Island, central Ryukyus, Japan. Crustaceana, **76**: 1211-1218.

NG, P. K. L. & EARL OF CRANBROOK, 2014. Fossil brachyuran crabs from the Jambusan Caves (Bau, Sarawak), collected by A.H. Everett in 1878-1879. Scripta geologica, **147**: 289-307.

PRETZMANN, G., 1968. Die Familie Trichodactylidae (Milne Edwards, 1853) Smith, 1870 (Vorläufige Mitteilung). Entomologisches Nachrichtenblatt, **15**(7-8): 70-76.

PRETZMANN, G., 1972. Some fossil chelae of river crabs and the distribution of *Potamon* in the Eastern Mediterranean and Western Asia. Thalassia jugoslavica, **8**: 71-73.

RABETT, R., G. BARKER, C. O. HUNT, T. NARUSE, P. PIPER, E. RADDATZ, T. REYNOLDS, N. VAN SON, C. STIMPSON, K. SZABÓ, N. C. TÂŃ & J. WILSON, 2009. The Tràng An project: late-to-post-Pleistocene settlement of the lower Song Hong valley, North Vietnam. Journal of the Royal Asiatic Society, **19**: 83-109.

RATHBUN, M. J., 1898. A contribution to a knowledge of the freshwater crabs of America. — The Pseudothelphusinæ. Proceedings of the United States National Museum, **21**(1158): 507-537.

RODDAZ, M., W. HERMOZA, A. MORA, P. BABY, M. PARRA, F. CHRISTOPHOUL, S. BRUSSET & N. ESPURT, 2010. Cenozoic sedimentary evolution of the Amazonian foreland basin system. In: C. M. HOORN & F. P. WESSELINGH (eds.), Amazonia — landscape and species evolution: a look into the past: 61-88. (Blackwell, Hoboken, NJ).

RODRÍGUEZ, G., 1981. Decapoda. In: S. H. HURLBERT, G. RODRÍGUEZ & N. D. SANTOS (eds.), Aquatic biota of tropical South America, part 1: Arthropoda: 41-51. (Academic Press, San Diego, CA).

RODRÍGUEZ, G., 1982. Les crabes d'eau douce d'Amérique: Famille des Pseudothelphusidae. Faune tropicale, **22**. (ORSTOM, Paris).

RODRÍGUEZ, G., 1986. Centers of radiation of freshwater crabs in the Neotropics. In: R. GORE & K. L. HECK (eds.), Crustacean biogeography. Crustacean Issues, **4**: 51-67. (A. A. Balkema Press, Rotterdam, Boston).

RODRÍGUEZ, G., 1997. Trichodactylid crabs. In: R. F. KAY, R. H. MADDEN, R. L. CIFELLY & J. J. FLYNN (eds.), Vertebrate paleontology in the Neotropics. The Miocene fauna of La Venta, Colombia: 63-66. (Smithsonian Institution Press, Washington, DC).

RODRÍGUEZ, G. & H. DÍAZ, 1977. Note sur quelques restes de crabes d'eau douce (Pseudothelphusidae) provenant d'un "kjoekken-moedding" du Vénézuéla. Crustaceana, **33**: 107-108.

SALAS-GISMONDI, R., J. J. FLYNN, P. BABY, J. V. TEJADA-LARA, J. CLAUDE & P.-O. ANTOINE, 2016. A new 13 million year old gavialoid crocodylian from proto-Amazonian mega-wetlands reveals parallel evolutionary trends in skull shape linked to longirostry. PLoS ONE, **11**: e0152453.

SALAS-GISMONDI, R., J. J. FLYNN, P. BABY, J. V. TEJADA-LARA, F. P. WESSELINGH & P.-O. ANTOINE, 2015. A Miocene hyperdiverse crocodylian community reveals peculiar trophic dynamics in proto-Amazonian mega-wetlands. Proceedings of the Royal Society of London Series B: Biological Sciences, **282**: 20142490.

TEJADA-LARA, J. V., R. SALAS-GISMONDI, F. PUJOS, P. BABY, M. BENAMMI, S. BRUSSET, D. DE FRANCESCHI, N. ESPURT, M. URBINA, P.-O. ANTOINE & A. GOSWAMI, 2015. Life in proto-Amazonia: middle Miocene mammals from the Fitzcarrald Arch (Peruvian Amazonia). Palaeontology, **58**: 341-378.

TSANG, L. M., C. D. SCHUBART, S. T. AHYONG, J. C. Y. LAI, E. Y. C. AU, T.-Y. CHAN, P. K. L. NG & K. H. CHU, 2014. Evolutionary history of true crabs (Crustacea: Decapoda: Brachyura) and the origin of freshwater crabs. Molecular Biology and Evolution, **31**(5): 1173-1187.

VON STERNBERG, R. & N. CUMBERLIDGE, 2003. Autapomorphies of the endophragmal system in trichodactylid freshwater crabs (Crustacea: Decapoda: Eubrachyura). Journal of Morphology, **256**: 23-28.

WESSELINGH, F. P., C. M. HOORN, J. GUERRERO, M. E. RÄSÄNEN, L. R. PITTMANN & J. A. SALO, 2006. The stratigraphy and regional structure of Miocene deposits in western Amazonia (Peru, Colombia and Brazil), with implications for late Neogene landscape evolution. Scripta Geologica, **133**: 291-322.

WESSELINGH, F. P. & M. I. RAMOS, 2010. Amazonian aquatic invertebrate faunas (Mollusca, Ostracoda) and their development over the past 30 million years. In: C. M. HOORN & F. P. WESSELINGH (eds.), Amazonia — landscape and species evolution: a look into the past: 302-316. (Blackwell, Hoboken, NJ).

YEO, D. C. J., P. K. L. NG, N. CUMBERLIDGE, C. MAGALHÃES, S. R. DANIELS & M. R. CAMPOS, 2008. Global diversity of crabs (Crustacea: Decapoda: Brachyura) in freshwater. Hydrobiologia, **595**: 275-286.

First received 20 September 2016.
Final version accepted 31 October 2016.

[When citing this volume, please refer to Crustaceana 90 (2017) 771-1288]

Michael Türkay Memorial Issue

A NEW SQUAT LOBSTER SPECIES OF THE GENUS *MUNIDA* (DECAPODA, ANOMURA, MUNIDIDAE) FROM THE DEEP-SEA OFF THE RYUKYU ISLANDS, JAPAN

BY

TOMOYUKI KOMAI[1])

Natural History Museum and Institute, Chiba, 955-2 Aoba-cho, Chuo-ku, Chiba, 260-8682, Japan

ABSTRACT

Munida michaeli, a new species of deep-sea squat lobster (Munididae), is described and illustrated on the basis of a single male specimen collected off Okinawa Island, Ryukyu Islands, Japan, at depths of 641-650 m. It closely resembles *M. sacksi* Macpherson, 1993 known with certainty only from the Philippines, but different spination of the carapace dorsum and the greatly reduced armature of the flexor margins of the ambulatory dactyli distinguish the new species from the latter. With the addition of the present new species, 41 species of *Munida* Leach, 1820 are now known from Japanese waters.

ZUSAMMENFASSUNG

Hier wird *Munida michaeli*, eine neue Art der Tiefsee-Furchenkrebse (Munididae) beschrieben und illustriert. Die Beschreibung basiert auf dem Fund eines einzelnen männlichen Individuums, welches vor der Insel Okinawa entlang der Ryukyu-Inseln in einer Tiefe von 641-650 m gesammelt wurde. Das Tier ähnelt stark *M. sacksi* Macpherson, 1993, welche lediglich von den Philippinen sicher bekannt ist, jedoch unterscheidet sich die neue Art gegenüber letzterer durch Unterschiede im Bezahnungsmuster des Carapax-Vorderrandes, sowie durch die stark reduzierte Bedornung des Innenrandes der Dactyli der Laufbeine. Unter Einbezug der hier beschriebenen, neuen Art sind nun 41 Arten von *Munida* Leach, 1820 aus japanischen Gewässern bekannt.

INTRODUCTION

Munida Leach, 1820 is the most species-rich genus in the squat lobster family Munididae Ahyong, Baba, Macpherson & Poore, 2010, represented by about 280 species worldwide (Baba et al., 2008; updated by Cabezas et al., 2009, 2011; Macpherson, 2009, 2013; Hendrickx & Ayon-Parente, 2010; Komai, 2011a, b, 2012, 2014; Liu et al., 2013; Komai & Higashiji, 2016). The northwest Pacific

[1]) e-mail: komai@chiba-muse.or.jp

region is one of the hot spot areas harbouring species of the genus (Baba, 2005; Baba et al., 2009; Komai, 2012), and discovery of new species still continues. In this article, a new species, *M. michaeli*, is described and illustrated on the basis of a single male specimen collected from deep sea off Okinawa Island, Ryukyu Islands, Japan. It closely resembles *M. sacksi* Macpherson, 1993, known with certainty only from the Philippines, but can be easily distinguished from the latter by a number of morphological characters.

The holotype of the new species was collected during the research cruise by TRV "Toyoshio-maru" of the School of Applied Biological Science, Hiroshima University, to the Ryukyu Islands in 2006 (chief scientist Prof. Dr. Susumu Ohtsuka of Hiroshima University), using a small beam trawl with a 2 m span opening. It is deposited in the Natural History Museum and Institute, Chiba (CBM), Japan. The carapace length (cl), indicating specimen size, was measured along the midline from the level of the sinus between rostrum and supraocular spines to the posterior margin of the carapace. The lengths of articles of the cheliped are measured along the dorsomesial margin and those of ambulatory legs are along the extensor margin. The higher classification follows Ahyong et al. (2010). Terminology for carapace spines follows Baba et al. (2009).

TAXONOMY

Family MUNIDIDAE Ahyong, Baba, Macpherson & Poore, 2010

Genus *Munida* Leach, 1820

Munida michaeli sp. nov.

[New Japanese name: Ibara-chu-koshiori-ebi]

(figs. 1, 2)

Type material.— Holotype: CBM-ZC 13564, male (cl 14.9 mm), TRV "Toyoshio-maru", 2006-03 cruise, stn 8, NW of Cape Zanpa, Okinawa Island, Ryukyu Islands, 26°27.06′N 127°32.25′E to 26°27.50′N 127°32.10′E, 650-641 m, 27 May 2006, beam trawl, coll. M. Shimomura.

Description.— Carapace (excluding rostrum) (fig. 1A) 1.4 times as long as wide. Dorsal surface gently convex transversely; main transverse ridges mostly interrupted medially and/or laterally; numerous secondary striae intervening main ridges, mostly interrupted into some short, sometimes scale-like, striae on entire dorsum; most ridges and striae with row of short setae, but no iridescent setae. Gastric region slightly elevated, with 1 median and 3 pairs of epigastric spines becoming smaller laterally, and 1 additional spine posterior to median epigastric spine; postrostral carina low, but discernible. Cervical groove shallow, but distinct. One parahepatic, 4 branchial dorsal (arranged in obliquely longitudinal row along cervical groove) and 5 postcervical spines (arranged in irregular transverse row)

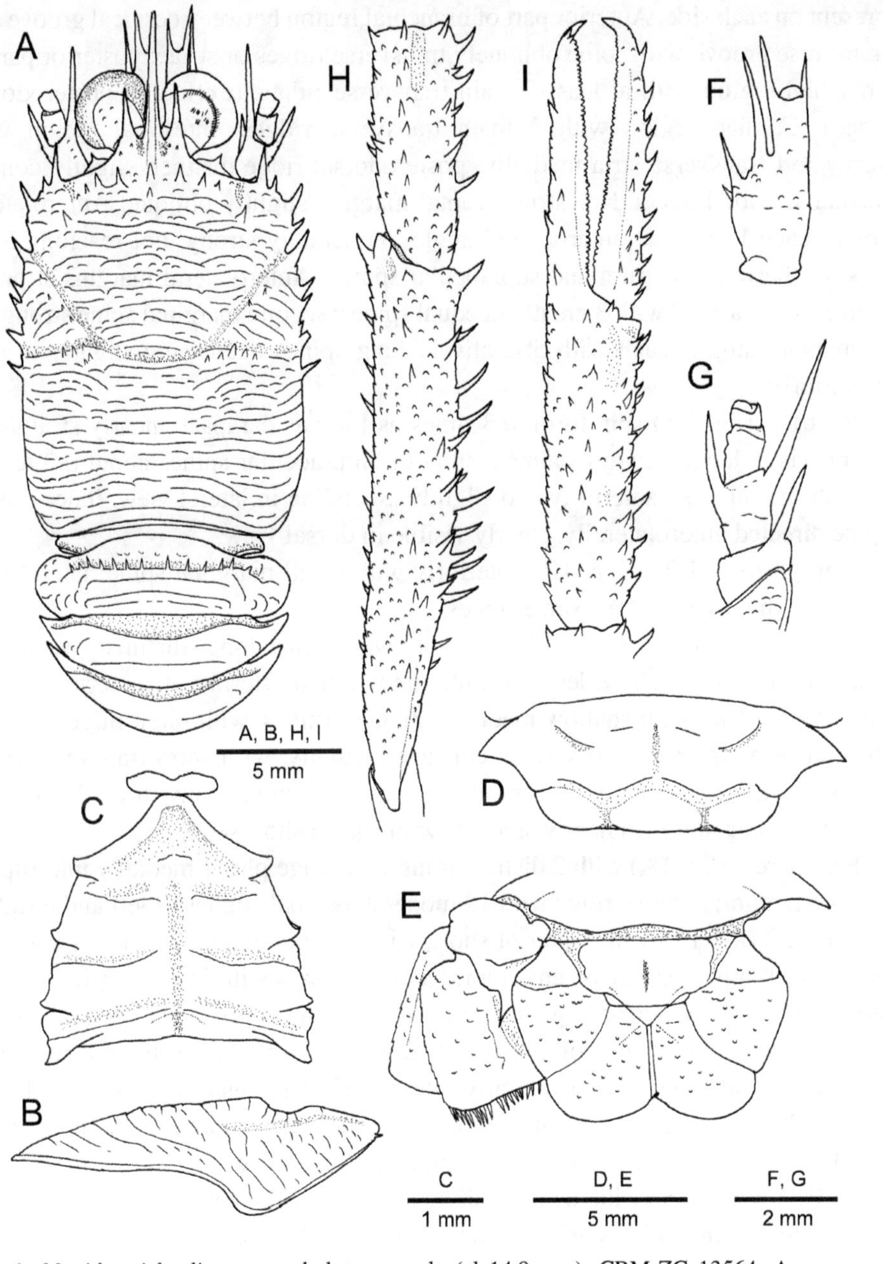

Fig. 1. *Munida michaeli* sp. nov., holotype, male (cl 14.9 mm), CBM-ZC 13564: A, carapace, pleomeres 1-4 and cephalic appendages, dorsal view (setae on carapace omitted; eye lash shown in right side); B, right pterygostomial flap, ventrolateral view (setae omitted); C, thoracic sternite, ventral view (setae omitted); D, pleomere 6, outer view; E, telson and left uropod, outer view; F, basal article of right antennular peduncle, ventral view (setae omitted); G, left antennal peduncle, ventral view (setae omitted); H, merus and carpus of left cheliped, dorsal view (setae omitted); I, chela of left cheliped, dorsal view (setae omitted).

present on each side. Anterior part of branchial region between cervical groove and transverse groove with some obliquely transverse ridges or striae; posterior part of branchial region with at least 5 main transverse ridges (excluding posterodorsal ridge). Cardiac region with 3 main transverse ridges. Intestinal region with very short transverse stria medially; posterodorsal ridge distinct, slightly convex medially, with 1 secondary stria. Frontal margins slightly oblique. Anterolateral spines each located at anterolateral angle, moderately strong, not reaching level of sinus between rostrum and supraocular spines. Anterolateral margins strongly convex, each armed with 3 greatly unequal spines; second marginal spine strongest. Branchial margins each with 5 relatively long spines decreasing slightly in size posteriorly.

Rostrum (fig. 1A) spiniform, 0.3 times as long as carapace, nearly straight in lateral view; lateral margins almost smooth. Supraocular spines about 0.3 length of rostrum, directed anteriorly and slightly ascending in lateral view. Inner orbital spine directed anterolaterally, clearly visible in dorsal view.

Pterygostomial flap (fig. 1B) anteriorly with small marginal spine; lateral face rugose with obliquely transverse ridges.

Thoracic sternite 3 (fig. 1C) 3.9 times wider than long, distinctly separated from sternite 4, much wider than anterior width of sternite 4; anterior margin faintly granulate, with shallow median notch. Sternite 4 with blunt ridge adjacent to anterior margin; ventral surface concave medially, with short transverse ridge narrowly interrupted medially. Sternites 5-7 smooth, without any striae. Transverse ridges dividing sternites nearly smooth, with sparse short setae.

Pleomere 2 (fig. 1A) with 2 distinct transverse ridge plus 1 medially interrupted stria on tergum; anterior ridge with 13 moderately small spines (7 left and 6 right). Pleomere 3 (fig. 1A) with 1 pair of short, widely separated striae and 1 transverse ridge on tergum, unarmed on anterior ridge. Pleomere 4 with 1 medially interrupted stria, unarmed on anterior ridge. Pleomere 5 with 2 medially interrupted transverse ridges on tergum. Pleura of pleomeres 2-5 each with few squamiform striae. Pleomere 6 (fig. 1D) with 1 laterally interrupted stria anterior to midlength and broadly W-shaped groove on posterior half, otherwise smooth; posterior margin faintly tri-lobed. All ridges or striae on pleomeres with fringe of short setae. Telson (fig. 1E) 1.5 times as wide as long, divided in 7 main plates; intermediate lateral and posteromedial plates with scattered, tiny scale-like tubercles or granules; anteromedian plate with shallow median groove, with division of small triangular plate posteriorly.

Eyes (fig. 1A) large. Cornea strongly dilated, corneal width 0.22 cl. Eyestalk widened distally; eyelashes very short.

Basal article of antennular peduncle (fig. 1A, F) moderately stout, length excluding distal spines 2.2 of width; distal spines unequal with mesial spine distinctly shorter than lateral; 2 lateral spines present, of them distal one located slightly distal to midlength of article, directed somewhat upward, extending beyond distolateral spine, proximal one about half length of distal spine; 1 small additional spine proximal to base of proximal spine; statocyst lobe weakly inflated, ventral surface rounded, with scattered scale-like tubercles and tufts of short setae.

Antennal peduncle (fig. 1A, G) moderately stout, not reaching distal corneal margin. Article 1 with moderately long distomesial spine slightly overreaching distal margin of article 2; distolateral angle unarmed. Article 2 with distomesial spine elongate, far overreaching distal end of article 4, distolateral spine nearly reaching distal end of article 4; surfaces with some scale-like tubercles and transverse striae. Articles 3 and 4 unarmed.

Maxilliped 3 (fig. 2A) moderately slender. Ischium distinctly longer than merus, with moderately large ventrodistal spine, extensor distal angle with minute spine; lateral surface with distinct longitudinal carina along extensor margin and few tiny scale-like tubercles. Merus narrowed distally, flexor margin only with 1 strong spine proximal to midlength; extensor distal margin unarmed; extensor margin and lateral face with several small, low tubercles. Exopod narrowing distally, distinctly overreaching distal margin of merus; lateral surface with scattered granules.

Cheliped (pereopod 1) (figs. 1H, I; 2B, C) elongate, about 3.7 times as long as carapace. Basis with small distolateral spine (fig. 2D). Ischium with row of small spines, increasing in size distally, on ventral surface. Merus with numerous scattered, tiny, scale-like tubercles and tufts of short to long plumose setae on surfaces (setae particularly numerous on mesial face), no iridescent setae; distal edges of these tubercles elevated, occasionally multidenticulate; dorsal surface with double row of long, slender spines mesially and single row of much smaller spines on midline to dorsolateral margin; ventrolateral margin with row of small spines and 1 strong spine at distal angle; mesial face with short row of long spines adjacent to dorsal margin in proximal 0.3 and row of shorter spines adjacent to ventral margin along entire length; ventral surface with row of small spines adjacent to mesial margin. Carpus 0.8 times as long as palm, approximately 3 times as long as distal width; surfaces with numerous scattered tubercles and tufts of short to long setae; dorsal surface with 3 main longitudinal rows of spines (spines on dorsomesial row strongest); lateral surface with few small spines in addition to tubercles; mesial surface with row of widely spaced, strong spines on midline; ventral surface with 2 strong spines in proximal half mesially and row of small spines extending to distal angle. Palm 3.3 times as long as wide; surfaces with numerous scattered tiny tubercles or granules and tufts of short setae; dorsal surface with 3 main rows of small spines, dorsal articular knob with 1 small spine; mesial face with row of small spines; ventral surface with row of small spines mesially and 1 spine at base of fixed finger, ventral articular knob

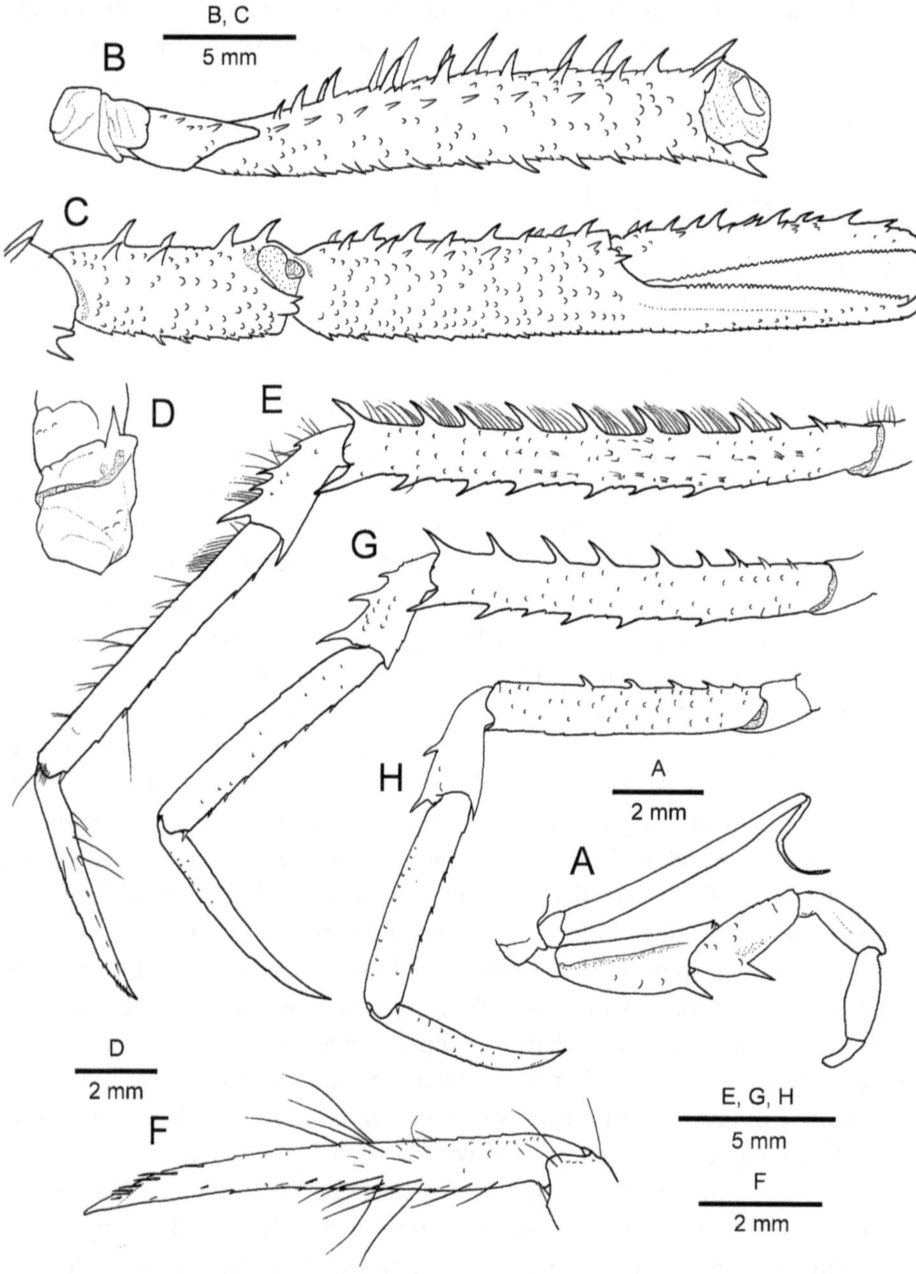

Fig. 2. *Munida michaeli* sp. nov., holotype, male (cl 14.9 mm), CBM-ZC 13564: A, right maxilliped 3, lateral view (setae omitted); B, ischium and merus of left cheliped, ventral view (setae omitted); C, carpus and chela of left cheliped, ventral view (setae omitted); D, coxa and ischium of left cheliped, ventral view; E, left pereopod 2, outer view; F, same, dactylus, outer view; G, pereopod 3, outer view (setae omitted); H, pereopod 4, outer view.

also with small spine. Fingers with tiny scattered tubercles, long simple setae and shorter plumose setae over entire surfaces; no hiatus between fingers. Fixed finger straight, terminating in curved, sharp claw; dorsal surface laterally with row of widely spaced, small spines; cutting edge straight, with row of minute, rounded or subacute denticles. Dactylus nearly straight, almost as long as palm, terminating in sharp, curved claw crossing tip of fixed finger; dorsal surface with 2 proximomesial spines and median row of smaller spines in proximal half; mesial face ventrally with row of strong spines extending nearly to tip; ventral surface mesially with row of small spines in proximal half.

Ambulatory legs (pereopods 2-4) relatively long and slender, successively becoming shorter. Pereopod 2 (fig. 2E), when anteriorly extended, distinctly overreaching tip of rostrum by tip of mero-carpal articulation, about 2.6 of carapace length. Merus 1.2 times as long as carapace; extensor margin with row of numerous setae and row of 11 spines increasing in length distally (proximalmost spine tiny); outer surface with scattered, minute, scale-like tubercles or granules and tufts of very short setae; outer flexor margin with 7 spines generally increasing in size distally and intervening minute acute granules; inner flexor margin with row of 6 (left) or 8 (right) spines increasing in size distally (those spines generally smaller than those of outer flexor margin). Carpus with prominent extensor distal spine followed by 3 smaller spines on extensor margin, flexor distal margin produced in prominent spine; lateral face with row of minute, low, scale-like tubercles adjacent to extensor margin. Propodus with sparse tufts or individual long simple setae on extensor margin; outer surface with some minute tubercles or granules; inner surface with row of numerous plumose setae adjacent to extensor margin; flexor margin with row of 8 evenly spaced spiniform setae, all but distalmost one arising from low protuberances, distalmost one without support by contiguous acute spine. Dactylus (fig. 2F) 0.8 times as long as propodus, very slender (9.7 times longer than greatest width), slightly curved distally, bearing sparse short to long stiff setae, flexor margin faintly sinuous, with 5 widely spaced, microscopic spiniform setae, distal 0.3 unarmed.

Pereopod 3 (fig. 2G) generally similar to pereopod 2 in general armature and setation, though distinctly shorter; merus 0.7 times as long as that of pereopod 2, with 8 (left) or 9 (right) spines on extensor margin, outer flexor margin with 5 (right) or 6 (left) spines, inner flexor margin with only 1 subterminal spine; carpus with 3 spines on extensor margin; propodus with 8 spiniform setae on flexor margin; dactylus with 5 microscopic spiniform setae on flexor margin, distal 0.3 unarmed. Pereopod 4 (fig. 2H) not reaching anterolateral angle of carapace by mero-carpal articulation; merus about half length of that of pereopod 2, extensor margin with 5 spines (extensor distal angle unarmed), outer flexor margin only with small distolateral spine, inner flexor margin unarmed; carpus with 2 spines on extensor margin and 1 small spine on flexor distal margin; propodus with 7 spiniform setae on flexor margin (including distolateral one), and with 1 additional spiniform seta on flexor distal margin mesially; dactylus with 3 microscopic spiniform setae on flexor margin.

Merus of pereopod 5 slightly rugose on lateral surface, outer flexor margin sharply carinate.

Epipods absent from all pereopods.

Gonopods 1 and 2 present.

Uropodal exopod (fig. 1E) with lateral margin minutely denticulate, with microscopic spiniform setae distally; distal margin truncate, without spiniform setae; outer surface with longitudinal row of minute scale-like tubercle adjacent to lateral margin, none of these tubercles with spiniform setae. Endopod (fig. 1E) with scattered, tiny scale-like tubercles on outer surface, none of these tubercles with spiniform setae; lateral margin minutely denticulate; distal margin truncate, with row of short to long spiniform setae. Uropodal protopod (fig. 1E) with slender spine on inner distal margin.

Coloration in life.— Not known.

Distribution.— Known only from the type locality, NW of Cape Zanpa, Okinawa Island, at depths of 641-650 m.

Remarks.— The holotype is generally in good condition, but the left cheliped and distal two articles of the right pereopod 2 are missing; right pereopods 2 and 3 are detached from the body, though still preserved. Although only the unique holotype is available for examination, the establishment of new taxon is warranted by the morphological evidence, as discussed below.

The new species closely resembles *Munida sacksi*, currently known only from the Philippines in sharing the following key characters: five spines on branchial margin of carapace; carapace with numerous secondary striae between main ridges; thoracic sternite 4 subtriangular, its anterior margin narrowly contiguous with sternite 3; thoracic sternites 4 to 7 smooth, without striae or granules; rostrum spiniform; pleomere 2 with at least eight spines on anterior ridge, while pleomeres 3 and 4 unarmed; eyes large, strongly dilated; basal article of antennular peduncle with distomesial spine distinctly shorter than distolateral spine; article 2 of antennal peduncle with distomesial spine overreaching distal end of article 4; and dactylus of pereopod 2 very slender, unarmed on distal 0.3 of flexor margin. Nevertheless, *M. michaeli* sp. nov. is characteristic in having a greatly reduced armature of the flexor margins of ambulatory dactyli. For example, the pereopod 2 dactylus bears only five microscopic spiniform setae on the flexor margin (fig. 2D). In *M. sacksi*, there are eight small but well-developed spiniform setae on the flexor margin of pereopod 2 dactylus (cf. Macpherson, 1993, fig. 6f). Furthermore, the anterior branchial region of the carapace bears four small spines in *M. michaeli* (fig. 1A), rather than being unarmed in *M. sacksi* (cf. Macpherson, 1993, fig. 6a); there are five postcervical spines on each side in *M. michaeli* (fig. 1A), but only one in *M. sacksi* (cf. Macpherson, 1993, fig. 6a); pleomere 2 has more numerous spines on the anterior ridge in *M. michaeli* than in *M. sacksi* (13 versus 8); the basal article of the antennal peduncle is unarmed at the distolateral angle in *M. michaeli* (fig. 1F), but is armed with a spine in *M. sacksi* (cf. Macpherson, 1993, fig. 6c). Unfortunately, the morphology of the cheliped remains unknown for *M. sacksi*, because the chelipeds were missing in the holotype, the only known representative of the species.

Munida delicata Macpherson, 2004, previously confounded with *M. sacksi* and now known from New Caledonia, Vanuatu, Fiji and Tonga (Macpherson, 2004), can be easily distinguished from *M. michaeli* by the characters above cited for *M. sacksi*. The much fewer secondary striae on the carapace and less spinose carpus and chela of pereopod 1 also separate *M. delicata* (cf. Macpherson, 2004, fig. 6A, E) from *M. michaeli*.

Komai (2012) briefly reviewed species of *Munida* known from Japanese waters, recording 37 species. Subsequently, two new species and one newly recorded species have been added (Komai, 2013; Komai & Higashiji, 2016). With the addition of the present new species, 41 species of the genus are now known from Japanese waters. From adjacent Taiwan, 28 species are recorded (Baba et al., 2009; Cabezas et al., 2011), of which only 12 species are common between the two areas. There is little doubt that ongoing study by the author and future investigations will reveal the presence of more unknown species in Japanese waters.

Etymology.— This new species is dedicated to the late Prof. Dr. Michael Türkay for his immense contributions to the taxonomy and systematics of decapod crustaceans.

ACKNOWLEDGEMENTS

I thank Dr. Michitaka Shimomura (Kitakyushu Museum of Natural History and Human History) and Prof. Dr. Susumu Ohtsuka (Hiroshima University) for making the interesting collections of decapods from cruises by TRV "Toyoshio-maru" available for study. Thanks are extended also to two anonymous referees for reviewing the manuscript.

REFERENCES

AHYONG, S. T., K. BABA, E. MACPHERSON & G. C. B. POORE, 2010. A new classification of the Galatheoidea (Crustacea: Decapoda: Anomura). Zootaxa, **2676**: 57-68.

BABA, K., 2005. Deep-sea chirostylid and galatheid crustaceans (Decapoda: Anomura) from the Indo-West Pacific, with a list of species. Galathea Reports, **20**: 1-317.

BABA, K., E. MACPHERSON, C.-W. LIN & T.-Y. CHAN, 2009. Crustacean fauna of Taiwan: squat lobsters (Chirostylidae and Galatheidae): ix + 311 pp. (National Taiwan Ocean University, Keelung).

BABA, K., E. MACPHERSON, G. C. B. POORE, S. T. AHYONG, A. B. P. CABEZAS, C.-W. LIN, M. NIZINSKI, C. RODRIGUES & K. E. SCHNABEL, 2008. Catalogue of squat lobsters of the world (Crustacea: Decapoda: Anomura — families Chirostylidae, Galatheidae and Kiwaidae). Zootaxa, **1905**: 1-220.

CABEZAS, P., C.-W. LIN & T.-Y. CHAN, 2011. Two new species of the deep-sea squat lobster genus *Munida* Leach, 1820 (Crustacea: Decapoda: Munididae) from Taiwan: morphological and molecular evidence. Zootaxa, **3036**: 26-38.

CABEZAS, P., E. MACPHERSON & A. MACHORDOM, 2009. Morphological and molecular descriptions of new species of squat lobster (Crustacea: Decapoda: Galatheidae) from the Solomon and Fiji Islands (south-west Pacific). Zoological Journal of the Linnean Society, **156**: 465-493.

HENDRICKX, M. E. & M. AYON-PARENTE, 2010. A new species of *Munida* Leach (Decapoda, Galatheidae) from off the west coast of Baja California, Mexico. In: C. H. J. M. FRANSEN, S. DE GRAVE & P. K. L. NG (eds.), Studies on Malacostraca. Crustaceana Monographs, **14**: 305-314.

KOMAI, T., 2011a. A new species of the squat lobster genus *Munida* (Decapoda: Anomura: Galatheidae) from the north Pacific off Japan. Bulletin of the National Museum of Nature and Science, (Supplement) **5**: 101-108.

KOMAI, T., 2011b. Squat lobsters of the genus *Munida* (Crustacea: Decapoda: Anomura: Munididae) from the Ogasawara Islands, with descriptions of four new species. Memoirs of the National Museum of Nature and Science, **47**: 339-365.

KOMAI, T., 2012. Squat lobsters of the genus *Munida* Leach, 1820 (Crustacea: Decapoda: Anomura: Munididae) from the Sagami Sea and Izu Islands, central Japan, with descriptions of 10 new species. Natural History Research, **12**: 1-69.

KOMAI, T., 2014. A new bathyal species of the squat lobster genus *Munida* Leach, 1820 (Crustacea: Decapoda: Anomura: Munididae) from Japan. Natural History Research, **13**: 11-18.

KOMAI, T. & T. HIGASHIJI, 2016. New records of the squat lobster genus *Munida* Leach, 1820 (Crustacea: Decapoda: Anomura: Munididae) from deep-water off Okinawa Islands, Ryukyu Islands, Japan, with description of a new species. Zootaxa, **4109**: 542-554.

LIU, X., R. LIN & D. HUANG, 2013. A new species of deep-sea squat lobster of the genus *Munida* Leach, 1820 (Crustacea: Decapoda: Anomura: Munididae) from a hydrothermal field in the southwestern Indian Ocean. Zootaxa, **3734**: 380-384.

MACPHERSON, E., 1993. Crustacea Decapoda: species of the genus *Munida* Leach, 1820 (Galatheidae) collected during the MUSORSTOM and CORINDON cruises in the Philippines and Indonesia. In: A. CROSNIER (ed.), Résultats des Campagnes MUSORSTOM, **10**. Mémoires du Muséum National d'Histoire Naturelle, Paris, **156**: 421-442.

MACPHERSON, E., 1994. Crustacea Decapoda: studies on the genus *Munida* Leach, 1820 (Galatheidae) in New Caledonia and adjacent waters with descriptions of 56 new species. In: A. CROSNIER (ed.), Résultats des Campagnes MUSORSTOM, **12**. Mémoires du Muséum National d'Histoire Naturelle, Paris, **161**: 421-569.

MACPHERSON, E., 2004. Species of the genus *Munida* Leach, 1820 and related genera from Fiji and Tonga (Crustacea: Decapoda: Galatheidae). In: B. A. MARSHALL & B. RICHER DE FORGES (eds.), Tropical deep-sea benthos, **23**. Mémoires du Muséum National d'Histoire Naturelle, Paris, **191**: 231-292.

MACPHERSON, E., 2009. New species of squat lobsters of the genera *Munida* and *Raymunida* (Crustacea, Decapoda, Galatheidae) from Vanuatu and New Caledonia. Zoosystema, **31**: 431-451.

MACPHERSON, E., 2013. New species and new occurrences of squat lobsters (Crustacea, Decapoda, Munididae, Eumunididae) from French Polynesia. In: S. T. AHYONG, T.-Y. CHAN, L. CORBARI & P. K. L. NG (eds.), Tropical deep-sea benthos, **27**. Mémoires du Muséum National d'Histoire Naturelle, Paris, **204**: 287-309.

First received 7 October 2016.
Final version accepted 19 October 2016.

[When citing this volume, please refer to Crustaceana 90 (2017) 771-1288]

Michael Türkay Memorial Issue

NEW RECORDS OF KING CRABS (DECAPODA, ANOMURA, LITHODIDAE) FROM SOUTHERN BRAZIL

BY

L. LIANOS[1,2]), M. C. MOLLEMBERG[2]), D. J. M. LIMA[1,2]) and W. SANTANA[1,2,3])

[1]) Programa de Pós-Graduação em Zoologia, Universidade Estadual Paulista "Júlio de Mesquita Filho", 18618-970 Botucatu, São Paulo, Brazil

[2]) Universidade do Sagrado Coração, Pró-reitora de Pesquisa e Pós-Graduação, Laboratório de Sistemática Zoológica, Rua Irmã Arminda, 10-50, Jd. Brasil, 17011-160 Bauru, São Paulo, Brazil

ABSTRACT

Until now only six species of lithodids have been found occasionally in the Brazilian coast. Most king crabs are accidentally caught during fisheries targeting other species. Herein we report new records for *Lithodes confundens* and *Neolithodes* aff. *asperrimus* from the south of Brazil, from the states of Rio Grande do Sul and Santa Catarina, respectively. *Lithodes confundens* is found in the western Atlantic with a previous northern limit in the Province of Buenos Aires, Argentina. *Neolithodes* aff. *asperrimus* is known to occur only along the western coast of Africa in the eastern Atlantic and this is the first record of the species in the western Atlantic.

Key words. — Lithodoidea, South America, western Atlantic, eastern Atlantic, distribution

ZUSAMMENFASSUNG

Bislang wurden nur sechs Arten der Lithodidae vereinzelt an der brasilianischen Küste gefunden. Die meisten Königskrabben werden zufällig von Fischern gefangen, welche auf andere Arten abzielen. Wir berichten hier über neue Nachweise von *Lithodes confundens* und *Neolithodes* aff. *asperrimus*, jeweils von den Staaten Rio Grande do Sul und Santa Catarina im Süden Brasiliens. *Lithodes confundens* ist im Westatlantik verbreitet. Seine ehemalige nördliche Verbreitungsgrenze lag in der Provinz von Buenos Aires, Argentinien. *Neolithodes* aff. *asperrimus* tritt bekannterweise an der Westküste Afrikas im Ostatlantik auf und dies ist somit der Erstnachweis dieser Art im Westatlantik.

[3]) Corresponding author; e-mail: willsantana@gmail.com

INTRODUCTION

The family Lithodidae currently comprises 10 genera and 123 species of some of the larger anomurans, with a high economical value (McLaughlin et al., 2010; Stevens & Lovrich, 2014). Lithodids can be found in subtidal waters in high latitudes to the deep sea in most of the world's oceans, although predominantly in colder water. This family has a greater diversity in the North Pacific and a limited distribution in tropical and subtropical areas (Zaklan, 2002; Hall & Thatje, 2009; Stevens & Lovrich, 2014).

Currently 28 species are recorded from the Atlantic, i.e., five *Neolithodes* A. Milne-Edwards & Bouvier, 1894, seven *Lithodes* Latreille, 1806 and 16 *Paralomis* White, 1856. These three genera are distributed from Greenland to the southwestern Atlantic sector of the Southern Ocean and the Barents Sea of the Cape Peninsula (Macpherson, 1988; Anosov et al., 2015). In South America, eleven species of lithodids are found in the western Atlantic south of 30°S, mainly on the continental shelf or the shelf break (Sotelano et al., 2013).

From Brazil lithodids are rarely recorded, as they are only accidentally caught during fisheries targeted at other species. Six species of lithodids have been recorded from Brazilian waters: *Lithodes manningi* Macpherson, 1988; *Neolithodes agassizii* (Smith, 1882); *Paralomis arethusa* Macpherson, 1994; *P. cubensis* Chace, 1939; *P. formosa* Henderson, 1888; and *P. grossmani* Macpherson, 1988 (Melo, 1999; Serejo et al., 2007; Stevens & Lovrich, 2014).

Herein we report the occurrence of *Lithodes confundens* and *Neolithodes* aff. *asperrimus* collected in Rio Grande do Sul and Santa Catarina states in the southern region of Brazil, extending the distribution of each species to 3°N and 32°W, respectively.

MATERIAL AND METHODS

The specimens are deposited in the collections of Centro de Estudos Marinhos do Atlântico Sul (CENEMAR), Rio Grande do Sul, and in the Laboratório de Sistemática Zoológica (LSZ), Universidade do Sagrado Coração, São Paulo, Brazil. The terminology used mostly follows Macpherson & Chan (2008). Carapace length (CL) was measured from the base of the rostrum to the posterior margin of the carapace, and carapace width (CW) without lateral spines. Measurements are given in mm.

SYSTEMATIC ACCOUNT

Family LITHODIDAE Samouelle, 1819
Genus *Lithodes* Latreille, 1806
Lithodes confundens Macpherson, 1988
(fig. 1)

Material examined.— Western Atlantic, Brazil, Rio Grande do Sul, off Chuí, F/V "Águia", Hélio de Navegantes coll., 18.viii.2005, 430 m, female CL 142 mm, CW 135 mm (LSZ 0127).

Diagnosis.— Rostrum short, bifid, somewhat upwardly directed, with anterior projection extending slightly beyond corneas, not reaching the distal end of antennal peduncle. Each branchial region armed dorsally with approximately 30 small spines or spiniform granules. Walking legs with numerous small spines and granules (based on Macpherson, 1988).

Distribution.— Previously reported from the western Atlantic, between 36 and 54°S, off the Province of Buenos Aires, Argentina, southern Patagonia, up to the eastern entrance of the Strait of Magellan and on the Burdwood Bank and south of the Falkland Islands, and in the eastern Pacific from Punta Arenas (Chile); on muddy bottoms from the intertidal to a depth of 600 m (Macpherson, 1988; Lovrich et al., 2002; Zaklan, 2002; Sotelano et al., 2013; Anosov et al., 2015; Pérez-Barros et al., 2015). The present record is from Brazil, Rio Grande do Sul, off Chuí; 33°S; 430 m.

Remarks.— The specimen studied herein generally conforms to the original description of *Lithodes confundens* (cf. Macpherson, 1988), however differences can be found in the ratio of the carapace length versus the length of the third walking leg. According to Macpherson (1988) and Anosov et al. (2015), the merus of the third walking leg is equal to or slightly less than the carapace length (excluding corneas), while in the specimen analysed here the carapace length is twice the merus length of the third walking leg (fig. 1) (see Macpherson, 1988, plate 11, female allotype CL 75 mm, and Anosov et al., 2015, fig. 2, male CL 83 mm). *Lithodes confundens* is closely related to *L. santolla* (Molina, 1782), as they are the only two species in this genus with a short anterior projection of the rostrum, revealing the basal spine dorsally. However, the two species can be distinguished by the size and number of spines on the carapace, chelipeds, and walking legs, which are always smaller and much more numerous in *L. confundens* (see Macpherson, 1988 for details). Despite these differences highlighted by Macpherson (1988), Pérez-Barros et al. (2015) indicated that morphology alone can be misleading in the identification of *L. confundens* and *L. santolla*. However, they also suggest that northern specimens are mostly like *L. confundens* and *L. santolla* has a restricted distribution to Puerto Montt and the Beagle Channel,

Fig. 1. *Lithodes confundens* Macpherson, 1988, female, CL 142 mm, CW 135 mm, Rio Grande do Sul (LSZ 0127): A, dorsal habitus; B, cephalothorax and abdomen, ventral view; C, cephalothorax, front; D, right cheliped, lateral view; E, left cheliped, lateral view. Scale bars: A, 30 mm; B-E, 20 mm.

which conforms to our identification. Pérez-Barros et al. (2015) have suggested that the use of the number of spines as a taxonomic character for *Lithodes* should be revised.

Genus *Neolithodes* A. Milne-Edwards & Bouvier, 1894

Neolithodes aff. **asperrimus** Barnard, 1947

(fig. 2)

Material examined.— Western Atlantic, Brazil, Santa Catarina, 2005, between 250-500 m, male CL 168 mm, CW 136 mm, dry specimen (CENEMAR CAR 0116).

Diagnosis.— Carapace, chelipeds and walking legs with numerous dispersed small spines distributed. Walking legs with reduced spines tending to be either thick granules or very short blunt spines; male third walking leg approximately 2.5 times CL, merus generally shorter than CL, dactyl of walking legs flattened dorsoventrally, ventral surface with a median cambered region in the distal half (based on Macpherson, 1988).

Distribution.— Eastern Atlantic, between 20°N-34°S and 6°W-18°E, from Mauritanian waters to Cape Point in the south, in Saldanha Bay and along

Fig. 2. *Neolithodes asperrimus* Barnard, 1947, male, CL 168 mm, CW 136 mm, Santa Catarina (CENEMAR CAR 0116): A, dorsal habitus; B, cephalothorax and abdomen, ventral view; C, cephalothorax, front; D, dactylus of third walking leg, ventral view; E, right cheliped, lateral view; F, left cheliped, lateral view. Scale bars: A-B, 50 mm; C, 10 mm; D, 40 mm; E-F, 20 mm.

Ivory Coast; on muddy bottoms between depths of 250-2000 m (Barnard, 1947; Macpherson, 1988; Muñoz & Garcia-Isarch, 2013). New record from western Atlantic, Brazil, Santa Catarina.

Remarks.— *Neolithodes asperrimus* Barnard, 1947 has only been recorded along the western coast of Africa in the eastern Atlantic. This is the first record of this species from the western Atlantic, and the eighth species of Lithodidae found in Brazil. *Neolithodes* aff. *asperrimus* recorded here is very similar to the material described by Macpherson (1988: 39, fig. 16, pls. 3A; male CL 141 mm) as *N. asperrimus* Barnard, 1947. However, some differences can be noted: (i) propodus of the third walking leg is about 9-10 times longer than high (7 times longer than high in *N. asperrimus* Barnard, 1947); (ii) dactyl of the third walking leg is approximately 2.5 times smaller than merus (dactyl is approximately 3.5 times smaller than merus in *N. asperrimus* Barnard, 1947); (iii) merus length of the third walking leg is similar to carapace length (merus is shorter than the carapace length in *N. asperrimus* Barnard, 1947); and (iv) merus of the walking legs are longer than propodus (merus shorter than propodus in *N. asperrimus*

Barnard, 1947). *Neolithodes asperrimus* seems to show a high degree of variation in the number and size of the carapace spines, a characteristic well known to be ontogenetically variable in lithodids (see Macpherson, 1988, pl. 3A and 3B). In large adults, the spines appear to be shorter and are commonly broken, as in the specimen studied here. Although we observed some differences in the specimen from Brazil compared to those of *N. asperrimus* from Africa, we have only one dry specimen. The large size of *N. asperrimus* and its rarity in museum collections inhibits loaning material and thus impairs a full comparison. Unfortunately, the specimen studied here is dried, also hindering a molecular comparison as well. *Neolithodes asperrimus* is closely related to *N. agassizii* (Smith, 1882) from the western Atlantic; however, it can be differentiated by several characters (see Macpherson, 1983, 1988 for additional information). *Neolithodes* aff. *asperrimus* can be quickly differentiated from *N. agassizii* by the length of the third walking leg, approximately 2.5 times the carapace length in *N.* aff. *asperrimus*, versus length of the third walking leg being more than three times the carapace length in *N. agassizii*. *Neolithodes* currently comprises only 12 species, of which only *N. grimaldii* has been recorded from both sides of the Atlantic (Macpherson, 1988). The amphi-Atlantic distribution recorded here for *N. asperrimus* is uncommon in decapod crustaceans but can be seen in other decapod groups, such as *Alpheus intrinsecus* Spence Bate, 1888 (Christoffersen, 1979), *Oncopagurus bicristatus* (A. Milne-Edwards, 1880) and *O. gracilis* (Henderson, 1888) (cf. Lemaitre, 2014). The specimen reported from Brazilian waters also increases the bathymetrical distribution of this species, which was previously only recorded between depths of 600 and 2000 m, which is now extended to between 250 and 2000 m.

ACKNOWLEDGEMENTS

We dedicate this work to the great carcinologist Michael Türkay (1948-2015), an example for those who want to study crustaceans. We thank José Carlos Tarasconi (CENEMAR) for providing working space and access to his collections. Thanks also to Marcos Tavares (Museu de Zoologia da Universidade de São Paulo) for his kind assistance throughout the execution of this work. LL and DL thank the Conselho Nacional de Desenvolvimento Científico e Tecnológico — CNPq for their fellowships (Nos 132450/2016-0 and 162247/2013-3). WS thanks Fundação de Amparo à Pesquisa do Estado de São Paulo — FAPESP for providing support throughout the research grant number 2013/1201-0. We thank the Universidade do Sagrado Coração — USC for supporting studies on the systematics of decapod crustaceans. This paper greatly benefited from comments from two anonymous reviewers and from Sammy De Grave (Oxford University Museum of Natural History, Oxford, U.K.).

REFERENCES

ANOSOV, S. E., V. A. SPIRIDONOV, T. V. NERETINA, E. F. URYUPOVA & D. SCHEPETOV, 2015. King crabs of the western Atlantic sector of Antarctic and adjacent areas: new records, molecular barcode data and distribution (Crustacea: Decapoda: Lithodidae). Polar Biology, **38**: 231-249.

BARNARD, K. H., 1947. XLI — Descriptions of new species of South African decapod Crustacea, with notes on synonymy and new records. Annals and Magazine of Natural History, Series 11, **13**: 361-392.

CHRISTOFFERSEN, M. L., 1979. Decapod Crustacea: Alpheoidea. Campagnes de la Calypso au large des côtes Atlantiques de l'Amérique du Sud (1961-1962). I. Résultats Scientifiques de la Campagne de la Calypso, **11**: 297-377.

HALL, S. & S. THATJE, 2009. Four new species of the family Lithodidae (Decapoda: Anomura) from the collections of the National Museum of Natural History, Smithsonian Institution. Zootaxa, **2302**: 31-47.

LEMAITRE, R., 2014. A worldwide taxonomic and distributional synthesis of the genus *Oncopagurus* Lemaitre, 1996 (Crustacea: Decapoda: Anomura: Parapaguridae), with descriptions of nine new species. The Raffles Bulletin of Zoology, **62**: 210-301.

LOVRICH, G. A., M. PERRONI, J. H. VINUESA, F. TAPELLA, A. CHIZZINI & M. C. ROMERO, 2002. Occurrence of *Lithodes confundens* (Decapoda: Anomura) in the intertidal of the southwestern Atlantic. Journal of Crustacean Biology, **22**: 894-902.

MACPHERSON, E., 1983. Crustaceos decapodos capturados em las costas de Namibia. Resultados Expediciones Cientificas, **11**: 3-79.

MACPHERSON, E., 1988. Revision of the family Lithodidae Samouelle, 1819 (Crustacea: Decapoda: Anomura) in the Atlantic Ocean. Monografías de Zoología Marina, **2**: 1-153.

MCLAUGHLIN, P. A., T. KOMAI, R. LEMAITRE & D. L. RAHAYU, 2010. Annotated checklist of anomuran decapod crustaceans of the world (exclusive of the Kiwaoidea and families Chirostylidae and Galatheidae of the Galatheoidea) part I — Lithodoidea, Lomisoidea and Paguroidea. The Raffles Bulletin of Zoology, **23**: 5-107.

MELO, G. A. S., 1999. Manual de identificação dos Crustacea Decapoda do litoral brasileiro: Anomura, Thalassinidea, Palinuridea, Astacidea. (Editora Plêiade, São Paulo).

MUÑOZ, I. & E. GARCÍA-ISARCH, 2013. New occurrences of lithodid crabs (Crustacea: Decapoda: Lithodidae) from the coasts of Africa, with the description of a new species of *Paralomis* White, 1856. Zootaxa, **3670**: 45-54.

PÉREZ-BARROS, P., V. A. CANFALONIERI, K. PASCHKE & G. A. LOVRICH, 2015. Incongruence between molecular and morphological characters in the southern king crabs *Lithodes santolla* and *Lithodes confundens* (Decapoda: Anomura). Polar Biology, **38**: 2097-2107.

SEREJO, C. S., P. S. YOUNG, I. C. CARDOSO, C. TAVARES, C. RODRIGUES & T. C. ALMEIDA, 2007. Abundância, diversidade e zonação dos crustáceos no talude da costa central do Brasil (11°-22°S) coletados pelo Programa REVIZEE/Score Central: prospecção pesqueira. In: P. A. S. COSTA, G. OLAVO & A. S. MARTINS (eds.), Biodiversidade da fauna marinha profunda na costa central brasileira: 133-162. (Museu Nacional, Rio de Janeiro).

SOTELANO, M. P., M. F. GOWLAND-SAINZ, M. J. DIEZ & G. A. LOVRICH, 2013. Distribution of *Lithodes confundens* Macpherson, 1988 (Decapoda, Anomura) along the Atlantic continental shelf of southern South America. Crustaceana, **86**: 246-252.

STEVENS, B. G. & G. A. LOVRICH, 2014. King crabs of the world: species and distributions. In: B. G. STEVENS (ed.), King crabs of the world: biology and fisheries management: 1-29. (CRC Press, London).

ZAKLAN, S. D., 2002. Review of the family Lithodidae (Crustacea: Anomura: Paguroidea): distribution, biology, and fisheries. In: A. J. PAUL, E. G. DAWE, R. ELNER, G. S. JAMIESON, G. H. KRUSE, R. S. OTTO, B. SAINTE-MARIE, T. C. SHIRLEY & D. WOODBY (eds.), Crabs in cold water regions: biology, management, and economics: 751-845. (University of Alaska Sea Grant, Fairbanks, AK).

First received 2 September 2016.
Final version accepted 21 February 2017.

[When citing this volume, please refer to Crustaceana 90 (2017) 771-1288]

Michael Türkay Memorial Issue

A NEW SPECIES OF THE GENUS *TYPHLAMPHIASCUS* (COPEPODA, HARPACTICOIDA, MIRACIIDAE) FROM THE SOUTH CHINA SEA

BY

LIN MA[1,2,3]) and XINZHENG LI[1,2,3,4])

[1]) Institute of Oceanology, Chinese Academy of Sciences, 7 Nanhai Road, Qingdao 266071, P.R. China

[2]) University of Chinese Academy of Sciences, Beijing 100039, P.R. China

[3]) Laboratory for Marine Biology and Biotechnology, 1 Wenhai Road, Aoshanwei, Qingdao 266237, P.R. China

ABSTRACT

A new species, *Typhlamphiascus tuerkayi* sp. nov. (Copepoda, Harpacticoida, Miraciidae), is described from specimens collected in the South China Sea. This new species differs from its congeners by the combined characters of the number of chitinous lamellae on the male P1, the apomorphic setal formulae of the swimming legs, and the setal number on P5. The new species is most similar to *T. ovale* Wells & Rao, 1987 and *T. higginsi* Chullasorn, 2009. It differs from the latter two species by the following features: the caudal ramus about three times as long as broad, P2 exp-3 with three outer spines, P3 enp-3 with two inner setae, P5 basoendopod in the female with four setae, P5 exopod in the female about three times as long as its greatest width, antennule of male 9-segmented, the inner edge of the basis in male P1 with eight chitinous lamellae, P5 exopod in male with four setae.

Key words. — Crustacea, new species, taxonomy, meiobenthos

ZUSAMMENFASSUNG

Eine neue Art, *Typhlamphiascus tuerkayi* sp. nov. (Copepoda, Harpacticoida, Miraciidae), wird hier anhand von im Südchinesischen Meer gesammelten Exemplaren beschrieben. Diese neue Art unterscheidet sich von ihren Verwandten durch eine Merkmalskombination der Anzahl chitinisierter Lamellen am männlichen P1, die apomorphischen Muster der Setae an den Schwimmbeinen und die Anzahl der Setae an P5. Die neue Art ähnelt *T. ovale* Wells & Rao, 1987 und *T. higginsi* Chullasorn, 2009 am stärksten. Sie unterscheidet sich von letzteren durch die folgenden Merkmale: der Kaudalramus ist ungefähr drei mal so lang wie breit, P2 exp-3 besitzt drei Außendornen, P3 enp-3 hat zwei innere Setae, P5 Basoendopod im Weibchen hat vier Setae, weiblicher P5 Exopod ist ungefähr dreimal so lang wie dessen größte Breite, männliche Antennule hat neun Segmente, der Innenrand der Basis des männlichen P1 hat acht chitinisierte Lamellen, P5 Exopod des Männchens hat vier Setae.

[4]) Corresponding author; e-mail: lixzh@qdio.ac.cn

INTRODUCTION

The genus *Typhlamphiascus* Lang, 1944 belongs to the large family Miraciidae Dana, 1846. Miraciids inhabit a wide range of sediment types from flocculent mud to coralline debris, primarily in intertidal and shallow subtidal areas of continental shelves (Chullasorn, 2009). The genus *Typhlamphiascus* comprises 17 species, of which *T. confuses* (Scott T., 1902) and *T. lamellifer* (Sars G.O., 1911) contain two subspecies, respectively (Walter et al., 2016). Species of the genus *Typhlamphiascus* can inhabit deep, sublittoral muddy bottoms, the littoral zone, algal cover and surface layers of the northern Atlantic and Pacific coast (Por, 1963).

Early taxonomic studies on the benthic copepods in China have usually focused on freshwater species (Dai & Song, 1979), and the marine species were carried on very late, just a few species having been reported (Gee & Mu, 2000; Mu & Gee, 2000; Mu & Huys, 2002, 2004; Huys & Mu, 2008; Ma & Li, 2011). Most records of benthic copepods from the China seas are mainly from the Bohai Gulf and Yellow Sea. No *Typhlamphiascus* species has previously been reported from tropical waters in the China seas.

In 2007, several specimens of harpacticoid copepods were sorted from benthic samples collected off Hainan Island, South China Sea, by RV "Kexue-I" during the cruise of a project of Chinese offshore marine investigation and assessment. These specimens proved to belong to an unknown species of *Typhlamphiascus*, which is herein described as a new species.

MATERIAL AND METHODS

The type specimens were collected from the South China Sea, fixed in 10% formalin. The specimens were extracted from benthic samples using a 38 μm sieve and the colloidal silica Ludox TM-50 suspension centrifugation flotation method. Specimens were preserved in 75% alcohol. For their identification, the specimens were cleared in lactic acid and observed under a light microscope. Before dissection, the habitus was drawn and the whole body length was measured while being temporarily mounted in lactophenol. Specimens were dissected in lactic acid, mounted on slides in lactophenol, and subsequently sealed with nail polish. Observations, dissection and drawings were made using a compound microscope. The habitus was drawn at 400× magnification; appendages were drawn at 1000× magnification, with an oil immersion lens.

The terminology follows Huys et al. (1996). Abbreviations used in the text and figures are: aes, aesthetasc; exp, exopod; exp-1(-2-3), the first (second, third) segment of the exopod; enp, endopod; enp-1(-2-3), the first (second, third) segment of the endopod; P1-P6, swimming legs 1-6. Body length was measured from the anterior margin of the rostrum to the posterior margin of the caudal rami. Type material is deposited in the Marine Biological Museum, Chinese Academy of Sciences, Qingdao, China (MBM, MBMCAS).

SYSTEMATIC PART

Order HARPACTICOIDA Sars, 1903
Family MIRACIIDAE Dana, 1846
Subfamily DIOSACCINAE Sars, 1906
Genus *Typhlamphiascus* Lang, 1944

Typhlamphiascus tuerkayi sp. nov.
(figs. 1-7)

Material examined.— Holotype: female, dissected on four slides (MBM 189121), South China Sea, 18°35.8071′N 110°43.4366′E, soft mud, 30.1 m depth, 19 October 2007. Allotype: male, dissected on three slides (MBM 189222) same collection data as holotype. Paratypes: one female and one male, preserved in alcohol (MBM 189223), same collection data as holotype; one male (MBM 189224), 21°17.7426′N 110°51.1891′E, 15.5 m depth, soft mud, 9 October 2007; one male (MBM 189225), 20°01.2523′N 111°09.9323′E, 50.2 m depth, muddy sand, 15 October 2007; one female (MBM 189226), 19°34.8292′N 111°31.7719′E, 102 m depth, muddy sand, 17 October 2007; one female (MBM 189227), 18°33.2539′N 110°22.8292′E, 49.9 m depth, soft mud, 7 October 2007; two female, one male (MBM 189228), 21°01.9355′N 111°04.2526′E, 26.2 m depth, soft mud, 10 October 2007. Collectors: J.-B. Wang, L.-M. Shuai, J. Zhou, Q.-X. Han & L. Ma.

Etymology.— The species is named in honour of the German carcinologist, the late Professor Michael Türkay, for his great contributions to the knowledge of crustacean taxonomy.

Diagnosis.— Antennule 8-segmented in female, with one aesthetasc on the fourth segment and terminal segment each. Antennary basis without seta, exopod 3-segmented. Mandibular gnathobase with row of nine teeth and one spiniform seta, exopod 2-segmented with one seta, respectively, endopod 1-segmented with seven setae. Maxillulary arthrite of praecoxa with nine spines at distal margin, coxal endite with three setae. Maxillary syncoxa bearing three endites with three, two, and three spinulose setae, respectively. Maxillipedal syncoxa with four setae. P1 exp-3 without inner seta, with three outer spines; P1 enp-3 with one inner seta, without outer spine. P2 and P3 exp-3 with one inner seta and three outer spines, respectively; P2 enp-3 with one inner seta and one outer spine; P3 enp-3 with two inner setae and one outer spine. P4 exp-3 with two inner setae and three outer spines; P4 enp-3 with one inner seta and one outer spine. Right and left legs of P5 separated; P5 basoendopod in female with four endopodal setae, in male with two setae. Inner edge of the basis of male P1 with eight chitinous lamellae. Caudal ramus about three times as long as broad.

Description.— Female (based on holotype and one paratype). Habitus (figs. 1A, 8A). Total length of holotype female (body plus caudal rami, excluding caudal setae): 920 μm. Body slender, almost linear, boundary between prosome and urosome inconspicuous, dorsal and lateral surfaces with scattered sensilla. Prosome 4-segmented: cephalothorax (including two thoracic somites bearing maxilliped and P1) and three articulated somites bearing P2 to P4. Urosome (fig. 1B) 5-segmented, comprising of posterior thoracic somite, genital double-somite, and three abdominal somites. Urosomites with rows of hyaline frills on dorsal edge respectively, excluding genital double-somite; urosomites ornamented with hyaline frills on ventral side except on first urosomite. Anal operculum large, furnished with rows of spinules dorsally and ventrally above the caudal edge. Caudal ramus (figs. 1A, 8B) about three times as long as broad, carrying seven setae: three outer setae, smooth; two medial setae, well developed, slightly pinnate; two inner setae, slender, all ramal setae located terminally.

Rostrum (fig. 2A) demarcated from cephalothorax, expanded at base, almost triangular with pair of sensilla positioned on either side about 1/5 the distance from acute tip.

Antennule (figs. 2B, 8C) with eight segments; the first segment and second segment longest; armature: 1, 11, 6, 4 + aes, 1, 4, 4, 5 + aes.

Antenna (fig. 2C) biramous, with small unornamented coxa. Allobasis with spinules near coxa. Exopod 3-segmented, with 1-0-(1 + 3) setae; exp-1 long, smooth; exp-2 clearly shorter than exp-1; exp-3 with row of spinules. Endopod 1-segmented, with eight spines on lateral edge; lateral armature comprising single spinulose seta; apical armature comprising seven elements: one pinnate spine, three geniculate setae, one spinulose seta and two slender setae.

Mandible (fig. 2D). Gnathobase with five large and strong teeth, seven small teeth and one seta. Basis with two rows of spinules and three setae. Exopod 2-segmented, with two and one setulose setae, respectively. Endopod 1-segmented, with two lateral and five terminal setae.

Maxillule (fig. 2E) with praecoxa and coxa demarcated. Arthrite with nine apical spines and two juxtaposed setae on surface. Coxal endite with three setae. Basis with row of spinules, six setae. Exopod 1-segmented, with two pinnate setae. Endopod 1-segmented, with four pinnate setae.

Maxilla (fig. 2F). Syncoxa with spinules along outer margin; with three endites bearing three, two, and three setae, respectively. Allobasis with one claw and two setae. Endopod 2-segmented, enp-1 with two setae, enp-2 with four setae.

Maxilliped (fig. 2G) subchelate. Syncoxa with four setae located at distal margin. Basis with two rows of spinules and two pinnate setae. Endopod 2-segmented; enp-1 with two setae, enp-2 with one strong claw.

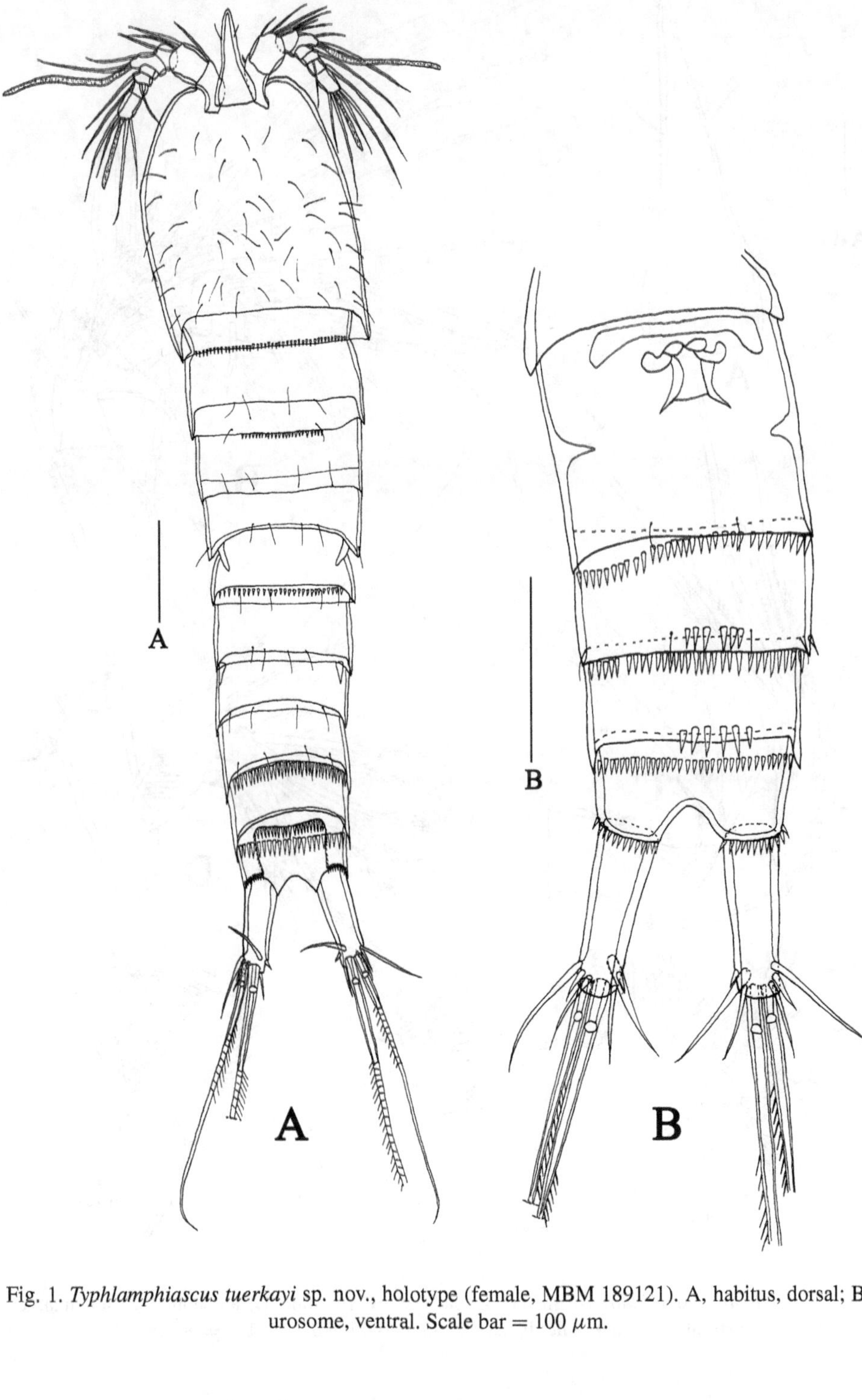

Fig. 1. *Typhlamphiascus tuerkayi* sp. nov., holotype (female, MBM 189121). A, habitus, dorsal; B, urosome, ventral. Scale bar = 100 μm.

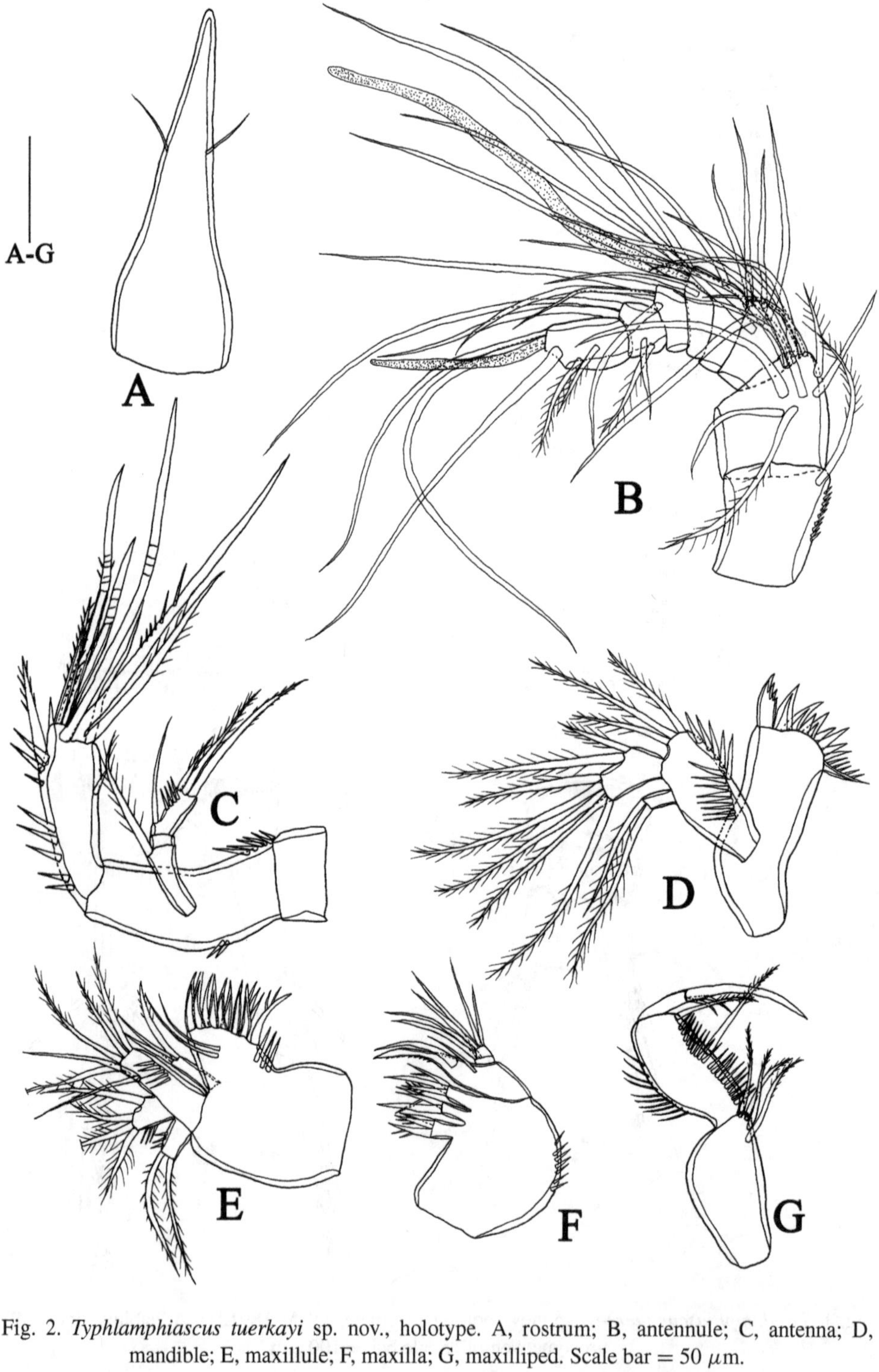

Fig. 2. *Typhlamphiascus tuerkayi* sp. nov., holotype. A, rostrum; B, antennule; C, antenna; D, mandible; E, maxillule; F, maxilla; G, maxilliped. Scale bar = 50 μm.

P1 (fig. 3A): Coxa with complex patterns of denticles on anterior surface, row of setules on inner margin. Basis bearing single outer and strong inner spinulose spines, terminal margin with spinules, inner margin with setules. Exopod and endopod 3-segmented, outer and distal margins with spinules, exopod short, about as long as enp-1.

P2-P4 (figs. 3B, 4A-B) with exopods and endopods 3-segmented, intercoxal sclerites with two lateral blunt projections. Coxae of almost rectangular shape with few rows of spinules on each. Basis with small and slender outer pinnate setae (spinulose seta in P2), without inner setae, ornamented with long spinules on anterior surface, distal margin with denticles; endopod about as long as exopod in P2 and P3, shorter in P4.

Setal formulae of female P1-P4 as follows:

	Exp	Enp
P1	0-1-0, 2, 3	1-1-1, 2, 0
P2	1-1-1, 2, 3	1-2-1, 2, 1
P3	1-1-1, 2, 3	1-1-2, 2, 1
P4	1-1-2, 2, 3	1-1-1, 2, 1

Right and left P5 (fig. 5A) not fused medially, baseoendopod and exopod separated. Endopodal lobe prominent, with two spinulose, one long and one slender setae. Exopod about three times as long as greatest width, ornamented with denticles and six bare setae.

Male (based on allotype and one paratype) differs from female as follows:

Body (figs. 6A, 8D) slightly shorter than female holotype, total length of allotype male (body plus caudal rami, excluding caudal setae): 660 μm. Urosome (fig. 6B) 6-segmented, genital somite and the first abdominal somite separate, urosomite 4 with two rows of spinules ventrally. Caudal ramus (figs. 6A, 8E) about 1.5 times as long as broad. Spermatophore (fig. 6A) ovoid visible inside genital somite.

Antennule (figs. 5B, 8F) 9-segmented, haplocer. Armature: 1, 10, 6, 5 + aes, 1, 2, 2, 4, 6 + aes., geniculation between segments 6 and 7.

Antenna, mandible, maxillule, maxilla, maxilliped similar to female holotype.

P1 (fig. 7A). Basis bearing eight unequal chitinous lamellae on the inner edge. Other characters as in female holotype.

P2 (fig. 7B) with protopod and exopod as in female holotype. Endopod 2-segmented; enp-1 with one inner seta, long spinules on outer margin; enp-2 modified as common in the genus, with one inner seta, two terminal setae, one stout spine and one seta on outer margin.

P4 (fig. 7C) as in female holotype, except enp-3 with two inner setae.

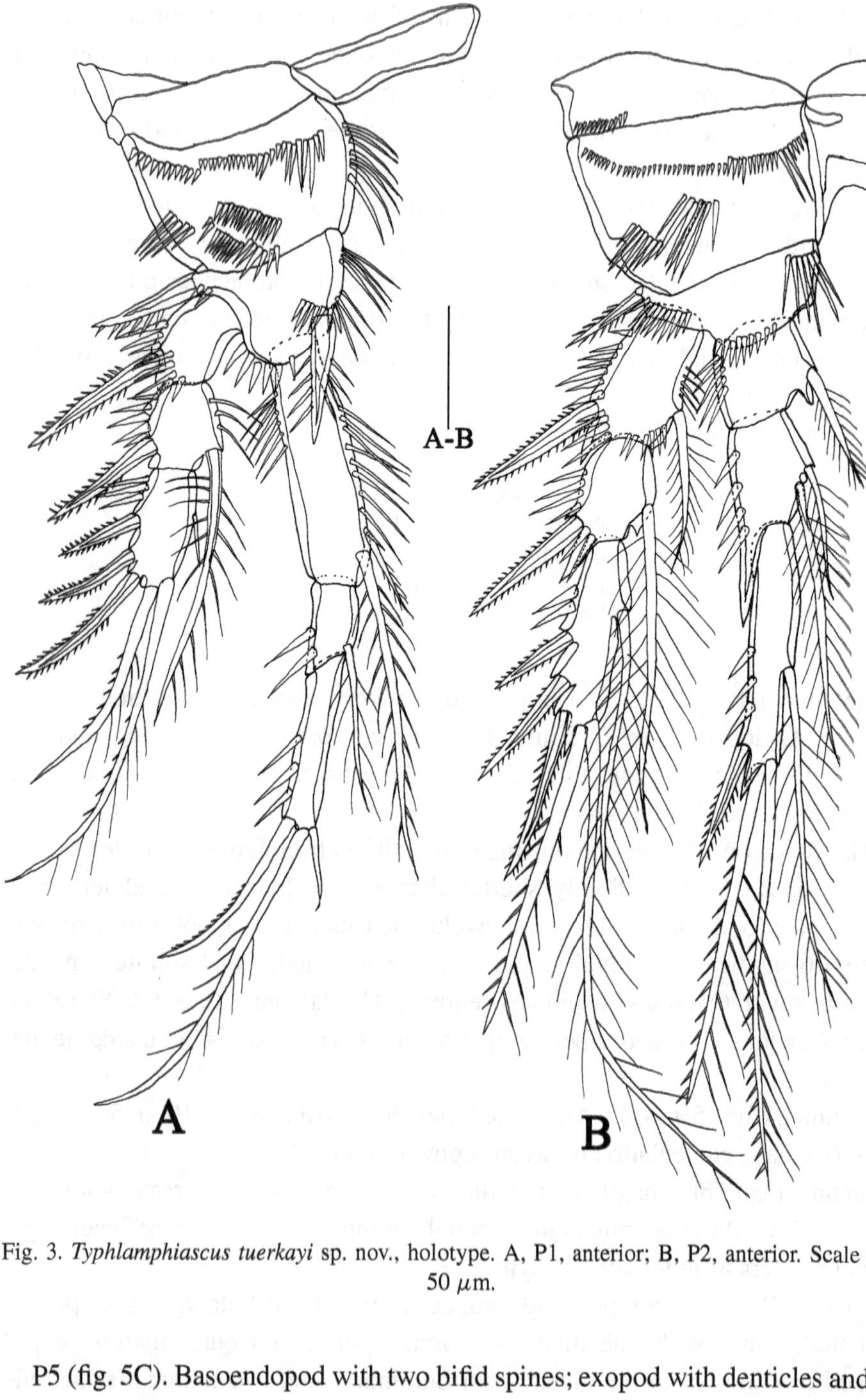

Fig. 3. *Typhlamphiascus tuerkayi* sp. nov., holotype. A, P1, anterior; B, P2, anterior. Scale bar = 50 μm.

P5 (fig. 5C). Basoendopod with two bifid spines; exopod with denticles and four unequal setae.

P6 (fig. 6B) unconfluent, each with three setae.

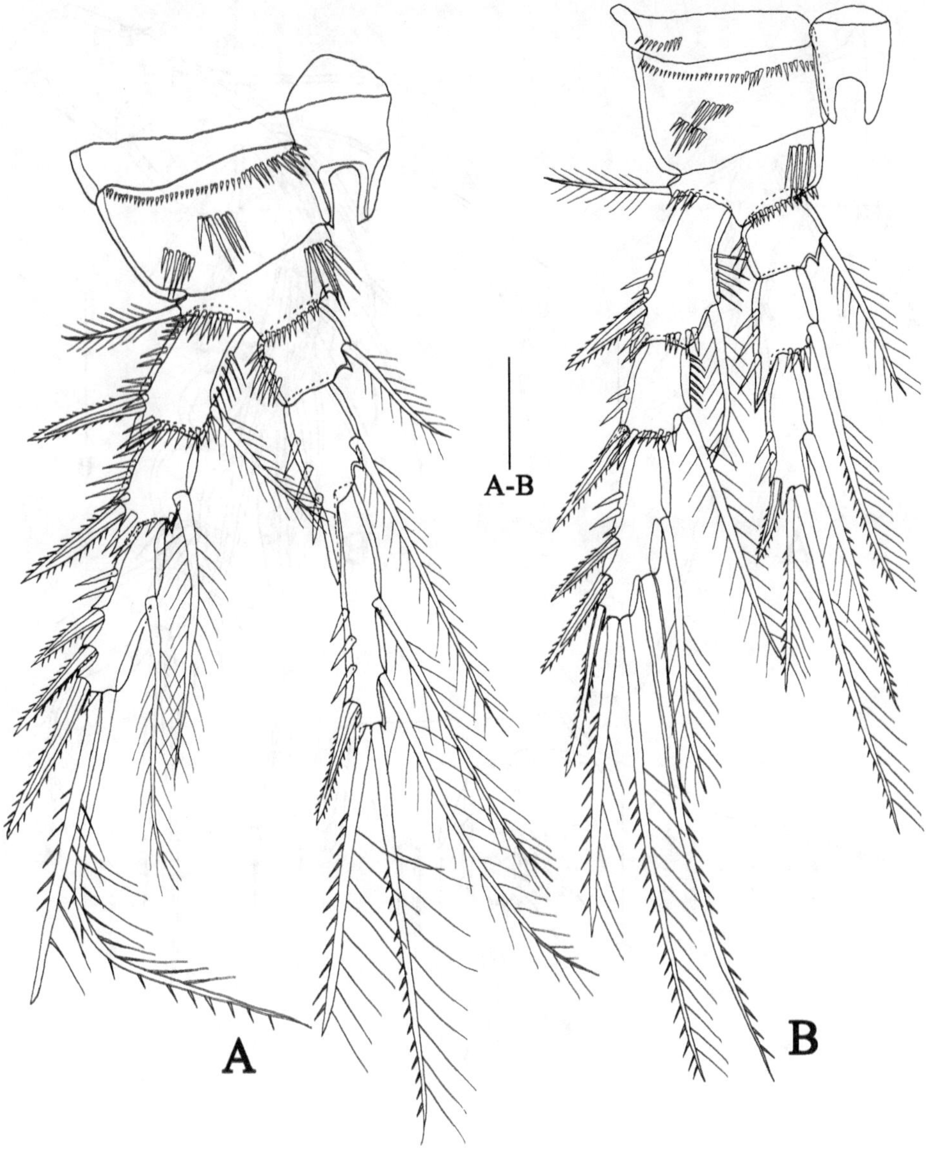

Fig. 4. *Typhlamphiascus tuerkayi* sp. nov., holotype. A, P3, anterior; B, P4, anterior. Scale bar = 50 μm.

Variability.— Most morphological features are conservative, except body length and the length/width ratio of caudal ramus. Body length of female varies from 518 to 920 μm and male from 497 to 660 μm. The length/width ratio of caudal ramus changes from 1.83 to 3 in female and 1.5 to 1.75 in male.

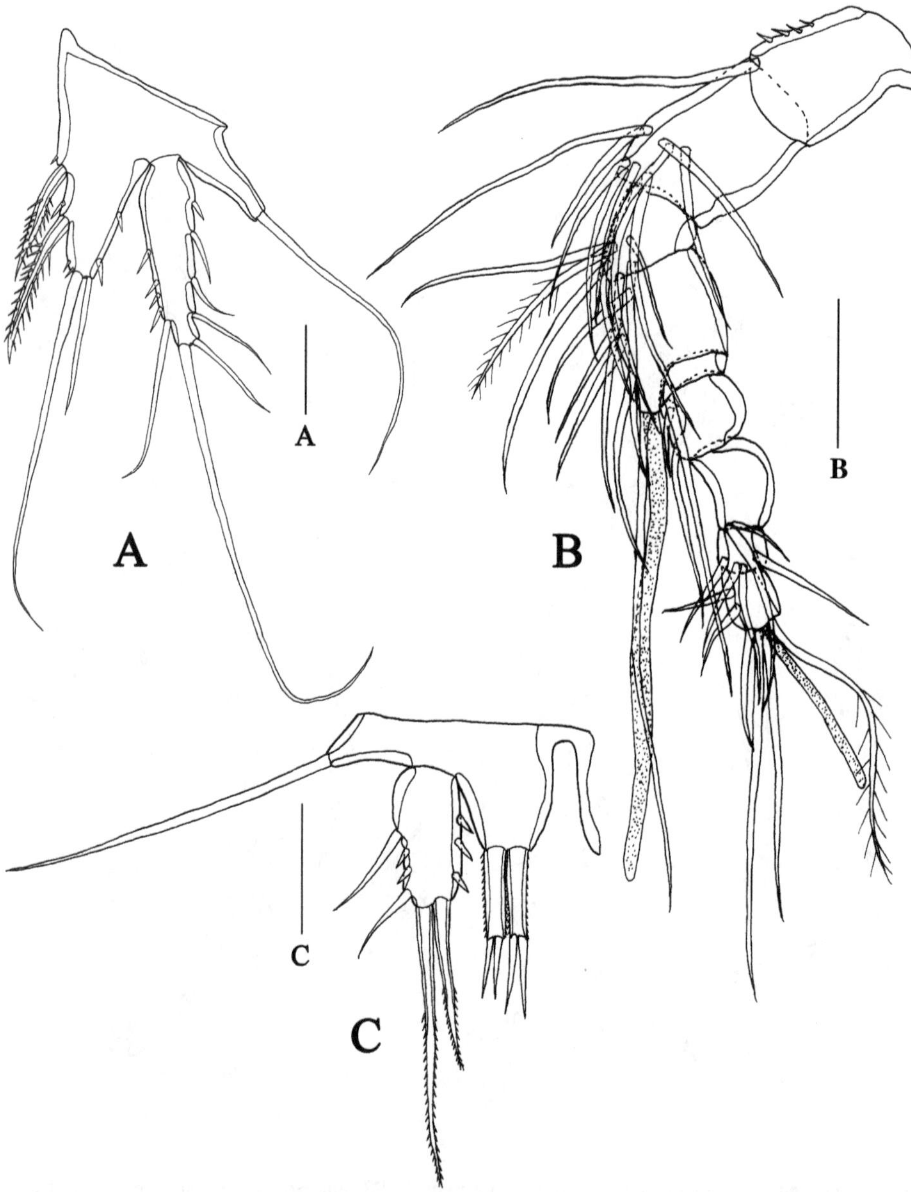

Fig. 5. *Typhlamphiascus tuerkayi* sp. nov., holotype. A, P5, anterior. Allotype (male, MBM 189222). B, antennule; C, P5, anterior. Scale bars = 50 μm.

DISCUSSION

In a revision of *Typhlamphiascus*, Por (1963) recognized two groups of species, based on the presence of two or single setae on the inner side of P4 enp-3. Though one of these species, *Typhlamphiascus accraensis* Scott T., 1894, was

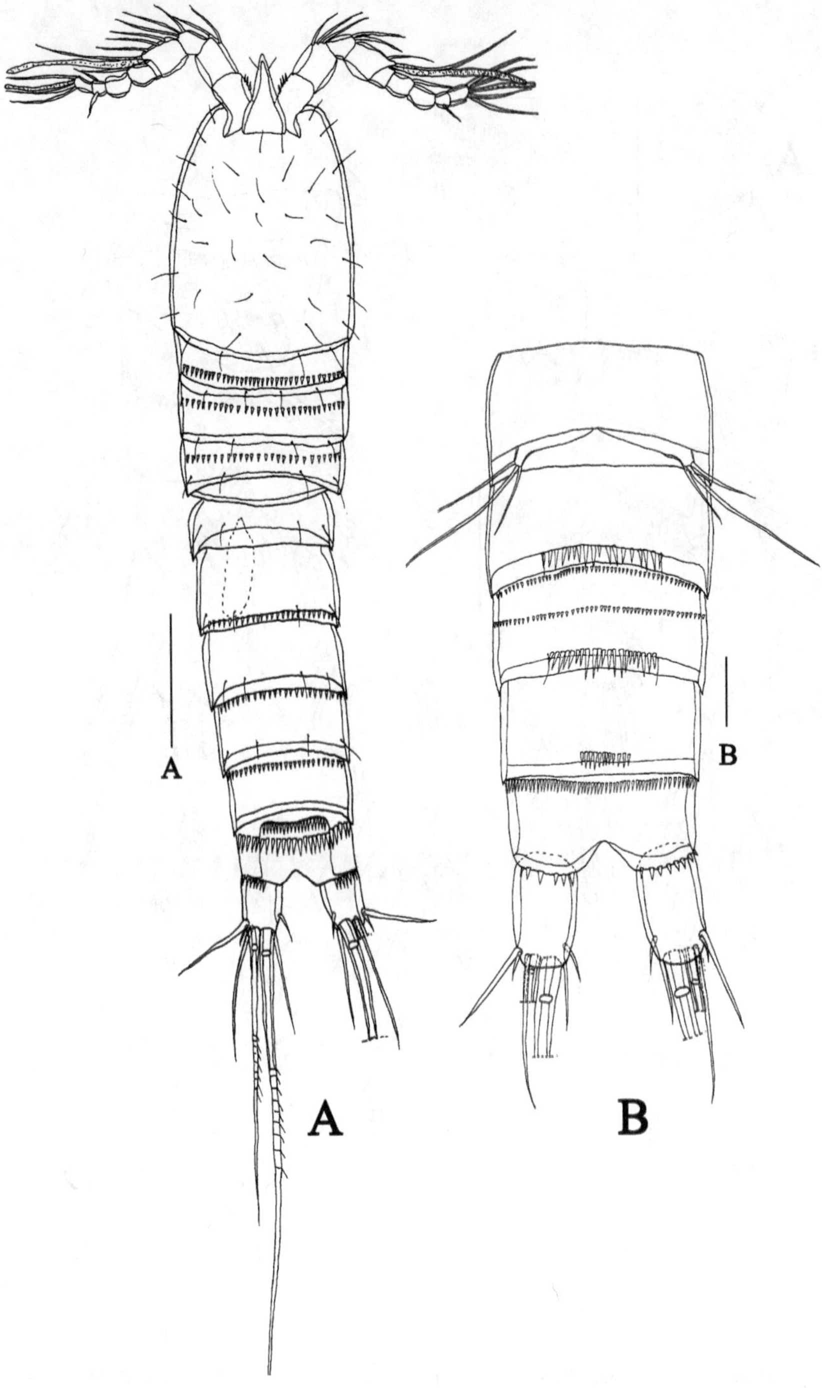

Fig. 6. *Typhlamphiascus tuerkayi* sp. nov., allotype. A, habitus, dorsal; B, urosome, ventral. Scale bars = 100 μm (A), 50 μm (B).

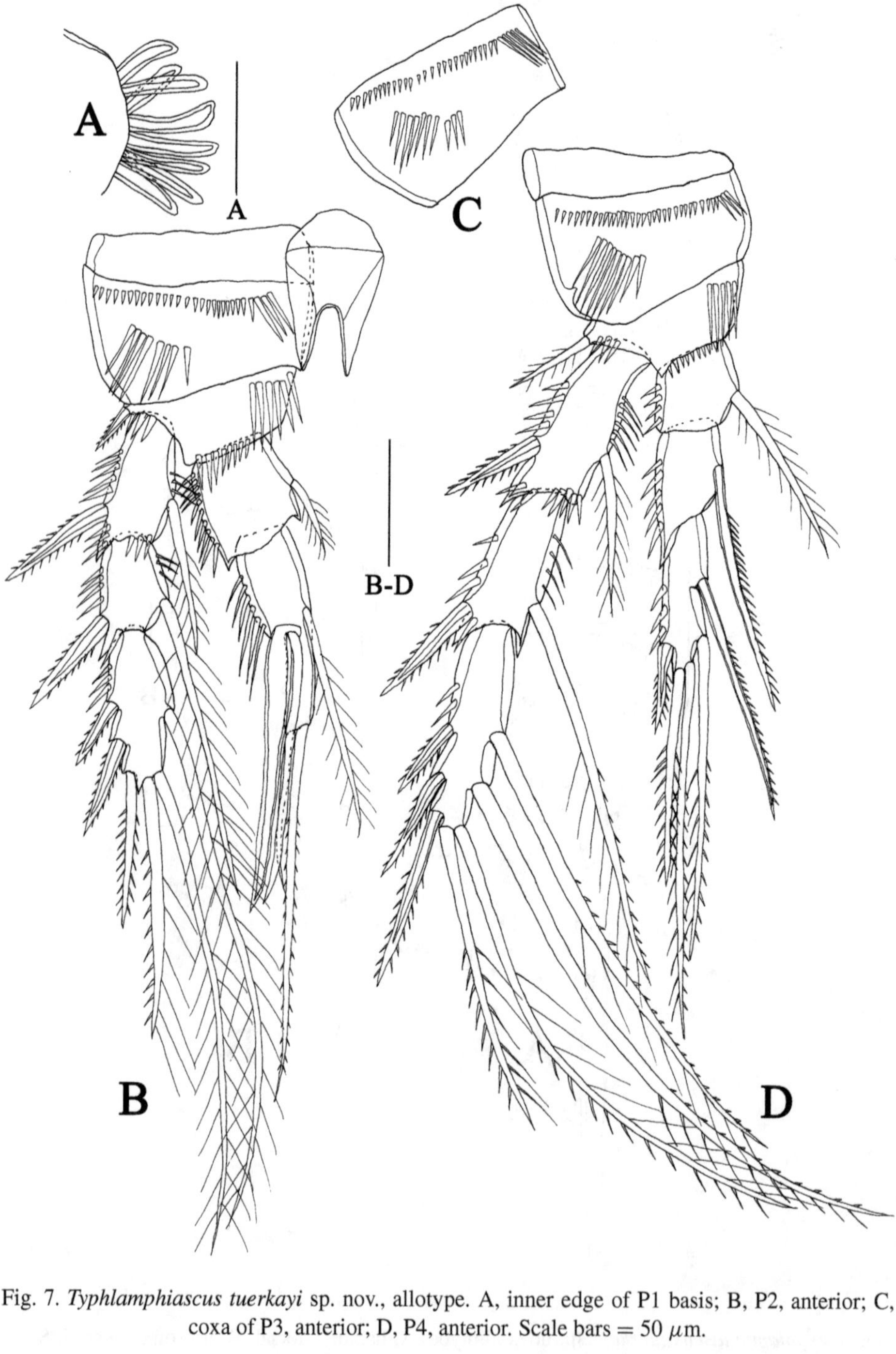

Fig. 7. *Typhlamphiascus tuerkayi* sp. nov., allotype. A, inner edge of P1 basis; B, P2, anterior; C, coxa of P3, anterior; D, P4, anterior. Scale bars = 50 μm.

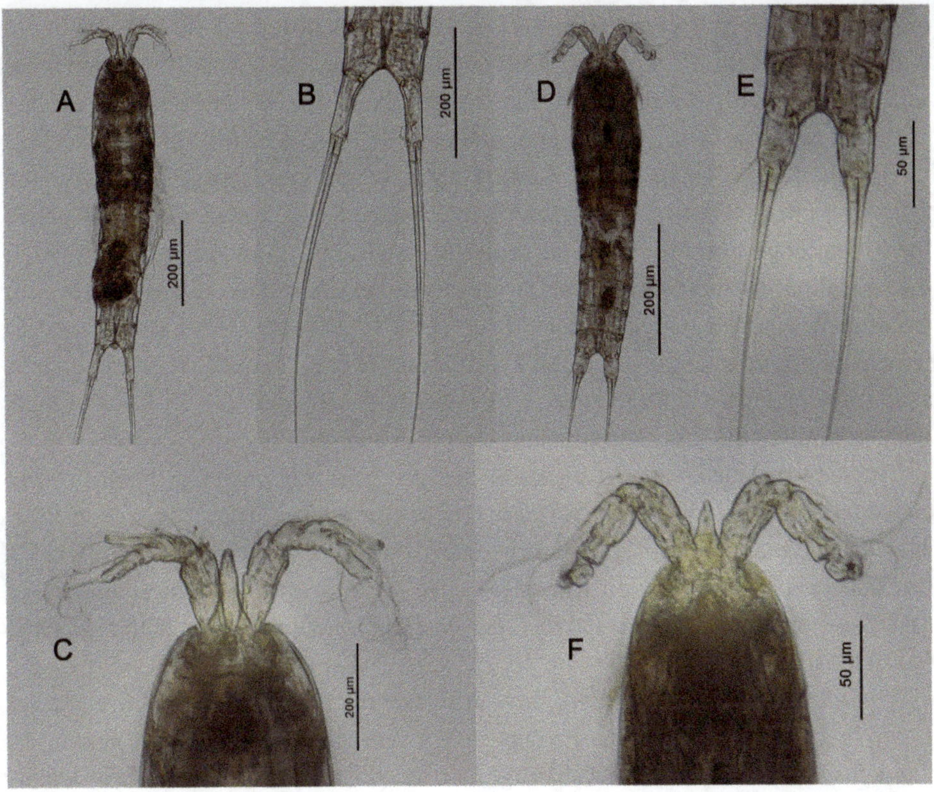

Fig. 8. *Typhlamphiascus tuerkayi* sp. nov., paratype (female, MBM 189222). A, habitus, dorsal; B, caudal ramus, dorsal; C, antennule, dorsal. Paratype (male, MBM 189228). D, habitus, dorsal; E, caudal ramus, dorsal; F, antennule, dorsal.

initially assigned to group 2 by Por (1963), the original description and figures (Scott T., 1984) describe and illustrate two setae on the inner side of P4 enp-3. Accordingly, we transfer *T. accraensis* to Por's group 1. Lang (1965) added *T. lutincola* Soyer, 1963 and *T. drachi* Soyer, 1964 to this group, and *T. pectinifer* Lang, 1965 and *T. unisetosus* Lang, 1965 to group 2. Therefore, group 1 now includes: *T. accraensis*, 1894, *T. blanchardi* (Scott T., 1895), *T. bouligandi* Soyer, 1971, *T. drachi*, *T. gracilicaudatus* (Thompson I.C. & Scott A., 1903), *T. gracilis* Por, 1963, *T. lamellifer capensis* Kunz, 1975, *T. lutincola*, *T. typhloides* (Sars G.O., 1911); while group 2 includes *T. confusus erythraeicus* Por, 1963, *T. confusus gullmaricus* Por, 1963, *T. higginsi* Chullasorn, 2009, *T. lamellifer lamellifer* (Sars G.O., 1911), *T. latifurca* Por, 1968, *T. longifurcatus* Rouch, 1962, *T. ovale* Wells & Rao, 1987, *T. pectinifer*, *T. typhlops* (Sars G.O., 1906) and *T. unisetosus*.

Typhlamphiascus tuerkayi sp. nov. bears only one setae on the inner side of P4 enp-3, so belongs to Por's (1963) group 2, wherein it most closely resembles *T. ovale* and *T. higginsi*. These three species share the following features: female urosomites 2-4 ornamented with hyaline frills on their ventral side; P1-P4 exp-2 with one inner seta; P2-P4 exp-1 with one inner seta; P1-P4 enp-1 with one inner seta; P1 enp-3 with three setae; P2 enp-2 of female with two inner setae; P5 exopod of female with six setae; P5 basoendopod of male with two bifid spines.

However, *Typhlamphiascus tuerkayi* sp. nov. differs from *T. ovale* by the following characteristics: anal operculum furnished with rows of spinules above the caudal edge both dorsally and ventrally (anal operculum simple in *T. ovale*); the caudal ramus about three times as long as broad (almost as long as in *T. ovale*); maxilliped syncoxa with four setae (two setae in *T. ovale*); P5 basoendopod in female with four setae (five setae in *T. ovale*); P5 exopod in male about three times as long as greatest width (about two times as long as in *T. ovale*); P1 basis of male without knob at the inner proximal corner (with a chitin knob in *T. ovale*).

The new species can be distinguished from *T. higginsi* by the following characteristics: P2 exp-3 and P3 exp-3 with three inner setae (two setae in *T. higginsi*); P3 enp-3 with two inner setae (only one seta in *T. higginsi*); P5 basoendopod in female with four setae (five setae in *T. higginsi*); antennule of male 9-segmented (8-segmented in *T. higginsi*); P5 exopod in male with four setae (six setae in *T. higginsi*).

Typhlamphiascus tuerkayi sp. nov. differs from other species of the second group by bearing five setae on P3 enp-3 (four setae in *T. unisetosus*, *T. pectinifer*, *T. confusus erythraeicus*, *T. confusus gullmaricus*, *T. latifurca*, *T. lamellifer lamellifer*); and four setae on P5 basoendopod of female (five setae in *T. typhlops* and *T. longifurcatus*).

To date, 15 species and 4 subspecies are currently recognized as valid in the genus. *Typhlamphiascus* can be identified with the specific key suggested below (amended from Soyer, 1963, 1964; and Wells, 2007).

1. Caudal ramus with an aberrant inner seta ..... *Typhlamphiascus typhloides* (Sars G.O., 1911)
– Caudal ramus with normal inner seta 2

2. Distal end of exterior margin in P5 exopod forming into a hook-like process .... *Typhlamphiascus blanchardi* (Scott T., 1895)
– Distal end of exterior margin in P5 exopod normal 3

3. Female P5 exopod with 4 setae .... *Typhlamphiascus longifurcatus* Rouch, 1962
– Female P5 exopod with more than 4 setae 4

4. Female P5 exopod with 5 setae 5
– Female P5 exopod with 6 setae 8

5. Female P5 baseoendopod with 4 setae 6
– Female P5 baseoendopod with 5 setae 7

6. P4 enp-3 with 2 inner setae .... *Typhlamphiascus gracilicaudatus* (Thompson I.C. & Scott A., 1903)
– P4 enp-3 with single inner seta .... *Typhlamphiascus tuerkayi* sp. nov.

7. Male P1 basis with 3 chitinous lamellae on inner edge .. *Typhlamphiascus drachi* Soyer, 1964

– Male P1 basis with 7 chitinous lamellae on inner edge . . . *Typhlamphiascus gracilis* Por, 1963

8. Female A1 9-segmented . . . . *Typhlamphiascus accraensis* Scott T., 1894
– Female A1 8-segmented. 9

9. P4 enp-3 with 2 inner setae 10
– P4 enp-3 with single inner seta 12

10. P1 exopod much shorter than P1 enp-1; male P1 without chitinous lamella . . . . *Typhlamphiascus bouligandi* Soyer, 1971
– P1 exopod extends to about the end of P1 enp-1 11

11. Caudal ramus at least two times as long as wide; fourth innermost seta of P5 baseoendopod as long as third innermost seta . . . . *Typhlamphiascus lutincola* Soyer, 1963
– Caudal ramus at most two times as long as wide; fourth innermost seta of P5 baseoendopod about half as long as third innermost seta . . . *Typhlamphiascus lamellifer capensis* Kunz, 1975

12. P3 enp-3 with 4 setae 13
– P3 enp-3 with 5 setae 14

13. P3 exp-3 with 5 setae . . . . *Typhlamphiascus higginsi* Chullasorn, 2009
– P3 exp-3 with 6 setae. . . . . *Typhlamphiascus unisetosus* Lang, 1965

14. P4 enp-2 without inner seta . . . . *Typhlamphiascus typhlops* (Sars G.O., 1906)
– P4 enp-2 with single inner seta 15

15. P5 exopod oval, the outer two setae of P5 baseoendopod reduced . . . . *Typhlamphiascus ovale* Wells & Rao, 1987
– P5 exopod not of this form, the outer two setae of P5 baseoendopod not reduced 16

16. Caudal ramus lamelliform. 17
– Caudal ramus sublinear or straight-rectangular 18

17. P5 exopod narrow oblong, the fifth innermost seta of P5 baseoendopod as long as the fourth innermost seta. . . . . *Typhlamphiascus pectinifer* Lang, 1965
– P5 exopod oblong oval, the fifth innermost seta of P5 baseoendopod about half as long as the fourth innermost seta . . . . *Typhlamphiascus lamellifer lamellifer* (Sars G.O., 1911)

18. Caudal ramus at least two times as long as wide . . . . *Typhlamphiascus confusus gullmaricus* Por, 1963
– Caudal ramus at most 1.7 times as long as wide 19

19. Urosomite 4 without armature . . . . *Typhlamphiascus confusus erythraeicus* Por, 1963
– Urosomite 4 with armature. . . . . *Typhlamphiascus latifurca* Por, 1968

ACKNOWLEDGEMENTS

This study was supported by the National Natural Science Foundation of China (No. 41206148) and the IOCAS Funding (No. 2012IO060104). This study was also funded partly by the Scientific and Technological Innovation Project Financially Supported by Qingdao National Laboratory for Marine Science and Technology (No. 2015ASKJ01). Thanks to the fellows in our team for their assistance in collecting samples.

REFERENCES

CHULLASORN, S., 2009. A review of *Typhlamphiascus* Lang, 1944 (Copepoda: Harpacticoida: Miraciidae) with a new species *Typhlamphiascus higginsi* from Phuket Island, Thailand. Zoological Studies, **48**: 493-507.

DAI, A. Y. (Tai, A. Y.) & Y. Z. SONG, 1979. Harpacticoida. In: C. J. SHEN (S. J. Shen) (chief ed.), Fauna Sinica. Crustacea. Freshwater Copepoda: 164-300. (Science Press, Beijing). [In Chinese.]

GEE, J. M. & F. H. MU, 2000. A new genus of Cletodidae (Copepoda; Harpacticoida) from the Bohai Sea, China. Journal of Natural History, **34**: 809-822.

HUYS, R., J. M. GEE, C. G. MOORE & R. HAMOND, 1996. Marine and brackish water harpacticoid copepods, part 1. Synopses of the British Fauna (new ser.), **51**: 1-viii, 1-352.

HUYS, R. & F. H. MU, 2008. Description of a new species of *Onychostenhelia* Itô (Copepoda, Harpacticoida, Miraciidae) from the Bohai Sea, China. Zootaxa, **1706**: 51-68.

LANG, K., 1965. Copepoda Harpacticoida from the Californian Pacific coast. Kungl. Svenska Vetenskapsakademiens Handlingar, **10**: 1-560.

MA, L. & X. Z. LI, 2011. *Delavalia qingdaoensis* sp. nov. (Harpacticoida, Miraciidae), a new copepod species from Jiaozhou Bay, Yellow Sea. Crustaceana, **84**: 1085-1097.

MU, F. H. & J. M. GEE, 2000. Two new species of *Bulbamphiascus* (Copepoda: Harpacticoida: Diosaccidae) and a related new genus, from the Bohai Sea, China. Cahiers de Biologie Marine, **41**: 103-135.

MU, F. H. & R. HUYS, 2002. New species of *Stenhelia* (Copepoda, Harpacticoida, Diosaccidae) from the Bohai Sea (China) with notes on subgeneric division and phylogenetic relationships. Cahiers de Biologie Marine, **43**: 179-206.

MU, F. H. & R. HUYS, 2004. Canuellidae (Copepoda, Harpacticoida) from the Bohai Sea, China. Journal of Natural History, **38**: 1-36.

POR, F. D., 1963. A comparative study in the genus *Typhlamphiascus* Lang (Copepoda: Harpacticoida). Arkiv för Zoologi, **16**: 189-206.

SCOTT, T., 1894. Report on Entomostraca from the Gulf of Guinea, collected by John Rattray, B.Sc. Transactions of the Linnean Society of London, **6**: 1-161.

WALTER, T. C., G. BOXSHALL & R. HUYS, 2016. *Typhlamphiascus* Lang, 1944. In: T. C. WALTER & G. BOXSHALL (eds.), World of Copepods database, available online at http://www.marinespecies.org/aphia.php?p=taxdetails&id=115329.

First received 12 May 2016.
Final version accepted 7 April 2017.

[When citing this volume, please refer to Crustaceana 90 (2017) 771-1288]

Michael Türkay Memorial Issue

A NEW SPECIES OF SQUAT LOBSTER OF THE GENUS *MUNIDA* (GALATHEOIDEA, MUNIDIDAE) FROM THE RED SEA

BY

E. MACPHERSON[1,5]), L. BEUCK[2]), C. RODER[3]) and C. R. VOOLSTRA[4])

[1]) Centro de Estudios Avanzados de Blanes (CEAB-CSIC), C. acc. Cala San Fracesc 14, E-17300 Blanes, Girona, Spain

[2]) Senckenberg am Meer, Marine Research Department, Südstrand 40, D-26382 Wilhelmshaven, Germany

[3]) Alfred-Wegener Institut Helmholtz-Zentrum für Polar- und Meeresforschung, am Binnenhafen 1117, D-27498 Helgoland, Germany

[4]) Red Sea Research Center, King Abdullah University of Science and Technology, Thuwal, Saudi Arabia

ABSTRACT

During a deep-water expedition to the Red Sea in 2013, an unusual specimen of squat lobster belonging to the genus *Munida* was collected off Thuwal, Saudi Arabia, at a depth of 320 m. This specimen is unique in having the pterygostomial flap visible from the dorsal side, the feature linking it to two eastern Pacific species, *M. bapensis* Hendrickx, 2000 and *M. macrobrachia* Hendrickx, 2003. The new species (*M. tuerkayi*) is readily distinguished from the eastern Pacific species by having the gastric region with numerous instead of less numerous spines, by having sternite 7 with three distinct carinae on each side, and by having the antennular basal article with two distal spines subequal instead of different in size. *Munida tuerkayi* was found associated with live colonies of the scleractinian coral *Eguchipsammia fistula* (Alcock, 1902).

Key words. — Anomura, taxonomy, deep-water coral, new species, *Munida*

ZUSAMMENFASSUNG

Während einer Tiefwasserexpedition im Roten Meer im Jahr 2013 wurde vor Thuwal (Saudi-Arabien) ein ungewöhnliches Exemplar eines Furchenkrebses, zugehörig zur Gattung *Munida*, in einer Tiefe von 320 m gesammelt. Dieses Tier zeichnet sich durch eine einzigartige Pterygostomialklappe aus, welche von der Dorsalseite zu sehen ist, und stellt somit die Verbindung her zu den zwei ostpazifischen Arten *M. bapensis* Hendrickx, 2000 und *M. macrobrachia* Hendrickx, 2003. Die neue Art (*M. tuerkayi*) ist leicht von den ostpazifischen Arten zu unterscheiden durch zahlreiche (anstatt weniger zahlreiche) Stacheln in der Gastralregion, durch drei deutlich sichtbare Kiele auf jeder Seite vom Sternit 7 und durch zwei fast gleich lange (anstatt unterschiedlich großer) Distalstacheln am Basalglied der 1. Antenne. *Munida tuerkayi* wurde zusammen mit lebenden Kolonien der Steinkoralle *Eguchipsammia fistula* (Alcock, 1902) gefunden.

[5]) Corresponding author; e-mail: macpherson@ceab.csic.es

INTRODUCTION

The Red Sea is considered one of the “hot spots” in marine biodiversity of the world, and previous studies on the decapod crustaceans of the area have demonstrated the existence of a very rich fauna, with a high proportion of endemic species (Spiridonov & Apel, 2007; Werding & Hiller, 2007; DiBattista et al., 2016). Among the decapods, the squat lobsters (Chirostyloidea and Galatheoidea) are considered one of the most diverse groups (Baba et al., 2008; Macpherson & Robainas-Barcia, 2015).

The genus *Munida* Leach, 1820, is one of the most diverse taxa of squat lobsters of the family Munididae Ahyong et al., 2010, and is represented in the Red Sea by five species (Baba et al., 2008; Macpherson & Baba, 2011; Macpherson et al., 2017): *M. babai* Tirmizi & Javed, 1976, *M. dispar* Macpherson & Baba, 1993, *M. eudora* Macpherson & Baba, 1993, *M. janetae* Tirmizi & Javed, 1992, and *M. roshanei* Tirmizi, 1966. During a cruise to the Red Sea (Saudi Arabia) in May 2013, a new species of *Munida* was collected from a bathyal depth of the central Red Sea, and associated with deep-sea corals (fig. 1). In the present study, we describe and illustrate this new species and characterize its environment.

The sample was collected during Leg 6 of the KAUST Red Sea Expedition (KRSE) 2013 at a depth of 320 m off Thuwal, Saudi Arabia (Red Sea) on board the RV “Aegaeo”. The expedition aimed to investigate deep-sea coral habitats in the central and northern Saudi Arabian Red Sea. Sampling was undertaken via a scoop used by a manipulator arm of the remotely operated vehicle (ROV) “Max Rover”. The live organisms were stored in seawater (+21°C) for subsequent macroscopic and microscopic documentation, prior to fixation in ethanol (⩾96% with 1% MEK ethylalcohol). Via timecode, the video material was linked to the ROV’s navigation track. Environmental conditions in 320 m depth were derived from four CTD downcasts (stations KRSE2013-6-4-2/1, KRSE2013-6-4-2/9, KRSE2013-6-4-2/15, KRSE2013-6-4-2/16), carried out in that working area off Thuwal. The material is deposited at the Senckenberg Research Institute and Natural History Museum, Frankfurt am Main (SMF) (see collection database SeSam: http://sesam.senckenberg.de). The terminology used follows Baba et al. (2009). The size of the specimen is indicated by the postorbital carapace length. Measurements of appendages are taken in dorsal (pereopod 1), lateral (antennule, pereopods 2-4) and ventral (antenna) midlines. Abbreviations used are: G, gonopod; Mxp3, maxilliped 3; P1, pereopod 1; P2-4, pereopods 2-4.

Fig. 1. Overview map of the Red Sea, showing the sampling site of *Munida tuerkayi* n. sp. in red (22°17.826′N 38°53.808′E, 320 m water depth).

TAXONOMY

MUNIDIDAE Ahyong, Baba, Macpherson & Poore, 2010

Munida tuerkayi n. sp.

(figs. 2-4)

Material examined.— Holotype: off Thuwal (Red Sea), RV "Aegaeo", sta. KRSE 2013-6-4-1/4d, ROV dive no. 6, 22°17.826′N 38°53.808′E, 10 May 2013, 320 m: male, 7.3 mm (SMF 49694).

Description.— Carapace: 1.3 times as long as broad, dorsal ridges with dense short non-iridescent setae and few scattered, long iridescent and non-plumose setae. Epigastric region with 2 transverse rows of spines, each with 4 spines flanking longitudinal row of 3 spines in midline behind rostral spine; protogastric region with 3 pairs of small spines laterally. Two branchial dorsal spines and 3 postcervical spines on each side, both transversally arranged. Frontal margins slightly oblique. Orbit visible in dorsal view, with spinose ventral crest and 2 small spines dorsoposterior to it. Lateral margins straight and subparallel. First

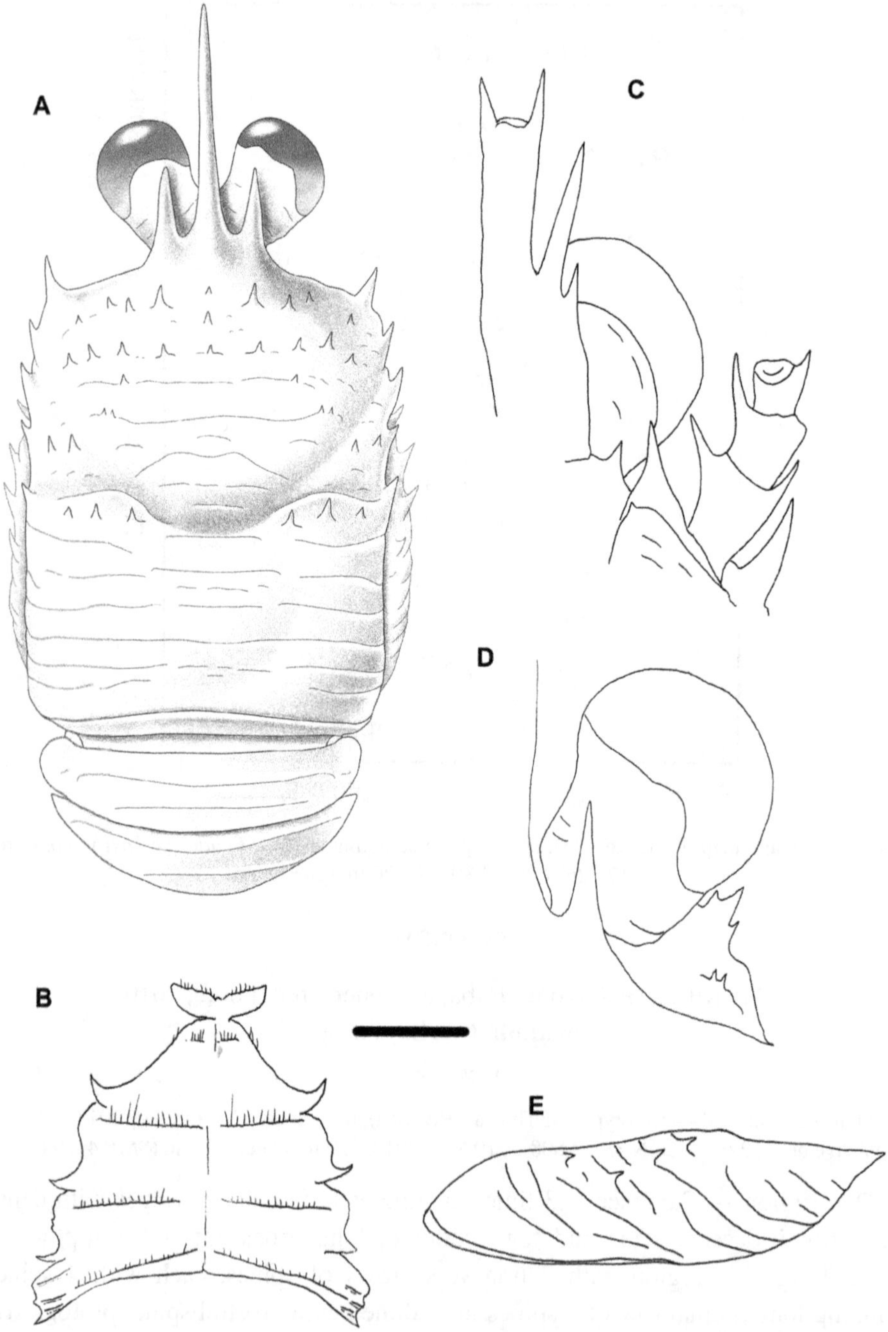

Fig. 2. *Munida tuerkayi* n. sp., holotype, male 7.3 mm (SMF), Red Sea. A, Carapace and abdomen, dorsal view; B, sternal plastron; C, cephalic region, showing antennular and antennal peduncles, ventral view; D, anterior part of carapace, showing ocular peduncle and orbit, right, dorsal; E, right pterygostomial flap, lateral view. Scale: A, E = 2.0 mm; B, C, D = 1.0 mm.

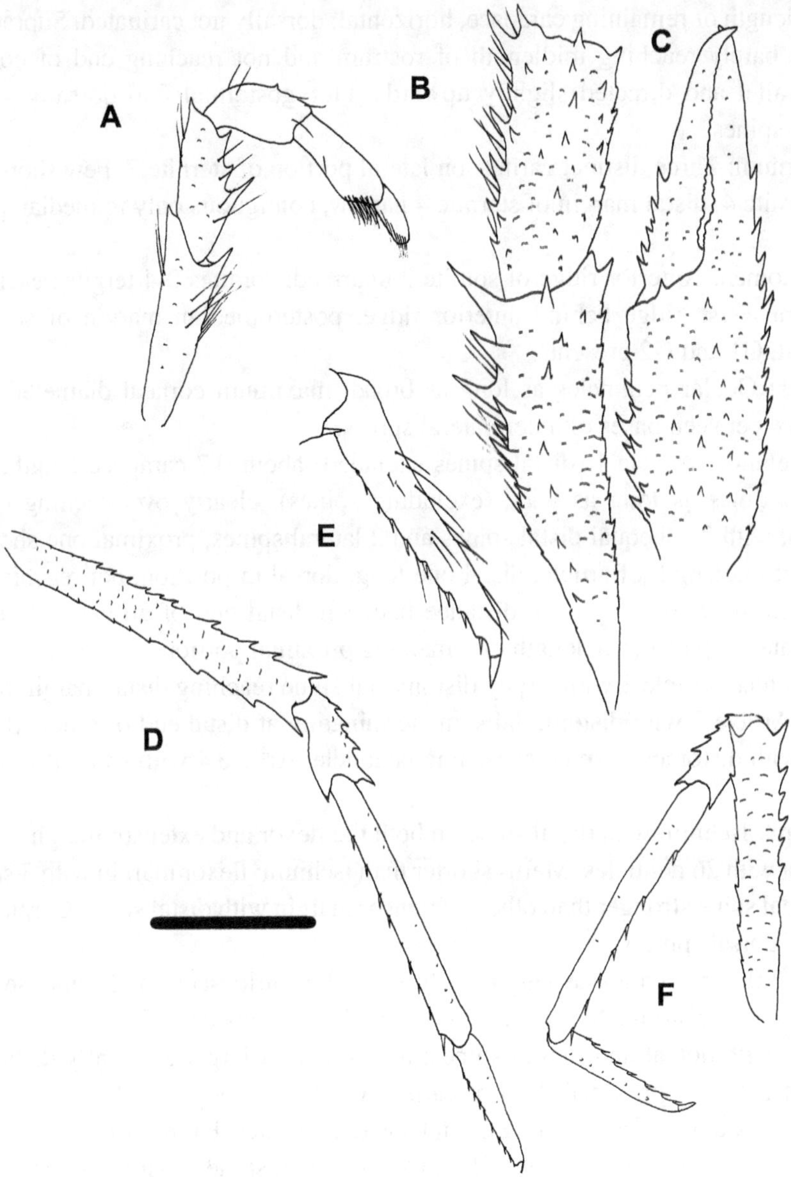

Fig. 3. *Munida tuerkayi* n. sp., holotype, male, 7.3 mm (SMF), Red Sea. A, Right Mxp3, lateral view; B, right P1 merus and carpus, dorsal view; C, right P1 palm and fingers, dorsal view; D, right P2, lateral view; E, dactylus of right P2, lateral view; F, left P3, lateral view. Scale: A, E = 1.0 mm; B, C, D, F = 2 mm.

lateral spine at anterolateral angle, long, clearly not reaching level of sinus between rostrum and supraocular spines; one small spine in front of anterior branch of cervical groove. Branchial margins with 3 spines. Rostrum spiniform, about 0.6

times length of remaining carapace, horizontal, dorsally not carinated. Supraocular spines barely reaching midlength of rostrum and not reaching end of corneae, subparallel and directed slightly upwards. Pterygostomial flap dorsally visible, with 4 spines.

Sternum: Three distinct carinae on lateral portion of sternite 7. Few short striae on sternite 4; distal margin of sternite 4 narrow, contiguous only to median part of sternite 3.

Abdomen: Anterior ridge of somite 2 unarmed; somites 2-4 tergites each with one transverse ridge behind anterior ridge; posteromedian margin of somite 6 straight. G1 and G2 present.

Eyes: Ocular peduncles as long as broad, maximum corneal diameter 0.4 × distance between bases of anterolateral spines.

Antennule: Article 1 (distal spines excluded) about 0.7 carapace length, elongate, 3 times as long as wide (excluding spines), clearly overreaching end of cornea, with 2 subequal distal spines and 2 lateral spines, proximal one short, located at midlength of article, distal one long, dorsal in position, not reaching end of article (excluding spines); distance between distal end of article and base of dorsolateral spine 0.8 × length of remaining proximal portion.

Antenna: Article 1 with strong distomesial spine reaching distal margin of article 2. Article 2 with distomesial spine terminating at distal end of article 3. Article 3 with distomesial spine exceeding peduncle. Article 4 with small distolateral spine.

Mxp3: Ischium with distal spine on both the flexor and extensor margins; crista dentata with 26 denticles. Merus shorter than ischium; flexor margin with 3 spines, proximal spine stronger than others; extensor margin with distal spine. Carpus with minute dorsal spine.

P1: 3.0 times carapace length, with some long iridescent and plumose setae along mesial margin. Merus 1.1 times length of carapace, 1.5 times as long as carpus, with dorsal and mesial spines; mesial spines large, especially distal one, but not reaching proximal third of carpus. Carpus as long as palm, 2.7 times as long as broad, with spines along mesial and dorsal sides. Palm 2.1 times as long as broad, with some small dorsal spines; row of lateral spines continuing on to fixed finger, and row of mesial spines continuing on to proximal half of movable finger. Fingers as long as palm.

P2-3 (P4 missing): Moderately long and slender, with numerous plumose and iridiscent setae along extensor margin. P2 2.4 times carapace length. Meri shorter posteriorly (P3 merus 0.8 length of P2 merus); P2 merus 0.9 × length of carapace, 7.5 times as long as broad, 1.3 times as long as P2 propodus; P3 merus 6 times as long as broad, as long as P3 propodus. Extensor margins of P2-3 meri with row of 7-10 proximally diminishing spines; flexor margins distally with some spines

Fig. 4. *Munida tuerkayi* sp. nov., colour in life.

followed proximally by several eminences. Carpi with 4 extensor marginal spines on P2-4; lateral surface with several granules sub-paralleling extensor margin on P2-3; flexor margin with distal spine. Propodi 8.6-9.0 times as long as broad; extensor margin unarmed; flexor margin with 7-8 slender, movable spines. Dactyli slender, length 0.5 × that of propodi; flexor margin with 8-9 movable spinules along entire length.

Colour.— Ground colour of carapace and abdomen whitish; spines, striae, and ridges reddish. Rostral and supraocular spines whitish, with lateral margins reddish. Sternite 3, and distal portion of sternite 4, reddish. P1-4 whitish with red bands; P1 fingers with proximal and distal parts whitish; P2-3 dactyli whitish. See fig. 4.

Etymology.— Named after our friend and colleague, Michael Türkay, for his important contributions to carcinology. The species name thus is a noun in the genitive singular.

Distribution and habitat.— Off Thuwal (Red Sea) at 320 m, associated with live colonies of the scleractinian coral *Eguchipsammia fistula*, which occur patchily and partly buried on soft sediment (fig. 5), at warm temperatures of 21.6°C, salinity of 40.5 PSU, at a pH value of 8.3, oxygen values of 8.4%, and under highly oligothrophic conditions (see also Roder et al., 2013; Roik et al., 2015).

Remarks.— *Munida tuerkayi* belongs to a group of species with the pterygostomial flap visible from the dorsal side. This character is shared with the genus *Pleuroncodes* as well as with two species of *Munida* from the eastern Pacific: *M. bapensis* Hendrickx, 2000, from the Gulf of California, 620 m, and *M. macrobrachia* Hendrickx, 2003, from southern California, 540-612 m.

Munida tuerkayi is easily distinguished from these two species by the following differences:

Fig. 5. ROV images of the sampling site (sta. KRSE2013-6-4-1/4d), in 320 m water depth. Images by courtesy of HCMR, Greece. A, Overview image of biotope with live colonies of *Eguchipsammia fistula* (Alcock, 1902) occurring patchily on, and partly buried in, soft sediment; B, detailed image with live *E. fistula* colonies; note the scoop (sampling tool), used by the manipulator arm of the ROV.

— The dorsal surface of the gastric region has numerous spines in *M. tuerkayi*, whereas these spines are less numerous in the other two species.
— The pterygostomial flap has some spines on the surface below the branchial region in *M. tuerkayi*, whereas it is unarmed in the other species (although *M. macrobrachia* has some spines on the anterior part).
— The abdominal somites are unarmed in the new species and *M. macrobrachia*, whereas the somites 2-4 have some spines on the anterior ridge in *M. bapensis*.
— Some spines on the posterior ridge of the carapace in *M. macrobrachia* and *M. tuerkayi*, whereas there is a row of spines in *M. bapensis*.
— There are three distinct carinae on the lateral portion of sternite 7 in the new species, whereas these lateral portions are smooth (*M. macrobrachia*) or with numerous granules (*M. bapensis*) in the other two species.
— Antennular article 1 has two subequal distal spines in the new species, whereas these spines are clearly different in size in *M. bapensis* and longer in *M. macrobrachia*. Furthermore, this article is clearly more elongated in *M. tuerkayi* than in the other two species.
— Antennal article 3 has a strong distomesial spine in *M. tuerkayi*, whereas this spine is absent in the other two species.

ACKNOWLEDGEMENTS

We thank the late Michael Türkay for his friendship, support, fruitful discussions, and efforts in the improvement of the crustacean collections. We also express our sincere thanks to the captain, his crew, and the shipboard staff of KRSE 2013 Leg 6 on RV "Aegaeo". Thanks are also due to two anonymous reviewers for their valuable comments. We thank Frank Mallon for the CTD casts. Jürgen Titschack (JT) is thanked for his support in documenting sampled live organisms. LB received funds from EU-FP7 Projects CoralFISH (contract no. 213144), HERMIONE (contract no. 226354), and the Hessian initiative for the development of scientific and economic excellence (LOEWE) at the Biodiversity and Climate Research Centre (BiK-F), Frankfurt a. M., Germany.

REFERENCES

AHYONG, S. T., K. BABA, E. MACPHERSON & G. C. B. POORE, 2010. A new classification of the Galatheoidea (Crustacea: Decapoda: Anomura). Zootaxa, **2676**: 57-68.

BABA, K., E. MACPHERSON, C. W. LIN & T. Y. CHAN, 2009. Crustacean fauna of Taiwan. Squat lobsters (Chirostylidae and Galatheidae): ix + 311. (National Taiwan Ocean University, Keelung).

BABA, K., E. MACPHERSON, G. C. B. POORE, S. T. AHYONG, A. BERMUDEZ, P. CABEZAS, C. W. LIN, M. NIZINSKI, C. RODRIGUES & K. E. SCHNABEL, 2008. Catalogue of squat lobsters of the world (Crustacea: Decapoda: Anomura — families Chirostylidae, Galatheidae and Kiwaidae). Zootaxa, **1905**: 1-220.

DIBATTISTA, J. D., J. H. CHOAT, M. R. GAITHER, J. P. A. HOBBS, D. F. LOZANO CORTES, R. F. MYERS, G. PAULAY, L. A. ROCHA, R. J. TOONEN, M. W. WESTNEAT & M. L. BERUMEN, 2016. On the origin of endemic species in the Red Sea. Journal of Biogeography, **43**: 13-30.

HENDRICKX, M. E., 2000. The genus *Munida* Leach (Crustacea, Decapoda, Galatheidae) in the eastern tropical Pacific, with description of two new species. Bulletin de l'Institut Royal des Sciences Naturelles de Belgique, **70**: 163-192.

HENDRICKX, M. E., 2003. The temperate species of the genus *Munida* Leach (Crustacea, Decapoda, Galatheidae) in the east Pacific, with the description of a new species and additional records for tropical-subtropical species. Bulletin de l'Institut Royal des Sciences Naturelles de Belgique, **73**: 115-136.

MACPHERSON, E. & K. BABA, 1993. Crustacea Decapoda: *Munida japonica* Stimpson, 1858, and related species (Galatheidae). In: A. CROSNIER (ed.), Résultats des Campagnes MUSORSTOM, Vol. 10, **156**: 381-420. (Mémoires du Muséum national d'Histoire naturelle, Paris).

MACPHERSON, E. & K. BABA, 2011. Chapter 2. Taxonomy of squat lobsters. In: G. C. B. POORE, S. T. AHYONG & J. TAYLOR (eds.), The biology of squat lobsters: 39-71. (CSIRO Publishing, Melbourne, VIC and CRC Press, Boca Raton, FL).

MACPHERSON, E. & A. ROBAINAS-BARCIA, 2015. Species of the genus *Galathea* Fabricius, 1793 (Crustacea, Decapoda, Galatheidae) from the Indian and Pacific Oceans, with descriptions of 92 species. Zootaxa, **3913**: 1-335.

MACPHERSON, E., P. C. RODRIGUEZ-FLORES & A. MACHORDOM, 2017. New sibling species and new occurrences of squat lobsters (Crustacea, Decapoda) from the western Indian Ocean. European Journal of Taxonomy, **343**: 1-61.

RODER, C., M. L. BERUMEN, J. BOUWMEESTER, E. PAPATHANASSIOU, A. AL-SUWAILEM & C. R. VOOLSTRA, 2013. First biological measurements of deep-sea corals from the Red Sea. Scientific Reports, **3**: DOI:10.1038/srep02802.

ROIK, A., T. RÖTHIG, C. RODER, P. J. MÜLLER & C. R. VOOLSTRA, 2015. Captive rearing of the deep-sea coral *Eguchipsammia fistula* from the Red Sea demonstrates remarkable physiological plasticity. PeerJ, **3**: e734.

SPIRIDONOV, V. A. & M. APEL, 2007. A new species and new records of deep-water Calappidae (Crustacea: Decapoda) from the Indian Ocean with a key to the *Mursia* Desmarest, 1823 species of the region. Journal of Natural History, **41**: 2851-2890.

TIRMIZI, N. M., 1966. Crustacea: Galatheidae. The John Murray Expedition 1933-34. Scientific Reports, **11**: 167-234.

TIRMIZI, N. M. & W. JAVED, 1976. A new species of *Munida* from the Indian Ocean with a redescription of a syntype of *Munida spinulifera* Miers, 1884 (Decapoda, Galatheidea). Crustaceana, **31**: 81-89.

TIRMIZI, N. M. & W. JAVED, 1992. Two new species of *Munida* Leach, 1820 (Decapoda, Anomura, Galatheidae) from the Indian Ocean. Crustaceana, **62**: 312-318.

WERDING, B. & A. HILLER, 2007. The Porcellanidae (Crustacea: Decapoda: Anomura) of the Red Sea with description of a new species of *Petrolisthes*. Zootaxa, **1460**: 1-24.

First received 29 April 2016.
Final version accepted 16 August 2016.

[When citing this volume, please refer to Crustaceana 90 (2017) 771-1288]

Michael Türkay Memorial Issue

A NEW GENUS AND SPECIES OF FRESHWATER CRAB (DECAPODA, PSEUDOTHELPHUSIDAE) FROM THE TAPAJÓS RIVER, A SOUTHERN TRIBUTARY OF THE AMAZON RIVER IN BRAZIL

BY

CÉLIO MAGALHÃES[1])

Instituto Nacional de Pesquisas da Amazônia, Av. André Araújo, 2936,
69.067-375 Manaus, AM, Brazil

ABSTRACT

A new genus and species of pseudothelphusid crab of the tribe Kingsleyini Ortmann, 1897, from the Tapajós River, a southern tributary of the Amazon River basin in the state of Pará, Brazil, is described and illustrated. *Michaelthelphusa tuerkayi* gen. et sp. nov. is characterized by the male first gonopod possessing marginal and lateral sutures fused at the distal third of the stem and the merged portion strongly twisted towards the lateral side; an apex transversally expanded along caudal-cephalic axis, with a well-developed, but low, somewhat depressed, apical plate; apical plate with spine-like outgrowth cross overlapping the spine-like projection of the mesial process. Moreover, the relationship of the new genus and species with other pseudothelphusid genera from the Amazon basin is discussed.

Key words. — Amazon, Brachyura, Crustacea, Kingsleyini, Neotropical region, taxonomy

ZUSAMMENFASSUNG

Hier wird eine neue Gattung und Art einer pseudothelphusiden Krabbe der Gattungsgruppe Kingsleyini Ortmann, 1897 beschrieben und illustriert. Diese stammt aus dem Tapayós-Fluss, einem südlichen Ausläufer des Amazonasbeckens im Staat Pará, Brasilien. *Michaelthelphusa tuerkayi* gen. et sp. nov. ist durch randständige und seitliche Nähte an den ersten männlichen Gonopoden, welche im distalen Drittel des Stammes verschmelzen, sowie durch eine starke Eindrehung dieses verschmolzenen Abschnittes in Richtung der Lateralseite gekennzeichnet. Weiterhin dehnt sich der Apex transversal entlang der caudal-cephalischen Achse aus. Die Apikalplatte ist gut ausgeprägt, aber niedrig und etwas eingedrückt und trägt einen dornartigen Auswuchs, welcher den dornartigen Vorsprung des Mesialfortsatzes quer überlappt. Ferner wird das Verwandtschaftsverhältnis der neuen Gattung und Art zu anderen pseudothelphusiden Gattungen aus dem Amazonasbecken diskutiert.

[1]) e-mail: celiomag@inpa.gov.br

INTRODUCTION

The Pseudothelphusidae is a highly diverse family of true freshwater crabs that is distributed in the Neotropical region from Mexico to Peru and Brazil, and up to 2014 comprised 34 genera and more than 260 known species (Cumberlidge et al., 2014). Six genera and 22 species occur partially or exclusively in Brazil (Magalhães et al., 2016; Pedraza et al., 2016; Pinheiro & Santana, 2016). They are mostly found in the Amazon River basin, although two species were found as far east as the state of Ceará, in northeastern Brazil (Magalhães et al., 2005; Pinheiro & Santana, 2016). However, there are still many gaps in the knowledge about the distribution and composition of the Brazilian pseudothelphusids, since large areas of the Amazon region have not yet been adequately surveyed. Magalhães & Pereira (2007) compiled a list of the species for the northern tributaries of the Amazon River that drain the Guyana Shield. With regards to the southern tributaries, only one list has been compiled for the Xingu River (Magalhães, 2003). This list was recently updated with the description of a new species from that drainage (Pedraza et al., 2015). Pseudothelphusid crabs were also described from the drainages of other right bank tributaries such as the Tocantins River (Magalhães, 2005; Pedraza & Tavares, 2015) and Madeira River (Magalhães, 2009; Magalhães & Türkay, 2010; Magalhães et al., 2014).

The Tapajós River is the fifth largest sub-basin of the Amazon; it is a clearwater river that arises from the hard crystalline basement of the Brazilian Shield and flows northwards for approximately 1900 km (Goulding et al., 2003). Draining the granite and sandstone tablelands of the Brazilian Shield, its upper course is marked by many rapids and cataracts that extend to the vicinities of Itaituba (Hales & Petry, 2015). The brachyuran fauna of this river system is poorly known; only four species have been unequivocally recorded from the basin: one Pseudothelphusidae, *Brasiliothelphusa tapajoense* Magalhães & Türkay, 1986 (see Magalhães, 1986), and four Trichodactylidae, *Sylviocarcinus devillei* H. Milne Edwards, 1853, *S. pictus* (H. Milne Edwards, 1853), *Trichodactylus borellianus* Nobili, 1896 and *Valdivia serrata harttii* (Rathbun, 1905) (see Rathbun, 1906; Bott, 1969; Magalhães & Türkay, 1996). In fact, *B. tapajoense* was the first pseudothelphusid crab described from a southern tributary of the Amazon River (Magalhães & Türkay, 1986). In the present paper I introduce a new genus to accommodate a hitherto unknown pseudothelphusid species from the Tapajós River, which is hereby described, illustrated and compared with allied taxa.

The specimens examined in this study are deposited at the Instituto Nacional de Pesquisas da Amazônia, Manaus (INPA), Museu Paraense Emilio Goeldi, Belém (MPEG), Museu de Zoologia, Universidade de São Paulo, São Paulo (MZUSP) and Forschungsinstitut und Naturmuseum Senckenberg, Frankfurt am Main (SMF).

Other abbreviations used are: carapace width (cw), measured across the carapace at its widest point; carapace length (cl), measured along the midline, from the frontal to the posterior margin; carapace height (ch), the maximum height of the cephalothorax, measured as the distance between the dorsal and ventral edges of the shell; frontal width (fw), the width of the front measured along its upper border; male first (G1) and second (G2) gonopods; third maxilliped (Mxp3); cheliped (P1); pereiopods 2 to 5 (P2-P5); sternal sulci between adjacent thoracic sternites (s1/2-s7/8). Geographic coordinates inserted between brackets were taken from Google Earth. Line drawings were made with the aid of a stereoscopic microscope (Wild M8) equiped with a drawing tube, then scanned at 1200 dpi and edited in Adobe Photoshop CS 2 software at 1200 dpi. The plates were mounted in CorelDraw X3 software. Measurements, in millimetres, were taken with a calliper and are given between parentheses after the number of specimens examined. Terminology for describing the morphology of the G1 was adapted from Smalley (1964) and Pedraza et al. (2015).

TAXONOMY

Family PSEUDOTHELPHUSIDAE Ortmann, 1983
Tribe Kingsleyini Bott, 1970
Michaelthelphusa gen. nov.

Diagnosis.— Third maxilliped with exopod strongly reduced, 0.17 times length of outer margin of ischium. G1 stout; marginal and lateral sutures strongly twisted towards lateral side, fused at distal third of stem, well-developed mesial process bearing strong, conical spine-like projection distally; apex transversally expanded along caudal-cephalic axis, with a well-developed, but low, depressed apical plate, with subdistal lobe, spine-like outgrowth cross overlapping spine-like projection of mesial process, and 2 superimposed distal lobes; field of apical spines weakly developed, situated in shallow depression along laterocephalic side enclosed by lateral and distal margins of apical lobes; opening of spermatic channel located subdistally.

Type species.— *Michaelthelphusa tuerkayi* sp. n., present designation.

Etymology.— The genus is named after Prof. Dr. Michael Türkay (see under etymology of the species) in combination with “Thelphusa”, a common suffix associated with freshwater crabs. The gender is feminine.

Remarks.— The present new genus belongs to the tribe Kingsleyini Bott, 1970 because it has the following characters: the exopod of the third maxilliped is less than 0.2 length of the ischium; the basis of the G1 is distinctly elongated dorso-ventrally; lateral side of the G1 has a field of apical spines; the mesial lobe of the

G1 bears a spine-like projection; and the marginal process of the G1 is distinct (Rodriguez, 1982).

Michaelthelphusa gen. nov. appears to be closely related to *Kingsleya* Ortmann, 1897 with regards to the G1 morphology. Similar to *Kingsleya*, the new genus bears an apical plate with two superimposed distal lobes, an apical field of spines situated laterally, and the distal portion of the mesial process produced into a spine-like projection. However, the fusion of the marginal and lateral sutures at the distal third of the G1, followed distomesially by a strong twist towards the lateral side, as well as the depressed, low apical plate and the crisscross arrangement of the subdistal spine-like projection of the apical plate and that of the distal portion of the mesial process, appears to be unique enough to justify the erection of a new generic taxon to accommodate this new species from the Tapajós River. These characters are not evident in the other genera of Kingsleyini, particularly when compared to the species that occur in the Amazon basin (see below under material examined). In *Fredius* spp., *Kingsleya* spp., *Microthelphusa* spp., and in the monotypic genera *Brasiliothelphusa tapajoense* Magalhães & Türkay, 1986 and *Prionothelphusa eliasi* Rodríguez, 1908, the marginal and lateral sutures run roughly parallel in a more or less straight direction along the mesial side of the G1. The lateral suture fails to reach the distal part of the mesial surface in these genera, only in the present new genus does the marginal suture tend to meet the lateral one as a consequence of the strong twist of the distal part of the G1. As a consequence of this convoluted distal part, the marginal process is displaced towards the lateral side in *Michaelthelphusa* gen. nov., and is not situated on the mesial side as is usual in the other genera.

The somewhat depressed structure of the apical plate is also peculiar to this new genus. In the other five genera compared herein, the apical plate shows a somewhat compressed arrangement and is usually higher (i.e., more distally projected) in relation to the size of the mesial process than that of *Michaelthelphusa* gen. nov., with the exception of the apical plate of *K. castrensis* Pedraza, Martinelli Filho & Magalhães, 2015 (however, in the latter species' the other characters are within those that are diagnostic for that genus). The direction of the spine-like distal portion of the mesial process can vary within the species compared here, either pointing towards the mesial or the caudal side; spine-like outgrowth in the apical plate can also be found in *Brasiliothelphusa* spp. (see Magalhães & Türkay, 1986, 2010); *Kingsleya celioi* Pedraza & Tavares, 2015 (see Pedraza & Tavares, 2015: 448, fig. 3A), *K. junki* Magalhães, 2003 (see Pedraza et al., 2015), *Microthelphusa lipkei* Magalhães, 2010, *M. bolivari* Rodríguez, 1980, *M. somanni* (Bott, 1967), *M. wymani* (Rathbun, 1905) (see Magalhães, 2010), *Fredius reflexifrons* (Ortmann, 1897), *F. fittkaui* (Bott, 1967) (see Magalhães & Rodríguez, 2002) and *F. buritizatilis* Magalhães & Mantelatto, 2014 in Magalhães et al.

(2014) (see Magalhães et al., 2014). However, none of them bear the crisscross arrangement seen in the present new genus.

Michaelthelphusa tuerkayi sp. nov.
(fig. 1)

Material examined.— Holotype: male (cw 43.0, cl 26.6, ch 14.5, fw 12.7), INPA 2186, Brazil, Pará, left bank of Tapajós River between Buburé and São Luiz do Tapajós (approximately 04°36′S 56°17′W), 24 Oct 1991, C. Magalhães & L. Moraes leg. Paratypes: 1 male (cw 45.4, cl 27.9), 1 female (cw 49.4, cl 30.3), INPA 2187, data as in holotype; 1 male (cw 31.7, cl 20.9), MPEG 1741, data as in holotype; 1 male (cw 40.9, cl 25.6), 1 female (cw 34.0, cl 22.1), MZUSP 34490, data as in holotype. Additional material: 2 spent females (cw 47.9, cl 29.4; cw 53.2, cl 31.5), INPA 2189, data as in holotype; 1 spent female (cw 47.9, cl 29.1), 1 female with youngs (cw 52.6, cl 31.8), INPA 2190, data as holotype, J. Zuanon et al. leg.

Comparative material examined.— *Brasiliothelphusa dardanelosensis* Magalhães & Türkay, 2010: male holotype (INPA 1474); *B. tapajoense* Magalhães & Türkay, 1986: male paratype (INPA 150); *Fredius buritizatilis* Magalhães & Mantelatto, in Magalhães, Sanches, Pileggi & Mantelatto, 2014: male holotype (INPA 1891); *F. denticulatus* (H. Milne Edwards, 1853): male (INPA 457); *F. fittkaui* (Bott, 1967): male (INPA 856); *F. reflexifrons* (Ortmann, 1897): male (INPA 159); *F. ykaa* Magalhães, 2009: male holotype (INPA 1465); *K. castrensis* Pedraza, Martinelli Filho & Magalhães, 2015: male holotype (INPA 2010); *K. gustavoi* Magalhães, 2005: male holotype (INPA 1320); *K. junki* Magalhães, 2003: male (INPA 2012); *K. latifrons* (Randall, 1840): male (INPA 289); *K. siolii* Bott, 1967: male (INPA 340); *K. ytupora* Magalhães, 1986: male holotype (INPA 369); *Microthelphusa bolivari* Rodríguez, 1980: male (INPA 1749); *M. lipkei* Magalhães, 2010: male holotype (INPA 1748); *M. somanni* (Bott, 1967): male holotype (SMF 2880a); *M. wymani* (Rathbun, 1905): male (INPA 356); and *Prionothelphusa eliasi* Rodríguez, 1980: male (INPA 390).

Diagnosis.— See diagnosis of genus.

Description of the holotype.— Carapace outline ellipsoid, widest medially (cb/cl 1.68); dorsal surface smooth, nearly flat, regions poorly defined (fig. 1A). Two distinct gastric pits, close to each other, on metagastric region. Cervical grooves shallow, narrow, nearly straight, faint proximally, distal end failing to reach anterolateral margin. Postfrontal lobules small, almost indistinct; median groove indistinct. Surface of carapace between front and postfrontal lobules smooth, with very slight anterior inclination. Upper border of front smooth, strongly angulate, slightly convex in dorsal view, median notch absent; lower border carinate, slightly sinuous in both frontal and dorsal view, a little more projected anteriorly than upper one, except medially. Upper orbital margin with minute papillae on the outer side, smooth on inner side; lower orbital margin slightly crenulate; exorbital angle very low, obtuse (fig. 1C). Anterolateral margin of carapace with very shallow depression just behind exorbital angle, followed posterolaterally by set of faint, minute, rounded teeth increasing in size from anterior to posterior portion; posterolateral margin smooth, marked by faint suture. Epistome narrow; epistomial tooth triangular, deflexed, with carinate, smooth borders. Suborbital and subhepatic regions of carapace branchiostegal

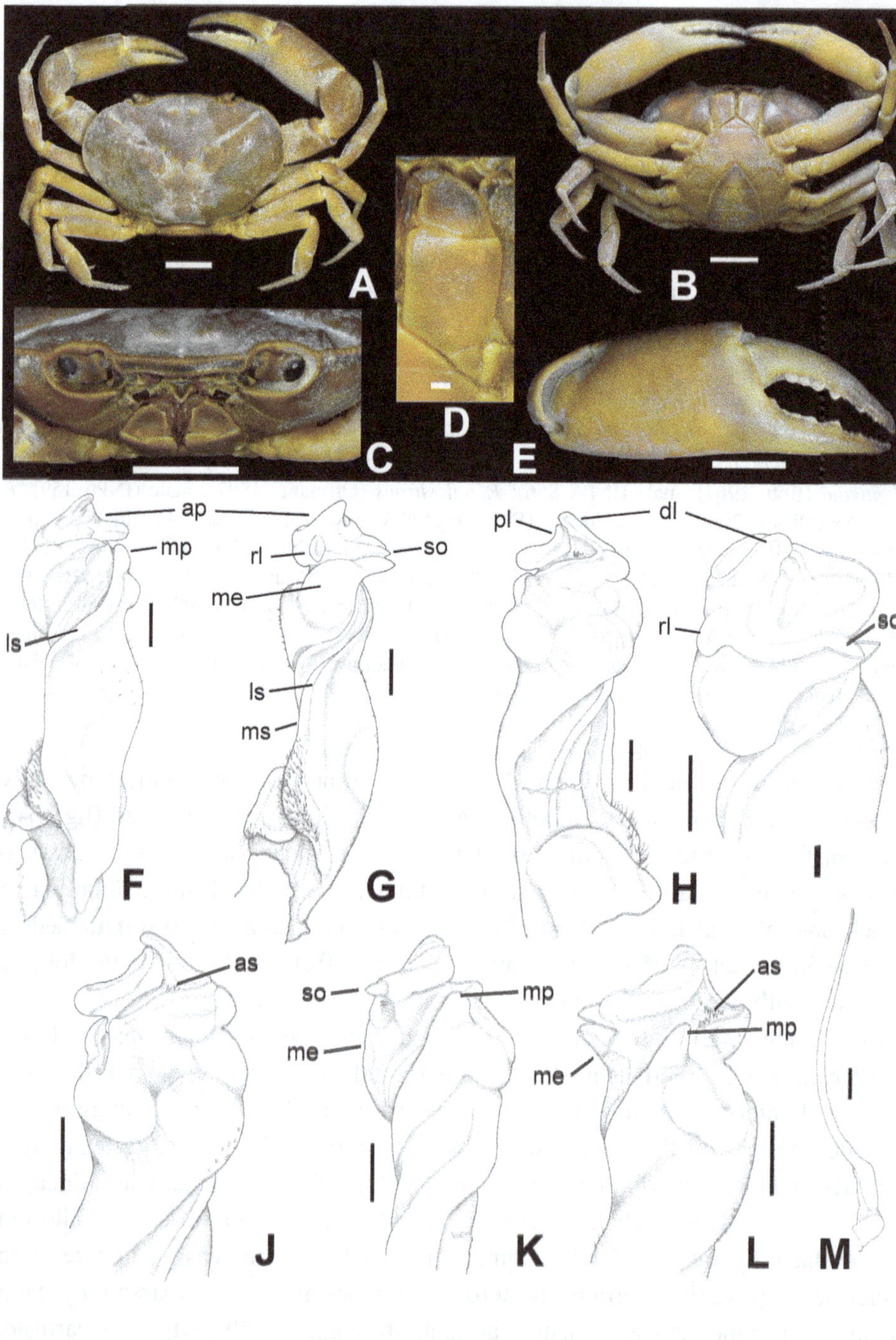
A
B
C
D
E
ap
mp
ls
rl
so
me
ms
pl
dl
as
F
G
H
I
J
K
L
M

walls smooth; pterygostomial regions with narrow, mostly smooth, weakly pilose patches only along outer borders of buccal cavity (fig. 1B).

Mxp3 ischium with nearly straight outer margin, inner margin straight; outer margin of merus broadly rounded; inner surface of palp covered with short setae; exopod of third maxiliped subtriangular, rudimentary, very short, narrow, 0.17 times length of outer margin of ischium (fig. 1D). Aperture of efferent branchial channel wide, upper margin subrectangular, smooth (fig. 1C).

First pereiopods (fig. 1A) moderately heterochelous, similarly armed, right P1 largest. Larger cheliped with merus subtriangular in cross section; superior margin rounded, with irregular row of low tubercles, fainter distally; mesial margin lined by longitudinal row of rounded, low teeth, slightly increasing in size distally; inferior lateral margin rounded, smooth; distal margin arched, smooth. Carpus inner margin with 1 prominent median spine, nearly smooth proximally (with only 3 faint, very low teeth) and smooth distally; outer margin rounded, smooth. Palm (fig. 1E) relatively narrow (length/breadth 1.22), smooth on both outer, inner sides. Fingers of larger chela (fig. 1E) distinctly gaping, those of smaller chela moderately gaping, tips not crossing; fixed finger with large tuberculiform teeth proximally, teeth smaller distally. Dactylus distinctly arched, slightly longer then palm (dactylus/palm = 1.16, measured dorsally), upper and outer surface of dactylus nearly smooth, with longitudinal, irregular rows of faint dark spots distally, distomedian portion slightly darker than proximal portion. Propodal finger with smooth surfaces, with few longitudinal, irregular rows of faint, dark spots. P2-P5 slender, dactylus/propodus and dactylus/merus ratios (left side measured), respectively, as follows: P2 = 1.46 and 0.77; P3 = 1.35 and 0.73; P4 = 1.37 and 0.74; P5 = 1.32 and 0.75. P2-5 with dactyli shorter than propodi, bearing 5 longitudinal rows of sharp, corneous spines, increasing in size distally, 2 faint grooves on proximal external surface.

Thoracic sternum slightly longer than broad. Thoracic sternites III and IV completely fused, except for lateral small notches; s4/5, s5/6 and s6/7 distinct, interrupted medially, failing to reach midline of thoracic sternum by a short distance; s7/8 complete, reaching midline. Midline of thoracic sternum marked by a deep

---

Fig. 1. *Michaelthelphusa tuerkayi* gen. et sp. n., male (cw 43.0 mm, cl 26.6 mm), holotype, INPA 2186. A, habitus, dorsal view; B, habitus, ventral view; C, frontal view; D, left third maxilliped, ventral view; E, right larger chela, outer view; F, left first gonopod, caudal-mesial view; G, same, mesial-cephalic view; H, same, cephalic view; I, same, distal part, apical-mesial view; J, same, distal part, lateral-cephalic view; K, same, distal part, caudal view; L, same, distal part, lateral view; M, left second gonopod, caudal-lateral view. Abbreviations: ap, apical plate; as, field of apical spines; dl, distal lobe of the apical plate; ls, lateral suture; me, mesial process; mp, marginal process; ms, marginal suture, pl, proximal lobe of the apical plate, rl, subdistal lobe of the apical plate; so, spine-like outgrowth of apical plate. Scale bars: A-C, E = 10 mm; D = 5 mm; F-M = 1 mm.

groove between sternites VII and VIII, deeper at junction with s7/8. Episternites 4-6 triangular posteriorly, episternite 7 posteriorly truncate. Sterno-abdominal cavity strongly concave, with few, scattered pubescence. Penis noticeably long, emerging from nearby coxo-sternal condyle articulation, located in shallow depression on sternite 8, proximally thick, abruptly tapering distally.

All abdominal somites free; lateral margins of somites fringed by line of short setae. Telson subtriangular, lateral margins nearly straight, fringed by line of short setae, tip rounded (fig. 1B).

G1 (fig. 1F-L) stout, nearly straight in caudal view, slightly arched in mesial view, apical processes somewhat depressed, low. Marginal suture slightly sinuous proximally, merging with lateral suture at approximately the distal third of stem, merged portion strongly twisted towards lateral side; marginal suture with dense patch of short setae along crested proximal portion; lateral suture distinctly crested in relation to marginal suture along median portion of stem. Marginal process well developed, distally rounded in cephalic view, distinctly projected over distal opening of spermatic channel, not overreaching field of apical spines area. Mesial process well developed, slightly longer than apex, proximal portion rounded, distal portion produced into strong, conical spine-like projection pointing in the caudal direction; mesial process separated from apex by deep, narrow groove on mesial surface, closely juxtaposed laterally. Apex transversally expanded along caudal-cephalic axis, roughly circular in apical-mesial view, with well-developed but somewhat depressed apical processes, consisting of subdistal rounded lobe on mesial-cephalic side, spine-like outgrowth on mesial margin cross overlapping spine-like distal portion of mesial process, pointing in the mesial-caudal direction, caudal margin rounded, low apical plate; apical plate with 2 superimposed distal lobes, proximal lobe short, situated subdistally on lateral side and folded under distal lobe in mesial view; distal lobe slightly longer than proximal one, distally rounded. Field of apical spines weakly developed, as curved, narrow patch of minute spines, situated in shallow depression along lateral-cephalic side, enclosed laterally by margins of apical lobes, proximally by distal margin of cephalic side of apex, distally by juxtaposition of apical lobes. Opening of spermatic channel located subdistally near base of the marginal process.

G2 (fig. 1M) slightly arched, approximately as long as G1, flagellum slender, regularly tapering, tip flattened, provided with short spinules on sternal surface.

Etymology.— The generic and the specific epithet names are dedicated to the distinguished German zoologist, Prof. Dr. Michael Türkay, for his exceptional contribution to the study of crustaceans worldwide. It is also a personal tribute to Michael as he was a mentor, a dear friend and a scientific collaborator for more than 30 years since we first met in 1984 (Magalhães, 2016). My first publication on freshwater crabs, and our first collaboration, was the description of a new genus and species from the Tapajós River (Magalhães & Türkay, 1986), and I have intentionally chosen a new taxon from the same area for this tribute.

Type locality.— Brazil, state of Pará, Tapajós River, municipality of Itaituba, between Buburé and São Luiz do Tapajós (to date, only known from the type locality).

Ecological notes.— The crabs were collected in granitic outcrops on the left bank of the Tapajós River in shallow, slow flowing waters. The specimens were hand-collected from under and among rocks during the early evening. Parental care by females was observed since they carried the young crabs in the sternoabdominal cavity.

Remarks.— The relationships of *Michaelthelphusa tuerkayi* sp. nov. and allied species have already been discussed under the genus. Very little is known about its distributional range as the species has been known only from the type locality. This is the second pseudothelphusid species described from the Tapajós River after *Brasiliothelphusa tapajosense* Magalhães & Türkay, 2010. It is noteworthy that, although their occurrence sites are less than 50 km apart, their distribution areas may not overlap. *B. tapajosense* was recorded from Monte Cristo and Fordlândia (Magalhães & Türkay, 1986), in the lower course of the river, which is characterized by a mouth-lake conformation bordered by high cliffs and low beaches (Goulding et al., 2003), and the specimens were collected "in the mud of river bank" (Magalhães & Türkay, 1986). On the other hand, specimens of *M. tuerkayi* sp. nov. were only found in rapids with rocky substrate in the vicinity of Itaituba. These rapids mark the northern boundary of the Tapajós-Juruena ecoregion, where the Tapajós River leaves the outcrops of the Brazilian Shield and runs into the sedimentary basin of the Amazon valley (Hales & Petry, 2015). The two species occur in the same river but each may be restricted to their own freshwater ecoregions (sensu Abell et al., 2008).

ACKNOWLEDGEMENTS

I thank the Conselho Nacional de Desenvolvimento Científico e Tecnológico - CNPq for an ongoing Research Grant (Proc. 304736/2015-5), C. Flanagan for reviewing the English, and Peter Ng (National University of Singapore) and Tohru Naruse (University of the Ryukyus, Japan) for their valuable comments to the manuscript.

REFERENCES

Abell, R., M. L. Thieme, C. Revenga, M. Bryer, M. Kottelat, N. Bogutskaya, B. Coad, N. Mandrak, S. Contreras Balderas, W. Bussing, M. L. J. Stiassny,

P. SKELTON, G. R. ALLEN, P. UNMACK, A. NASEKA, R. NG, N. SINDORF, J. ROBERTSON, E. ARMIJO, J. V. HIGGINS, T. J. HEIBEL, E. WIKRAMANAYAKE, D. OLSON, H. L. LÓPEZ, R. E. REIS, J. G. LUNDBERG, M. H. SABAJ PÉREZ & P. PETRY, 2008. Freshwater ecoregions of the world: a new map of biogeographic units for freshwater biodiversity conservation. BioScience, **58**: 403-414.

BOTT, R., 1967. Flusskrabben aus Brasilien und benachbarter Gebiete. *Potamocarcinus* (*Kingsleya*) Ortmann 1897 (Crustacea, Decapoda). Senckenbergiana biologica, **48**: 301-312.

BOTT, R., 1969. Die Süsswasserkrabben Süd-Amerikas und ihre Stammesgeschichte. Eine Revision der Trichodactylidae und der Pseudothelphusidae östlich der Anden (Crustacea, Decapoda). Abhandlungen der Senckenbergischen Naturforschenden Gesellschaft, **518**: 1-94.

BOTT, R., 1970. Betrachtungen über die Entwicklungsgeschichte und Verbreitung der Süßwasserkrabben nach der Sammlung des Naturhistorischen Museums in Genf/Schweiz. Revue Suisse de Zoologie, **77**: 327-344, pls. 1-2.

CUMBERLIDGE, N., F. ÁLVAREZ & J. L. VILLALOBOS, 2014. Results of the global conservation assessment of the freshwater crabs (Brachyura, Pseudothelphusidae and Trichodactylidae): the Neotropical region, with an update on diversity. Zookeys, **457**: 133-157.

GOULDING, M., R. BARTHEM & E. FERREIRA, 2003. The Smithsonian atlas of the Amazon. (Smithsonian Institution Press, Washington, DC).

HALES, J. & P. PETRY, 2015. Freshwater ecoregions of the world. 320: Tapajos-Juruena. Available online at http://www.feow.org/ecoregions/details/320 (accessed 2 March 2016).

MAGALHÃES, C., 1986. Revisão taxonômica dos caranguejos de água doce brasileiros da família Pseudothelphusidae (Crustacea, Decapoda). Amazoniana, **9**: 609-636.

MAGALHÃES, C., 2003. The occurrence of freshwater crabs (Crustacea: Decapoda: Pseudothelphusidae, Trichodactylidae) in the rio Xingu, Amazon Region, Brazil, with description of a new species of Pseudothelphusidae. Amazoniana, **17**: 377-386.

MAGALHÃES, C., 2005. A new species of freshwater crab (Crustacea: Decapoda: Pseudothelphusidae) from the southeastern Amazon Basin. Nauplius, **12**: 99-107 [2004].

MAGALHÃES, C., 2009. A new species of freshwater crab of the genus *Fredius* Pretzmann, 1967 from the middle Amazon River Basin, Brazil (Crustacea: Decapoda: Pseudothelphusidae). Proceedings of the Biological Society of Washington, **122**: 81-86.

MAGALHÃES, C., 2010. A new species of freshwater crab of the genus *Microthelphusa* (Decapoda, Pseudothelphusidae) from a tepui in the Serra do Aracá, state of Amazonas, Brazil. In: C. FRANSEN, S. DE GRAVE & P. NG (eds.), Studies on Malacostraca: Lipke Bijdeley Holthuis memorial volume. Crustaceana Monographs, **14**: 453-460.

MAGALHÃES, C., 2016. A tribute to Michael Türkay (3 April 1948 – 9 September 2015): contributions and legacy of a lifelong study on New World decapods, and personal impressions. Nauplius, **24**: e2016012, 17 pp. DOI:10.1590/2358-2936e2016012.

MAGALHÃES, C., F. A. ABRUNHOSA, M. O. PEREIRA & M. A. MELO, 2005. New records of *Fredius denticulatus* (H. Milne Edwards, 1853) and *F. reflexifrons* (Ortmann, 1897), and the eastern limits of the distribution of pseudothelphusid crabs (Crustacea: Decapoda) in Brazil. Acta Amazonica, **35**: 93-96.

MAGALHÃES, C., M. R. CAMPOS, P. A. COLLINS & F. L. MANTELATTO, 2016. Diversity, distribution and conservation of freshwater crabs and shrimps in South America. In: T. KAWAI & N. CUMBERLIDGE (eds.), A global overview of the conservation of freshwater decapod crustaceans: 303-322. (Springer, Cham).

MAGALHÃES, C. & G. PEREIRA, 2007. Assessment of the decapod crustacean diversity in the Guayana Shield region aiming at conservation decisions. Biota Neotropica, **7**: 111-124.

MAGALHÃES, C. & G. RODRÍGUEZ, 2002. The systematic and biogeographical status of *Fredius reflexifrons* (Ortmann, 1897) and *Fredius fittkaui* (Bott, 1967) (Crustacea: Brachyura: Pseudothelphusidae) from the Amazon and Atlantic Guianas river basins. Acta Amazonica, **32**: 677-689.

MAGALHÃES, C., V. Q. A. SANCHES, L. G. PILEGGI & F. L. MANTELATTO, 2014. Morphological and molecular description of a new species of *Fredius* (Decapoda: Pseudothelphusidae) from Rondônia, southern Amazonia, Brazil. In: D. C. J. YEO, N. CUMBERLIDGE & S. KLAUS (eds.), Advances in freshwater decapod systematics and biology. Crustaceana Monographs, **19**: 101-114.

MAGALHÃES, C. & M. TÜRKAY, 1996. Taxonomy of the Neotropical freshwater crab family Trichodactylidae II. The genera *Forsteria*, *Melocarcinus*, *Sylviocarcinus*, and *Zilchiopsis* (Crustacea: Decapoda: Brachyura). Senckenbergiana biologica, **75**: 97-130.

MAGALHÃES, C. & M. TÜRKAY, 2010. A new freshwater crab of the genus *Brasiliothelphusa* Magalhães & Türkay, 1986 from Rio Aripuanã, southern Amazon Region, Brazil (Crustacea: Decapoda: Pseudothelphusidae). Nauplius, **18**: 103-108.

MILNE EDWARDS, H., 1853. Mémoire sur la famille des ocypodiens. Annales des Sciences Naturelle, Zoologie, **20**: 163-228.

NOBILI, G., 1896. Viaggio del Dott. A. Borelli nella Republica Argentina en el Paraguay. XIX. Crostacei Decapodi. Bollettino dei Musei di Zoologia ed Anatomia comparata della R. Università di Torino, **11**(222): 1-4.

ORTMANN, A., 1893. Die Dekapoden-Krebse des Strassbourg Museums, mit besonder berücksichtigung der von Herrn Dr. Döderlein bei Japan und bei den Liu-Kiu-Inseln gesammelten und zur Zeit im Strassburger Museum aufbewahrten Formen. VII Theil Abtheilung: Brachyura (Brachyura genuina Boas) II. Unterabtheilung: Cancroidea, 2 Section: Cancrinea, 1. Gruppe: Cyclometopa. Zoologische Jahrberichter, **7**: 411-495.

ORTMANN, A., 1897. Carcinologische Studien. Zoologische Jahrbücher, Abteilung für Systematik, Geographie and Biologie der Tiere, **10**: 258-372.

PEDRAZA, M., J. E. MARTINELLI FILHO & C. MAGALHÃES, 2015. A new species of the genus *Kingsleya* from rio Xingu, and range extension for *Kingsleya junki* (Crustacea: Decapoda: Pseudothelphusidae), freshwater crabs from southern Amazon Basin, Brazil. Zoologia, **32**: 41-46.

PEDRAZA, M. & M. TAVARES, 2015. A new species of freshwater crab of the genus *Kingsleya* Ortmann, 1897 (Crustacea: Brachyura: Pseudothelphusidae) from Amazonia, Brazil. Zootaxa, **4032**: 444-450.

RANDALL, J. W., 1840. Catalogue of the Crustacea brought by Thomas Nuttall and J. K. Townsend, from the west coast of North America and the Sandwich Islands, with descriptions of such species as are apparently new, among which are included several species of different localities, previously existing in the collection of the Academy. Journal of the Academy of Natural Sciences of Philadelphia, **8**: 106-147, pls. 3-7.

RATHBUN, M. J., 1905. Les crabes d'eau douce (Potamonidae). Nouvelles Archives du Muséum d'Histoire Naturelle, Paris, **7**(4): 159-321, pls. 11-22.

RATHBUN, M. J., 1906. Les crabes d'eau douce (Potamonidae). Nouvelles Archives du Muséum d'Histoire Naturelle, Paris, **8**(4): 33-122.

RODRÍGUEZ, G., 1980. Description préliminaire de quelques especes et genres nouveaux de Crabes d'eau douce de l'Amérique tropicale (Crustacea: Decapoda: Pseudothelphusidae). Bulletin du Muséum nationale d'Histoire naturelle, 4e sér., **2** Section A (3): 889-894.

SMALLEY, A., 1964. A terminology for the gonopods of the American river crabs. Systematic Zoology, **13**: 28-31.

First received 30 March 2016.
Final version accepted 16 December 2016.

[When citing this volume, please refer to Crustaceana 90 (2017) 771-1288]

Michael Türkay Memorial Issue

THE CORAL GENUS *CAULASTRAEA* DANA, 1846 (SCLERACTINIA, MERULINIDAE) AS A NEW HOST FOR GALL CRABS (DECAPODA, CRYPTOCHIRIDAE), WITH THE DESCRIPTION OF *LITHOSCAPTUS TUERKAYI* SP. NOV.

BY

SANCIA E. T. VAN DER MEIJ[1,2,3,4])

[1]) Oxford University Museum of Natural History, Parks Road, Oxford OX1 3PW, U.K.

[2]) Linacre College, St Cross Road, Oxford OX1 3JA, UK

[3]) Naturalis Biodiversity Center, Darwinweg 2, NL-2333 CR Leiden, The Netherlands

ABSTRACT

A new gall crab species is described from the stony coral genus *Caulastraea*, a new host coral genus for Cryptochiridae crabs. Specimens were collected during fieldwork off Kudat (Malaysian Borneo) and Okinawa (Japan). Further material was retrieved from the collections of the Institute of Oceanography in Nha Trang (Vietnam). The new species, here named *Lithoscaptus tuerkayi* sp. nov., is the tenth assigned to the genus in which it is tentatively placed. It can be separated from its congeners by the strongly demarcated, inverted V-shaped protuberance flanking the mesogastric region, with denuded spots on the anterolateral sides of this inverted V-shape, as well as the denuded area just below the frontal margin. Based on morphology and COI sequence data, the closest species to *L. tuerkayi* sp. nov. is determined to be *Lithoscaptus helleri* (Fize & Serène, 1957).

ZUSAMMENFASSUNG

Hier wird eine neue Gallenkrabben-Art von Steinkorallen der Gattung *Caulastrea*, einer neuen Wirtskorallengattung für Cryptochiridae beschrieben. Die Individuen wurden während Feldarbeiten vor Kudat (Malaysisch-Borneo) und Okinawa (Japan) gesammelt. Zusätzlich wurde weiteres Material aus den Sammlungen des Institutes für Ozeanographie in Na Thrang (Vietnam) begutachtet. Die neue Art, die hiermit als *Lithoscaptus tuerkayi* sp. nov. benannt wird, ist die zehnte innerhalb der Gattung, zu welcher sie vorläufig zugeordnet wurde. Sie kann von ihren Artverwandten durch den stark abgegrenzten, invers V-förmigen Fortsatz, welcher die Mesogastralregion flankiert und blanke Flecken an den anterolateralen Seiten der inversen V-Form aufweist, unterschieden werden. Basierend auf Daten ihrer Morphologie und CO1-Sequenzen ist *Lithoscaptus helleri* (Fize & Serène, 1957) die nächste verwandte Art zu *L. tuerkayi* sp. nov.

[4]) e-mail: Sancia.vanderMeij@oum.ox.ac.uk

INTRODUCTION

Coral-dwelling gall crabs are common components of coral reef communities, but are frequently overlooked because of their minute size and cryptic habitat (Hoeksema & Van der Meij, 2013). Scleractinian corals belonging to at least 66 genera act as hosts to cryptochirids (Castro, 2015; Van der Meij, unpubl.) and many reef communities still need to be searched for their presence. In their monograph on Cryptochiridae Fize & Serène (1957) mentioned *Caulastraea* as one of the coral genera that still needed to be explored for possible gall crab inhabitants. Fieldwork in Sabah (Malaysia) in 2012 and Okinawa (Japan) in 2016 recovered the first gall crab specimens inhabiting this coral genus. This new gall crab species, here named *Lithoscaptus tuerkayi* sp. nov., is the tenth species assigned to the genus.

MATERIAL AND METHODS

Gall crabs were collected in Malaysia (off Kudat, Sabah) during the Tun Mustapha Park Expedition in September 2012 and during fieldwork in Japan (Okinawa) in April 2016. Corals were searched for gall crabs, photographed and subsequently split with hammer and chisel. The crabs were preserved in 80% ethanol, after being photographed with a digital SLR camera equipped with a macro lens to register colour patterns. Crab specimens, including the holotype, were deposited in the Crustacea collection of Naturalis Biodiversity Center in Leiden, the Netherlands (formerly Rijksmuseum van Natuurlijke Historie, collection coded as RMNH.Crus.D), while a paratype and part of the Japanese material was deposited in the Oxford University Museum of Natural History in Oxford, United Kingdom (collection coded as OUMNH.ZC). Drawings of the new species were made with a stereomicroscope with camera lucida. Carapace lengths and widths were measured to the nearest 0.1 mm using an eyepiece graticule, with the crabs positioned on a level surface. Abbreviations used: CL, carapace length; CW, carapace width (at widest point); MXP3, third maxilliped; ovig., ovigerous; P, pereiopod. Carapace measurements are given as CL × CW (in mm).

Fize & Serène (1957) described 16 new gall crab species in their monograph on the gall crabs of Vietnam and deposited the types in the museum of the Institute of Oceanography in Nha Trang (Vietnam). The Nha Trang collections were studied by the author in October 2012 and were found to contain two undescribed specimens collected from *Caulastraea* corals. This material was identified as *Lithoscaptus tuerkayi* sp. nov. Collection numbers were mentioned on labels on the outside of the collection jars as well as on labels inside the vials. In case of discrepancy, the information on the labels inside the vial was accepted as most likely the 'correct' information for the specimen(s) concerned. The locality data 'Rte' (recolte) on the Vietnamese labels can be retrieved from Fize & Serène (1957).

In earlier studies (e.g., Van der Meij & Nieman, 2016), analyses of sequences from the 'barcoding gene' mitochondrial *cytochrome c oxidase subunit I* gene (*COI*, partially; Folmer et al., 1994) were used to infer phylogenetic relationships between examined taxa. In Van der Meij & Klaus (2015) a sequence of *Lithoscaptus tuerkayi* sp. nov. was used under the name "cf. *Lithoscaptus* 'Caula'". Based on this result the position of the new species within the Cryptochiridae was retrieved in the largest clade, containing all cryptochirid species inhabiting Merulinidae corals. Therefore, in this study a modified dataset of crab species mostly inhabiting Merulinidae corals was used to determine the closest relatives of *L. tuerkayi* sp. nov. based on COI mtDNA.

Sequences were obtained using a PCR protocol based on Qiagen DNeasy Kit manufacturer's protocol. DNA was isolated from the fifth pereiopod. Maceration took place overnight for approximately 18 h. The final elution step was performed with 100 μl elution buffer. Polymerase chain reaction was carried out with the following PCR conditions (2.5 μl PCR buffer (10×), 0.5 μl dNTPs (10 mM), 1.0 μl primer FW (0.1 μM), 1.0 μl primer RV (0.1 μM), 0.3 μl *Taq* polymerase (5 U/μl), 18.7 μl MilliQ and 1.0 μl DNA template). Thermal cycling was as follows: initial denaturation at 95°C for 5 min, followed by 39 cycles of 95°C for 5 s, 47°C for 1 min and 72°C for 1 min. This was followed by 10 min at 72°C. The sequences ($n = 24$) were aligned using the ClustalW algorithm on the Guidance2 server (Sela et al., 2015), resulting in an alignment score of 1.00. The gall crab *Opecarcinus cathyae* Van der Meij, 2014 was used as the outgroup (Van der Meij & Nieman, 2016). Prior to the model-based phylogenetic analysis, the best-fit model of nucleotide substitution (11 schemes) was based on the Akaike Information Criterion (AIC) calculated with jModeltest (Posada, 2008), resulting in GTR + I + G as the best model. A maximum likelihood analysis (1000 bootstraps) was run in Phyml 3.1 (Guindon et al., 2010) in the Seaview platform (Gouy et al., 2010).

SYSTEMATIC PART

Family CRYPTOCHIRIDAE Paul'son, 1875
Genus *Lithoscaptus* A. Milne-Edwards, 1862
Lithoscaptus tuerkayi sp. nov.
(figs. 1-3)

Cf. *Lithoscaptus* 'Caula' Van der Meij & Klaus, 2015: fig. 1.

Type locality.— Banggi Outer NE Reef, off Kudat, Sabah, Malaysia (TMP20: 07°22′53″N 117°22′24″E).

Coral host holotype.— *Caulastraea furcata* Dana, 1846.

DNA barcoding.— COI sequence (partially, Folmer et al., 1994) of RMNH.Crus.D.54336b has been deposited in GenBank under accession number KU745732.

Type material.— Holotype. RMNH.Crus.D.54336, 1 ovig. female (3.3 × 2.9). Paratype. OUMNH.ZC.2016-01-016 (ex-lot RMNH.Crus.D.54336), 1 ovig. female (3.3 × 2.3). Collected by the author on 14 September 2012 at 9 m depth.

Non-type material.— RMNH.Crus.D.54336b (collected with holotype and paratype), 1 ovig. female (3.8 × 2.9), slightly damaged; RMNH.Crus.D.54920, SE Banggi Dangers, off Kudat, Sabah, Malaysia (TMP16: 07°11′33″N 117°23′38″E), 10 m depth, 12 September 2012, 2 ovig. females (4.1 × 2.8, 3.8 × 2.5), leg. S. E. T. van der Meij; RMNH.Crus.D.54316, Patanunan Island, off Kudat, Sabah, Malaysia (TMP36: 07°05′59″N 117°05′21″E), 10 m depth, 19 September 2012, 1 non-ovig. female (3.0 × 2.1), leg. S. E. T. van der Meij; RMNH.Crus.D.54262, Tanjung Kapor, off Kudat, Sabah, Malaysia (TMP52: 06°56′21″N 116°50′25″E), 4 m depth, 16 September 2012, 1 non-ovig. female (3.9 × 3.2), leg. B. T. Reijnen; RMNH.CRUS.D.57180, Hedo Point I, Okinawa, Japan (RYU08: 26°52′37″N 128°15′36″E), 20 m depth, 21 April 2016, 1 ovig. female (3.0 × 2.0), leg. S. E. T. van der Meij; OUMNH.ZC.2016-01-030, Oura Bay — AoSango, Okinawa, Japan (RYU05: 26°32′14″N 128°04′34″E), 7 m depth, 18 April 2016, 1 ovig. female (3.0 × 1.9), leg. S. E. T. van der Meij. All specimens were collected from *Caulastraea furcata* Dana, 1846, with the exception of RMNH.Crus.D.57180, which was collected from *Caulastraea tumida* Matthai, 1928.

Description (female holotype).— Carapace (fig. 1A) rectangular, longer than wide, CL 1.4 times as long as CW, widest near midlength. Dorsal surface about 45° convex in lateral view, deflected anteriorly; denuded areas are present close to frontal margin, anterior half of carapace with an inverted V-shaped protuberance flanking mesogastric region, inverted V-shape flanked by denuded areas on anterolateral sides; sharp tubercles on anterior half of carapace, small,

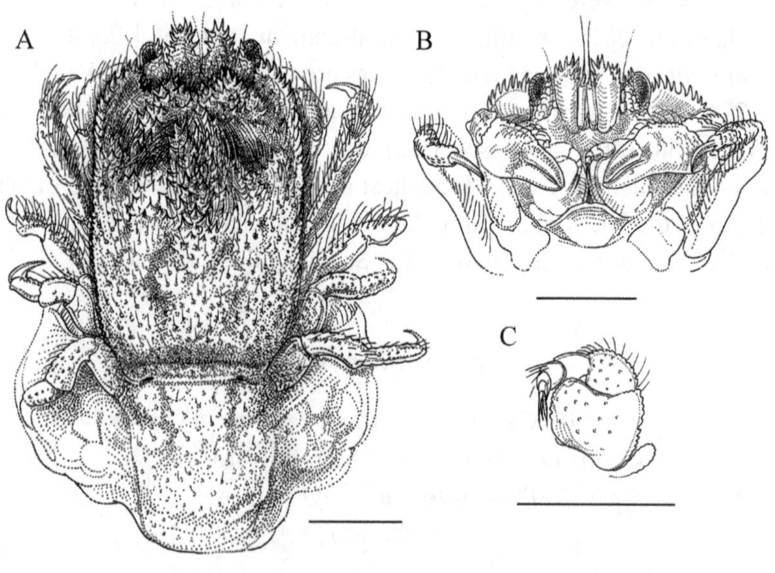

Fig. 1. Ovigerous female holotype (3.3 mm × 2.9 mm) of *Lithoscaptus tuerkayi* sp. nov. (RMNH.Crus.D.54336): A, habitus, dorsal view; B, anterior cephalothorax, ventral view; C, MXP3. Scale bars = 1 mm.

rounded granules on posterior half; cardiointestinal region slightly inflated; simple setae all over carapace, mostly on mesogastric region. Frontal margin armed with anteriorly directed spines; the internal orbital angle does not extent beyond the external orbital angle (fig. 1B). Pterygostomial region fused to carapace.

Eyestalk exposed dorsally, granular, small spines on mesial margin. Cornea on anterolateral portion of eyestalk (fig. 1A, B).

Dorsal surface margins of antennular peduncle with sharp tubercles, not inflated, apex extending beyong tip of eyes. Ventral surface smooth (fig. 1B). Distal antennule segments with long, simple setae.

MXP3 (fig. 1C) exopod rectangular; ischium squaroid, smooth with few small rounded tubercles, mesial and distal margins slightly granulated; merus rounded, surface smooth with few small rounded tubercles, extensor margin granulated, with simple setae; carpus, propodus smooth, with few simple setae, dactylus with bundle of setae.

P1 (cheliped, figs. 1B, 2A) slender; merus length twice height, smooth, with small tubercles on distodorsal surface; carpus with small tubercles on dorsal surface; propodus length 2.5 times height, smooth, with small tubercles on dorsal surface; fingers slender, mesial surface of fingers smooth, cutting edge entire, tips of fingers not crossing.

P2 (fig. 2B) coarser than P1; ischium without setae; merus stout, rectangular, smooth, with sharp tubercles on distal half of dorsal surface, simple setae on dorsal and lateral surfaces; joint between merus, carpus not extending more than at right angle; carpus mostly smooth with rounded tubercles and simple setae on dorsal surface; propodus same length as carpus, surface mostly smooth with rounded tubercles and simple setae on dorsal surface; dactylus half-length of propodus, smooth, sharp, curved ventrally.

P3 (fig. 2C) ischium without setae; merus length 1.5 times height, rectangular, smooth, with sharp tubercles and setae on distal half of dorsal surface; joint

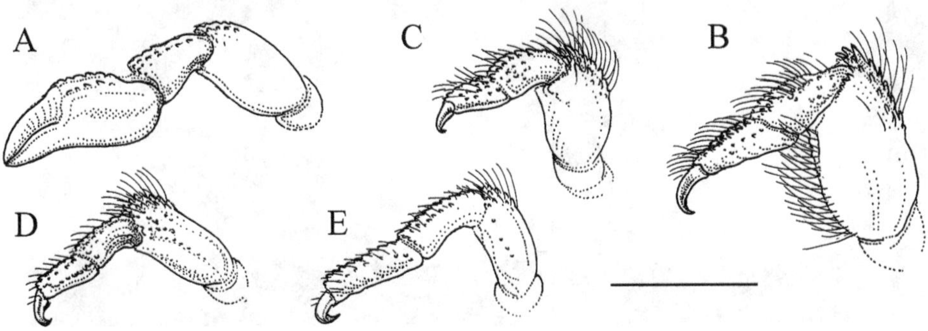

Fig. 2. Ovigerous female holotype of *Lithoscaptus tuerkayi* sp. nov. (RMNH.Crus.D. 54336): A, left P1 (cheliped); B, left P2; C, left P3; D, left P4; E, left P5. Scale bar = 1 mm.

between merus, carpus not extending more than at right angle; carpus length 2.5 times height, tubercles and setae on dorsal surface; propodus length twice height, rounded tubercles and scattered simple setae on dorsal surface; dactylus short, sharp, curved ventrally.

P4 (fig. 2D) similar to P3, merus less rounded; ischium without setae; merus length approximately 2.3 times height; carpus and propodus of similar length, twice height; dactylus short, sharp, curved ventrally.

P5 (fig. 2E) slender; ischium without setae; merus, carpus, propodus of about equal length, all with short simple setae and tubercles on dorsal surface; dactylus short, sharp, curved ventrally.

P3, P4 decreasing in size from P2 (fig. 2C, D).

Abdomen enlarged, lateral margins fringed with setae (fig. 1A).

Gonopore (vulva) reniform, size about half the height of sternite 6.

Male.— The male of *Lithoscaptus tuerkayi* sp. nov. is unknown (but see paragraph on Nha Trang collections hereafter).

Colour.— *Lithoscaptus tuerkayi* is overall off-white, without striking colours. In all the available photographs sediment is trapped between the setae on the carapace, obscuring any colour (patterns). The concavities in the carapace (next to inverted V-shaped depression) are brownish, the antennular peduncles and MXP3 have a soft green marbled pattern. Eyes brownish-red with white patch laterally (fig. 3A, B).

Host coral.— The coral genus *Caulastraea* belongs to the large Indo-Pacific coral family Merulinidae. Within this family *Caulastraea* forms a monophyletic group with the genera *Oulophyllia*, *Mycedium* and *Pectinia*, based on molecular

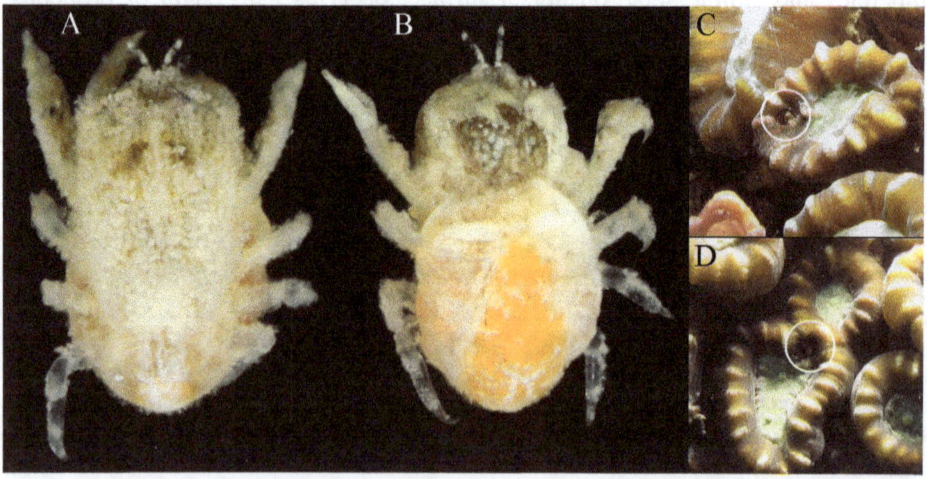

Fig. 3. A-B, Colour in life of *Lithoscaptus tuerkayi* sp. nov. (OUMNH.ZC.2016-01-030); C-D, Dwellings (circled) in *Caulastraea furcata* at SE Banggi Dangers, Sabah, Malaysia, at 10 m depth.

and morphological data (Huang et al., 2014). There are currently five valid species in the genus: *Caulastraea connata* (Ortmann, 1892), *C. curvata* Wijsman-Best, 1972, *C. echinulata* (H. Milne Edwards & Haime, 1849), *C. furcata* Dana, 1846 (type species) and *C. tumida* Matthai, 1928. The corals from which *L. tuerkayi* sp. nov. was collected are identified as *C. furcata* and *C. tumida*. Dwellings have been found in between the septa (fig. 3C) or in between the space between two (closely) adjacent corallites (fig. 3D). Surprisingly, two other gall crab species have been found to inhabit *Caulastraea* as well, but their identities are not yet known. Because of the paucity of records from *Caulastraea* the extent of the associations (i.e. host specificity and host switch patterns) are unclear. *Caulastraea* is one of the only recorded host corals with more than two associated gall crab species. The only other known examples so far are the mushroom corals *Lithophyllon scabra* and *L. repanda* (see Van der Meij et al., 2015). In the mushroom corals there was a common gall crab inhabitant and single records from the other two crab species. Such single records are possibly the results of crab larvae which needed to settle, without having their 'most preferred' host coral available.

Nha Trang collections.— In the collections of the Institute of Oceanography in Nha Trang two vials are present with the numbers E.39492 (Rte 1663) and E.39515 (Rte 1668). This material was collected from respectively *Caulastraea* sp. and *Caulastraea furcata*. The vials contain a label with the name "*Troglocarcinus gordoni*"; however, this name has never been used in a published species description. The vial numbered E.39492 contains a male, in good condition, and vial E.39515 contains an ovigerous female, also in good condition (fig. 4). The material of the ovigerous female is in agreement with the description of *L. tuerkayi* sp. nov.; the identity of the male specimen could not be confirmed.

Placement in genus.— The placement of *L. tuerkayi* sp. nov. in the genus *Lithoscaptus* is somewhat tentative, as was also the case with *L. prionotus* Kropp, 1994 and *L. semperi* Van der Meij, 2015. Molecular reconstructions of relationships within the Cryptochiridae show that *Lithoscaptus* is paraphyletic (Van der Meij & Reijnen, 2014; Van der Meij & Nieman, 2016). The characters defining the genus need to be redefined and the species confirmed or reassigned to a new genera.

Comparisons.— Nine species of *Lithoscaptus* are currently recognised (Ng et al., 2008: 212; Davie, 2015). The new species can be separated from all *Lithoscaptus* species by the strongly demarcated inverted V-shaped protuberance on the carapace, flanked by denuded areas, as well as the denuded spots close to the front margin. In addition, *L. tuerkayi* sp. nov. can be distinguished from *L. nami* (Fize & Serène, 1957), *L. tri* (Fize & Serène, 1956) and *L. pardalotus* Kropp, 1995 by not having the internal orbital angle extending beyond the external orbital angle. *Lithoscaptus tuerkayi* sp. nov. is morphologically most similar to *L. helleri*

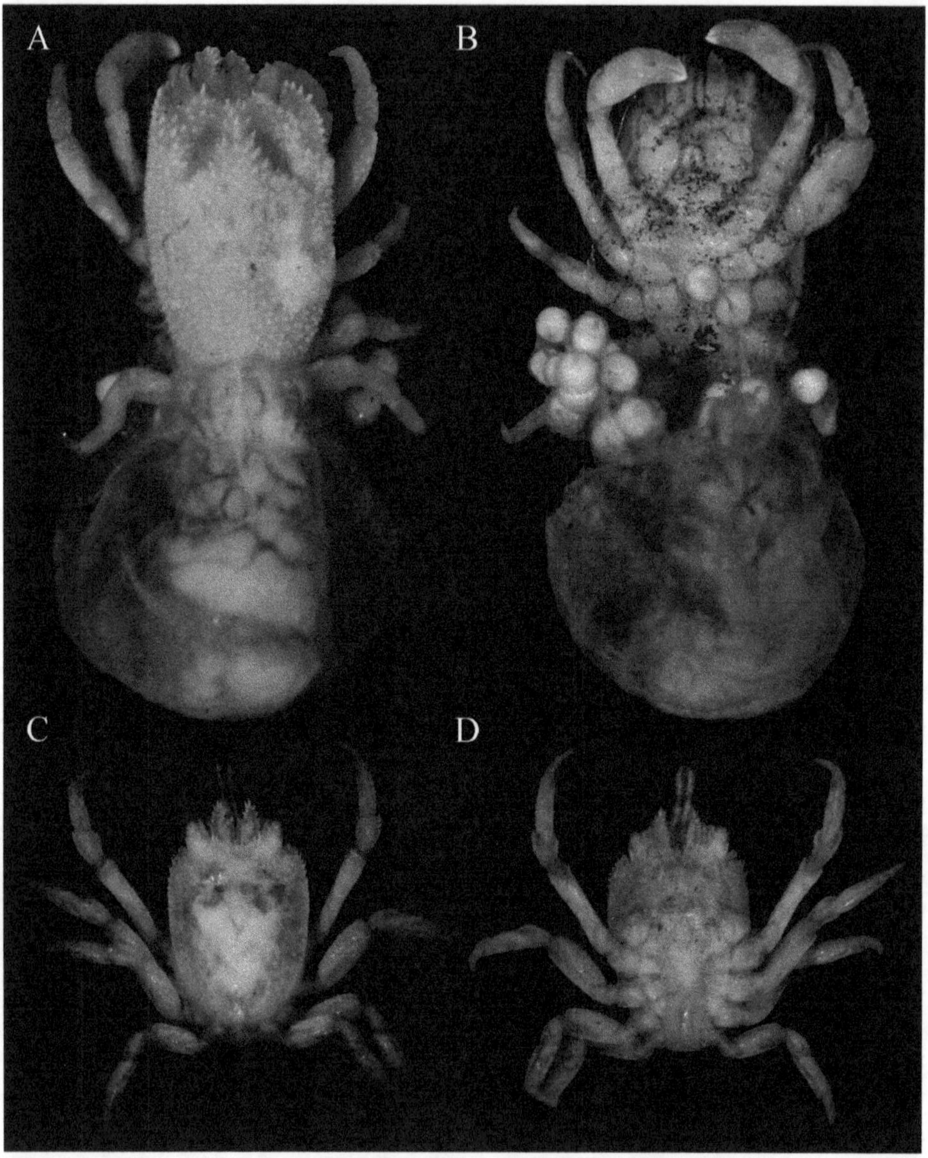

Fig. 4. Specimen of *Lithoscaptus tuerkayi* sp. nov. (labelled as *Troglocarcinus gordoni*) in the Nha Trang collections. A-B, dorsal and ventral view of ovigerous female (E.39515); C-D, dorsal and ventral view of male (E.39492). Photographs not to scale.

(Fize & Serène, 1957), but there are noticeable differences apart from the denuded areas on the carapace of the new species. The demarcation and tuberculation of the mesogastric region is more prominent in *L. helleri* and this species has a pattern of deep furrows and elevations on the carapace, which are absent or only weakly present in *L. tuerkayi* sp. nov. (Fize & Serène, 1957: fig. 25B). Live specimens of

L. helleri have an overall greyish-blue hue with striking yellow markings, contrary to *L. tuerkayi* sp. nov. which is off-white.

Molecular analyses.— Based on COI data the closest species to *Lithoscaptus tuerkayi* sp. nov. is *L.* cf. *helleri* (fig. 5), an inhabitant of *Oulophyllia* corals. *Caulastraea* and *Oulophyllia* form a monophyletic group within the Scleractinia, together with the coral genera *Mycedium* and *Pectinia*, based on molecular and morphological data (Huang et al., 2014). The type species of the genus is *L. paradoxus*, which clusters with several other *Lithoscaptus* species and a *Xynomaia* species (fig. 5).

Remarks.— It is not uncommon in gall crabs for the males to be unknown (e.g., Kropp, 1994). In Cryptochiridae the holotypes are based on females, contrary to most crab families. This discrepancy is probably related to the sedentary life style of the females, which allows for easier collecting. There is a striking sexual dimorphism between males and females of most species, with males being smaller and thus more difficult to collect. There is some uncertainty about the identity of the Nha Trang material, since several vials were found to have multiple labels with conflicting information. It is, however, likely that the male in vial E.39492 (fig. 4C, D) represents the only known male specimen of *L. tuerkayi* sp. nov.

Distribution.— *Lithoscaptus tuerkayi* sp. nov. is so far known from Tun Mustapha Park, northern Borneo, Malaysia, and Okinawa, Japan. Based on the Fize & Serène collections in Nha Trang the distribution of *L. tuerkayi* sp. nov. also includes Nha Trang, Vietnam. The coral genus *Caulastraea* is widely distributed on reefs of the Indo-Pacific, recorded as far east as the Pitcairn Islands in the southern hemisphere, but absent eastwards from Hawaii (USA) in the north (Huang et al., 2014).

Etymology.— This species is named after Prof. Dr. Michael Türkay in recognition of his major contributions to crustacean research.

ACKNOWLEDGEMENTS

The Tun Mustapha Park Expedition (TMPE) 2012 was jointly organized by WWF-Malaysia, Universiti Malaysia Sabah (UMS), Sabah Parks and Naturalis Biodiversity Center. TMPE was funded by the Ministry of Science, Technology and Innovation (MOSTI) and USAID Coral Triangle Support Partnership (CTSP). The research permits were granted by the Economic Planning Unit, Prime Minister's Department Malaysia and Sabah Biodiversity Centre. I am grateful to James Reimer (University of the Ryukyus) for making the fieldwork in Okinawa possible. Funding was provided by the TREUB-maatschappij (Society for the Advancement of Research in the Tropics), JJ ter Pelkwijkfonds (Naturalis) and the John

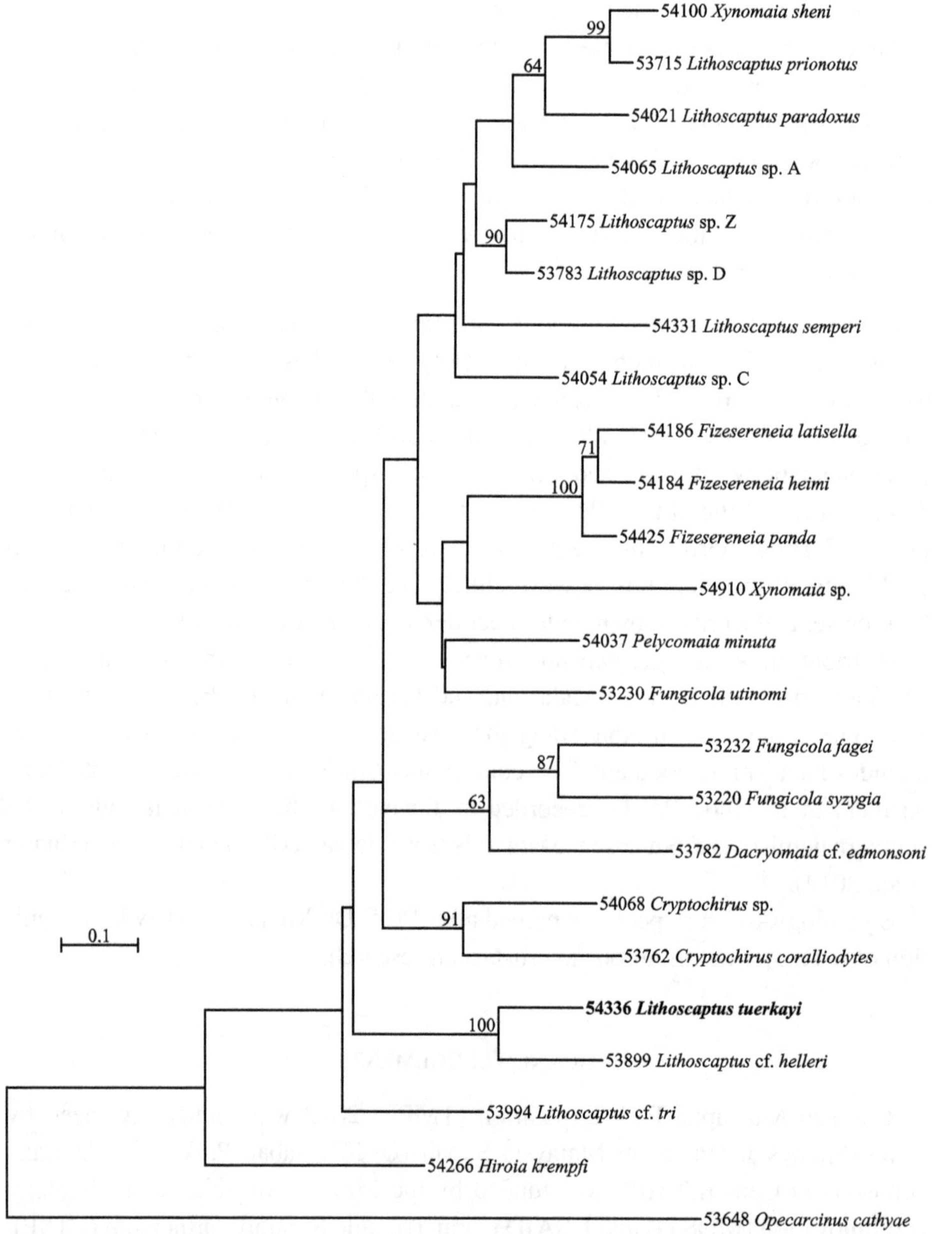

Fig. 5. Phylogeny reconstruction of the Cryptochiridae, based on *COI*, indicating the position of *Lithoscaptus tuerkayi* sp. nov. (in bold). Topology and support values derived from maximum likelihood analysis in Phyml 3.1. The numbers preceding the species names refer to collection numbers of the specimens in Naturalis Biodiversity Center (collection coded as RMNH.Crus.D.).

Fell Oxford University Press Research Fund. Many thanks to Bùi Quang Nghị and Nguyễn Thị Mỹ Ngân (Institute of Oceanography, Nha Trang, Vietnam) for facilitating my visit to the institute and their indispensable help with locating the Fize & Serène material. Bastian Reijnen (Naturalis) collected material used in this study and assisted with photography. Inge van Noortwijk (Naturalis) made the beautiful line drawings. The COI barcode was produced as part of the Naturalis Barcoding Project.

REFERENCES

CASTRO, P., 2015. Symbiotic Brachyura. In: P. CASTRO, P. DAVIE, D. GUINOT, F. R. SCHRAM & J. C. VON VAUPEL KLEIN (eds.), Decapoda: Brachyura, treatise on zoology — anatomy, taxonomy, biology, **9C-I**: 543-581 (Chapter 71-10). (Brill, Leiden).

DANA, J. D., 1846-1849. United States exploring expedition during the years 1838-1842. Zoophytes, **7**: 1-740. (Lea and Blanchard, Philadelphia, PA).

DAVIE, P., 2015. *Lithoscaptus*. World Register of Marine Species, available online at http://marinespecies.org/aphia.php?p=taxdetails&id=439231 (accessed 20 February 2016).

FIZE, A. & R. SERÈNE, 1956. Note préliminaire sur quatre espèces nouvelles d'Hapalocarcinidés avec quelques remarques au sujet du *Cryptochirus rugosus* Edmonson. Bulletin de la Société Zoologique de France, **45**: 379-382.

FIZE, A. & R. SERÈNE, 1957. Les Hapalocarcinidés du Viet-Nam. Mémoires de l'Institut Océanographique de Nhatrang, **10**: 1-202, pls. 1-18.

FOLMER, O., M. BLACK, W. HOEH, R. LUTZ & R. VRIJENHOEK, 1994. DNA primers for amplification of mitochondrial cytochrome c oxidase subunit I from diverse metazoan invertebrates. Molecular Marine Biology and Biotechnology, **3**: 294-297.

GOUY, M., S. GUINDON & O. GASCUEL, 2010. SeaView version 4: a multiplatform graphical user interface for sequence alignment and phylogenetic tree building. Molecular Biology and Evolution, **27**: 221-224.

GUINDON, S., J.-F. DUFAYARD, V. LEFORT, M. ANISIMOVA, W. HORDIJK & O. GASCUEL, 2010. New algorithms and methods to estimate maximum-likelihood phylogenies: assessing the performance of PhyML 3.0. Systematic Biology, **59**: 307-321.

HOEKSEMA, B. W. & S. E. T. VAN DER MEIJ, 2013. Gall crab city: an aggregation of endosymbiotic crabs inhabiting a colossal colony of *Pavona clavus*. Coral Reefs, **32**: 59.

HUANG, D., F. BENZONI, H. FUKAMI, N. KNOWLTON, N. D. SMITH & A. F. BUDD, 2014. Taxonomic classification of the reef coral families Merulinidae, Montastraeidae and Diploastraeidae (Cnidaria: Anthozoa: Scleractinia). Zoological Journal of the Linnean Society, **171**: 277-355.

KROPP, R. K., 1994. The gall crabs (Crustacea: Decapoda: Brachyura: Cryptochiridae) of the Rumphius expeditions revisited, with descriptions of three new species. Raffles Bulletin of Zoology, **42**: 521-538.

KROPP, R. K., 1995. *Lithoscaptus pardalotus*, a new species of coral-dwelling gall crab (Crustacea: Brachyura: Cryptochiridae) from Belau. Proceedings of the Biological Society of Washington, **108**: 637-642.

MATTHAI, G., 1928. A monograph of the recent meandroid Astraeidae. Catalogue of the madreporarian corals in the British Museum. Natural History, **7**: 1-288, pls. 1-72.

MILNE-EDWARDS, A., 1862. Faune carcinologique de l'île de la Réunion. In: L. MAILLARD (ed.), Notes sur l'île de la Réunion. Vol. Annexe F: 1-16. (Paris, France).

MILNE EDWARDS, H. & J. HAIME, 1849. Recherches sur les polypiers. Quatrième mémoire. Monographiedes astréides(1). Tribu II. Astréens (Astreinae). Annales Des Sciences Naturelles, Série 3, Zoologie, **11**: 233-312.

NG, P. K. L., D. GUINOT & P. J. F. DAVIE, 2008. Systema Brachyurorum: Part I. An annotated checklist of extant brachyuran crabs of the world. The Raffles Bulletin of Zoology, **17**(Suppl.): 1-286.

ORTMANN, A., 1892. Koralriffe von Dar-es-Salaam und Umgegend. Zoologische Jahrbüchern, Abteilung für Systematik, Biologie und Biogeographie der Tiere, **6**: 631-670, pl. 29.

PAUL'SON, O., 1875. Studies on Crustacea of the Red Sea with notes regarding other seas. Part I. Podophthalmata and Edriophthalmata (Cumacea): 1-144, i-xiv, pls. 1-22. [In Russian. English translation by the Israel Program for Scientific Translations, Jerusalem, 1961.]

POSADA, D., 2008. jModelTest: phylogenetic model averaging. Mol. Biol. Evol., **25**: 1253-1256.

SELA, I., H. ASHKENAZY, K. KATOH & T. PUPKO, 2015. GUIDANCE2: accurate detection of unreliable alignment regions accounting for the uncertainty of multiple parameters. Nucleic Acids Research, **43** (Web Server issue): W7-W14.

VAN DER MEIJ, S. E. T., 2014. A new species of *Opecarcinus* Kropp and Manning, 1987 (Crustacea: Brachyura: Cryptochiridae) associated with the stony corals *Pavona clavus* (Dana, 1846) and *P. bipartita* Nemenzo, 1980 (Scleractinia: Agariciidae). Zootaxa, **3869**: 44-52.

VAN DER MEIJ, S. E. T., 2015. A new gall crab species (Brachyura, Cryptochiridae) associated with the free-living coral *Trachyphyllia geoffroyi* (Scleractinia, Merulinidae). ZooKeys, **500**: 61-72.

VAN DER MEIJ, S. E. T., C. H. J. M. FRANSEN, L. R. PASMAN & B. W. HOEKSEMA, 2015. Phylogenetic ecology of gall crabs (Cryptochiridae) as associates of mushroom corals (Fungiidae). Ecology and Evolution, **5**: 5770-5780.

VAN DER MEIJ, S. E. T. & S. KLAUS, 2015. Origin and diversification of dwelling gall crabs (Cryptochiridae). In: S. E. T. VAN DER MEIJ, Evolutionary diversification of coral-dwelling gall crabs (Cryptochiridae): 79-88. (Ph.D. Thesis, Leiden University, Leiden).

VAN DER MEIJ, S. E. T. & A. M. NIEMAN, 2016. Old and new DNA unweave the phylogenetic position of the eastern Atlantic gall crab *Detocarcinus balssi* (Monod, 1956) (Decapoda: Cryptochiridae). Journal of Zoological Systematics and Evolutionary Research, **54**: 189-196.

VAN DER MEIJ, S. E. T. & B. T. REIJNEN, 2014. The curious case of *Neotroglocarcinus dawydoffi* (Decapoda, Cryptochiridae): unforeseen biogeographic patterns resulting from isolation. Systematics and Biodiversity, **12**: 503-512.

WIJSMAN-BEST, M., 1972. Systematics and ecology of New Caledonian Faviinae (Coelenterata–Scleractinia). Bijdragen tot de Dierkunde, **42**: 3-90.

First received 22 February 2016.
Final version accepted 27 September 2016.

[When citing this volume, please refer to Crustaceana 90 (2017) 771-1288]

Michael Türkay Memorial Issue

SUNDATHELPHUSA MIGUELITO, A NEW SPECIES OF FRESHWATER CRAB FROM THE SOUTHERN PHILIPPINES (BRACHYURA, GECARCINUCIDAE)

BY

JOSE C. E. MENDOZA[1,3]) and E. Y. SY[2])

[1]) Lee Kong Chian Natural History Museum, Faculty of Science, National University of Singapore, 2 Conservatory Drive, 117377 Singapore

[2]) Philippine Center for Terrestrial and Aquatic Research, 1198 Benavidez Street, No. 1202, Tondo, Manila 1012, Philippines

ABSTRACT

A new species of the freshwater crab genus *Sundathelphusa* Bott, 1969 (Gecarcinucidae) is described from the island of Mindanao in the southern Philippines. The Philippine Islands host the greatest species diversity of this genus, although only four species were previously known from Mindanao. The new species is most similar to *S. mistio* (Rathbun, 1904) and *S. montanoanus* (Rathbun, 1904) in the general carapace outline and leg proportions, but it can be distinguished by features of the carapace, thoracic sternum and gonopods. With this discovery, there are now 29 species of *Sundathelphusa* known from the Philippines.

Key words. — Biodiversity, Decapoda, Lake Sebu, Michael Türkay, Mindanao, South Cotabato, taxonomy

ZUSAMMENFASSUNG

Hier wird eine neue Art der Süßwasserkrabbengattung *Sundathelphusa* Bott, 1969 (Gecarcinucidae) von der Insel Mindanao (südliche Philippinen) beschrieben. Die Philippinen beherbergen bezüglich dieser Gattung die größte Artenvielfalt, obwohl bislang nur vier Arten aus Minadao bekannt waren. Die neue Art ähnelt am ehesten *S. mistio* (Rathbun, 1904) und *S. montanoanus* (Rathbun, 1904) durch das generelle Profil ihres Carapax und ihre Beinproportionen. Sie kann jedoch durch Merkmale des Carapax, des thorakalen Sternums und der Gonopoden unterschieden werden. Durch diese Entdeckung sind nun 29 Arten der Gattung *Sundathelphusa* von den Philippinen bekannt.

[3]) Corresponding author; e-mail: jcmendoza@nus.edu.sg

INTRODUCTION

Sundathelphusa Bott, 1969 is a Southeast Asian genus of freshwater gecarcinucid crab, distributed in the islands of Borneo, Sulawesi, the Lesser Sundas and the Philippines (Ng, 2015). The Philippine Islands host the greatest species diversity of this genus, with 28 out of 37 species at the latest count. Nine of these species were described only in the last six years, as interest in Philippine freshwater crabs is on the upswing due to their high diversity and conservation value (see Husana et al., 2009, 2014, 2015; Mendoza & Naruse, 2010; Ng, 2010; Stasolla et al., 2015). Only four species, however, are known from the large southern Philippine island of Mindanao, namely *S. mistio* (Rathbun, 1904) (type locality Mindanao, no specific locality), *S. montanoanus* (Rathbun, 1904) (type locality: Mindanao, no specific locality); *S. woltereckі* (Balss, 1937) (type locality: Lake Lanao, Lanao del Sur Province), and *S. hades* Takeda & Ng, 2001 (type locality: Latay Cave, Rosario, Agusan del Sur Province) (viz. Mendoza & Naruse, 2010). A fifth species, from the municipality of Lake Sebu, South Cotabato Province, in southern Mindanao, is described here. High-resolution photographs of the types of *S. mistio* and *S. montanoanus*, including those of the gonopods, are also provided here for comparison.

Specimens examined are deposited in the Crustacean Reference Collection of the National Museum, Manila, Philippines (NMCR); Museum national d'Histoire naturelle, Paris, France (MNHN); U.S. National Museum of Natural History, Smithsonian Institution, Washington, D.C., U.S.A. (USNM); and Zoological Reference Collection of the Lee Kong Chian Natural History Museum (former Raffles Museum of Biodiversity Research), National University of Singapore, Singapore (ZRC). Measurements provided are of the carapace width and length, respectively, in millimeters. The terminology essentially follows that of Ng (1988) and Ng & Sket (1996). The following abbreviations are used: coll. for collected by; G1 and G2 for male first and second gonopods, respectively; P1 for the chelipeds; and P2-P5 for the first to fourth ambulatory legs, respectively.

TAXONOMY

Superfamily GECARCINUCOIDEA Rathbun, 1904

Family GECARCINUCIDAE Rathbun, 1904

Sundathelphusa Bott, 1969

Sundathelphusa miguelito sp. nov.

(figs. 1-4)

Material examined.— Holotype, male, 26.1 × 20.5 mm (NMCR-39999), small stream in Barangay Lake Seloton, Municipality of Lake Sebu, South Cotabato Province, Mindanao Island, Philippines, coll. I. Galang, Dec. 2012.

Paratypes: 1 male, 23.4 × 18.9 mm, 2 females, 25.0 × 20.6 mm, 31.4 × 25.1 mm (ZRC 2016.0666), same collection information as holotype.

Comparative material.— *Sundathelphusa mistio* (Rathbun, 1904), holotype, male, 31.8 × 25.8 mm (MNHN-IU-2014-11829 (= MNHN-B5295), central Mindanao, Philippines, coll. J. Montano, Oct.-Nov. 1880 (fig. 5). *Sundathelphusa montanoanus* (Rathbun, 1904), lectotype (here designated), male, 38.8 × 29.9 mm (MNHN-IU-2014-11830 (= MNHN-B5304); paralectotypes (here designated), 1 male, 35.9 × 27.8 mm, 2 females, 35.1 × 27.3 mm, 36.8 × 28.9 mm (USNM 29996); all from Mindanao, Philippines, coll. J. Montano, 1880 (figs. 6, 7).

Diagnosis.— Carapace with dorsal surface mostly smooth, branchial regions striated, inflated; without well-defined epigastric cristae, low, interrupted postorbital cristae; cervical grooves shallow, broad. Frontal median triangle distinct, dorsal margin cristate, not contiguous with lateral margins. Exorbital angle broadly triangular, external margin gently convex. Epibranchial tooth nearly obliterated, not separated from exorbital angle by notch or slit. Male pleonal somite 6 subrectangular, slightly longer than broad, lateral margins straight, parallel. G1 short, relatively slender for genus; terminal segment about 0.21-0.28 times total length, subconical, aperture small. G2 terminal segment short, 0.22 times total length.

Description.— Carapace (figs. 1A, 2A, 3A) transversely subovate in outline, width-to-length ratio about 1.21-1.27, widest at level of mesobranchial region; dorsal surface mostly smooth, sparsely pitted, strongly convex longitudinally, moderately convex transversely; branchial, sub-branchial regions distinctly inflated; cardiac, intestinal regions gently convex; epigastric region without well-defined cristae, rather with grouping of discrete pits forming rugose area; postorbital cristae low, interrupted at midpoint, not reaching anterolateral margin; lateral regions with weak striation; shallow median groove between epigastric regions, cervical groove moderately deep, broad, H-shaped gastric groove shallow. Front moderately broad, about 0.20-0.21 times greatest carapace width, gently deflexed ventrally; anterior margin appearing straight in dorsal view; frontal median triangle (figs. 1C, 2B, 3C, J) with distinct margins, dorsal margin cristate, straight or gently curving outward, not fusing with lateral margins. Exorbital tooth broadly triangular, external margin weakly convex, distinctly longer than internal margin; epibranchial tooth very small, triangular, without notch separating it from external orbital tooth. Anterolateral margin weakly cristate, strongly convex; posterolateral margin nearly straight, converging gradually towards posterior margin of carapace; central region of posterior margin slightly concave. Orbital margins cristate, supraorbital margin smooth, infraorbital margin finely granulate. Suborbital, sub-branchial, pterygostomial regions rugose, with rows of small granules. Posterior margin of epistome (figs. 1C, 2B, 3D, K) with protruding, triangular median lobe with rounded tip, separated from sinuous lateral lobes by thin sutures.

Fig. 1. *Sundathelphusa miguelito* sp. nov., holotype, male, 26.1 × 20.5 mm (NMCR-39999). A, dorsal habitus; B, cephalothorax, ventral view; C, cephalothorax, anterior view; D, right (minor) chela, external view; E, left (major) chela, external view. Scale bars: A = 10.0 mm; B-E = 5.0 mm.

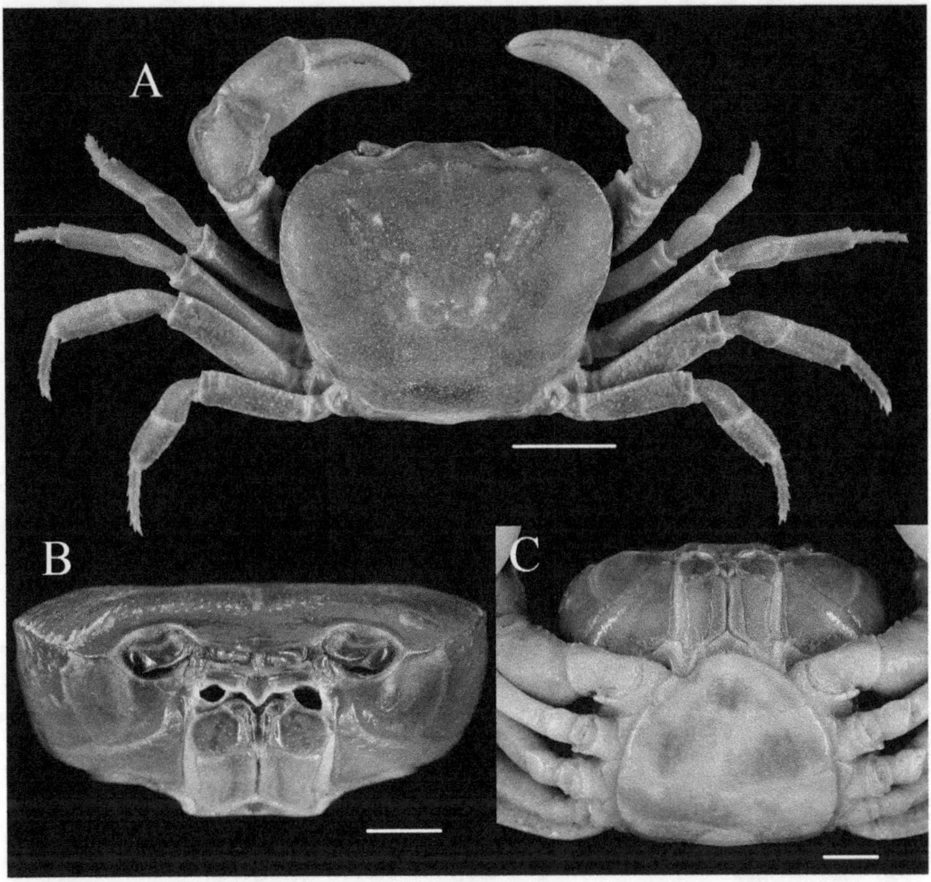

Fig. 2. *Sundathelphusa miguelito* sp. nov., paratype, female, 31.4 × 25.1 mm (ZRC 2016.0666). A, dorsal habitus; B, cephalothorax, anterior view; C, cephalothorax, ventral view. Scale bars: A = 10.0 mm; B, C = 5.0 mm.

Eyes (figs. 1C, 2B) large, occupying almost entire orbit, corneas well developed. Mandibular palp with two segments, terminal segment bilobed, inner lobe about one-third longer than outer lobe, both lobes broad, flattened. Third maxilliped (fig. 3B) sparsely setose; ischium subrectangular, with distinct, oblique, submedian sulcus closer to mesial margin; merus subquadrate, distal (anterior) margin slightly sinuous, almost straight, lateral margin convex, lateral, mesial and proximal margins raised, central region depressed; exopod slender, appearing distally tapering when pressed against lateral margins of endopod, external margin slightly sinuous, distal tip reaching level of mid-length of merus, with small subtriangular projection on subdistal inner margin, flagellum well developed, reaching mesial margin of merus.

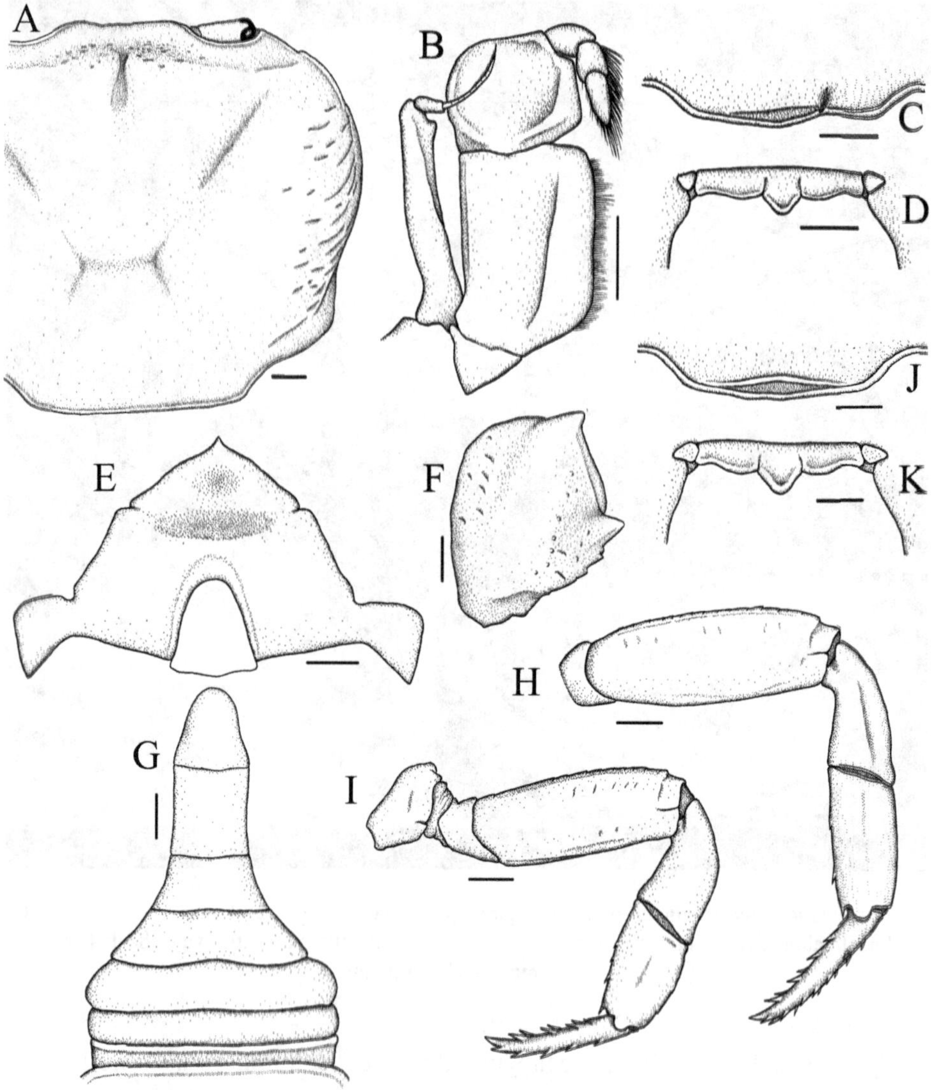

Fig. 3. *Sundathelphusa miguelito* sp. nov., holotype, male, 26.1 × 20.5 mm (NMCR-39999) (A-I); paratype, female, 31.4 × 25.1 mm (ZRC 2016.0666) (J, K). A, carapace, dorsal view; B, right third maxilliped, external view; C, J, frontal triangle, anterior view; D, K, epistome, anterior view; E, anterior thoracic sternum, ventral view; F, carpus of left cheliped (P1), dorsal view; G, pleon (unfolded), external view; H, right P4, dorsal view; I, right P5, dorsal view. Scale bars: A-K = 2.0 mm.

Male thoracic sternum (figs. 1B, 3E) broad, generally smooth except for slightly pitted anterior region; sternites 1-4 fused, traces of sutures between sternites 2, 3 and 3, 4 seen as deep transverse depressions; sternite 4 with straight lateral margins, slightly raised area around anterior portion of sterno-abdominal cavity. Sterno-abdominal cavity deep; press-button on sternite 5, midway between sutures.

Chelipeds, P1 (fig. 1A, D, E), distinctly asymmetric (in males), robust; merus lower (ventral) surface gently striated with rows of fine granules continuing onto posterior margin, anterior margin ratcheted with short transverse rows of fewer, larger granules; upper surface of carpus (fig. 3F) lightly striated, with small scattered pits, inner angle with short, pointed tooth, followed by two or three large granules proximally; palm of major chela inflated, with smooth dorsal, external, internal surfaces, inferior margin distinctly convex and with distinct junction with fixed finger; palm of minor chela smaller, less inflated; fingers sparsely pitted, slender, distal tips pointed, crossing, gape moderately wide, cutting edges armed with several small teeth interspersed with slightly larger teeth, no molariform teeth, dactylus longer than superior margin of palm.

Ambulatory legs, P2-P5 (figs. 1A, 2A, 3H, I), relatively short, P3 longest, total length (coxa-dactylus) about 1.3 times maximum carapace width, P5 shortest, total length (coxa-dactylus) about 1.1 times maximum carapace width; anterior margins of meri weakly serrated, with low subdistal angle not distinctly formed into tooth or spine, posterior margins smooth, subcristate only on P4, P5; carpi short, with longitudinal submedian ridge on dorsal surface except for P5, widened distally; propodi flattened, anterior, posterior margins with spiniform setae; dactyli slender, slightly curved at distal tip, subequal in length to propodi, with several spiniform, marginal setae.

Male pleon (figs. 1B, 3G) shaped like inverted 'T', moderately broad for the genus; somite 1 thin, sinuous; somites 2-5 subtrapezoidal, progressively narrowing distally; somite 6 rectangular, proximal width equal to distal width, median length slightly greater than maximum width, lateral margins nearly straight, slightly divergent distally; telson subtriangular, apex rounded, lateral margins straight, basal width subequal to median length, distal tip reaching to level of posterior quarter of P1 coxae in ventral view.

G1 (fig. 4A, B, D, E) short, stout at base, tapering distally, subterminal segment generally straight, mesial and lateral margins gently sinuous, widely divergent proximally; terminal segment distinctly curving laterally, about 0.21 and 0.28 times total length (measured from mesial and lateral margins, respectively), subconical, distal tip narrow. G2 (fig. 4C) slender, tapering distally; terminal segment short, about 0.20 times total length.

Female morphology.— Female resembling male in all morphological aspects except for sexual characters. Anterior thoracic sternum setose; vulvae large, subcircular, positioned on antero-mesial part of sternite 6 close to suture with sternite 5. Chelipeds (fig. 2A) relatively less robust than in the male, size difference between major, minor chelae much less pronounced. Pleon (fig. 2C) broad,

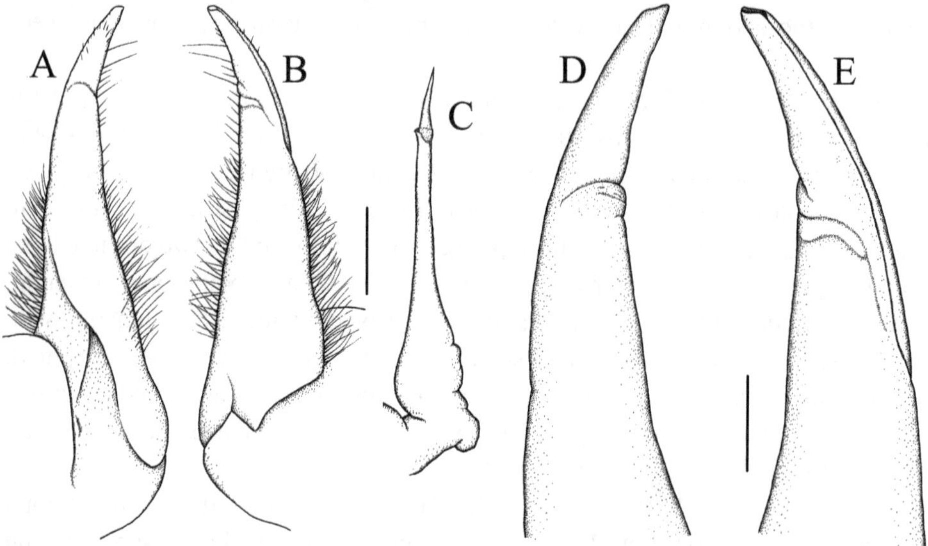

Fig. 4. *Sundathelphusa miguelito* sp. nov., holotype, male, 26.1 × 20.5 mm (NMCR-39999). Left G1: A, external view; B, internal view; D, apex (setae not included), external view; E, apex (setae not included), internal view. Left G2: C, external view. Scale bars: A-C = 1.0 mm; D, E = 0.5 mm.

covering nearly the entire thoracic sternum, lateral margins pronouncedly convex; telson broadly subtriangular, basal width 2.5 times median height, tip widely rounded, lateral margins gently convex.

Etymology.— This species is dedicated to Prof. Michael Türkay for his exemplary contribution to the field of carcinology. The epithet, *miguelito*, is from Spanish/Filipino usage which means "little Miguel/Michael" and is commonly used with endearment. It is used here as a noun in apposition.

Habitat and ecology.— The crabs were collected in a small, shallow stream approximately one meter in width with substrate consisting of rocks and gravel. During the day crabs were observed hiding underneath rocks near cascading water.

Utilisation.— According to locals interviewed by the second author, the new species is regularly collected and sold as food by ambulant fish vendors to nearby municipalities of the province. Crabs are sold by the bagful (ca. 200 g per bag). The type material was likewise purchased from such vendors by Mr. Ivan Galang and provided to the second author.

Remarks.— The *Sundathelphusa* of Mindanao are very poorly known, and an examination of the literature reveals that these species are only known from the type material, and the taxonomy, in some instances, can be somewhat confused. This is compounded by, with the exception of *S. hades*, the incompatibility of early descriptions and illustrations to modern species concepts. Moreover, two of the four species, do not have their type localities specified beyond "Mindanao" in the literature (e.g., *S. mistio* and *S. montanoanus*; see discussion below). Husana et al. (2015) have highlighted this lack of specificity in locality data and the scarcity of specimens available for examination, and we believe these are the main obstacles to understanding the diversity of freshwater crabs in the Philippines. To overcome this problem, a two-pronged approach is recommended, which entails (1) the systematic re-examination of all type material from the Philippines, many of which are deposited in European and American museums, and (2) a renewed and intensive collection effort in suitable habitats throughout the Philippines to provide fresh material for morphological examination as well as tissue for molecular phylogenetic analyses.

Rathbun (1904: 302, fig. 33, pl. 13 fig. 3) gave a brief description for *Potamon* (*Potamon*) *mistio*, based on one medium-sized male ("un mâle de médiane taille") from Mindanao, collected by a certain "M. Montano" (we assume "M." in this case is for the French honorific, "monsieur"), which she designated as the type. She also listed two broken male specimens ("deux mâles cassés") collected by "M. Marche" from the "Montagnes au nord-ouest de Tayabas". High-resolution photographs of the male holotype (MNHN-IU-2014-11829) have been examined by the first author (fig. 5), as well as the actual specimen of one of the males collected by "Marche" (USNM 29995).

The former specimen (holotype) is accompanied by several labels, one of which has the words "Museum Paris" printed on it, and the following handwritten entry: "Potamon (Potamon) mistis Rathbun 1♂ Type! (phot.-abd.fig.) Philippines centre de Mindanao Montano 611_81". Another handwritten label extends the numerical information thus: "611/1881". The collector, "Montano", is almost certainly the French anthropologist, Dr. Joseph Montano, who went on a scientific survey in the Philippines in 1879-1881; he also published a travelogue, "*Voyage aux Philippines et en Malaisie*" (Montano, 1886), in which he writes an account of, among others, his travels through the interior of Mindanao, including an ascent of Mount Apo, Mindanao's highest peak, in October and November, 1880. Montano (1886) details very clearly his north-to-south journey through the center of Mindanao, which included travel on rivers which could be suitable habitat for *S. mistio*. Based on the available information, we, therefore, interpret the labels to mean that the holotype was collected by Joseph Montano from central Mindanao in the year 1880. We make no definite conclusions regarding the numbers "611-81" or "611/1881", although we suspect this may refer to the date the specimen was received by the museum.

The latter specimen has the antero-dorsal portion of the carapace and the left epibranchial tooth intact, as well as the dismembered pereopods (two chelipeds and five ambulatory legs), the pleon and one G1. While the carapace, pleon, and G1 bear some similarity with the holotype, the recorded locality, "Tayabas", is suspect,

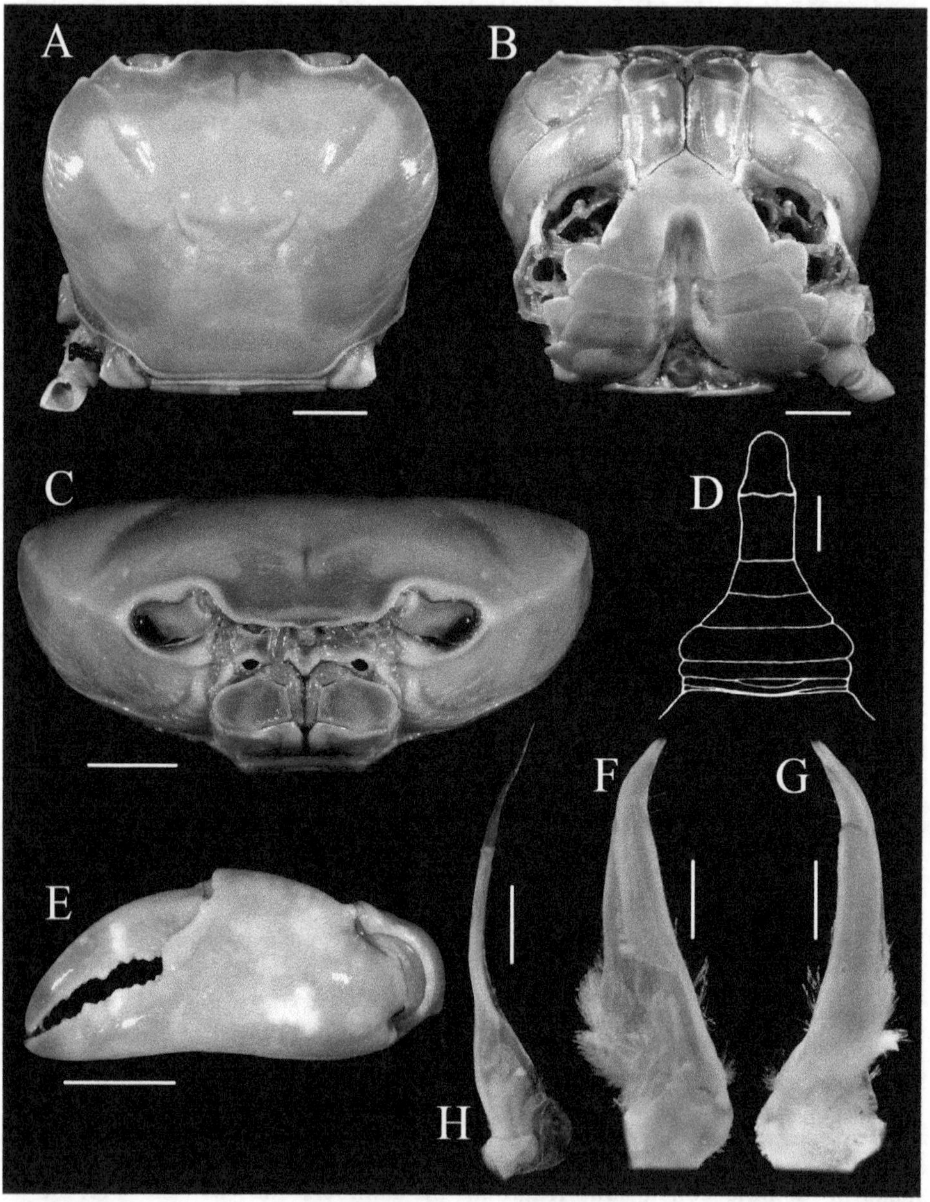

Fig. 5. *Sundathelphusa mistio* (Rathbun, 1904), holotype, male, 31.8 × 25.8 mm (MNHN-IU-2014-11829). A, carapace, dorsal view; B, cephalothorax, ventral view; C, cephalothorax, anterior view; D, pleon (unfolded), external view [adapted from Rathbun, 1904]; E, left (major) chela, external view; F, left G1, external view; G, left G1, internal view; H, left G2 external view. Scale bars: A-E = 5 mm; F-H = 1 mm.

as it is the old name of a province in Luzon Island, which has since been split up into the present-day provinces of Aurora and Quezon. In the handwritten labels accompanying the specimen (including one in Rathbun's handwriting), however, the locality is spelled "Bayabas". Although there is a small municipality of that name in present-day Surigao del Sur Province, in northeastern Mindanao, we cannot ascertain whether this is the same place Marche collected from. Moreover, "Marche" is most probably the French naturalist and explorer, Antoine-Alfred Marche, who was in the Philippines in the 1880s, but his explorations were mostly focused on the islands of Luzon (including the province of Tayabas), Marinduque, Catanduanes, and Palawan, not Mindanao (viz., Marche, 1887). As taxonomists in the early 20th century sometimes lumped similar-looking specimens from disparate islands as one species, we cannot entirely discount that the specimens may have come from a locality outside of Mindanao.

Rathbun (1904: 309) also gave a brief description for *Potamon* (*Potamon*) *montanoanus*, basically a differential diagnosis vis-à-vis *P. (P.) mistio*, based on 6 male and 6 female specimens also collected by "Montano" from "Mindanao". The first author has examined one male and two female syntypes (USNM 29996), as well as a detailed photograph of the male syntype figured by Rathbun (1904, fig. 36, pl. 13 fig. 7) (MNHN-IU-2014-11830), here designated as the lectotype. Unlike *S. mistio*, however, there are no further indications in the literature nor in the labels accompanying the specimens regarding a more specific locality. Again, one can only go by Montano's (1886) detailed itinerary in Mindanao to try and deduce the likely type locality. Unfortunately, while there are several likely candidate localities mentioned in Montano's travelogue, he does not seem to take note of any crabs collected, but it can be safely assumed that the type material was collected at around the same time (1880, the year he was in Mindanao) as the holotype of *S. mistio*. As such, the type locality for *S. montanoanus* cannot be ascertained with greater specificity than Mindanao Island for the time being.

Among its congeners in Mindanao, *S. miguelito* sp. nov. is most similar to *S. mistio* and *S. montanoanus* in the general carapace outline, the proportions of the carapace and ambulatory legs and the general form of the G1. These two species are very poorly figured in the literature, and the G1 and G2 morphology is unknown. We here provide additional illustrations to facilitate better comparison. The two other species from Mindanao are very easily distinguishable from the new species by their morphology. *Sundathelphusa hades* is a troglobitic crab, with the usual morphological adaptations for cave life such as highly elongated ambulatory legs, reduced eyes/corneas, and reduced pigmentation (Takeda & Ng, 2001, figs. 1, 2). *Sundathelphusa woltereckі* has a flat, subquadrate carapace, as well as slender, elongated ambulatory legs (Balss, 1937, figs. 17, 18; Bott, 1970, pl.14 figs. 64-66).

Sundathelphusa miguelito sp. nov. differs from *S. mistio* in the following morphological features: (1) the carapace is more globose, with the epibranchial regions more inflated and the lateral margins more arcuate (figs. 1A, C, 3A) (vs. less globose in *S. mistio*; fig. 5A, C); (2) the dorsal margin of frontal triangle is better defined, cristate (figs. 1C, 2B, 3C, J) (vs. poorly defined, not cristate in *S. mistio*; fig. 5C); (3) the exorbital angle is more obtuse, the external margin of the exorbital tooth is longer and slightly convex (figs. 1A, 2A, 3A) (vs. exorbital angle relatively acute, external margin of exorbital tooth shorter and slightly concave in *S. mistio*; fig. 5A); (4) the epibranchial tooth is smaller and without a distinct notch separating it from the exorbital tooth (figs. 1A, 2A, 3A) (vs. epibranchial tooth larger and more prominent, separated from exorbital tooth by a notch in *S. mistio*; fig. 5A); (5) the anterior thoracic sternum is relatively shorter, with two deep transverse depressions marking the divisions between the second and third, and between the third and fourth sternites (figs. 1B, 3E) (vs. relatively longer, without deep depressions in *S. mistio*; fig. 5B); (6) the inferior border of the palm of the major chela of the male is more pronouncedly convex (fig. 1E) (vs. gently convex in *S. mistio*; fig. 5E); (7) the ambulatory legs are shorter relative to the carapace (figs. 1A, 2A) (vs. relatively longer in *S. mistio*, cf. Rathbun, 1904, pl. 13 fig. 3); (8) the subterminal segment of the G1 is more slender and the terminal segment is less curved (fig. 4A, B, D, E) (vs. subterminal segment stouter, especially at the base, and curvature of terminal segment more pronounced; fig. 5F, G); and (9) the terminal segment of the G2 is distinctly relatively shorter, about 1/5 total length (fig. 4C) (vs. longer, about 1/3 total length, in *S. mistio*; fig. 5H).

The new species differs from *S. montanoanus* in the following morphological features: (1) the carapace is relatively less globose, with the branchial regions less inflated and more prominently striated, and the cervical groove shallower (figs. 1A, C, 2A, B, 3A) (vs. carapace relatively more globose, branchial regions much more inflated but smoother, cervical furrow deeper in *S. montanoanus*; fig. 6A, C); (2) the exorbital tooth is more obtuse and its external margin is gently convex (figs. 1A, 3A) (vs. exorbital tooth more pointed, external margin straight in *S. montanoanus*; figs. 6A, 7A); (3) the epibranchial tooth is smaller and less distinct (figs. 1A, 2A, 3A) (vs. comparatively larger, particularly in the male types, in *S. montanoanus*; fig. 6A); (4) somite 6 of the male pleon is rectangular, lateral margins parallel (figs. 1B, 3G) (vs. somite 6 trapezoidal, lateral margins divergent distally; fig. 6B, also Rathbun, 1904, fig. 36b); (5) the male telson is more broadly rounded at the tip and its median height is equal to its basal width (figs. 1B, 3G) (vs. tip of telson narrower, median height greater than basal width; fig. 6B, also Rathbun, 1904, fig. 36b); (6) the female telson is relatively more broadly rounded, with lateral margins convex (fig. 2C) (vs. female telson more triangular, lateral margins straight or sinuous; fig. 7C, D); (7) the G1 is more slender (fig. 4A, B, D, E) (vs. G1 stouter,

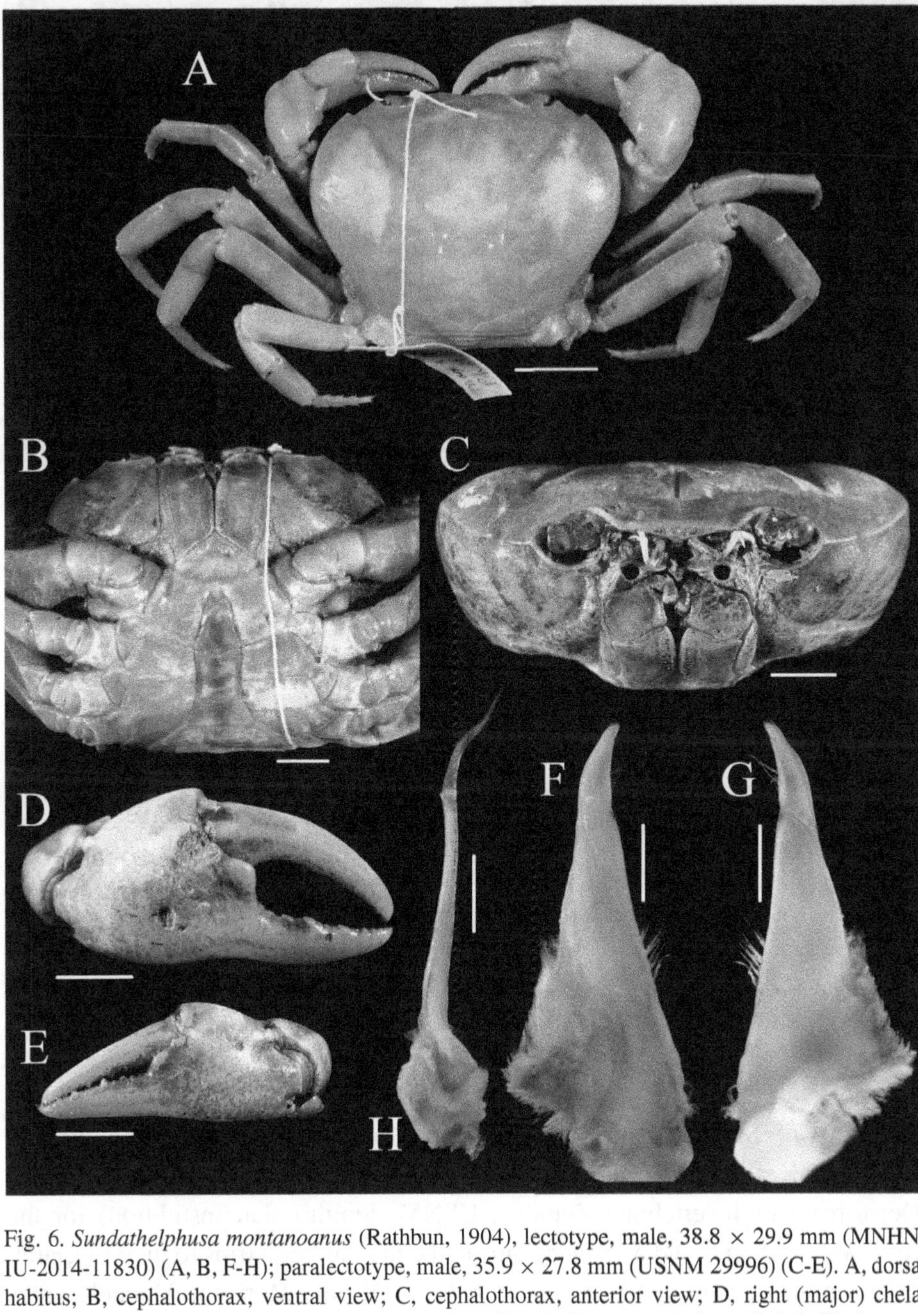

Fig. 6. *Sundathelphusa montanoanus* (Rathbun, 1904), lectotype, male, 38.8 × 29.9 mm (MNHN-IU-2014-11830) (A, B, F-H); paralectotype, male, 35.9 × 27.8 mm (USNM 29996) (C-E). A, dorsal habitus; B, cephalothorax, ventral view; C, cephalothorax, anterior view; D, right (major) chela, external view; E, left (minor) chela, external view; F, left G1, external view; G, left G1, internal view; H, left G2, external view. Scale bars: A = 10.0 mm; B-E = 5.0 mm; F-H = 1.0 mm.

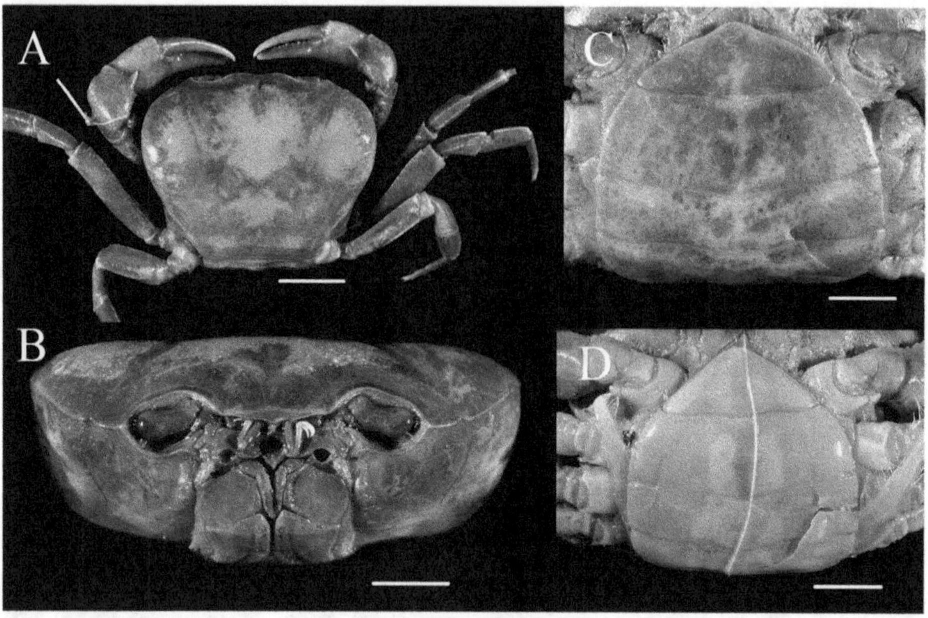

Fig. 7. *Sundathelphusa montanoanus* (Rathbun, 1904), paralectotype females (USNM 29996), 36.8 × 28.9 mm (A-C), 35.1 × 27.3 mm (D). A, dorsal habitus; B, cephalothorax, anterior view; C, D, pleon, ventral view. Scale bars: A = 10.0 mm; B-D = 5.0 mm.

especially at base of subterminal segment; fig. 6F, G); and (8) the terminal segment of the G2 shorter, about 1/5 total length (fig. 4C) (vs. terminal segment longer, about 1/3 total length; fig. 6H).

ACKNOWLEDGEMENTS

We would like to thank Célio Magalhães for his kind invitation to contribute to this special issue; Tohru Naruse (University of the Ryukyus, Japan) for providing the photos of the types deposited in the MNHN, Paris; Peter Ng (National University of Singapore) for kindly sharing his data on the type and other old material of Philippine freshwater crabs; Rafael Lemaitre and Karen Reed (Department of Invertebrate Zoology, USNM, Smithsonian Institution) for the warm welcome and kind assistance during the first author's trips to their museum; Paula Martin-Lefevre and Laure Corbari (MNHN) and Marivene Manuel-Santos (NMCR) for assistance with the catalogue numbers; and finally Tohru Naruse and Peter Ng for their critical review and helpful comments for the improvement of this article.

REFERENCES

BALSS, H., 1937. Potamiden (Dekapoda Brachyura) der Philippinen und des Malayischen Archipels. International Revue der gesamten Hydrobiologie und Hydrographie, **34**(3-5): 143-187.

BOTT, R., 1969. Flußkrabben aus Asien und ihre Klassifikation (Crustacea, Decapoda). Senckenbergiana Biologica, **50**: 359-366.

BOTT, R., 1970. Die Süßwasserkrabben von Europa, Asien, Australien und ihre Stammesgeschichte. Abhandlungen der Senckenbergischen Naturforschenden Gesellschaft, **526**: 1-338, pls. 1-58.

HUSANA, D. E. M., T. KASE & J. C. E. MENDOZA, 2015. Two new species of the freshwater crab genus *Sundathelphusa* Bott, 1969 (Crustacea: Brachyura: Gecarcinucidae) from Negros Island, Philippines. Raffles Bulletin of Zoology, **63**: 226-236.

HUSANA, D. E. M., T. NARUSE & T. KASE, 2009. Two new cavernicolous species of the genus *Sundathelphusa* from western Samar, Philippines (Decapoda: Brachyura: Parathelphusidae). Journal of Crustacean Biology, **29**: 419-427.

HUSANA, D. E. M., M. YAMAMURO & P. K. L. NG, 2014. Two new species of freshwater crabs of the genus *Sundathelphusa* Bott, 1969 (Decapoda: Brachyura: Gecarcinucidae) from caves in Luzon, Philippines. Zootaxa, **3815**: 565-574.

MARCHE, A.-A., 1887. Luçon et Palaouan, six années de voyages aux Philippines: i-xx, 1-406. (Libraires Hachette, Paris).

MENDOZA, J. C. E. & T. NARUSE, 2010. A new species of riverine crab of the genus *Sundathelphusa* Bott, 1969 (Crustacea: Brachyura: Gecarcinucidae) from northeastern Luzon, Philippines. Philippine Journal of Science, **139**: 61-70.

MONTANO, J., 1886. Voyage aux Philippines et en Malaisie: i-viii, 1-351. (Libraires Hachette, Paris).

NG, P. K. L., 1988. The freshwater crabs of Peninsular Malaysia and Singapore: 1-156, pls. 1-4. (Department of Zoology, National University of Singapore and Shinglee Press, Singapore).

NG, P. K. L., 2010. On the identity of *Para-Bary-Thelphusa grapsoides longipes* Balss, 1937, with description of a new species from the Philippines (Brachyura: Gecarcinucidae). In: C. H. J. M. FRANSEN, S. DE GRAVE & P. K. L. NG (eds.), Studies on Malacostraca: Lipke Bijdeley Holthuis memorial volume. Crustaceana Monographs, **14**: 561-571. (Brill, Leiden).

NG, P. K. L., 2015. On a new species of *Sundathelphusa* Bott, 1969 (Crustacea: Brachyura: Gecarcinucidae) from southeastern Borneo. Raffles Bulletin of Zoology, **63**: 494-502.

NG, P. K. L. & B. SKET, 1996. The freshwater crab fauna (Crustacea: Decapoda: Brachyura) of the Philippines. IV. On the collection of Parathelphusidae from Bohol. Proceedings of the Biological Society of Washington, **109**: 695-706.

RATHBUN, M. J., 1904. Les crabes d'eau douce (Potamonidae). Nouvelles Archives du Muséum d'Histoire Naturelle, 4e ser., **6**: 225-311, pls. 9-18.

STASOLLA, G., A. ABARCHI & G. INNOCENTI, 2015. *Sundathelphusa spelaeophila*, a new species of cavernicolous crab from Samar, Philippines (Decapoda: Brachyura: Gecarcinucidae). Raffles Bulletin of Zoology, **63**: 448-453.

TAKEDA, M. & P. K. L. NG, 2001. The freshwater crab fauna of the Philippines: VI. A new cavernicolous crab from Mindanao. Zoological Science, **18**: 1123-1127.

First received 15 October 2016.
Final version accepted 7 December 2016.

[When citing this volume, please refer to Crustaceana 90 (2017) 771-1288]

Michael Türkay Memorial Issue

ANKERIUS GRUSOCURARE, A NEW SPECIES OF APHANODACTYLIDAE (DECAPODA, BRACHYURA) FROM IRIOMOTE ISLAND, RYUKYU ISLANDS, JAPAN

BY

TOHRU NARUSE[1,3]) and DAISUKE UYENO[2])

[1]) Tropical Biosphere Research Center, Iriomote Station, University of the Ryukyus, 870 Uehara, Taketomi, Okinawa 907-1541, Japan

[2]) Graduate School of Science and Engineering, Kagoshima University, 1-21-35 Korimoto, Kagoshima, Kagoshima 890-0065, Japan

ABSTRACT

A new species of aphanodactylid genus *Ankerius* Ng, Rahayu & Shih, 2016, is described from Iriomote Island, Ryukyu Islands, Japan. *Ankerius grusocurare* n. sp. is distinct from the only congener, *A. aenigmaticus*, by the shape of the lateral corner of the front (= supraorbital tooth), proportionally wider carapace, and the fewer number of the teeth along the upperflexor margin of the ambulatory meri. The present study brings the number of aphanodactylid taxa to 13 species.

ZUSAMMENFASSUNG

Hier wird eine neue Art der Gattung *Ankerius* Ng, Rahayu & Shih, 2016 [Aphanodactylidae] beschrieben, welche auf der Insel Iriomote, Ryukyu-Inseln, Japan gesammelt wurde. *Ankerius grusocurare* n. sp. unterscheidet sich von der einzigen weiteren Art dieser Gattung, *A. aenigmaticus*, durch die Form der Lateralkante der Stirn (= Supraorbitalzahn), durch einen proportional breiteren Carapax und die kleinere Anzahl an Zähnen entlang des oberen Flexor-Randes der Laufbein-Meri. Diese Studie erhöht die Zahl der Aphanodactylidae auf 13 Arten.

INTRODUCTION

Aphanodactylidae Ahyong & Ng, 2009, is a relatively small Indo-west Pacific family, currently known by 12 species in 6 genera (Ng et al., 2008, 2016; Ahyong & Ng, 2009; Naruse et al., 2009; Ng & Naruse, 2009; Naruse & Maenosono, 2012; Ng & Rahayu, 2016; Ng et al., 2016). Although many species are associated with terebellid worms (Tesch, 1918; Konishi & Noda, 1999; Ahyong & Ng, 2009; Ng &

[3]) Corresponding author; e-mail: naruse@lab.u-ryukyu.ac.jp

Naruse, 2009; Ng & Rahayu, 2016), an undescribed species of the genus *Ankerius* Ng, Rahayu & Shih 2016, was collected from a tube of chaetopterid worm at off Iriomote Island, Ryukyu Islands, Japan. The present study describes this species.

The material examined is deposited in the Ryukyu University Museum, Fujukan (RUMF), University of the Ryukyus, Okinawa, Japan. The following abbreviations are used: CL, carapace length; CW, carapace width; Mxp3, third maxilliped; P1-5, first to fifth pereopods.

TAXONOMY

Family APHANODACTYLIDAE

Ankerius Ng, Rahayu & Shih, 2016

Ankerius grusocurare n. sp. (figs. 1, 2)

Material examined.— Holotype, RUMF-ZC-4358, ovigerous female, CL 8.1 × CW 9.6 mm, Nazuni, off Sonai of Iriomote Island, Ryukyu Islands, Japan, 20 m, from a tube of *Chaetopterus* sp. (RUMF-ZO-17), coll. D. Uyeno, 12 Aug. 2016.

Description.— Carapace subcircular, vaulted, widest at posterior quarter (fig. 1A), CW 1.19 times CL, dorsal surface setose, lateral margins more heavily pubescent than midportion. Front deflexed, with median groove; frontal margin medially produced as triangular lobe, up-turned distally, lateral corner (supraorbital tooth) rounded, slightly produced (fig. 1C). Supraorbital margin entire; external orbital angle inconspicuous; infraorbital margin entire, inner orbital tooth acute. Lateral margin of carapace convex, entire (fig. 1B), with line of tiny, randomly aligned granules along anterolateral margin that joining epimeral sulcus from around above base of P3. Suborbital, pterygostomial, subbranchial regions not granulated, lower pterygostomial, subbranchial regions setose. Milne-Edwards aperture distinct. Epistome posterior margin with strong median triangular projection. Sublateral portions weakly produced, posteriorly continued to endostomial ridge (fig. 1C).

Thoracic sternites 1, 2 fused, anterolateral margin serrated; sternites 2, 3 demarcated by horizontal groove, sternite 3 short, wide, lateral margin upturned; sternites 3, 4 fused, laterally demarcated by short fissure. Vulva subovoid, anterior margin on mesial end of suture of sternites 5/6.

Eyes short, stout, cornea much narrower than peduncle (fig. 1C). Antennular fossa relatively high, large. Antenna set vertically, entering to orbit (fig. 1C).

Mxp3s (fig. 1C) leaving wide rhomboidal gap between them. Ischium as long as merus, mesial to proximal margins produced triangularly; merus sub-oval; carpus, attached to distal median portion of merus; carpus, propodus, dactylus sylindrical,

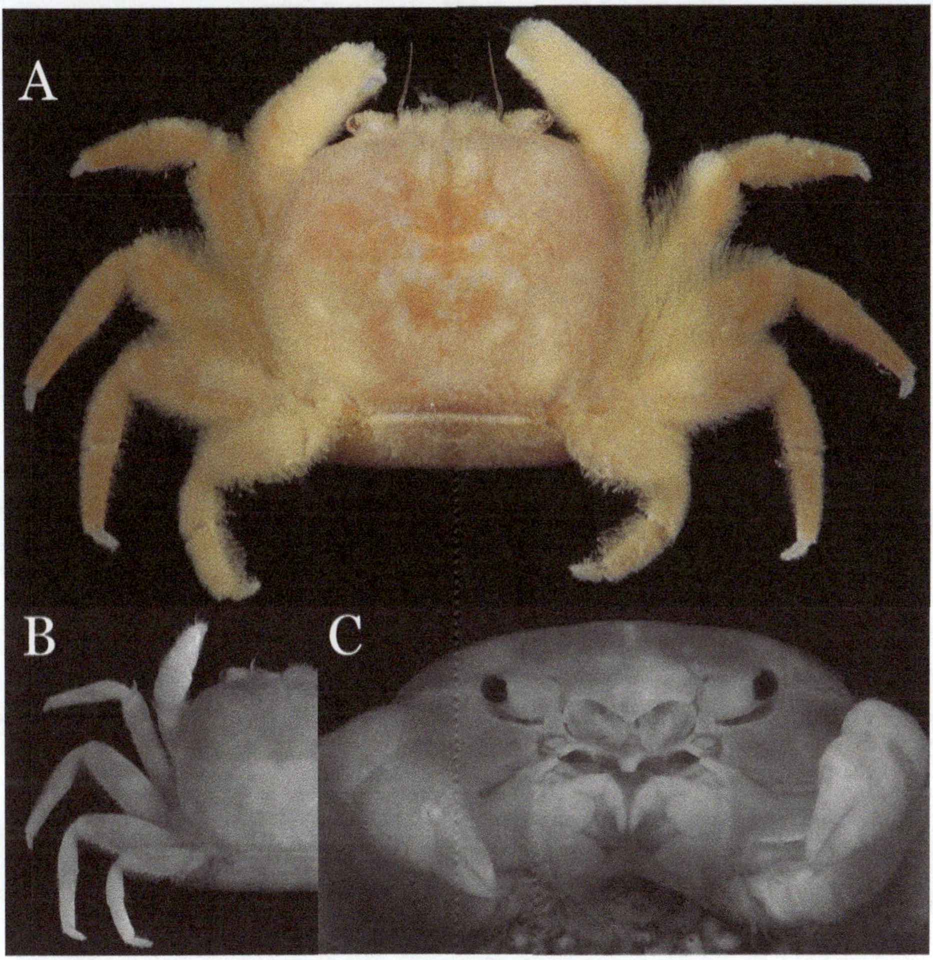

Fig. 1. *Ankerius grusocurare* n. sp., holotype, RUMF-ZC-4358. A, habitus in live colour, dorsal view; B, left half, setae denuded; C, cephalothorax, anterior view.

subequal in length, articulating end-to-end. Exopod slender, reaching distal third of lateral margin of merus, with a long flagellum (fig. 2A).

Chelipeds, ambulatory legs heavily setose, except for fingers tips of chelipeds and ambulatory dactyli (fig. 1A). Cheliped meri triangular in cross-section, outer-lower margin lined with small teeth. Carpus with outer surface rough, inner angle rounded. Chela (fig. 2B) relatively small, as high as carpus, palm outer surface sparsely granulated, upper, inner surface smooth, no ridge. Fingers much shorter than palm, no gap when closed, distally crossing, occlusal margin densely lined with small teeth.

Ambulatory legs (figs. 1A, B and 2C) slender, long, P3 longest, meri with row of distinct teeth over length on upper flexor margin, numbers of teeth of left/right

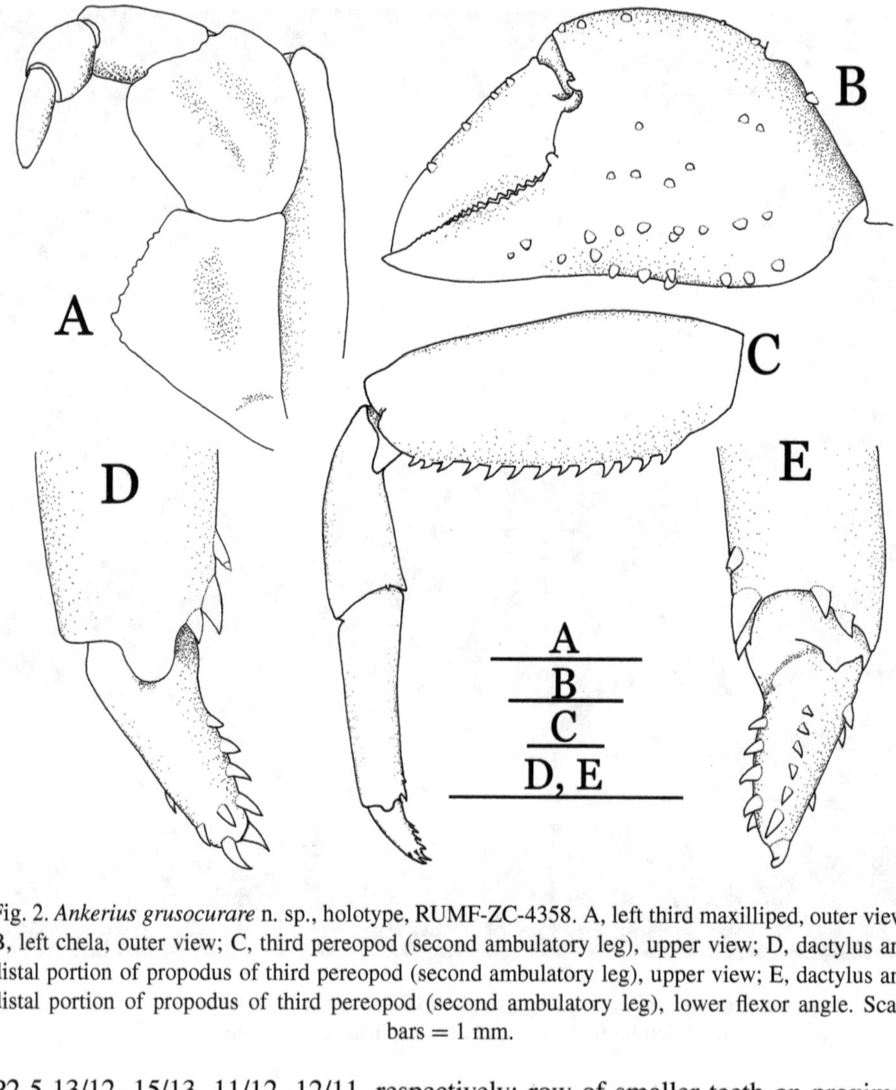

Fig. 2. *Ankerius grusocurare* n. sp., holotype, RUMF-ZC-4358. A, left third maxilliped, outer view; B, left chela, outer view; C, third pereopod (second ambulatory leg), upper view; D, dactylus and distal portion of propodus of third pereopod (second ambulatory leg), upper view; E, dactylus and distal portion of propodus of third pereopod (second ambulatory leg), lower flexor angle. Scale bars = 1 mm.

P2-5 13/12, 15/13, 11/12, 12/11, respectively; row of smaller teeth on proximal half of lower flexor margin. Propodus with 2 pairs of sharp teeth on upper, lower corners of distoflexor margin, sometimes with a few additional teeth proximally (fig. 2C-E). Dactylus (fig. 2C-E) much shorter than respective propodus, e.g., about 0.3 times in P3; flexor margin with double row of sharp teeth, two distal teeth on upper flexor row largest, terminal one with corneous distal tip; small irregular teeth present, especially on extensor surface of P4, P5.

Etymology.— The species name *"grusocurare"* is derived from a combination of the Latin "grus" (= crane) and "curare" (= curator, to take care of), which alludes to Japanese characters "鶴飼" ("crane" and "keeping animal") that has a similar pronunciation to "Türkay" and was used on his name card (fig. 3).

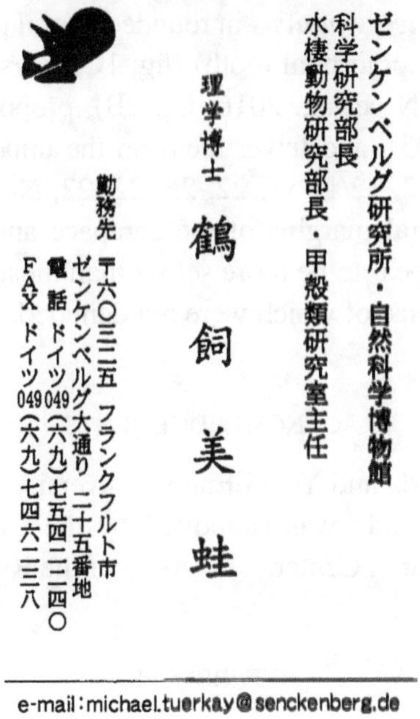
ゼンケンベルグ研究所・自然科学博物館
科学研究部長
水棲動物研究部長・甲殻類研究室主任

理学博士 鶴 飼 美 蛙

勤務先 〒六〇三二五 フランクフルト市
ゼンケンベルグ大通り 二十五番地
電 話 ドイツ049（六九）七五四二二四〇
FAX ドイツ049（六九）七四六二三八

e-mail：michael.tuerkay@senckenberg.de

Fig. 3. Name card in Japanese of the late Prof. Dr. Michael Türkay.

Ecological note.— *Ankerius grusocurare* was collected from a tube of *Chaetopterus* sp. (fig. 4). Habitat of the worm was the sea floor of a depth of about 20 m and composed of sands and buried dead corals.

Remarks.— *Ankerius grusocurare* n. sp. is morphologically very close to the only congener, *A. aenigmaticus* Ng, Rahayu & Shih, 2016. The new species,

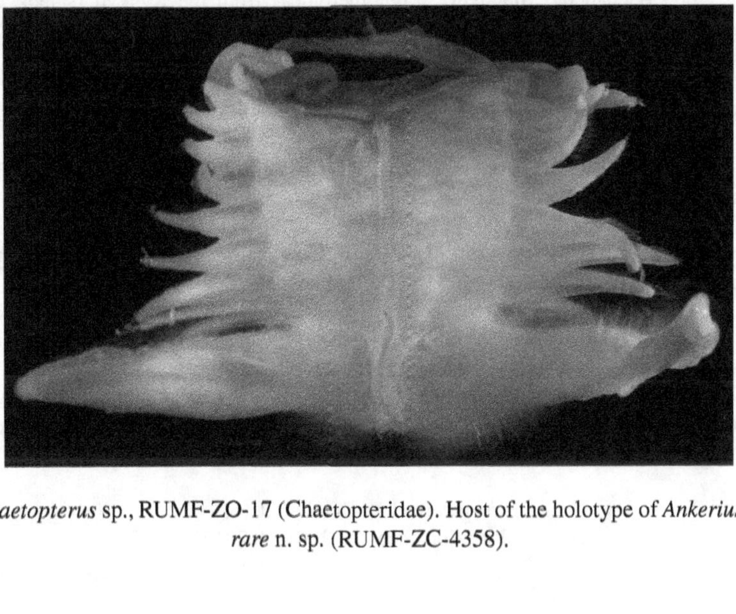

Fig. 4. *Chaetopterus* sp., RUMF-ZO-17 (Chaetopteridae). Host of the holotype of *Ankerius grusocurare* n. sp. (RUMF-ZC-4358).

however, differs from *A. aenigmaticus* in rounded lateral projections of the anterior margin of the front (= supraorbital tooth) (fig. 1C) (versus projected triangularly in *A. aenigmaticus*; see Ng et al., 2016, fig. 2B), proportionally wider carapace (CW/CL 1.19 versus 1.03), and fewer teeth on the upper flexor margin of P2-5 meri (13/12, 15/13, 11/12, 12/11 vs. 22, 25, 22, 22, Ng et al., 2016: fig. 4A, C, E, G). Furthermore, lateral margins of the carapace and ambulatory meri of *A. grusocurare* (fig. 1A) appear to be more setose than those of *A. aenigmaticus* (see Ng et al., 2006, fig. 1), setae of which were not denuded, P.K.L. Ng, pers. comm.).

ACKNOWLEDGEMENTS

We thank Ryuta Yoshida and Yuuki Endo (University of the Ryukyus) for their help in sampling. This study was supported by the Collaborative Research of Tropical Biosphere Research Center, University of the Ryukyus (FY2016).

REFERENCES

AHYONG, S. T. & P. K. L. NG, 2009. Aphanodactylidae, a new family of thoracotreme crabs (Crustacea: Brachyura) symbiotic with polychaete worms. Zootaxa, **2289**: 33-47.

KAMMERER, C. F., 2006. Notes on some preoccupied names in Arthropoda. Acta Zootaxon. Sinica, **31**: 269-271.

KONISHI, K. & H. NODA, 1999. A new species of commensal crab genus *Aphanodactylus* (Crustacea: Brachyura: Pinnotheridae) from the Yaeyama Islands, southern Japan. Publ. Seto Mar. Biol. Lab., **38**: 223-229.

NARUSE, T., Y. FUJITA & P. K. L. NG, 2009. A new genus and new species of symbiotic crab (Crustacea: Brachyura: Pinnotheroidea) from Okinawa, Japan. Zootaxa, **2053**: 59-68.

NARUSE, T. & T. MAENOSONO, 2012. A new genus and species of Aphanodactylidae Ahyong & Ng, 2009 (Decapoda, Brachyura, Pinnotheroidea) from the Ryukyu Islands, Japan. In: H. KOMATSU, J. OKUNO & K. FUKUOKA (eds.), Studies on Eumalacostraca: a homage to Takeda Masatsune. Crustaceana Monographs, **17**: 225-231.

NG, P. K. L., D. GUINOT & P. J. F. DAVIE, 2008. Systema Brachyurorum: Part I. An annotated checklist of extant brachyuran crabs of the world. Raffles Bull. Zool., **17**(Suppl.): 1-286.

NG, P. K. L. & T. NARUSE, 2009. On the identity of *Pinnixa brevipes* H. Milne Edwards, 1853, and a new species of *Aphanodactylus* Tesch, 1918 (Crustacea: Decapoda: Brachyura: Pinnotheroidea) from the Philippines. Raffles Bull. Zool., **20**(Suppl.): 283-290.

NG, P. K. L. & D. L. RAHAYU, 2016. On the genera *Selwynia* Borradaile, 1903, and *Gandoa* Kammerer, 2006, with descriptions of two new species from Papua New Guinea and French Polynesia (Crustacea: Decapoda: Brachyura: Aphanodactylidae). Zootaxa, **4092**(3): 339-370.

NG, P. K. L., D. L. RAHAYU & H.-S. SHIH, 2016. *Ankerius aenigmaticus*, a new genus and new species of aphanodactylid crab symbiotic with polychaetes from the Red Sea coast of Saudi Arabia (Crustacea: Decapoda: Brachyura: Aphanodactylidae). Zootaxa, **4179**(2): 271-278.

TESCH, J. J., 1918. Decapoda Brachyura II. Goneplacidae and Pinnotheridae. Siboga Exped. Monogr., **39c1**: 149-295.

First received 18 October 2016.
Final version accepted 2 March 2017.

[When citing this volume, please refer to Crustaceana 90 (2017) 771-1288]

Michael Türkay Memorial Issue

REPRODUCTIVE ASPECTS OF THE SHRIMP *CUAPETES AMERICANUS* (KINGSLEY, 1878) (CARIDEA, PALAEMONIDAE) FROM BOCAS DEL TORO, PANAMA

BY

MARIANA NEGRI[1]), TATIANA MAGALHÃES[1]), NATÁLIA ROSSI[1]), DARRYL L. FELDER[2]) and FERNANDO L. MANTELATTO[1,3])

[1]) Laboratory of Bioecology and Crustacean Systematics (LBSC), Department of Biology, Faculty of Philosophy, Sciences and Letters at Ribeirão Preto (FFCLRP), University of São Paulo (USP), São Paulo, Brazil

[2]) Department of Biology, University of Louisiana at Lafayette, 410 East St. Mary Boulevard, Billeaud Hall, Room 108, Lafayette, LA 70503, U.S.A.

ABSTRACT

The shrimp *Cuapetes americanus* is widely distributed in shallow waters of the western Atlantic Ocean, inhabiting sandy and rocky bottoms, macroalgae, seagrass beds, and mangrove roots. As the life cycle and ecology of this species remain poorly known, we investigated some reproductive traits of this species at Bocas del Toro, Panama. We collected and examined 167 individuals (63 males, 58 non-ovigerous females and 46 ovigerous females). Although males numerically dominated the larger size classes, ovigerous females were significantly larger than males and non-ovigerous females. All 46 ovigerous females were used to calculate embryo loss and embryo volume increase. Only females with eggs at an initial developmental stage ($n = 25$) were used to estimate fecundity and reproductive output. A mean of 75.84 ± 40.16 embryos ($N_e = 32\text{-}208$) occurred per female, a value near those reported for other species of Pontoniinae. A positive correlation between fecundity and female body size was observed, but it lacked significance. Likewise, a significant correlation between female body size and embryo volume was not detected. Embryo volume increased 35%, and females lost 26% of embryos on average during the incubation period, though significant changes were measurable only during early embryonic development. Mean reproductive output in the analysed reproductive event (0.22 ± 0.11, $R = 0.02\text{-}0.53$) suggested a rate of energy investment in reproduction higher than average, as compared to females of most other decapods. A decrease in relative reproductive output was observed with the increase of female size, indicating that larger females were investing proportionally less energy in reproduction than smaller ones, when measured in relation to the body size, but this correlation also lacked significance. We review how these initial insights on reproduction of *C. americanus* at Bocas del Toro compare to those of other populations in various localities.

Key words. — Embryo loss and volume, fecundity, reproductive output, size class distribution

[3]) Corresponding author; e-mail: flmantel@usp.br

ZUSAMMENFASSUNG

Die Garnele *Cuapetes americanus* ist weit verbreitet auf Sand- und Felsböden, Makroalgen, Seegraswiesen und Mangrovenwurzeln im Flachwasser des Westatlantiks. Da Lebenszylus und Ökologie dieser Art schlecht bekannt sind, haben wir einige Reproduktionsmerkmale dieser Art in Bocas del Toro, Panama untersucht. Wir sammelten und untersuchten 167 Individuen (63 Männchen, 58 nicht ovigere und 46 ovigere Weibchen). Obgleich die Männchen von ihrer Anzahl her die größeren Größenklassen dominierten, waren die ovigeren Weibchen signifikant größer als die Männchen und nicht ovigeren Weibchen. An allen 46 ovigeren Weibchen wurden Embryoverlust und embryonaler Volumenzuwachs untersucht. Nur bei Weibchen mit Eiern im Anfangsstadium der Entwicklung ($n = 25$) wurden Fruchtbarkeit und Reproduktionsleistung untersucht. Gemittelt traten 75.84 ± 40.16 Embryos ($N_e = 32$-208) pro Weibchen auf, ein Wert vergleichbar mit denen anderer Arten der Pontoniinae. Eine Positivkorrelation zwischen Fruchtbarkeit und weiblicher Körpergröße wurde beobachtet, jedoch ohne statistische Signifikanz. Ebenso wurde keine signifikante Korrelation zwischen weiblicher Körpergröße und Embryonalvolumen festgestellt. Durchschnittlich nahm das Embryonalvolumen um 35% zu und die Weibchen verloren 26% der Embryos während der Brutzeit, wobei nur während der frühen Embryonalentwicklung signifikante Veränderungen festgestellt werden konnten. Die mittlere Reproduktionsleistung im amanysierten Reproduktionsereignis (0.22 ± 0.11, $R = 0.02$-0.53) wies auf ein überdurchschnittliches Energieinvestitionsniveau im Vergleich mit Weibchen der meisten anderen Dekapoden hin. Es wurde eine Abnahme der relativen Reproduktionsleistung bei Größenzunahme der Weibchen beobachtet, die darauf hindeutet dass größere Weibchen proportional weniger Energie in die Reproduktion investieren als kleinere. Diese Korrelation ist jedoch ebenfalls statistisch nicht signifikant. Wir überprüfen, inwiefern sich diese ersten Erkenntnisse zur Reproduktion von *C. americanus* in Bocas del Toro mit denen anderer Populationen an verschiedenen Orten vergleichen lassen.

INTRODUCTION

Cuapetes americanus (Kingsley, 1878) is a small shrimp originally described as a member of the genus *Anchistia* Dana, 1852 (Kingsley, 1878, 1899), but for many years treated within the genus *Periclimenes* Costa, 1844 (Rathbun, 1902; Holthuis, 1951; Escobar-Briones & Villalobos-Hiriart, 2003) and briefly thereafter in *Kemponia* Bruce, 2004 (see Bruce, 2004). Only recently was it reassigned to the genus *Cuapetes* Clark, 1919, a taxon resurrected by Okuno (2009). *Cuapetes americanus* is endemic to the western Atlantic Ocean (Holthuis, 1951; Román-Contreras & Martínez-Mayén, 2010), with records from USA (Beaufort, North Carolina to Texas) (Holthuis, 1951; Felder & Chaney, 1979) to Brazil (Pará to São Paulo) (Coelho & Ramos, 1972; Markham et al., 1990; Escobar-Briones & Villalobos-Hiriart, 2003; Román-Contreras & Martínez-Mayén, 2010).

Populations of *Cuapetes americanus* are commonly found free living on seagrass beds, macroalgae, roots of mangroves, or rocky and sandy bottoms (Markham et al., 1990; Campos-Vázquez, 2000; Wehrtmann & Vargas, 2003; Román-Contreras & Martínez-Mayén, 2010; Bruce, 2012). However, they have also been reported to occur in association with anemones and crinoids or living in shallow water sponges together with snapping shrimps *Synalpheus longicarpus* (Herrick, 1891) and *S. townsendi* Coutière, 1909 (Wehrtmann & Vargas,

2003; Bruce, 2004), although these may be passive associations (Bruce, 2004). Despite the wide distribution and high abundance of *C. americanus* in the western Atlantic (Gurney, 1943), little information exists regarding its biology. Most data are restricted to regional faunal inventories and expedition collections (Chace, 1972; Coelho & Ramos, 1972; Ledoyer, 1986; Nizinski, 2003; Román-Contreras & Martínez-Mayén, 2010; Santos et al., 2012), caridean catalog (De Grave & Fransen, 2011), and morphological taxonomy (Gurney, 1943; Holthuis, 1951; Bruce, 2004; Okuno, 2009).

Some aspects of reproduction in *C. americanus* have been previously investigated for populations from Puerto Rico (Bauer, 1989, 1991), Florida (Corey & Reid, 1991), and Mexico (Martínez-Mayén & Román-Contreras, 2014). The present study aimed to describe fecundity, embryo volume and loss, investment per reproductive event, and size-class distributions in previously unstudied populations of *C. americanus* at Bocas del Toro (Central America, Panama). Our objective was to compare present findings to those for previously studied populations in search of possible explanations for differences and similarities in reproductive performance.

MATERIAL AND METHODS

Sampling and measurements

Sampling of *Cuapetes americanus* was conducted using manually pushed nets in shallow waters with depths of no more than 1 m in *Thalassia* seagrass meadows surrounding Cólon Island, inside Almirante Bay (09°19′N 82°15′W), Bocas del Toro, Panama, in August 2011. A total of 167 individuals (63 males, 58 non-ovigerous females, and 46 ovigerous females) were collected and preserved in 80% ethyl alcohol. Once identified according to Abele & Kim (1986), they were counted and separated by sex, which was determined by the presence (male) or absence (female) of the appendix masculina on the second pair of pleopods. Ovigerous females were identified by the presence of embryos attached to pleopods. Body size was measured as carapace length (CL) from the postorbital margin to the posterior margin of the carapace. This measurement was obtained under a stereomicroscope Leica M205C® with an attached camera and a calibrated micrometer eyepiece (0.01 mm accuracy). Embryos were measured from images captured by the same stereomicroscope and processed using the Leica Application Suite software (LAS) (0.01 mm accuracy).

Female reproductive analyses

Embryos were gently removed from the abdomen of 46 ovigerous females using fine forceps, counted, and classified according to criteria proposed by Wehrtmann

(1990): Stage I, embryos with uniform yolk and without visible eyespot; Stage II, eyespot enlarged and barely visible; stage III, eyes clearly visible and free abdomen.

Only data from females with newly formed embryos (stage I) were used to estimate fecundity (N_e = number of embryos) (Kuris, 1991; Mantelatto & Fransozo, 1997), since loss and predation of embryos during incubation is a well-documented phenomenon in decapods (Kuris, 1991). Data from females carrying embryos in stage II and III were used only to estimate the increase in embryo volume and embryo loss during the incubation period.

Embryo volume (Ev) was calculated based on 15 embryos arbitrarily separated from each brood mass in stage I, II, and III. We measured the shortest (a_1) and longest (a_2) axes of each embryo and calculated volumes in accordance with the formula for oblate spheroids: $\mathrm{Ev} = (\pi(a_1)^2 a_2/6)$, where a_1 = shortest axis, a_2 = longest axis (Turner & Lawrence, 1979). After these measurements, embryos were returned to their respective brood masses for subsequent analyses.

Reproductive output (RO) was calculated on the bases of only females in stage I for one single brood per female and then was defined as investment per reproductive event. Females (without their embryos) and their embryos masses were dried separately at 60°C for 48 h and 24 h, respectively, after which they were weighed on an analytical scale (±1 mg). The formula RO = total dry mass of embryos/dry mass of female (which previously carried that mass of embryos, now without it) was applied to determine the reproductive output for each female. RO defined as the investment per brood and calculated as the percentage of the female body weight is used in several studies (Hines, 1988; Clarke et al., 1991; Lardies & Castilla, 2001; Terossi et al., 2010; Zimmermann et al., 2015).

Statistical analyses

A histogram of body size classes was constructed for males, ovigerous females, and non-ovigerous females. The number of classes was determined according to the following formula: $K = 1 + \log_2 n$, in which n represents the sample size and K is the number of classes (Sturges, 1926). Thus, size classes were established at 0.3 mm intervals. All the statistical analyses were performed with SigmaStat® Windows software and Excel. To verify the normality of data, the Kolmogorov-Smirnov test was applied. Unlike females (ovigerous and non-ovigerous), males showed a non-normal distribution in size. Student's t-test was used to compare average size between ovigerous and non-ovigerous females and a non-parametric Mann Whitney test was adopted to compare sizes between males and females (ovigerous and non-ovigerous separately).

Percentage of embryo loss and percentage of increase in embryo volume between different stages were calculated based on averages of each stage, and analysis of covariance (ANCOVA) was applied to compare the number of embryos and embryo volumes between the developmental stages, with CL as the covariate. Linear regression analysis was applied to evaluate the relationship between embryo mass dry weight and female dry weight, fecundity and female carapace length, embryo volume and female carapace length, and RO and female carapace length. Additionally, the Pearson correlation coefficient (r) and the correlation significance were estimated for all regressions ($\alpha < 0.05$). All statistical baselines followed Zar (1996).

Voucher specimens were deposited in the Collection of Crustacean Biology Department (CCDB), Faculty of Philosophy, Sciences and Letters of Ribeirão Preto (FFCLRP), University of São Paulo, Brazil (CCDB 4934).

RESULTS

Among the 46 ovigerous females of *Cuapetes americanus* examined, 25 were stage I, 10 stage II, and 11 stage III. The mean size of all ovigerous females was 2.07 ± 0.36 mm, ranging from 1.37 to 2.79 mm (table I). Although only males occupied the larger size classes, ovigerous females were significantly larger than males ($p<0.001$) and non-ovigerous females ($p<0.001$). No significant difference between male and non-ovigerous female sizes was found ($p = 0.139$) (table I; fig. 1).

N_e varied between 32 and 208 embryos (table II) with a significant decrease in the average number of embryos accompanying embryonic development (table I; 35%). The most significant loss occurred between the stages I and II (21%), while the loss of 10% between the stages II and III was not significant (table III). Removing the outlier value of 208, we obtained a positive low correlation between fecundity of females and their sizes, which was not significant though ($p>0.30$; $r = 0.23$). Despite this result, we observed a tendency for larger females to have a higher number of extruded embryos (fig. 2).

TABLE I

Average, maximum and minimum carapace length (CL) of sampled males, ovigerous females, and non-ovigerous females of *Cuapetes americanus* (Kingsley, 1878) from Bocas del Toro, Panama

	Carapace length		
	Average ± SD	Min	Max
Males	1.74 ± 0.63	1.05	3.98
Non-ovigerous females	1.73 ± 0.31	1.15	2.60
Ovigerous females	2.07 ± 0.36	1.37	2.79

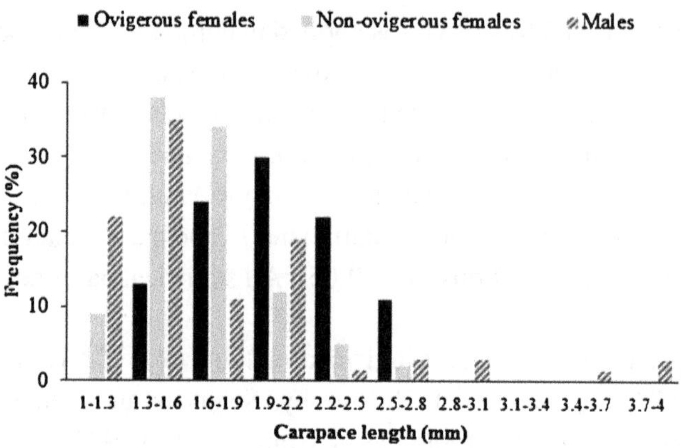

Fig. 1. Size-frequency distribution of sampled ovigerous females (n = 46), non-ovigerous females (n = 58) and males (n = 63) of *Cuapetes americanus* (Kingsley, 1878) from Bocas del Toro, Panama.

TABLE II

Number of embryos and embryo volume of sampled females of *Cuapetes americanus* (Kingsley, 1878) from Bocas del Toro, Panama

Stage	n	Number of embryos			Embryo volume (mm^3)		
		Average ± SD	Min	Max	Average ± SD	Min	Max
I	25	75.84 ± 40.16	32	208	0.005 ± 0.001	0.001	0.006
II	10	59.40 ± 34.57	10	115	0.007 ± 0.001	0.005	0.008
III	11	45.73 ± 33.47	12	131	0.008 ± 0.001	0.006	0.009

n, number of females with embryos at each stage of development.

TABLE III

Percentages of embryo loss and embryo volume increase between stages of development in a population of *Cuapetes americanus* (Kingsley, 1878) from Bocas del Toro (Panama) and values of ANCOVA test

Stages		Embryo loss	Embryo volume increase
I-II	%	21*	19*
	Slope	$F = 1.64$; $p = 0.2097$	$F = 0.17$; $p = 0.6813$
	y-intercept	$F = 5.52$; $p = 0.0252$	$F = 8.91$; $p = 0.0054$
II-III	%	10	8
	Slope	$F = 0.00$; $p = 0.9712$	$F = 0.00$; $p = 0.8747$
	y-intercept	$F = 3.77$; $p = 0.067$	$F = 1.89$; $p = 0.1856$
I-III	%	35*	26*
	Slope	$F = 0.90$; $p = 0.35$	$F = 0.02$; $p = 0.8914$
	y-intercept	$F = 9.58$; $p = 0.004$	$F = 28.22$; $p<0.0001$

*Significant % of loss or increase ($p<0.05$).

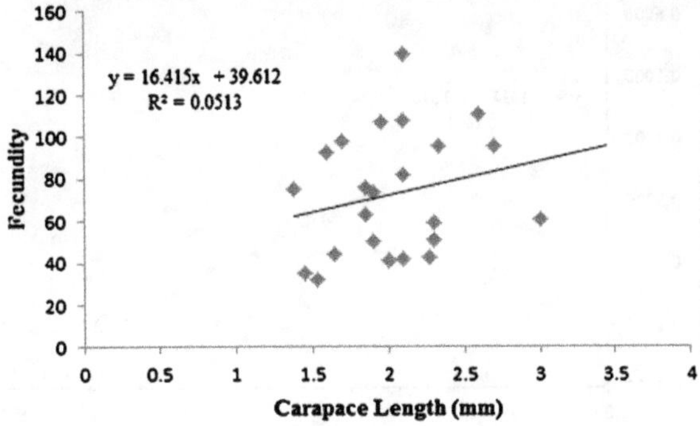

Fig. 2. Linear regression between fecundity and carapace length (CL) for sampled ovigerous females of *Cuapetes americanus* (Kingsley, 1878) from Bocas del Toro, Panama.

The low correlation between embryo volume (average = 0.006 ± 0.001 mm^3) (table II) and female body size was not significant ($p>0.05$; $r = 0.18$). During embryogenesis, there was an average increase in the embryo volume of approximately 26%. The major and significant increase (19%) was found in the early development of the embryos between stages I and II. In the final phase of embryo development (between stages II and III), the difference of 8% was not significant (table III).

Reproductive output of *C. americanus* females ranged from 0.02 to 0.53 with an average of 0.22 ± 0.11. Although the Pearson coefficient ($r = 0.30$) showed a moderate positive correlation between female and embryo dry weights, this correlation was not significant ($p = 0.11$) (fig. 3). The RO decreased with the increase of female body size (fig. 4), showing a moderate negative correlation ($r = 0.35$) but this also was not significant ($p = 0.06$).

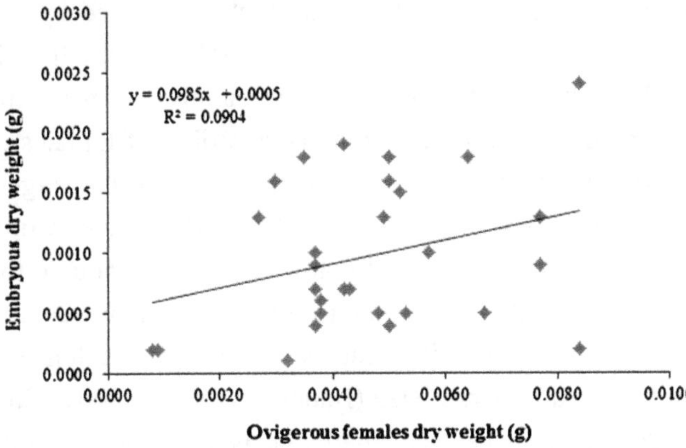

Fig. 3. Linear regression between the total dry weight of embryos and dry weight of ovigerous females of *Cuapetes americanus* (Kingsley, 1878) from Bocas del Toro, Panama.

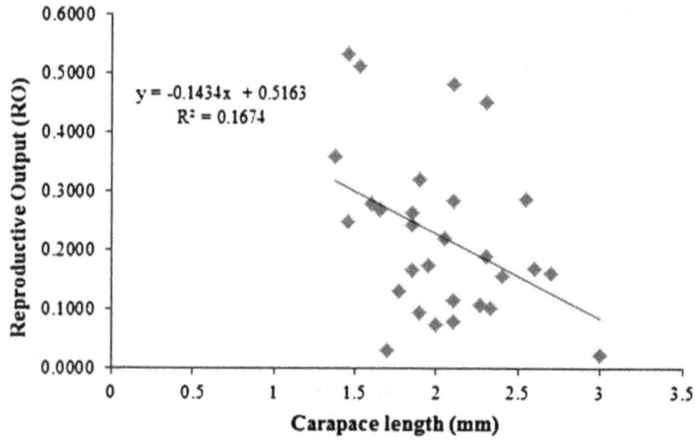

Fig. 4. Linear regression between reproductive output (RO) and carapace length (CL) of females for sampled ovigerous females of *Cuapetes americanus* (Kingsley, 1878) from Bocas del Toro, Panama.

DISCUSSION

Size

In many penaeid and caridean shrimp species males are smaller than females (Bauer, 1996). Generally, this difference in size is greater when males are compared to ovigerous females (e.g., Penha-Lopes et al., 2007). For *Cuapetes americanus* from Bocas del Toro, no size difference was found between males and non-ovigerous females but males were significantly smaller than ovigerous females. Small males and sexually monomorphic chelipeds are usually evidence of a 'pure search' mating system (Bauer & Abdalla, 2001). Males with such a mating system roam through the population searching for receptive females, and upon encounters, males transfer sperm in brief acts quickly followed by couple separation (Bauer & Abdalla, 2001; Correa & Thiel, 2003). So, small and agile males may be favoured, as they may search more efficiently for mates in the population. In addition, as there is no precopulatory guarding of females or aggressive behaviour against conspecifics, no selective pressure favours larger males with enlarged chelae (Bauer, 1996; Correa & Thiel, 2003). In other words, reproductive success depends on males' efficient search behaviour favoured by small-sized specimens with higher ability to find and inseminate as many receptive females as possible rather than competition for females with a strong defence performance, which could be favoured by larger size (Bauer & Abdalla, 2001; Correa & Thiel, 2003). Larger sizes could also elevate energy costs, given an increase of their vulnerability to predation and the lower efficiency of escape responses (Correa & Thiel, 2003).

Fecundity

Fecundity is usually deemed to be an essential indicator of reproductive performance (Corey & Reid, 1991). In an intraspecific comparison, the mean fecundity calculated for *C. americanus* from Bocas del Toro ($N_e = 75.84$) was smaller than the value found by Corey & Reid (1991) for a population from Florida, USA (29°05′N 83°03′W; 247.33 embryos). On the other hand, fecundity measured at Bocas del Toro was greater than that found by Bauer (1991) for a population from Puerto Rico (18°29′N 66°15′W; 56 ± 27 embryos) and greater than the value estimated by Martínez-Mayén & Román-Contreras (2014) for a population from Quintana Roo, Mexico (19°30′N 87°07′W) sampled in 2000 (47 ± 21 embryos). Fecundities of *C. americanus* from Bocas del Toro and from a sample obtained in Quintana Roo, México in 2001 (74 ± 57 embryos) were on the other hand very similar (Martínez-Mayén & Román-Contreras, 2014). Thus, while relationships between fecundity and latitude might be expected, with decreased brood sizes at higher latitudes (see Thorson, 1950; Clarke et al., 1991; Lardies & Castilha, 2001; Mantelatto et al., 2010; Terossi et al., 2010), it is not evident across the latitudinal differences represented by these particular populations, and suggests the likely influence of other factors.

The greater measure of fecundity reported by Corey & Reid (2001) from Florida could result from sampling bias. Fecundity is usually correlated positively with female body size in caridean shrimps (Bauer, 1991), so it is possible that only large ovigerous females were included in the analyses performed by Corey & Reid (2001). This might be suspected because their sample size was very small ($n = 9$) and the ovigerous females analysed by them also had a greater mean size (CL = 3.06 mm) than in other studies to which they can be compared. However, despite the moderate positive correlation between fecundity and female body size for the population from Bocas del Toro, this correlation was not significant which may reflect the low number of samples. Nevertheless, female body size is not the only variable that can explain the differences in fecundity among populations of *C. americanus*. For instance, the populations from Puerto Rico and Bocas del Toro were composed of ovigerous females with similar mean body sizes (CL = 2.07 and 2.1 mm, respectively) but the Puerto Rican population showed the lower value of fecundity. Furthermore, the ovigerous females from Quintana Roo (CL = 2.20 ± 0.23 mm in 2000 and 2.46 ± 0.47 mm in 2001) were larger than the ones of Bocas del Toro but their fecundity was lower than the value found in the present work.

Differences in the number of embryos among populations of *C. americanus* from different regions might simply be associated with variations in the environments occupied by them, with the size and number of embryos influenced by abiotic factors like temperature, salinity, photoperiod, and lunar cycle during embryogenesis (Thorson, 1950; Sastry, 1983; Clarke et al., 1991; Lardies & Castilla, 2001;

Litulo et al., 2005). However, if temperature and salinity are very similar among the compared regions, it is not reasonable to associate the differences in fecundity with variations in such environmental aspects (see Bauer, 1985; Mason & Zengel, 1996; Martínez-Mayén & Román-Contreras, 2014). Rather, latitudinal variations in food supply might partially determine differences in fecundity (Sastry, 1983; Martínez-Mayén & Román-Contreras, 2014). It is very plausible that periodic stress in shallow waters of Bocas del Toro (Carruthers et al., 2005; Guzmán et al., 2005) might result in episodes of natural eutrophication and modification of food supply in this region (D'Croz et al., 2005).

From an interspecific perspective, the fecundity observed for the studied population was in general smaller than the values obtained for other species of the family Palaemonidae (Bauer, 1991; Corey & Reid, 1991). However, that would be expected if other species of this family were characterized by specimens larger than *C. americanus*, as there usually is a tendency for females of larger-sized species to carry a greater number of embryos than those of smaller species (Bauer, 1991). On the other hand, species with smaller-sized females do sometimes produce relatively more embryos, for their size, than do females of larger species (Bauer, 1991). Some other species, mainly those belonging to the subfamily Pontoniinae, exhibit fecundity lower than or similar to *C. americanus*, but those species also tend to represent animal sizes smaller than or similar to those of *C. americanus* (cf. Corey & Reid, 1991). Other species of Pontoniinae with larger-sized females tend to have greater brood sizes (Omori et al., 1994).

Embryo volume and loss

The volume of embryos in the initial development phases of *C. americanus* from Bocas del Toro (0.005 mm^3) was smaller than in populations from Florida (0.034 mm^3), Puerto Rico (0.013 mm^3), and Quintana Roo (0.034 mm^3 and 0.028 mm^3 for 2000 and 2001, respectively). Considering a given energetic allocation to reproduction, embryo size is negatively correlated with fecundity (Steele & Steele, 1975). This assumption could explain the difference in the embryo volume between Bocas del Toro and Puerto Rico or Quintana Roo (only in 2000). If we consider that these three populations allocate a similar quantity of energy to reproduction, the population from Bocas del Toro with higher fecundity would be expected to have smaller embryos, as we in fact observed. However, this explanation could not be applied to values found for the Florida population, with greater embryo volume and fecundity (Corey & Reid, 1991), nor to the population from Quintana Roo sampled in 2001, which showed embryos greater than those for the population from Bocas del Toro even though they had a similar value for fecundity.

Compared to other species of the family Palaemonidae, including those of the subfamily Pontoniinae (Bauer, 1991; Corey & Reid, 1991), *C. americanus* from Bocas del Toro showed the smallest embryo volume. Bauer (1991) noted that the production of small-sized embryos requires a short incubation period, increasing the number of broods spawned per unit of time.

The embryo volume increase during embryogenesis is a consequence of water accumulation by uptake to the embryo or retention of water resulting from metabolic processes such as respiration (Pandian, 1970; Amsler & George, 1984; Penha-Lopes et al., 2007). Although a 26% volume increase was detected during the embryo development of *C. americanus* from Bocas del Toro, with the largest and statistically most significant increase occurring between the initial stages I and II, this value is much lower than those observed for the population from Quintana Roo (73.5% in 2000 and 78.5% in 2001) by Martínez-Mayén & Román-Contreras (2014) as well as in a number of other caridean shrimps (Lardies & Werhtmann, 2001). By comparison, volume increase in *Hippolyte obliquimanus* Dana, 1852 was found to be 100% (Terossi et al., 2010), 100% in *Latreutes pymoeus* Nobili, 1904 (Penha-Lopes et al., 2007), 82.75% in *Periclimenes siankaanensis* Martinez-Mayen & Roman-Contreras, 2006 (Martínez-Mayén & Román-Contreras, 2009), 161% in *Palaemon gravieri* (Yu, 1930) (Yu, 1930), and 47% in *Leander paulensis* Ortmann, 1897 (Zimmermann et al., 2015). In addition, as observed for other carideans, the embryo volume was not significantly correlated with the sizes of the ovigerous females analysed (Penha-Lopes et al., 2007).

The intra- and interspecific differences in the embryo volume increase might reflect differences in the duration of the embryonic development, as these two variables can be positively correlated, for example tropical seagrass species tend to have shorter incubation periods and smaller embryos than temperate and boreal species (Allen, 1966; Wear, 1974; Bauer, 1991). Incubation period decreases with increasing temperature, and it is suggested that short embryonic development is an adaptation of tropical species that yields an increase of the number of broods spawned per unit of time (Bauer, 1991). Alternatively, the short incubation period may also be a temperature-modulated metabolic consequence of differences in environmental conditions (Wear, 1974).

The mean loss of embryos of *C. americanus* from Bocas del Toro during embryonic development was slightly higher (35%) than the values found for the population of Quintana Roo (21.5% in 2000 and 27.5% in 2001; Martínez-Mayén & Román-Contreras, 2014). In addition, in contrast to the population of Quintana Roo, where the largest loss occurred in late development, *C. americanus* from Bocas del Toro showed the largest and significant loss of embryos in early developmental stages (between stages I and II). According to Kuris (1991), embryo losses are more intense at oviposition or advanced stages of embryogenesis. Two non-exclusive factors are pointed out as potential explanations for the loss at oviposition: incomplete fertilization and inadequate attachment of embryos to the pleopods (Oh & Hartnoll, 1999). Many factors may influence embryo loss during embryogenesis, including mechanical losses, aborted development, parental care, predation, and parasitism (Kuris, 1991). However, a large loss of embryos during incubation naturally occurs as consequence of their volume increase. The embryos' swelling can make them susceptible to the wear of their attachment structures and thus more likely to be lost (Oh & Hartnoll, 1999). Mechanical abrasions and physical injuries to the abdomen and pleopods are also well-known causes of embryo loss (Kuris, 1991). The first is usually associated with losses of embryos at the tip of pleopods. Even though some specializations can reduce damage to embryos by parasites and predators, embryos are also damaged and lost to these agents during incubation (Kuris, 1991). Grooming activities, however, may come with a cost, as some viable embryos may be removed along with those that are damaged or infested by parasites.

Reproductive output (RO)

Comparative analysis of RO shows that the high intraspecific variation noted for *C. americanus* (average = 22.60%; min = 2.30%, max = 53.30%), was also observed in populations of other caridean species (Terossi et al., 2010). The mean value found for *C. americanus* is higher than the average of 10% of the body weight proposed as the energy invested in reproduction by female decapods (Hines, 1988; Lardies & Wehrtmann, 1996; Hernáez-Bové, 2001; Hartnoll, 2006; Miranda et al., 2006; Thiel & Duffy, 2007; Torati & Mantelatto, 2008; Zimmermann et al., 2015). However, there are many reports of elevated RO values for other caridean shrimps from Latin America (see Terossi et al., 2010 for review).

The independence between the energy invested in the production of embryos and the ovigerous female body size is a common pattern for decapods (Wehrtmann & Andrade, 1998; Wehrtmann & Lardies, 1999). Since RO is calculated as the ratio between dried mass of embryos and dried mass of females without their embryos and larger females tend to carry a greater number of embryos (Bauer, 1991; Corey & Reid, 1991; Mantelatto & Fransozo, 1997; Thiel & Duffy, 2007), the RO value tends to remain constant. However, while not significant in the present study, there is a tendency for RO to decrease as female size increases (Terossi et al., 2010). Thus, larger females could be investing similar or less energy in reproduction than small ones, in relation to their size. These findings might indicate senescence of reproductive output in larger (older) females (Bauer, 1991). As observed, the quantity of produced embryos may be increasing with the increase of female body size, but the energetic content of these embryos may be lower.

This research provides initial insights on the reproduction of a previously understudied, but very common caridean species. It addresses fecundity, embryo volume, embryo loss, and RO of *C. americanus* from Bocas del Toro, while also making comparisons to other populations at the intraspecific and interspecific level. Although the studied population inhabits an area with relatively low seasonality throughout the year (Kaufmann & Thompson, 2005), a better comprehension of population structure and reproductive dynamics could likely be achieved by the analysis of periodically obtained samples over a longer period, taking into account the environmental instability of this coastal habitat due high annual rainfall and local terrestrial runoff that impacts the Bocas del Toro archipelago. It is nonetheless a starting pointing, providing a basis for studies to follow.

ACKNOWLEDGEMENTS

We are deeply grateful to Smithsonian Tropical Research Institute (STRI) in Bocas Del Toro, Panama for a Grant support enabling travel to Panama and facilities during the development of the course on Crustacean Biology and Taxonomy from July 31 to August 15, 2011 and that enabled us to collect *C. americanus* specimens. Special thanks are due to Rachel Collin (STRI), who secured funding for this course and assisted with logistics and collecting permits, and to Jenny Felder for her enthusiastic help during both field and lab activities. All authors thank students, assistants and STRI staff involved in the course hosted at STRI's Bocas del Toro Research Station. Additional support was provided to FLM by the Fundação de Amparo à Pesquisa do estado de São Paulo — FAPESP (Biota 2010/50188-8; Coleções Científicas 2009/54931-0) and the Conselho Nacional de Desenvolvimento Científico e Tecnológico — CNPq (Proc. 301359/2007-5; 302748/2010-6; 304968/2014-5). MN was supported by CNPq 123990/2010-16 and FAPESP 2012/06300-3 and is under continued support by a Post-Doctoral scholarship from CNPq (PROTAX 152377/2016-6). TM and NR were supported by the Coordenação de Aperfeiçoamento de Pessoal de Nível Superior — CAPES. NR was also supported by FAPESP 2009/12409-5. We thank the suggestions of the anonymous reviewers that improved the quality of the manuscript. For DLF this is contribution no. 187 of the University of Louisiana-Lafayette Laboratory for Crustacean Research. The authors are honoured to recognize the many achievements in the field of biology and taxonomy of decapod crustaceans by our esteemed colleague, Michael Türkay, and therefore would like to dedicate this contribution on reproductive aspects to him.

REFERENCES

ABELE, L. G. & W. KIM, 1986. An illustrated guide to the marine decapod crustaceans of Florida. State of Florida Department of Environmental Regulation Technical Series, **8**: 1-760. (The Florida State University, Tallahassee, FL).

ALLEN, J. A., 1966. The dynamics and interrelationships of mixed populations of Caridea found off the north-east coast of England. In: H. B. BARNES (ed.), Some contemporary studies in marine science: 45-66. (Allen & Unwin, London).

AMSLER, M. O. & R. Y. GEORGE, 1984. Seasonal variation in the biochemical composition of the embryos of *Callinectes sapidus* Rathbun. Journal of Crustacean Biology, **4**: 546-553.

BAUER, R. T., 1985. Diel and seasonal variation in species composition and abundance of caridean shrimps (Crustacea, Decapoda) from seagrass meadows on the north coast of Puerto Rico. Bulletin of Marine Science, **36**: 150-162.

BAUER, R. T., 1989. Continuous reproduction and episodic recruitment in nine shrimp species inhabiting a tropical seagrass meadow. Journal of Experimental Marine Biology and Ecology, **127**: 175-187.

BAUER, R. T., 1991. Analysis of embryo production in a caridean shrimp guild from a tropical seagrass meadow. In: A. WENNER & A. KURIS (eds.), Crustacean egg production. Crustacean Issues, **7**: 181-191. (A. A. Balkema, Rotterdam).

BAUER, R. T., 1996. A test of hypotheses on male mating systems and female molting in decapods shrimp, using *Sicyonia dorsalis* (Decapoda: Penaeoidea). Journal of Crustacean Biology, **16**: 429-436.

BAUER, R. T. & J. H. ABDALLA, 2001. Male mating tactics in the shrimp *Palaemonetes pugio* (Decapoda, Caridea): precopulatory mate guarding vs. pure searching. Ethology, **17**: 185-199.

BRUCE, A. J., 2004. A partial revision of the genus *Periclimenes* Costa, 1884 (Crustacea: Decapoda: Decapoda: Palaemonidae). Zootaxa, **582**: 1-26.

BRUCE, A. J., 2012. *Cuapetes uncinatus* sp. nov. (Crustacea: Decapoda: Pontoniinae) from Futuna Island, eastern Pacific Ocean. Cahiers de Biologie Marine, **53**: 151-154.

CAMPOS-VÁZQUEZ, C., 2000. Crustáceos asociados a macroalgas en Bajo Pepito, Isla Mujeres, Caribe mexicano. Revista de Biología Tropical, **48**: 361-364.

CARRUTHERS, T. J. B., P. A. G. BARNES, G. E. JACOME & J. W. FOURQUREAN, 2005. Lagoon scale processes in a coastally influenced Caribbean system: implications for the seagrass *Thalassia testudinum*. Caribbean Journal of Science, **41**: 441-445.

CHACE JR., F. A., 1972. The shrimps of the Smithsonian-Bredin Caribbean expeditions with a summary of the West Indian shallow-water species (Crustacea: Decapoda: Natantia). Smithsonian Contributions to Zoology, **98**: 1-179.

CLARKE, A., 1987. Temperature, latitude and reproductive effort. Marine Ecology Progress Series, **38**: 89-99.

CLARKE, C., C. C. E. HOPKINS & E. M. NILSSEN, 1991. Egg size and reproductive output in the deep-water prawn *Pandalus borealis* Kroyer, 1838. Functional Ecology, **5**: 724-730.

COELHO, P. A. & M. A. RAMOS, 1972. A constituição e a distribuição da fauna de decápodos do litoral leste da América do sul entre as latitudes de 5° N e 39° S. Trabalhos do Instituto Oceanográfico da Universidade Federal de Pernambuco, **13**: 133-236.

COLLIN, R., 2005. Ecological monitoring and biodiversity surveys at the Smithsonian Tropical Research Institute's Bocas Del Toro Research Station. Caribbean Journal of Science, **41**: 367-373.

COREY, S. & D. M. REID, 1991. Comparative fecundity of decapod crustaceans: I. The fecundity of thirty-three species of nine families of caridean shrimp. Crustaceana, **60**: 270-294.

CORREA, C. & M. THIEL, 2003. Mating systems in caridean shrimp (Decapoda: Caridea) and their evolutionary consequences for sexual dimorphism and reproductive biology. Revista Chilena de Historia Natural, **76**: 187-203.

D'CROZ, L., J. B. DEL ROSARIO & P. GÓNDOLA, 2005. The effect of fresh water runoff on the distribution of dissolved inorganic nutrients and plankton in the Bocas del Toro Archipelago, Caribbean Panama. Caribbean Journal of Science, **41**: 414-429.

DÍAZ, M. C., 2005. Common sponges from shallow marine habitats from Bocas del Toro region, Panama. Caribbean Journal of Science, **41**: 465-475.

ENGLISH, S., C. WILKINSON & V. BAKER, 1997. Survey manual for tropical marine resources (2nd ed.): 1-390. (Australia Institute of Marine Sciences, Townsville, QLD).

ESCOBAR-BRIONES, E. & J. L. VILLALOBOS-HIRIART, 2003. Deep-water caridean shrimps (Crustacea: Decapoda) from Banco Chinchorro and adjacent areas in the northern Caribbean. Bulletin of Marine Science, **73**: 99-122.

FELDER, D. L. & A. H. CHANEY, 1979. Decapod crustacean fauna of Seven and One-Half Fathom Reef, Texas: species composition, abundance and species diversity. Contributions in Marine Science, **22**: 1-29.

GURNEY, R., 1943. Notes on *Periclimenes americanus* and the growth of the cheliped in this and other palaemonid prawns. Annals and Magazine of Natural History, **10**: 495-504.

GUZMAN, H. M., P. A. G. BARNESI, C. E. LOVELOCK & I. C. FELLER, 2005. A site description of the CARICOMP Mangrove, Seagrass and Coral Reef Sites in Bocas del Toro, Panama. Caribbean Journal of Science, **41**: 430-440.

GUZMÁN, H. M. & C. A. GUEVARA, 1998. Arrecifes coralinos de Bocas del Toro, Panamá. I. Distribución, estructura y estado de conservación de los arrecifes continentales de la Laguna de Chiriquí y la Bahía Almirante. Revista de Biología Tropical, **46**: 601-622.

GUZMÁN, H. M. & C. A. GUEVARA, 2002. Annual reproductive cycle, spatial distribution, abundance, and size structure of *Oreaster reticulatus* (Echinodermata: Asteridea) in Bocas del Toro, Panama. Marine Biology, **141**: 1077-1084.

HARTNOLL, R. G., 2006. Reproductive investment in Brachyura. Hydrobiologia, **184**: 31-40.

HECK, K. L., K. W. ABLE, C. T. ROMAN & M. P. FAHAY, 1995. Composition, abundance, biomass, and production of macrofauna in a New England estuary: comparisons among eelgrass meadows and other nursery habitats. Estuaries, **18**: 379-389.

HERNÁEZ-BOVÉ, P., 2001. Producción y rendimiento reproductivo en *Petrolisthes granulosus* (Decapoda, Anomura, Porcellanidae) en diferentes localidades del norte de Chile: una comparación latitudinal. Investigaciones Marinas, **29**: 73-81.

HINES, A. H., 1988. Fecundity and reproductive output in two species of deep sea crabs, *Geryon fenneri* and *G. quinquedens* (Decapoda: Brachyura). Journal of Crustacean Biology, **8**: 557-562.

HOLTHUIS, L. B., 1951. A general revision of the Palaemonidae (Crustacea Decapoda Natantia) of the Americas. I. The subfamilies Euryrhynchinae and Pontoniinae. Allan Hancock Foundation Publication, Occasional Papers, **11**: 1-332.

HOWARD, R. K., 1984. The trophic ecology of caridean shrimps in an eelgrass community. Aquatic Botany, **18**: 155-174.

KAUFMANN, K. W. & R. C. THOMPSON, 2005. Water temperature variation and the meteorological and hydrographic environment of Bocas del Toro, Panama. Caribbean Journal of Science, **41**: 392-413.

KINGSLEY, J. S., 1878. List of the North American Crustacea belonging to the sub-order Caridea. Bulletin of the Essex Institute, **10**: 53-71.

KINGSLEY, J. S., 1899. Synopses of North-American invertebrates. III. The Caridea of North America. The American Naturalist, **33**(393): 709-720.

KURIS, A. M., 1991. A review of patterns and causes of crustacean brood mortality. In: A. WENNER & A. KURIS (eds.), Crustacean egg production: 117-141. (A. A. Balkema, Rotterdam).

LARDIES, M. A. & J. C. CASTILLA, 2001. Latitudinal variation in the reproductive biology of the commensal crab *Pinnaxodes chilensis* (Decapoda: Pinnotheridae) along the Chilean coast. Marine Biology, **139**: 1125-1133.

LARDIES, M. A. & I. S. WEHRTMANN, 1996. Aspects of the reproductive biology of *Petrolisthes laevigatus* (Guérin, 1835) (Decapoda, Anomura, Porcellanidae). Part I: reproductive output and chemical composition of eggs during embryonic development. Archive of Fishery Marine Research, **43**: 121-135.

LARDIES, M. A. & I. S. WEHRTMANN, 2001. Latitudinal variation in the reproductive biology of *Betaeus truncatus* (Decapoda: Alpheidae) along the Chilean coast. Ophelia, **55**: 55-67.

LEDOYER, M., 1986. Faune mobile des herbiers de phanérogames marines (*Halodule* et *Thalassia*) de la Laguna de Términos (Mexique, Campéche). I. Les Caridea (Crustacea Decapoda) et aperçu sur la faune globale. Anales del Instituto de Ciencias del Mar y Limnología, Universidad Nalional Autónoma de México, **13**: 147-170.

LITULO, C., A. MACIA & F. L. MANTELATTO, 2005. Fecundity and sexual maturity of the crab *Macrophthalmus depressus* (Brachyura: Ocypodidae) from Inhaca Island, Mozambique. African Journal of Aquatic Science, **30**: 179-183.

LOVELOCK, C. E., I. C. FELLER, K. L. MCKEE & R. THOMPSON, 2005. Variation in mangrove forest structure and sediment characteristics in Bocas del Toro, Panama. Caribbean Journal of Science, **41**: 456-464.

MANTELATTO, F. L., L. C. FERNANDES-GÓES, M. Z. FANTUCCI, R. BIAGI, L. M. PARDO & J. M. DE GÓES, 2010. A comparative study of population traits between two South American populations of the striped-legged hermit crab *Clibanarius vittatus*. Acta Oecologica, **36**: 10-15.

MANTELATTO, F. L. & A. FRANSOZO, 1997. Fecundity of the crab *Callinectes ornatus* Ordway, 1863 (Decapoda, Brachyura, Portunidae) from the Ubatuba region, São Paulo, Brazil. Crustaceana, **70**: 214-226.

MARKHAM, J. C., F. E. DONATH-HERNÁNDEZ, J. L. VILLALOBOS-HIRIART & A. C. DÍAZ-BARRIGA, 1990. Notes on the shallow-water marine Crustacea of the Caribbean Coast of Quintana Roo, Mexico. Anales del Instituto de Biología, Universidad Nacional Autónoma de México, Serie Zoología, **61**: 405-446.

MARTÍNEZ-MAYÉN, M. & R. ROMÁN-CONTRERAS, 2009. Reproduction of *Periclimenes siankaanensis* (Decapoda, Caridea, Palaemonidae) in Bahía De La Ascensión, Quintana Roo, Mexico. Crustaceana, **82**: 27-37.

MARTÍNEZ-MAYÉN, M. & R. ROMÁN-CONTRERAS, 2014. Sexual maturity, fecundity, and embryo loss in the pontoniine shrimp *Cuapetes americanus* (Kingsley, 1878) (Decapoda: Palaemonidae) in Bahía de la Ascensión, Quintana Roo, Mexico. Journal of Crustacean Biology, **34**: 342-348.

MASON JR., W. T. & S. A. ZENGEL, 1996. Foods of juvenile spotted seatrout in seagrasses at Seahorse Key, Florida. Gulf of Mexico Science, **14**: 89-104.

MIRANDA, I., A. L. MEIRELES, R. BIAGI & F. L. MANTELATTO, 2006. Is the abundance of the red brocade hermit crab *Dardanus insignis* (Decapoda: Anomura: Diogenidae) in the infralitoral region of southern Brazil determined by reproductive potential? Crustacean Research, **6**: 45-55.

NIZINSKI, M. S., 2003. Annotated checklist of decapod crustaceans of Atlantic coastal and continental shelf waters of the United States. Proceedings of the Biological Society of Washington, **116**: 96-157.

OH, C. W. & R. G. HARTNOLL, 1999. Brood loss during incubation in *Philocheras trispinosus* (Decapoda) in Port Erin Bay, Isle of Man. Journal of Crustacean Biology, **19**: 467-476.

OKUNO, J., 2009. *Cuapetes* Clark, 1919, a senior synonym of *Kemponia* Bruce, 2004 (Crustacea: Decapoda: Palaemonidae). Zootaxa, **2028**: 67-68.

OMORI, K., Y. YANAGISAWA & N. HORI, 1994. Life history of the caridean shrimp *Periclimenes ornatus* Bruce associated with a sea anemone in southwest Japan. Journal of Crustacean Biology, **14**: 132-145.

PANDIAN, T. J., 1970. Ecophysiological studies on the developing eggs and embryos of the European lobster *Homarus gammarus*. Marine Biology, **5**: 154-167.

PENHA-LOPES, G., P. TORRES, A. MACIA & J. PAULA, 2007. Population structure, fecundity and embryo loss of the sea grass shrimp *Latreutes pymoeus* (Decapoda: Hippolytidae) at Inhaca Island, Mozambique. Journal of Marine Biological Association of the United Kingdom, **87**: 879-884.

RATHBUN, M. J., 1902. Japanese stalk-eyed crustaceans. Proceedings of the United States National Museum, **26**(1307): 23-55.

RODRÍGUEZ, E., R. ALMANZA & R. ALVARADO, 1993. Situación biofísica y ambiental de la Provincia de Bocas del Toro. In: S. HECKADON-MORENO (ed.), Agenda Ecológica y Social para Bocas del Toro: 55-72. (Impresora Continental, Panama City).

ROMÁN-CONTRERAS, R. & M. MARTÍNEZ-MAYÉN, 2010. Palaemonidae (Crustacea: Decapoda: Caridea) from the shallow waters from Quintana Roo, Mexican Caribbean coast. Revista Mexicana de Biodiversidad, **81**: 43-45.

SANTOS, P. S., G. O. SOLEDADE & A. O. ALMEIDA, 2012. Decapod crustaceans on dead coral from reef areas on the coast of Bahia, Brazil. Nauplius, **20**: 145-169.

SASTRY, A. N., 1983. Ecological aspects of reproduction. In: F. J. VERNBERG & W. B. VERNBERG (eds.), The biology of Crustacea, **8**, Environmental adaptations: 179-270. (Academic Press, New York, NY).

SHERIDAN, P., 1997. Benthos of adjacent mangrove, seagrass and non-vegetated habitats in Rookery Bay, Florida, USA. Estuarine, Coastal and Shelf Science, **44**: 455-469.

STEELE, D. H. & V. J. STEELE, 1975. Egg size and duration of embryonic development i[n] Crustacea. Internationale Revue der gesamten Hydrobiologie und Hydrographie, **60**: 711-715

STURGES, H. A., 1926. The choice of a class interval. Journal of the American Statistic[al] Association, **21**(153): 65-66.

TEROSSI, M., I. S. WEHRTMANN & F. L. MANTELATTO, 2010. Interpopulation comparison reproduction of the Atlantic shrimp *Hippolyte obliquimanus* (Caridea: Hippolytidae). Jour[nal] of Crustacean Biology, **30**: 571-579.

THIEL, M. & J. E. DUFFY, 2007. The behavioral ecology of crustaceans: a primer taxono[my] morphology and biology. In: J. E. DUFFY & M. THIEL (eds.), Evolutionary ecology of so[cial] and sexual systems: crustaceans as model organisms: 3-28. (Oxford University Press, N[ew] York, NY).

THORSON, G., 1950. Reproductive and larval ecology of marine bottom invertebrates. Biolo[gical] Reviews of the Cambridge Philosophical Society, **25**: 1-45.

TORATI, L. S. & F. L. MANTELATTO, 2008. Uncommon mechanism of egg incubation i[n] endemic Southern hermit crab *Loxopagurus loxochelis*: how is this phenomenon relat[ed] egg production? Acta Zoologica, **89**: 79-85.

TURNER, R. L. & J. M. LAWRENCE, 1979. Volume and composition of echinoderm implications for use of eggs size in life-history models. In: S. E. STANCKY (ed.), Reprod[uctive] ecology of marine invertebrates: 25-40. (University of South Carolina Press, Columbia, S

WEAR, R. G., 1974. Incubation in British decapod Crustacea, and the effects of temperature rate and success of embryonic development. Journal of the Marine Biological Associat[ion of] the United Kingdom, **54**: 745-762.

WEHRTMANN, I. S., 1990. Distribution and reproduction of *Ambidexter panamense* and *monetes schmitti* in Pacific Costa Rica (Crustacea, Decapoda). Revista de Biología Tr[opical] **38**(2A): 327-329.

WEHRTMANN, I. S. & W. G. ANDRADE, 1998. Egg production in *Heterocarpus reec[i]* northern Chile, with a comparison between iced and living females. Ophelia, **49**: 71-82

WEHRTMANN, I. S. & M. A. LARDIES, 1999. Egg production of *Austropandalus grayi* (De[capoda:] Caridea, Pandalidae) from the Magellan Region, South America. Scientia Marina, **63**: 3[…]

WEHRTMANN, I. S. & R. VARGAS, 2003. New records and range extensions of shrimps (De[capoda:] Penaeoidea, Caridea) from the Pacific and Caribbean coasts of Costa Rica, Central A[…] Revista de Biología Tropical, **51**: 268-274.

YU, S. C., 1930. Note sur les crevettes chinoises appartenant au genre *Leander* Desm. avec description de nouvelle espèces. Bulletin de la Société Zoologique de France, **55**: 553-573.

ZAR, J. H., 1996. Biostatistical analysis (3rd ed.): 1-662. (Prentice Hall International, Upper Saddle River, NJ).

ZIMMERMANN, U., F. L. CARVALHO & F. L. MANTELATTO, 2015. The reproductive performance of the Red-Algae shrimp *Leander paulensis* (Ortmann, 1897) (Decapoda, Palaemonidae) and the effect of post-spawning female weight gain on weight-dependent parameters. Brazilian Journal of Oceanography, **63**: 207-216.

First received 2 June 2016.
Final version accepted 23 April 2017.

[When citing this volume, please refer to Crustaceana 90 (2017) 771-1288]

Michael Türkay Memorial Issue

ARCOTHERES BORRADAILEI (NOBILI, 1906) AND *PINNOTHERES RIDGEWAYI* SOUTHWELL, 1911: A REASSESSMENT OF CHARACTERS AND GENERIC ASSIGNMENT OF SPECIES TO *ARCOTHERES* MANNING, 1993 (DECAPODA, BRACHYURA, PINNOTHERIDAE)

BY

PETER K. L. NG[1]), PAUL F. CLARK[2]), SANTANU MITRA[3]) and APPUKUTTANNAIR BIJU KUMAR[4,5])

[1]) Lee Kong Chian Natural History Museum, Faculty of Science, National University of Singapore, 2 Conservatory Drive, Singapore 117377, Republic of Singapore

[2]) Department of Zoology, The Natural History Museum, Cromwell Road, London SW7 5BD, England

[3]) Crustacea Division, Zoological Survey of India, Fire Proof Spirit Building, 27 J. L. Nehru Road, Kolkata [Internationally known as: Calcutta] – 700016, West Bengal, India

[4]) Department of Aquatic Biology & Fisheries, University of Kerala, Kariavattom, Thiruvananthapuram – 695581, Kerala, India

ABSTRACT

The identity of the pea crab, *Arcotheres borradailei* (Nobili, 1906) is clarified. The holotype female is re-diagnosed, figured and its recent generic assignment to *Arcotheres* Manning, 1993, is discussed. A second poorly known species from India and Sri Lanka that resembles *A. borradailei*, *Pinnotheres ridgewayi* Southwell, 1911, is redescribed and assigned to *Arcotheres*. In addition, *Pinnotheres sinensis atrinae* Sakai, 1939, and *Pinnotheres cyclinus* Shen, 1932, from East Asia are re-examined and also referred to *Arcotheres*.

Key words. — *Arcotheres borradailei*, *Pinnotheres ridgewayi*, *Pinnotheres sinensis atrinae*, *Pinnotheres cyclinus*, identification, assignment to *Arcotheres*

ZUSAMMENFASSUNG

Hier wird die Identität des Muschelwächters *Arcotheres borradailei* (Nobili, 1906) geklärt. Der weibliche Holotyp wird erneut untersucht, gezeichnet und dessen jüngste Gattungszuordnung zu *Arcotheres* Manning, 1993 wird diskutiert. Eine zweite, wenig bekannte Art aus Indien und Sri Lanka, *Pinnotheres ridgewayi* Southwell, 1911, welche *A. borradailei* ähnelt, wird neu beschrieben und *Arcotheres* zugeordnet. Zusätzlich werden *Pinnotheres sinensis atrinae* Sakai, 1939 und *Pinnotheres cyclinus* Shen, 1932 aus Ostasien erneut untersucht und ebenfalls der Gattung *Arcotheres* zugeordnet.

[5]) Corresponding author; e-mail: bijupuzhayoram@gmail.com

INTRODUCTION

The pinnotherid *Pinnotheres borradailei* (Nobili, 1906) is associated with *Mya* clams (Myidae) as well as with the fan mussel *Pinna* (Pinnidae) and this pea crab has been reported from several localities in the western Indian Ocean. Originally established as a replacement name for *Pinnotheres tenuipes* Borradaile, 1903 (pre-occupied by *Pinnotheres tenuipes* Bürger, 1895), it was recently assigned to *Arcotheres* Manning, 1993, by Ng & Kumar (2015). The type of this species was compared with three pinnotherid specimens recently obtained from *Pinna* in southwestern India; and while the material superficially resembles *A. tenuipes*, it differs markedly in structures of the carapace and ambulatory legs. These specimens are here identified as a poorly known species, *Pinnotheres ridgewayi* (Southwell, 1911). The original description of this species from Sri Lanka is poor and the figures are inaccurate. On the basis of the present study, *Pinnotheres ridgewayi* is also referred to *Arcotheres*.

The purpose of the present paper is to re-diagnose *Arcotheres borradailei* and *A. ridgewayi*, compare them with allied taxa including *Pinnotheres sinensis atrinae* Sakai, 1939, and *Pinnotheres cyclinus* Shen, 1932, from East Asia, and revise the number of species assigned to *Arcotheres*.

MATERIAL AND METHODS

Abbreviations: CW × CL = carapace width and length measurements, respectively, in millimeters; coll. = collected by; MXP3 = third maxilliped; P2-P5 = second to fifth pereopods (first to fourth ambulatory legs), respectively. Specimens examined are deposited in the Department of Aquatic Biology & Fisheries, University of Kerala (DABFUK), India; Western Ghat Regional Centre, Zoological Survey of India (ZSI-K), Calicut, Kerala; Zoological Survey of India (ZSI), Kolkata [= Calcutta], West Bengal; and the Cambridge University Museum of Zoology (CUMZ), Cambridge, United Kingdom.

SYSTEMATICS

Superfamily PINNOTHEROIDEA De Haan, 1833
Family PINNOTHERIDAE De Haan, 1833
Arcotheres Manning, 1993

Remarks.— The nomenclature of *Arcotheres* is confused because the genus was established in the personal notes of Ernst Nauck and never published. Bürger (1895) cited Nauck's notes and discussed the name *Arcotheres* but did not consider it as a valid genus. This was a decision with which Rathbun (1918) agreed and considered the taxon (*Arcotheres* Nauck in Bürger, 1895) as a junior synonym of *Pinnotheres* Latreille [sic]. This genus was later reviewed by Manning (1993) who considered *Arcotheres* as valid, attributed the authority to Bürger and designated *Pinnotheres palaensis* Bürger, 1895, as the type species. Campos & Manning (2001: 167) subsequently noted that the author of *Arcotheres* should be Manning (1993) because he (Manning, 1993), unlike Nauck in Bürger (1895), had provided the first valid diagnosis of the genus.

Manning (1993) had originally diagnosed *Arcotheres* as possessing a spatulate MXP3 dactylus which is inserted proximally on the ventral margin of the propodus; and the dactyli of P4 and P5 being longer and different in structure compared to those of P2 and P3. Campos & Manning (2001) refined the diagnosis by stating that the MXP3 dactylus was digitiform and may extend to or slightly overreach the tip of the propodus; and that the P4 is also asymmetrical in length, with either left or right leg longer. *Arcotheres* currently contains 23 species, however, and considering the diversity of diagnostic characters (Ahyong & Ng, 2007; Ng et al., 2008; Ng & Kumar, 2015), this taxon may require further revision to establish monophyly of the genus.

Ng & Kumar (2015: 265) transferred *P. borradailei* to *Arcotheres*, briefly commenting that the structures of the MXP3 and P2-P5 were diagnostic of the genus. The MXP3 dactylus of *A. borradailei* is digitiform, and while the original figure has it almost reaching the tip of the propodus (Borradaile, 1903, fig. 113C; fig. 3B), it is actually slightly shorter (fig. 3C, D). Like in other *Arcotheres* species, the P5 dactyli of *A. borradailei* are much longer and more slender than those of P2-P4 (figs. 1A, 2, 3A). The P4 dactylus, however, is only slightly longer than those on P2 and P3 (figs. 1A, 2, 3A); and there is no distinct asymmetry between the left and right P4 (figs. 1A, 2, 3A). For the new species described below, the same general features are observed.

One generic character appears to be the structure of the P5 with basis-ischium of *A. borradailei* and the new species being elongated, cylindrical and much longer (figs. 2, 6D, H) than the short, subquadrate structure observed in other pinnotherids. This character has also been figured in most other *Arcotheres* species (see Manning, 1993; Campos, 2001; Ahyong & Ng, 2007).

Arcotheres borradailei (Nobili, 1906) (figs. 1-3, 8A)

Pinnotheres rouxi — ?Paul'son, 1875: 70, pl. 9 fig. 2-2c (not *Pinnotheres rouxi* H. Milne Edwards, 1853).

Pinnotheres tenuipes Borradaile, 1903: 431, fig. 113 (name pre-occupied by *Pinnotheres tenuipes* Bürger, 1895; now assigned to *Buergeres* Ng & Manning, 2003).

Pinnoteres borradailei Nobili, 1906a: 410 (replacement name for *Pinnotheres tenuipes* Borradaile, 1903); Nobili, 1906b: 306 (part?); Guinot, 1967: 279 (list).

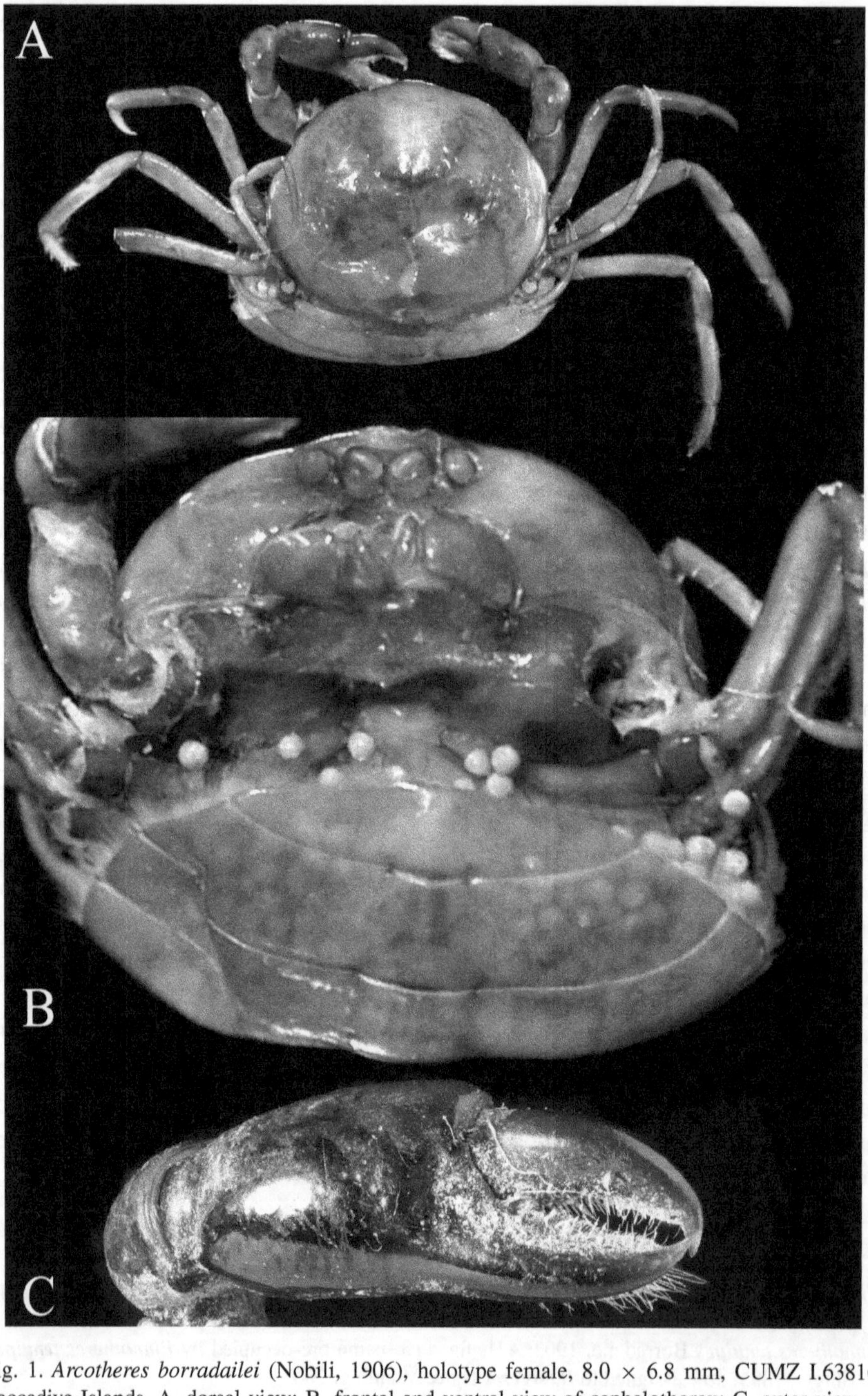

Fig. 1. *Arcotheres borradailei* (Nobili, 1906), holotype female, 8.0 × 6.8 mm, CUMZ I.63816, Laccadive Islands. A, dorsal view; B, frontal and ventral view of cephalothorax; C, outer view of right chela.

Fig. 2. *Arcotheres borradailei* (Nobili, 1906), holotype female, 8.0 × 6.8 mm, CUMZ I.63816, Laccadive Islands. A, left P2-P5; B, right P2-P5.

Pinnotheres borradailei — Laurie, 1915: 415 (list); Tesch, 1918: 248, 253, 287 (list and key); Silas & Alagarswami, 1967: 1197 (list); Schmitt et al., 1973: 40 (list); Ng et al., 2008: 250 (list).
Arcotheres borradailei — Ng & Kumar, 2015: 265.

Material examined.— Holotype: ovigerous female, 8.0 × 6.8 mm, CUMZ I.63816, Minikoi (present day = Minicoy), Lakshadweep Islands (previously known as the Laccadive, Minicoi and Aminidivi Islands), India, previously part of Maldive and Laccadive Islands, supposedly from *Mya*, coll. J. S. Gardiner.

Diagnosis.— Female: Carapace subcircular, slightly wider than long, width to length ratio 1.18; dorsal surface, lateral regions mostly smooth, glabrous; front broadly triangular, slightly projecting anteriorly beyond orbits, margin convex (figs. 1A, 3A, 8A). Eyes small, not visible in dorsal view (figs. 1A, 3A). MXP3 outer surface with scattered setae; propodus about 3 times as long as high, stout, longer than carpus; dactylus short, digitiform, inserted submedially on ventral margin of propodus, tip not reaching propodal apex; ischiomerus completely fused, relatively short, slightly longer than wide, mesial margin subangular at widest point; exopod slender, about half length of ischiomerus (figs. 1B, 3B-D). Chela short, dactylus and pollex shorter than palm; palm relatively stout, outer surfaces of palm, fingers mostly glabrous; fingers gently curved with sharp tip; dactylus

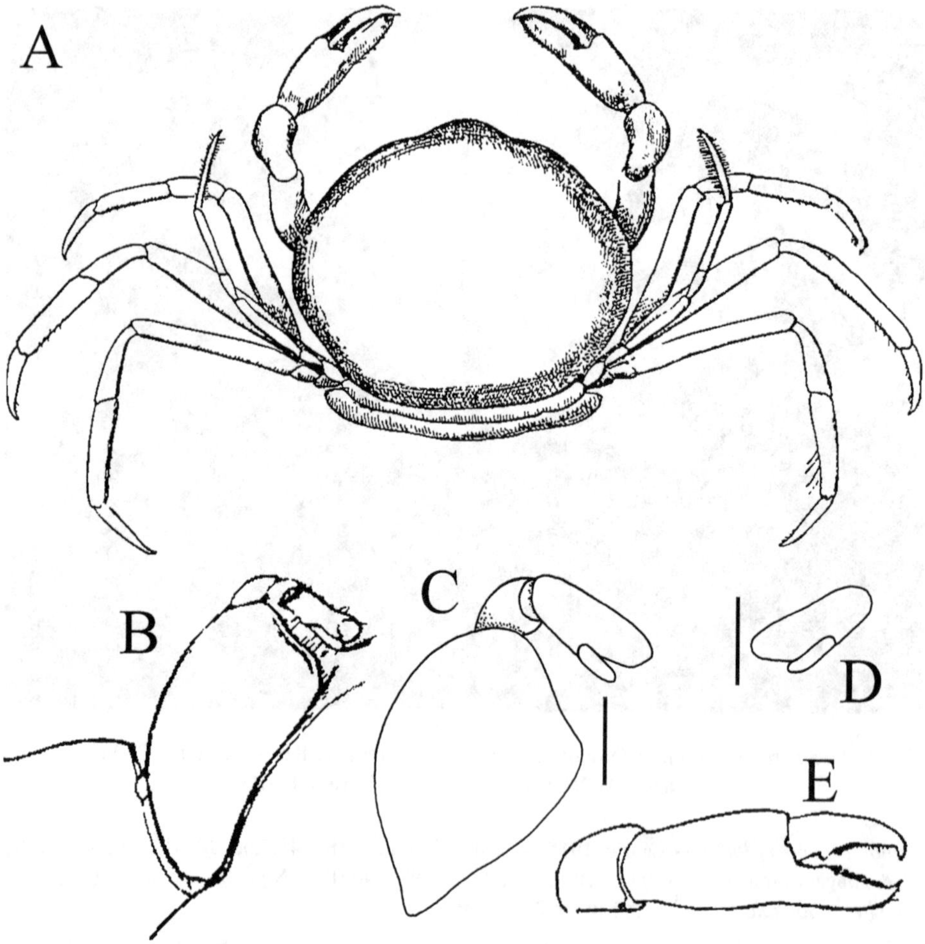

Fig. 3. *Arcotheres borradailei* (Nobili, 1906), holotype female, 8.0 × 6.8 mm, CUMZ I.63816, Laccadive Islands. A, dorsal view; B, C, right MXP3 (setae not drawn); D, propodus and dactylus of left MXP3 (setae not drawn); E, outer view of right chela. A, B, E after Borradaile (1903: fig. 113). Scales = 0.25 mm.

occlusal margin with large tooth on proximal third, covered by dense short setae; outer surface of proximal part of pollex with distinct long setae; pollex with ventral margin setose, occlusal margin with 1 larger and 1 smaller tooth on proximal third, lined with dense setae which partially covers margin (figs. 1A, C, 3A, E). Left and right ambulatory meri and dactyli almost symmetrical; P2-P4 ischiomerus subquadrate, that of P5 elongate, cylindrical; outer surface of ischium, merus and carpus glabrous; propodus mostly glabrous except for distal half or third of ventral margin, lined with scattered setae, without natatory setae; P2 dactylus slender, with scattered setae; P3 and P4 dactyli relatively wider with dense setae; P5 dactylus long, falciform with dense setae; relative lengths of meri P3 = P4 > P2 > P5; P2 dactylus 2/3 length of P3 dactylus, P3 dactylus slightly shorter than P4 dactyli, P5 dactylus longest (figs. 1A, 2, 3A). Pleon wide, extending to buccal region, covering bases of ambulatory legs; telson broadly semicircular, partially recessed into distal margin of somite 6 (fig. 1B).

Type locality.— Minikoi, Laccadives (Borradaile, 1903: 432), present day Minicoy, Lakshadweep Islands, India.

Distribution.— Red Sea (Paul'son, 1875); Périm Island, Red Sea; Djibouti, Somalia (Nobili, 1906b); Minicoy, Lakshadweep Islands, India (Borradaile, 1903).

Remarks.— The figure of the whole animal and MXP3 in Borradaile (1903, fig. 113) is schematic. The frontal margin of the specimen actually does not protrude as prominently anteriorly (fig. 1A vs fig. 3A). Furthermore, the MXP3 was drawn in situ, and a position, with the angles and proportions not accurately depicted (fig. 3B vs fig. 3C, D).

Paul'son (1875: 70) recorded "*Pinnotheres rouxi*" from the Red Sea without reference to a host. He commented that the P2-P5 dactylus vary, with one specimen having all the dactyli pubescent while the other has this character only for the P5. Although Nobili (1906b) referred the *P. rouxi* of Paul'son (1875) to *P. borradailei*, this decision is difficult to confirm and they may well represent different taxa. While his simple figure (Paul'son, 1875, pl. 9, fig. 2) at least shows a crab with a rounded carapace like *A. borradailei* s. str., the proportions of the ambulatory dactyli and MXP3 exopod appear to be different.

Arcotheres ridgewayi (Southwell, 1911) (figs. 4-7, 8B)

Pinnotheres borradailei – ?Lenz, 1910: 558, 576 (not *Pinnoteres borradailei* Nobili, 1906).
Pinnoteres ridgewayi Southwell, 1911: 226, pl. 3, figs. 1, 2, 2a.
Pinnotheres ridgewayi – Tesch, 1918: 250, 251 (list); Gravely, 1927: 146, pl. 23, fig. 37; Prasad & Tampi, 1957: 22, fig. 1 (larvae); Sankarankutty, 1966: 349, 350, 358, 360, figs. 23, 28; Silas & Alagarswami, 1967: 1177, 1183, 1208, 1219, 1223, 1225, 1227 (discussion); Ng et al., 2008: 251 (list).
Nepinnotheres ridgewayi – Ng & Kumar, 2015: 265 (discussion).

Material examined.— Lectotype (here designated): female (14.9 × 12.8 mm) (ZSI/WGRC/IR-INV 8102/10), from *Pinna bullata* Gmelin, 1791, pearl banks, Kondatchi Paar, Ceylon (= Sri Lanka), coll. T. Southwell, 1910. Paralectotypes (here designated): 2 females (14.4 × 11.9 mm, one with carapace in poor condition), same data as lectotype. Non-types: 1 non-ovigerous female (17.7 × 13.6 mm) (ZSI-K/WGRC/IR-INV 8235), coll. A. B. Kumar & R. Ravinesh, 2014; 1 ovigerous female (14.7 × 11.2 mm) (DABFUK), coll. R. Ravinesh, early 2016; 1 ovigerous female (16.6 × 12.3 mm) (DABFUK); all collected from *Pinna* (*Subitopinna*) *atropurpurea* Sowerby, 1825 (Pinnidae), 5-10 m depth, Kovalam, 8.3°N 77.2°E, Thiruvananthapuram, Kerala, southwestern India.

Diagnosis.— Carapace transversely subovate, distinctly wider than long, width to length ratio 1.30-1.35; dorsal surface, lateral regions mostly smooth, glabrous; front broadly triangular, distinctly projecting anteriorly beyond orbits, margin convex (figs. 5A, 8B). Eyes small, not visible in dorsal view (figs. 4B, 5A, B).

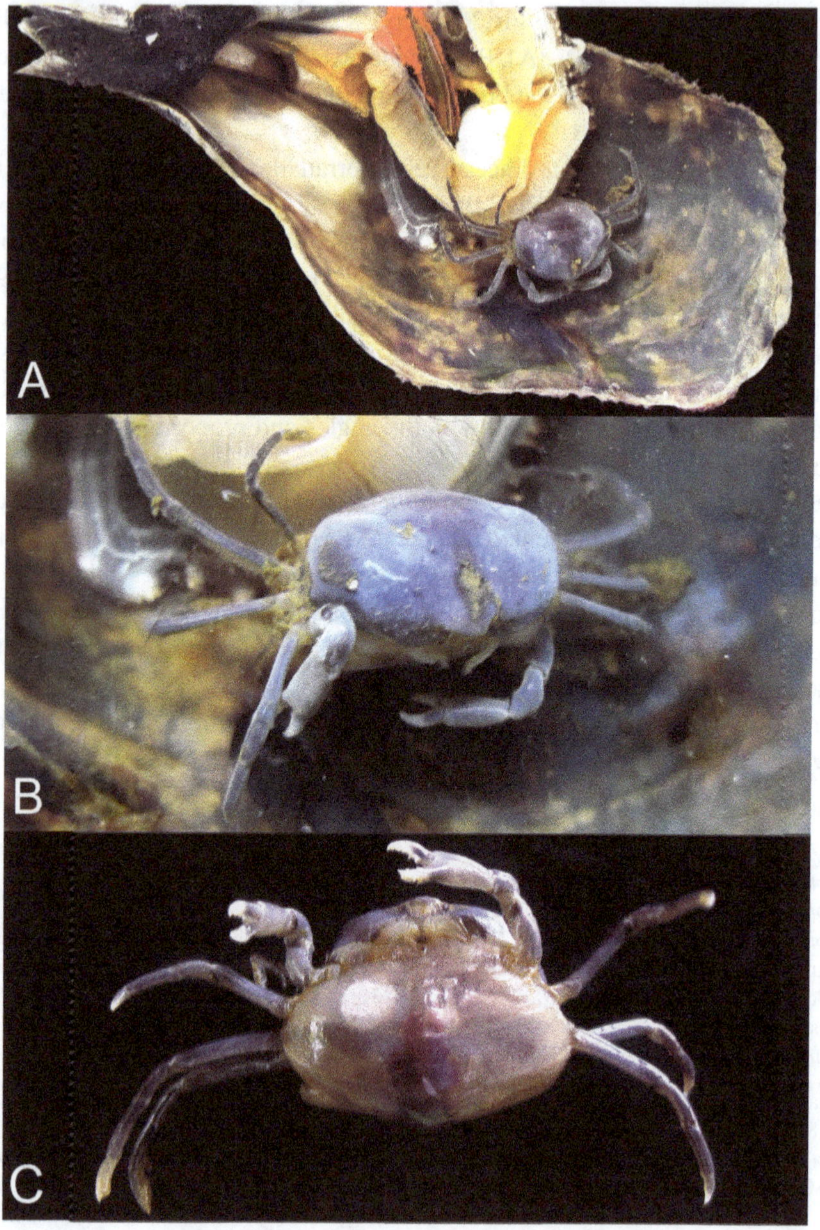

Fig. 4. *Arcotheres ridewayi* (Southwell, 1911), non-ovigerous female, 17.7 × 13.6 mm, ZSI-K/WGRC/IR-INV 8235, Kerala, India, colour in life. A, B, in situ in *Pinna*; C, ventral view.

MXP3 outer surface with scattered setae; propodus about 3 times as long as high, stout, longer than carpus; dactylus short, digitiform, inserted submedially on ventral margin of propodus, tip not reaching propodal apex; ischiomerus completely fused, relatively long, distinctly longer than wide, mesial margin

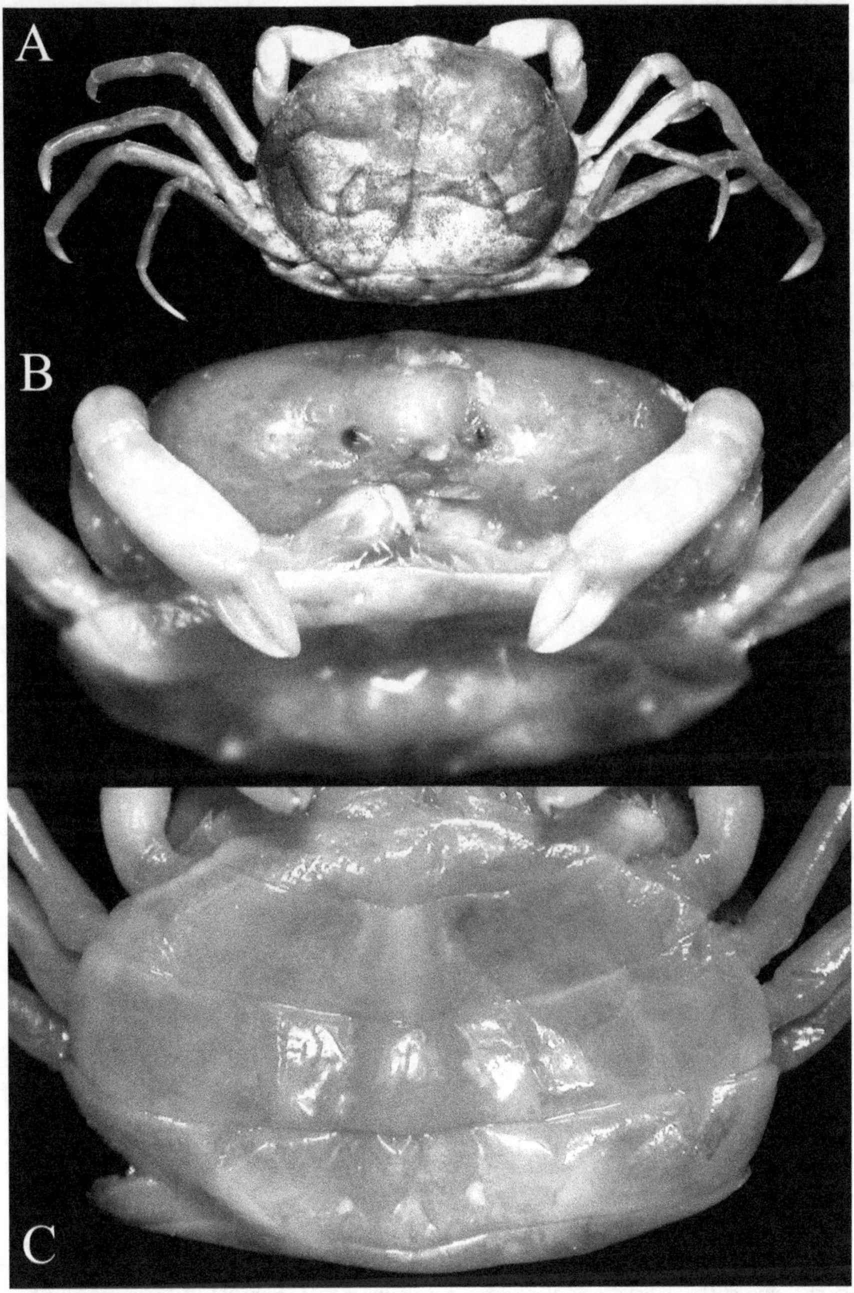

Fig. 5. *Arcotheres ridgewayi* (Southwell, 1911), non-ovigerous female, 17.7 × 13.6 mm, ZSI-K/WGRC/IR-INV 8235, Kerala, India. A, dorsal view; B, frontal view of cephalothorax; C, ventral view of cephalothorax showing pleon.

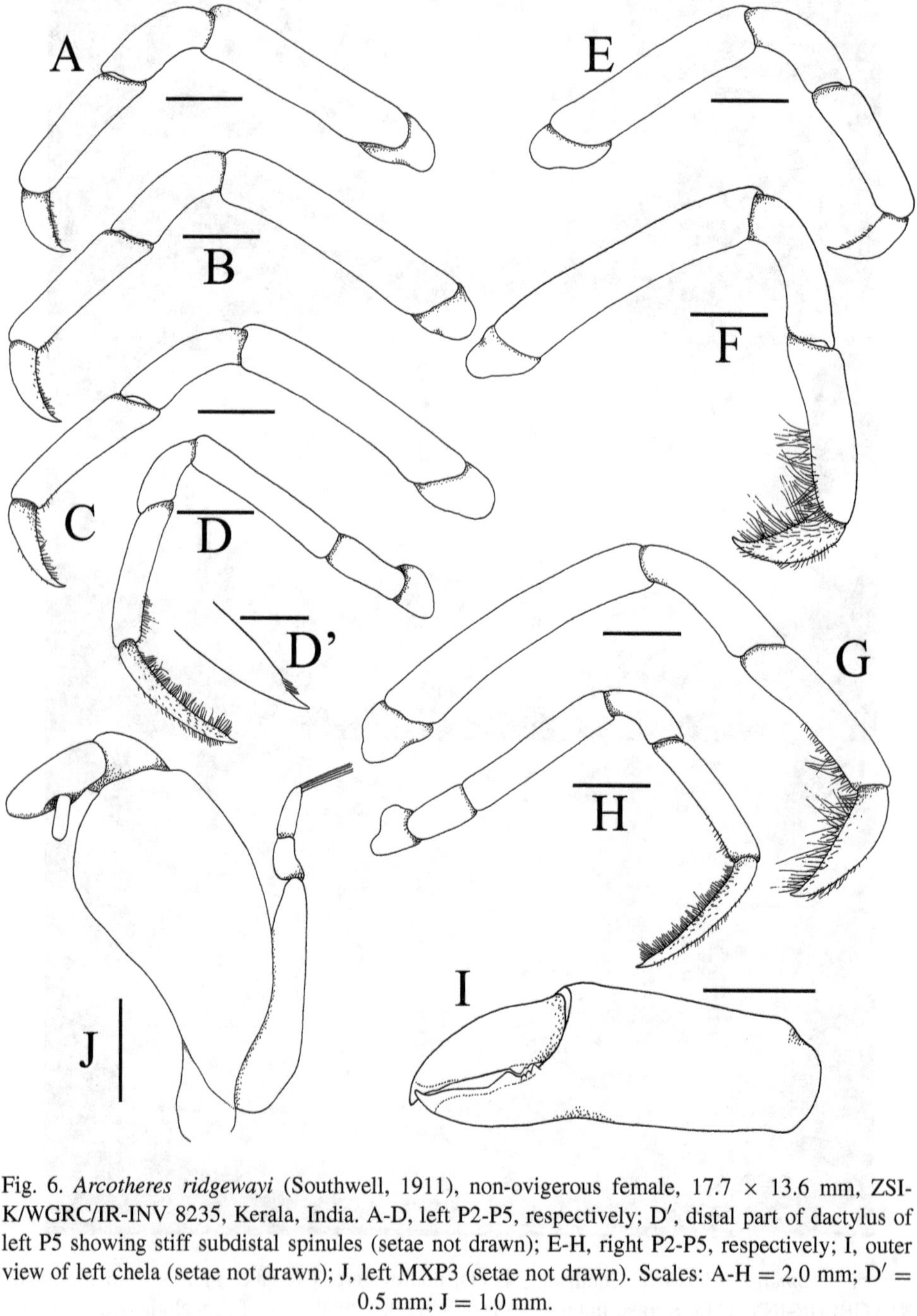

Fig. 6. *Arcotheres ridgewayi* (Southwell, 1911), non-ovigerous female, 17.7 × 13.6 mm, ZSI-K/WGRC/IR-INV 8235, Kerala, India. A-D, left P2-P5, respectively; D′, distal part of dactylus of left P5 showing stiff subdistal spinules (setae not drawn); E-H, right P2-P5, respectively; I, outer view of left chela (setae not drawn); J, left MXP3 (setae not drawn). Scales: A-H = 2.0 mm; D′ = 0.5 mm; J = 1.0 mm.

rounded at widest point; exopod slender, about half length of ischiomerus (fig. 6J). Chela short, dactylus and pollex shorter than palm; palm relatively stout, outer surfaces of palm, fingers mostly glabrous; outer surface of proximal part of pollex with scattered short setae; fingers gently curved with sharp tip; dactylus occlusal margin with large tooth on proximal third, partially covered by setae; pollex with ventral margin setose, occlusal margin with 1 large tooth on proximal third and 3 denticles posterior to, lined with setae which partially covers margin (figs. 4B, C, 5B, 6I). Left and right P2 and P3 meri similar in length; P2 and P2 dactyli relatively shorter, stouter, right P3 dactylus slightly broader than left P3 dactylus; P2-P4 ischiomerus subquadrate, that of P5 elongate, cylindrical; outer surface of ischiomerus, merus and carpus glabrous; P2 propodus glabrous; P3-P5 propodus mostly glabrous except for distal half or third of ventral margin, lined with scattered setae (longest on right P3), without natatory setae; P2 dactylus relatively shorter, stouter, with scattered setae; P3 and P4 dactyli relatively wider with dense setae, that on P3 dense with more long setae; P5 dactylus long, falciform with dense setae; right P4 merus and dactylus longer than left P4 merus and dactylus; left and right P5 symmetrical, P5 dactylus longest (fig. 6A-H). Pleon extremely wide, extending to buccal region, covering bases of ambulatory legs; telson broadly semicircular, partially recessed into distal margin of somite 6 (figs. 4C, 5C).

Variation.— The carapace of the lectotype female is intact but one of its P4 is damaged so the condition of its dactylus and propodus is not discernible. The two paralectotypes are in poor condition but their P2-P5 are more complete. They agree with the non-type specimens in all major characters. The carapace shapes of the specimens vary somewhat. The frontal margin of the smaller non-type female (14.7 × 11.2 mm, DABFUK) is relatively more projected anteriorly with the anterolateral margin less rounded (fig. 7A) and agrees very well with that of the lectotype female (cf. Southwell, 1911, fig. 2). In the larger non-type female (16.6 × 12.3 mm, DABFUK) (fig. 7B), the carapace appears to be proportionately wider, with the frontal and anterolateral margins more similar to that of the largest non-type female (17.7 × 13.6 mm, ZSI-K/WGRC/IR-INV 8235, width to length ratio 1.35; fig. 5A). The carapaces of the two more intact type specimens are somewhat shrivelled because of long preservation and the present measurements are almost certainly not reflective of their original state. As such, it is not surprising that their carapace width to length ratios (1.16 and 1.21) are smaller than those of the recent specimens (1.30-1.35). The chelae of all specimens are the same, without prominent setae on the outer surface. The proportions of the P2-P5 of the type specimens and the two smaller non-type females are similar to the largest female except that their P5 dactylus is slightly shorter and less slender (fig. 7A, B). This is also the case for the MXP3 of the smaller type and non-type females; their structures are similar to that of the largest female in proportions (fig. 6J) except

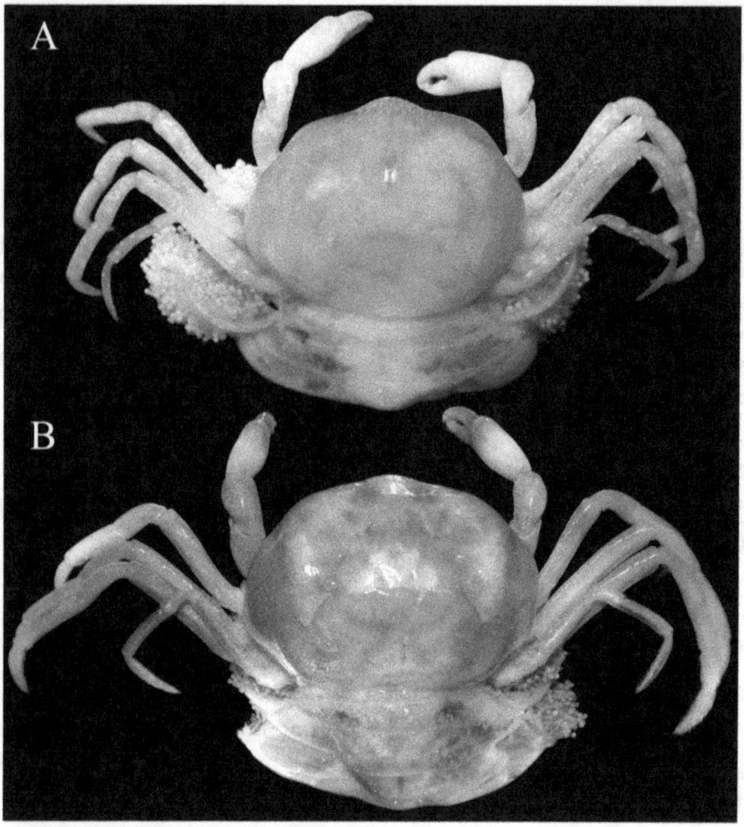

Fig. 7. *Arcotheres ridgewayi* (Southwell, 1911), dorsal views. A, ovigerous female, 14.7 × 11.2 mm, DABFUK, Kerala, India; B, ovigerous female, 16.6 × 12.3 mm, DABFUK, Kerala, India.

that the anteromesial angle of the ischiomerus is slightly more angular and less rounded; but still quite different from the condition in *A. borradailei* (fig. 3B, C).

Colour.— Dorsal surface of carapace, chelipeds and ambulatory legs purplish-blue except for white tips of fingers and dactyli; ventral surfaces dark purple (fig. 4).

Distribution.— Pemba Island, Tanzania (Lenz, 1910); Ceylon Pearl Banks, Sri Lanka (Southwell, 1911); Mandapam, Madras State, Palk Bay and Gulf of Manaar, Tamil Nadu, southeastern India (Gravely, 1927; Prasad & Tampi, 1957; Sankarankutty, 1966); and Thiruvananthapuram, Kerala, southwestern India (present data).

Remarks.— Nobili (1906b: 306) reported seven females of "*Pinnotheres borradailei*" from Red Sea *Pinna* and five females from an unspecified host mollusc from Djibouti. He commented that they agreed with the descriptions in Borradaile (1903) but the carapace seemed wider, although the measurements he gave of the two largest specimens (13.0 × 11.0 mm, 13.5 × 11.5 mm; no host specified) have width to length ratios of 1.18 and 1.17 respectively, not different from that of the holotype female (width to length ratio 1.18). As such, there is confusion over the precise measurements, or if the material comprises more than one species.

Lenz (1910: 558) also reported a female specimen obtained in *Pinna* from Pemba Island in Tanzania measuring 15.0 × 11.5 mm. The width to length proportion of this specimen is 1.30, distinctly broader than that of the holotype of *A. borradailei* and similar to that recorded here for *A. ridgewayi*; and as such the material of Lenz is provisionally referred to the latter taxon. Considering that Tanzania is over 4000 km southwest of India, it is possible that it belongs to another species, but this can only be ascertained once the material of Lenz is located and re-examined. In addition, Borradaile (1903) noted that his specimen of *A. borradailei* (as *Pinnotheres tenuipes*) probably came from a *Mya*; unlike the present material which was from *Pinna*.

Southwell (1911: 226, pl. 3 figs. 1, 2) described *Pinnotheres ridgewayi* on the basis of three ovigerous females (measured specimen 15.0 × 13.0 mm) obtained in *Pinna bullata* Gmelin, 1791, collected from Kondatchi Paar, Sri Lanka. Ng & Kumar (2015: 265) referred *Pinnotheres ridgewayi* to *Nepinnotheres* Manning, 1993, because the left and right P4 appeared to be symmetrical in length and structure (cf. Southwell, 1911: 226, pl. 3 figs. 1, 2) although the third maxilliped was not described or figured. The present material from Kerala agrees well with the description and figures in Southwell (1911) especially with respect to the structure of the carapace, chelipeds and female pleon. The carapace shape of the largest non-type female (17.7 × 13.6 mm, ZSI-K/WGRC/IR-INV 8235, fig. 5A) is relatively more transversely ovate but that of the smaller non-type female (14.7 × 11.2 mm, DABFUK, fig. 7A) matches that in fig. 2 of Southwell (1911, pl. 3). Furthermore, the ambulatory legs of the Kerala material generally agree with that figured in Southwell (1911, pl. 3 figs. 1, 2) in the relative proportions of the merus and especially in the relatively broad and highly setose dactylus. There are, however, two major discrepancies with the structures of the ambulatory legs. One is in the relative lengths of the merus and dactylus of P4. Southwell (1911, pl. 3 figs. 1, 2) figured the left and right P4 meri and dactyli as symmetrical; but in the Kerala specimens, one of the legs is always longer, with both the merus and dactylus prominently more elongated on one side than those on the other side (fig. 6C, G). Secondly, the P5 dactylus was figured as relatively short and stout (Southwell, 1911, pl. 3 figs. 1, 2) while in the Kerala material, it is distinctly more slender and elongate (fig. 6D, H).

The three type specimens of Southwell (1911) were re-examined for the present study. One specimen is in an extremely poor condition with the carapace badly damaged. The P5 dactyli of all three syntype specimens are elongate and slender, and the P4 meri and dactyli on one side are slightly longer than on the other, as in the Kerala specimens. The figures of the ambulatory legs illustrated by Southwell (1911) are therefore inaccurate. The condition of the third maxilliped, notably the form of the ischium as well as proportions and positions of the dactylus and propodus also agree. The Kerala material is here considered conspecific with *Pinnotheres ridgewayi*.

Southwell (1911) described the species without designating a holotype, and as such, all three of his specimens in ZSI are syntypes. The largest and most intact syntype specimen of *Pinnotheres ridgewayi* Southwell, 1911, a female measuring 14.9 × 12.8 mm (ZSI/WGRC/IR-INV 8102/10) is here selected as the lectotype of the species to stabilize the taxonomy of this and allied taxa.

Although the P4 asymmetry is not as marked as in some *Arcotheres* species, it is nevertheless present, and as such, *Pinnotheres ridgewayi* is here assigned to this genus. The inaccuracy of the figures by Southwell has precedence: in their paper describing a new pinnotherid, *Pinnotheres placunae*, from the Rann of Kutch in northwestern India, Hornell & Southwell (1909, figs. 5, 6) figured the female P4 as symmetrical. The first author has since obtained specimens from the type locality and nearby areas, and all the specimens of *Pinnotheres placunae* (now assigned to *Arcotheres*) have the P4 asymmetrical (unpubl. data). In addition, the P5 dactyli of actual *Arcotheres placunae* are proportionately longer than those figured by Hornell & Southwell (1909, figs. 5, 6). Interestingly, Gravely (1927, pl. 23 fig. 37) figured the P5 dactylus of *Pinnotheres ridgewayi* as being much longer and more slender than those of the other legs, although he still incorrectly depicted the left and right P4s as symmetrical.

Sankarankutty (1966, figs. 23, 28) figured the male first gonopod and male pleon of *A. ridgewayi* (as a *Pinnotheres*). Notable is that the male telson of *A. ridgewayi* is unusual, being trapezoidal in shape, with the distal margin much wider than the proximal one (Sankarankutty, 1966, fig. 28). Such a male telson is not typical for *Arcotheres* species and suggests that the specimen may need to be re-examined for further taxonomic clarification.

DISCUSSION

The two species treated in this paper, *A. borradailei* (Nobili, 1906) and *A. ridgewayi* (Southwell, 1911), are characterized by their smooth and strongly convex dorsal carapace surface, the frontal margin being broadly triangular with a gently convex tip, the anterolateral margin is evenly convex, the ambulatory legs are relatively long with the P5 dactylus elongate, slender, and distinctly longer than those on other legs.

The carapace shape of *A. ridgewayi* (figs. 7, 8B) is distinct in being prominently transversely ovate with a width to length ratio of 1.30-1.35 (vs more rounded

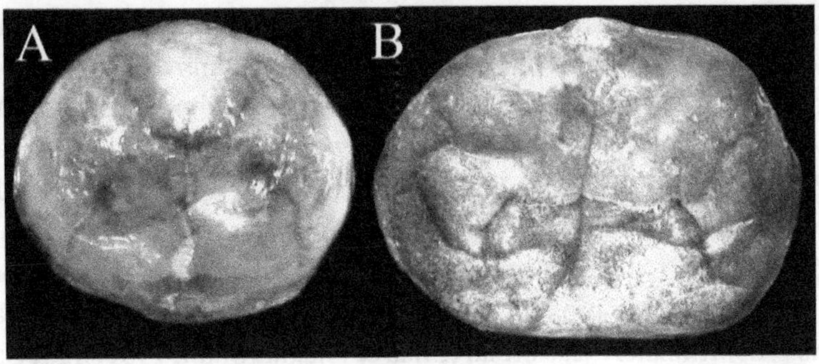

Fig. 8. Dorsal view of female carapace. A, *Arcotheres borradailei* (Nobili, 1906), holotype, 8.0 × 6.8 mm, CUMZ I.63816, Laccadive Islands; B, *A. ridgewayi* (Southwell, 1911), non-ovigerous female, 17.7 × 13.6 mm, ZSI-K/WGRC/IR-INV 8235, Kerala, India.

with a ratio of 1.18 for *A. borradailei*; fig. 8A). It must be noted, however, that the smallest non-type female of *A. ridgewayi* (14.7 × 11.2 mm, DABFUK) is still larger than that of the holotype of *A. borradailei* (8.0 × 6.8 mm, CUMZ I.63816). Considering their relatively poorly calcified carapace, the use of shape as a taxonomic character should be carefully considered and one may expect it to vary with size to some degree. Other than in the shape of the carapace, *A. borradailei* has distinct, dense setae on the outer surface of the cheliped dactylus, mainly along the proximal cutting edges (fig. 1C) (vs setae being short and sparse in *A. ridgewayi*; figs. 4B, 5B). In *A. ridgewayi*, the ischiomerus of MXP3 is proportionately longer, with the mesial margin of the ischiomerus of MXP3 more rounded at its widest point (fig. 6J) (vs ischiomerus relatively shorter with the mesial margin angular to subangular at its widest point in *A. borradailei*; figs. 1B, 3B, C). The P3 and P4 meri of *A. ridgewayi* (figs. 5A, 6B, C, F, G) are also proportionately more slender and longer (vs relatively shorter and stouter for *A. borradailei*; figs. 1A, 2, 3A).

Arcotheres borradailei and *A. ridgewayi*, which have broadly triangular frontal margins, can be separated from the following Indo-West Pacific *Arcotheres* species which have truncate to subtruncate frontal margins which may be gently sinuous: *A. arcophilus* (Bürger, 1895), *A. exiguus* (Bürger, 1895), *A. guinotae* Campos, 2001, *A. modiolicola* (Bürger, 1895), *A. rotundatus* (Bürger, 1895), *A. similis* (Bürger, 1895), *A. spinidactylus* (Gordon, 1936), *A. tivelae* (Gordon, 1936), and *A. winckworthi* (Gordon, 1936) (see Gordon, 1936; Ahyong & Ng, 2007). The carapace of *A. ridgewayi* closely resembles that of *A. latus* (Bürger, 1895) from the Philippines but the MXP3 of *A. latus* is proportionately shorter and the proportions of the ambulatory dactyli are different (cf. Bürger, 1895; Ahyong & Ng, 2007; PKLN, unpubl. data). From India, *A. ridgewayi* most closely resembles *A. casta* (Antony & Kutyamma, 1971) found in the clam *Meretrix casta* (Gmelin, 1791) (Veneridae) in the general carapace shape and relatively slender ambulatory legs.

Arcotheres casta can be distinguished, however, by its more truncate frontal margin (Silas & Alagarswami, 1967, fig. 1A; Antony & Kutyamma, 1971, text-fig. 1-1) (vs broadly triangular in *A. ridgewayi*; fig. 8B); the ischiomerus of the MXP3 is relatively shorter with the mesial part distinctly angular (Silas & Alagarswami, 1967, fig. 1B; Antony & Kutyamma, 1971, text-figs. 1-4) (vs more rounded in *A. ridgewayi*; fig. 6J); the left and right P4 are conspicuously asymmetrical in size and length (Silas & Alagarswami, 1967, fig. 2D; Antony & Kutyamma, 1971, text-figs. 2-3, 4) (vs less pronounced in *A. ridgewayi*; fig. 6C, G); the larger P3 dactylus is proportionately less broad (Silas & Alagarswami, 1967, fig. 2C; Antony & Kutyamma, 1971, text-figs. 2-3, 4) (vs broader in *A. ridgewayi*; fig. 6F); the P4 dactylus is relatively more slender (Silas & Alagarswami, 1967, fig. 2CD; Antony & Kutyamma, 1971, text-figs. 2-3, 4) (vs distinctly broad in *A. ridgewayi*; fig. 6G); and the P5 dactylus is proportionately shorter and stouter (Silas & Alagarswami, 1967, fig. 2E; Antony & Kutyamma, 1971, text-figs. 2-3, 4) (vs longer and more slender in *A. ridgewayi*; fig. 6D, H).

In contrast to the projected frontal margin and convex anterolateral margins of *A. borradailei* and *A. ridgewayi*, the frontal margin and anterior part of the carapace of *A. latifrons* (Bürger, 1895), *A. placunae* (Hornell & Southwell, 1909) and *A. rayi* Ahyong & Ng, 2007, are prominently flattened, with the margin almost straight (Ahyong & Ng, 2007, figs. 5A, 11A). *Arcotheres nudifrons* (Bürger, 1895) has a carapace shape which superficially resembles those of *A. borradailei* and *A. ridgewayi*, but in this species, the P2 and P3 dactyli are distinctly shorter, and the P5 dactylus is proportionately shorter and stouter (Ahyong & Ng, 2007, fig. 8A). Three other *Arcotheres* species each have characteristic carapaces: *A. palaensis* (Bürger, 1895) with a subhexagonal carapace (Ahyong & Ng, 2007, fig. 9A); *A. pernicola* (Bürger, 1895) with a rounded carapace with the frontal and anterolateral margins almost smoothly confluent (Ahyong & Ng, 2007, fig. 10F); and *A. rhombifer* (Bürger, 1895) which has the anterior half prominently less broad and appears somewhat protruding (Ahyong & Ng, 2007, fig. 12A). *Arcotheres coarctatus* (Bürger, 1895) has a distinctly rounded carapace with the frontal margin gently convex, and with the P4 dactylus conspicuously short (Ahyong & Ng, 2007, fig. 3A). *Arcotheres purpureus* is distinct from *A. borradailei* and *A. ridgewayi* in having a subquadrate carapace in which the front is almost confluent with the anterolateral margins and the ambulatory legs are long and slender, and with the P4 dactylus slender and longer than the P5 dactylus (Alcock & McArdle, 1903, pl. 62 fig. 6).

Ahyong & Ng (2007) discussed the taxonomy of *A. alcocki* (Rathbun, 1909) and suggested that this species was close to, if not synonymous with, *A. sinensis* (Shen, 1932). These two species are distinct, however, in their transversely ovate carapace in which the frontal margin is low and almost confluent with the rest of the anterolateral margin (see Ahyong & Ng, 2007, fig. 1A; Shen, 1932, text-fig. 78b). They can be separated from the two species discussed in the present study by the carapace shape.

On the basis of the longer P4 and P5 dactyli and the structure of the MXP3, two East Asian taxa can also be assigned to *Arcotheres*: *Pinnotheres sinensis atrinae* Sakai, 1939 (see Sakai, 1939, pl. 69 fig. 4; Sakai, 1976, text-fig. 309b, b′) and *P. cyclinus* Shen, 1932 (see Shen, 1932, text-fig. 80). Ng et al. (2008: 250) listed both as taxa of *Pinnotheres*. Both are here regarded as distinct *Arcotheres* species, with *A. atrinae* Sakai, 1939, here treated as a full species. *Arcotheres atrinae* and *A. cyclinus* can be distinguished from the two Indian Ocean species treated in this paper by the same characters discussed for *A. alcocki* and *A. sinensis*.

With regard to the confirmed mollusc host for the recent specimens, in the general Indian literature, the host species of *A. ridgewayi* has been broadly known as *Pinna bicolor* Gmelin, 1791 (Pinnidae), following the study of Rosewater (1961), which synonymized many different forms. In the recent detailed revision by Schultz & Huber (2013), however, the various forms are recognized as separate species, and the Indian species should be known as *Pinna* (*Subitopinna*) *atropurpurea* Sowerby, 1825. All three recent specimens of *A. ridgewayi* were found in this *Pinna* species. In a specimen of the related pinnid *Atrina* (*Atrina*) *vexillum* (Born, 1778) collected with the present *Pinna* material, a pair of *Afropinnotheres ratnakara* Ng & Kumar, 2015, were found instead. *Afropinnotheres ratnakara* was previously known from the mussel *Perna perna* (Linnaeus, 1758) (Mytilidae) (Ng & Kumar, 2015).

ACKNOWLEDGEMENTS

The present paper is written in honour of our great friend and colleague, the late Michael Türkay of the Senckenberg Museum, Frankfurt am Main, Germany. Michael always endeavoured to strengthen research ties between countries. One of his last hopes expressed to the first author was that more Indian colleagues should be engaged and encouraged to work with western-trained scientists; a wish which this paper nicely fulfils. The authors thank Matthew Lowe (CUMZ) for allowing them to examine the material in his museum. Phil Hurst of the Photographic Unit in The Natural History Museum in London took photographs of *A. borradailei* (fig. 1C). The fourth author thanks R. Ravinesh for assistance in collection and UGC-SAP for funding support. We also thank Shane Ahyong and Ernesto Campos for their many constructive suggestions which have substantially improved the paper. The first author is grateful to Tan Siong Kiat for advice on the taxonomy of the Pinnidae, and help in identifying the host species.

REFERENCES

AHYONG, S. T. & P. K. L. NG, 2007. The pinnotherid type material of Semper (1880), Nauck (1880) and Bürger (1895) (Crustacea: Decapoda: Brachyura). Raffles Bulletin of Zoology, Supplement, **16**: 191-226.

ALCOCK, A. & A. F. MCARDLE, 1903. Illustrations of the Zoology of the Royal Indian Marine Survey Ship Investigator, under the command of Commander T.H. Heming, R.N. Crustacea. — Part X, Plates LVI-LXVII. Published under the Authority of Captain W. S. Goodridge, R.N., C.I.E., Director of the Royal Indian Marine. Calcutta: Office of the Superintendent of Government Printing, India, 1902.

ANTONY, A. & V. J. KUTYAMMA, 1971. A new species of the pea-crab *Pinnotheres* Latreille (Crustacea: Brachyura: Pinnotheridae) from the clam *Meretrix casta* (Chemnitz). Bulletin of the Department of Marine Biology and Oceanography, University of Cochin, **5**: 59-68.

BORRADAILE, L. A., 1903. Marine crustaceans. V. The crabs of the catometope families. In: J. S. GARDINER (ed.), The fauna and geography of the Maldive and Laccadive Archipelagoes: being the account of the work carried on and of the collections made by an expedition during the years 1899 and 1900, **1**: 429-433.

BÜRGER, O., 1895. Ein Beitrag zur Kenntnis der Pinnotherinen. Zoologische Jahrbucher, Abtheilung fur Systematik, Geographie und Biologie der Thiere, **8**(3): 361-390, pls. 9, 10.

CAMPOS, E., 2001. A new crab species of the genus *Arcotheres* Manning, 1993, from Thailand (Crustacea, Brachyura, Pinnotheridae). Zoosystema, **23**(3): 493-497.

CAMPOS, E. & R. B. MANNING, 2001. Authorship and diagnosis of the genus *Arcotheres* Manning, 1993 (Crustacea: Brachyura: Pinnotheridae). Raffles Bulletin of Zoology, **49**(1): 167-170.

GORDON, I., 1936. On a few Indo-Pacific species of *Pinnotheres*, with special reference to asymmetry of the walking legs. Journal of the Linnean Society of London, Zoology, **40**: 163-180.

GRAVELY, F. H., 1927. Decapoda and Stomatopoda. In: The littoral fauna of Krusadai Island in the Gulf of Manaar. Bulletin of the Madras Government Museum, New Series (Natural History), **1**: 141-155, figs. 1, 2, pls. 20-26.

GUINOT, D., 1967. La faune carcinologique (Crustacea Brachyura) de l'Océan Indien occidental et de la mer Rouge. Catalogue, remarques biogéographiques et bibliographie. In: Réunion de Spécialistes C. S. A. sur les Crustacés, Zanzibar 1964. Mémoires de l'Institute fondamental d'Afrique Noire, 77: 235-352.

LAURIE, R. D., 1915. On the Brachyura. Reports on the marine biology of the Sudanese Red Sea, from collections made by Cyril Crossland, M.A., B. Sc., F.Z.S., XXI. Journal of the Linnaean Society of London, Zoology, **31**(209): 407-475, pls. 42-45.

LENZ, H., 1910. Crustaceen von Madagaskar, Ostafrika und Ceylon. Reise in Ostafrika in den Jahren 1903-1905, **2**: 539-576.

MANNING, R. B., 1993. Three genera removed from the synonymy of *Pinnotheres* Bosc, 1802 (Brachyura: Pinnotheridae). Proceedings of the Biological Society of Washington, **106**: 523-531.

NG, P. K. L., D. GUINOT & P. J. F. DAVIE, 2008. Systema Brachyurorum: Part I. An annotated checklist of extant brachyuran crabs of the world. Raffles Bulletin of Zoology, Supplement **17**: 1-286.

NG, P. K. L. & A. B. KUMAR, 2015. A new species of *Afropinnotheres* Manning, 1993 (Crustacea, Brachyura, Pinnotheridae) from southwestern India, the first record of the genus from the Indian Ocean, with a review of the Pinnotheridae of India and adjacent seas. Zootaxa, **3947**(2): 264-274.

NOBILI, G., 1906a. Diagnoses préliminaires de 34 espèces et variétés nouvelles, et de 2 genres nouveaux de Décapodes de la Mer Rouge. Bulletin du Muséum national d'Histoire naturelle, Paris, **11**(6)**1905**(1906): 393-411.

NOBILI, G., 1906b. Faune carcinologique de la Mer Rouge. Décapodes et Stomatopodes. Annales de Sciences Naturelles, Paris, Zoologie, (9) **4**(1-3): 1-347, pls. 1-11.

PAUL'SON, O., 1875. Studies on Crustacea of the Red Sea with notes regarding other areas. Part 1. Podophthalmata and Edriophthalmata (Cumacea). S. V. Kul'zhenko, Kiev, 144 p., 21 pls., with 21 tables. (English translation by F. D. Por, pp. 1-164, The National Science Foundation and Smithsonian Institution by The Israel Program for Scientific Translation, Washington D.C.).

PRASAD, R. R. & P. R. S. TAMPI, 1957. Notes on some decapod larvae. Journal of the Zoological Society of India, **9**: 22-39.

RATHBUN, M. J., 1918. The grapsoid crabs of America. Bulletin of the United States National Museum, **97**: 1-461.

ROSEWATER, J. R., 1961. The family Pinnidae in the Indo-Pacific. Indo-Pacific Mollusca, **1**(4): 175-226.

SAKAI, T., 1939. Studies on the crabs of Japan. IV. Brachygnatha, Brachyrhyncha: 365-741 + 1-23 [= Index] pp., figs. 1-129, pls. XLII-CXI, Tables 1-6. (Yokendo, Tokyo).

SAKAI, T., 1976. Crabs of Japan and the Adjacent Seas. In three volumes; English Text, xxix + 773 pp., Japanese Text, 461 pp., Plates volume, pp. 1-16, pls. 1-251. (Kodansha, Tokyo).

SANKARANKUTTY, C., 1966. On Decapoda Brachyura from the Gulf of Mannar and Palk Bay. In: Proceedings of the Symposium on Crustacea held at Ernakulam from January 12 to 15, 1965. Part I. Marine Biological Association of India, India, Symposium Series, **2**: 347-362, figs. 1-33, pls. I, II, map 1.

SCHMITT, W. L., J. C. MCCAIN & E. S. DAVIDSON, 1973. Decapoda. I. Brachyura. I. Fam. Pinnotheridae. In: GRUNER, H.-E. & L. B. HOLTHUIS (eds.), Crustaceorum Catalogus, pars **3**: 1-160.

SCHULTZ, P. W. & M. HUBER, 2013. Revision of the worldwide recent Pinnidae and some remarks of fossil European Pinnidae. Acta Conchyliorum, **13**: 1-164.

SHEN, C.-J., 1932. The brachyuran Crustacea of north China. Zoologica Sinica, (A) **9**(1): 1-320, pls. 1-10, 1 map.

SILAS, E. G. & K. ALAGARSWAMI, 1967. On an instance of parasitisation by the pea-crab (*Pinnotheres* sp.) on the backwater clam [*Meretrix casta* (Chemnitz)] from India, with a review of the work on systematics, ecology, biology and etiology of pea crabs of the genus *Pinnotheres*, Latreille, 1802. Proceedings of the Symposium on Crustacea held at Ernakulam, 1965, Symposium Series 2. Marine Biological Association of India, **3**: 1161-1227.

SOUTHWELL, T., 1911. Description of a new species of *Pinnoteres* [sic] and of female of *Pinnoteres* [sic] *margaritiferae*, Laurie. [Article 19]. Reports from the Ceylon Marine Biological Laboratory, **1**(5): 226-227, pl. 3.

TESCH, J. J., 1918. Decapoda Brachyura II. Goneplacidae and Pinnotheridae. Siboga Expeditie Monographie, **39**c1: 149-295, pls. 7-18.

First received 6 July 2016.
Final version accepted 11 October 2016.

[When citing this volume, please refer to Crustaceana 90 (2017) 771-1288]

Michael Türkay Memorial Issue

A NEW SPONGE ASSOCIATED SHRIMP SPECIES OF THE INDO-WEST PACIFIC GENUS *PARACLIMENAEUS* (DECAPODA, CARIDEA, PALAEMONIDAE)

BY

MACHTELD ODIJK and CHARLES H. J. M. FRANSEN[1])
Department of Taxonomy and Systematics, Naturalis Biodiversity Center, P.O. Box 9517, 2300 RA Leiden, The Netherlands

ABSTRACT

A new sponge-associated species, *Paraclimenaeus michaeli* sp. nov. from Sulawesi, Indonesia, and Singapore, is described and illustrated. The new species can be distinguished from its congenerics by: (1) the downwards curved rostrum with 2 acute dorsal teeth subdistally, (2) the absence of subdistal teeth on the fixed finger of the major second chela, (3) the pairs of dorsal telson spines at about 0.33 and 0.66 of the telson length, (4) the caudal fan sparsely setose, and (5) the second pereiopods without subdistal teeth on fixed finger. To evaluate its relationship with congenerics and species of the related genera *Apopontonia* and *Climeniperaeus*, a phylogenetic analysis based on the 16S mitochondrial gene is presented. A key for the identification of the species in *Apopontonia*, *Climeniperaeus* and *Paraclimenaeus* is proposed.

Key words. — Palaemonidae, *Paraclimenaeus michaeli* sp. nov., sponge associate, identification key, phylogenetic analysis

ZUSAMMENFASSUNG

Hier wird eine neue Schwamm-assoziierte Art, *Periclimenaeus michaeli* sp. nov. aus Sulawesi, Indonesien und Singapur beschrieben und illustriert. Die neue Art kann von Verwandten gleicher Gattung wie folgt unterschieden werden: (1) nach unten gekrümmtes Rostrum mit zwei spitzen, subdistalen Zähnen, (2) Fehlen subdistaler Zähne am unbeweglichen Finger der großen, zweiten Chela, (3) Paare dorsaler Telsonstachel mit ungefähr 0,33 bis 0,66-facher Telsonlänge, (4) Kaudalfächer etwas rauh und (5) zweite Pereiopoden ohne subdistale Zähne am unbeweglichen Scherenfinger. Um das Verwandtschaftsverhältnis zu Arten innerhalb derselben Gattung und zu den Gattungen *Apopontonia* und *Climeniperaeus* zu bewerten, wird eine phylogenetische Analyse basierend auf der mitochondrialen 16S-Gensequenz gezeigt. Ferner wird ein Identifikationsschlüssel für die Arten der Gattungen *Apopontonia*, *Climeniperaeus* und *Paraclimenaeus* vorgeschlagen.

[1]) Corresponding author; e-mail: charles.fransen@naturalis.nl

INTRODUCTION

The genus *Paraclimenaeus* was erected by Bruce in 1988 to receive the species *Periclimenaeus fimbriatus* Borradaile, 1915. Together with *Apopontonia* Bruce, 1976 and *Climeniperaeus* Bruce, 1996, it forms a group of sponge associated palaemonid genera which have several morphological characters in common: rostrum well developed, dentate, lateral carinae expanded, hepatic spine absent, orbit well developed, maxillipeds with well-developed exopods, fourth thoracic sternite without median process, ambulatory dactyls compressed, biunguiculate, corpus ventrally dentate, uropodal exopod laterally multispinose (Bruce, 2009).

Bruce (2009) revised this group of genera and emended the definitions of each genus. The group at present comprises (De Grave & Fransen, 2011): *Apopontonia falcirostris* Bruce, 1976; *Climeniperaeus orbitospinatus* (Bruce, 1969); *C. truncatus* (Rathbun, 1906); *C. truncoideus* (Chace & Bruce, 1993); *Paraclimenaeus dentimanus* (Mitsuhashi & Chan, 2008); *P. fimbriatus* (Borradaile, 1915); *P. seticauda* (Bruce, 2008); and *P. spinicauda* (Bruce, 1969).

During fieldwork in SW and N Sulawesi, Arni P.H.M. Janssen and Nicole J. de Voogd collected endofauna of selected sponge species. From the interior canals of these sponges, various species belonging to the three above-mentioned genera were collected. In *Melophlus sarasinorum* Thiele, 1899, an unknown species of *Paraclimenaeus* was encountered. In 2013 a similar specimen was collected and photographed from Singapore (Anker & De Grave, 2016: 409, fig. 77, as *Climeniperaeus* sp.). Sammy De Grave compared our morphological description with the features of their specimen and concluded that the Singapore specimen is conspecific with the Sulawesi specimens. This new species is described and figured herein.

A molecular phylogeny is reconstructed to determine the relationships within the group. A key to the species in the three genera is provided.

MATERIAL AND METHODS

The present material was collected during fieldwork at SW Sulawesi in 2000 and preserved in ethanol (80%). Eggs or pleopods were selected for DNA analyses of partial 16S sequences (table I). The sponge associated symbiotic palaemonid *Orthopontonia ornata* (Bruce, 1970), the sponge associated *Periclimenaeus djiboutensis* Bruce, 1970, and the compound ascidian associated *Periclimenaeus storchi* Bruce, 1989, were selected as outgroups.

The material is deposited in Museum Zoologicum Bogoriense, Research Center for Biology, Indonesian Institute of Sciences, Cibenong, Indonesia (MZB-LIPI) and Naturalis Biodiversity Center, formerly Rijksmuseum van Natuurlijke Historie

TABLE I

Sequences obtained from Naturalis Biodiversity Center with species name, voucher registration number, sampling locality, host, and GenBank accession numbers

Species	Voucher	Locality	Host	16S GenBank accession No.
Paraclimenaeus michaeli spec. nov.	RMNH.CRUS.D.50212	SW Sulawesi	Porifera: *Melophlus sarasinorum*	KY511239
Paraclimenaeus michaeli spec. nov.	RMNH.CRUS.D.50216	SW Sulawesi	Porifera: *Melophlus sarasinorum*	KY511240
Paraclimenaeus michaeli spec. nov.	RMNH.CRUS.D.50218	SW Sulawesi	Porifera: *Melophlus sarasinorum*	KY511241
Paraclimenaeus spinicauda (Bruce, 1969)	RMNH.CRUS.D.49827	SW Sulawesi	Porifera: unid. sponge	KY511242
Paraclimenaeus spinicauda (Bruce, 1969)	RMNH.CRUS.D.49828	SW Sulawesi	Porifera: unid. sponge	KY511243
Paraclimenaeus spinicauda (Bruce, 1969)	RMNH.CRUS.D.50225	SW Sulawesi	Porifera: *Melophlus sarasinorum*	KY511244
Paraclimenaeus spinicauda (Bruce, 1969)	RMNH.CRUS.D.50229	SW Sulawesi	Porifera: *Melophlus sarasinorum*	KY511245
Paraclimenaeus spinicauda (Bruce, 1969)	RMNH.CRUS.D.50234	SW Sulawesi	Porifera: *Melophlus sarasinorum*	KY511246
Paraclimenaeus dentimanus (Mitsuhashi & Chan, 2008)	RMNH.CRUS.D.50246	SW Sulawesi	Porifera: *Melophlus sarasinorum*	KY511247
Paraclimenaeus dentimanus (Mitsuhashi & Chan, 2008)	RMNH.CRUS.D.50248	SW Sulawesi	Porifera: *Melophlus sarasinorum*	KY511248
Paraclimenaeus dentimanus (Mitsuhashi & Chan, 2008)	RMNH.CRUS.D.50255	SW Sulawesi	Porifera: *Melophlus sarasinorum*	KY511249
Climeniperaeus orbitospinatus (Bruce, 1969)	RMNH.CRUS.D.50188	SW Sulawesi	Porifera: *Ircinia* sp.	KY511250
Climeniperaeus orbitospinatus (Bruce, 1969)	RMNH.CRUS.D.50197	SW Sulawesi	Porifera: *Melophlus sarasinorum*	KY511251
Climeniperaeus orbitospinatus (Bruce, 1969)	RMNH.CRUS.D.50223	SW Sulawesi	Porifera: *Melophlus sarasinorum*	KY511252
Apopontonia falcirostris Bruce, 1976	RMNH.CRUS.D.50178	N Sulawesi	Porifera: *Leucosolenia* sp.	KY511253
Apopontonia falcirostris Bruce, 1976	RMNH.CRUS.D.50181	N Sulawesi	Porifera: *Mycale* aff. *cockburniana*	KY511254
Apopontonia falcirostris Bruce, 1976	RMNH.CRUS.D.50185	S Sulawesi	Porifera: *Mycale* aff. *cockburniana*	KY511255
Periclimenaeus djiboutensis Bruce, 1970	RMNH.CRUS.D.50125	SW Sulawesi	Porifera: *Hippospongia amata*	KY511256
Periclimenaeus storchi Bruce, 1989	RMNH.CRUS.D.50154	SW Sulawesi	Ascidiacea: *Didemnum molle*	KY511257
Orthopontonia ornata (Bruce, 1970)	RMNH.CRUS.D.50272	N Sulawesi	Porifera: *Melophlus sarasinorum*	KY511258

(RMNH), Leiden, The Netherlands. Post-orbital carapace length (pocl) is used as the standard measurement of size.

Drawings were made using a Zeiss Wild M5 and Olympus CH2 compound stereo microscope, both with drawing tube. Drawings were mounted using Adobe Photoshop CS6 software.

Total genomic DNA was extracted with the NucleoMag® 96 Tissue kit (Macherey-Nagel, Düren, Germany) using a Thermo Scientific KingFisher™ Flex Magnetic Particle Processor. Extraction was performed with 200 μl lysis buffer and 20 μl proteinase K. Incubation took place overnight at 56°C.

A 443 bp fragment of the 16S mitochondrial ribosomal gene was amplified with 16Sar and 16Sbr rDNA primers (Palumbi et al., 1991). 2.5 μl of 10 × PCR buffer (containing 15 mM $MgCl_2$, Qiagen, Hilden, Germany); 2.0 μl dNTPs (2.5 mM, Amersham, Amersham, U.K.); 1.0 μl of both forward and reverse primer; 1.0 μl $MgCl_2$ (25 mM, Qiagen); 2.0 μl glycerol (50%); 0.5 μl *Taq* polymerase (5 U/ml, Qiagen); 12.0 μl MilliQ; and 3 μl of DNA extract.

The PCR program used is as follows: 94°C for 1 min; 94°C for 5 s; 47°C for 1 min; 72°C for 1 min; last three steps repeated 39 times; then 72°C for 5 min and 4.0°C forever. PCRs and cycle sequence reactions were performed in a PTC-200 (Biozym, Landgraaf, The Netherlands).

Positive PCR products of about 500-600 kb were gel purified after cutting out the product under low intensity UV-light, using either Nucleospin® Extract II PCR clean-up Gel extraction (Macherey-Nagel) or PCR purification kit (250 μl; Qiagen), both according to standard protocol and diluted in the last step in 30 μl elution buffer.

Cycle sequence reactions were performed with BigDye Terminator 1.1 cycle sequencing kit (Applied Biosystems, Foster City, CA, U.S.A.). Cycle sequence reaction mixes were 2 μl Big Dye; 2 μl Enzyme dilution buffer; 2.6 μl MilliQ; 3 μl template. The cycle sequence program was as follows: 96°C for 10 s; 50°C for 5 s; 60°C for 4 min, these steps repeated 39 times, and then 4°C. Cycle sequence reaction products were purified using Sephadex G-50 Auto Seq Columns (Amersham) according to standard protocol. Samples were electrophoresed with an ABI PrismTM 377 DNA Sequencer overday using the 36E-2400 program, and results were analysed using ABI-377 collection software (Perkin-Elmer, Waltham, MA, U.S.A.).

Twenty 16S sequences were obtained (table I). From the chromatograms, forward and reverse sequences were contra-aligned and consensi were made using Sequencher 4.2 (Gene Codes, Ann Arbor, MI, U.S.A.). Consensi were checked for contamination using BLAST from Genbank (National Center for Biotechnology Information, NCBI, Bethesda, MD, U.S.A.). Alignments were made using MUSCLE in AliView (vers.1.18, Larsson, 2014) with default settings.

The best-fitting model for sequence evolution (TIM3 + G) was determined by jModelTest 2.3.1. (Darriba et al., 2012), selected by the AIC (Akaike Information Criterion), and subsequently applied to the maximum likelihood (ML) analyses with PAUP* 4.0b10 (Swofford, 2003). ML trees were constructed using a heuristic search with tree-bisection-reconnection (TBR) branch-swapping algorithm. For ML analyses, tree reliability was estimated with a bootstrap test of phylogeny comprising 1000 replications. All codon positions, as well as non-coding sites, were examined. Gaps and missing data were ignored. Bayesian inferences were estimated in MrBayes 3.1.2 (Ronquist & Huelsenbeck, 2003). The programme was run for 5×10^6 generations with a burnin set to 25% for each marker using the most complex GTR + I + G model.

SYSTEMATIC PART

Family PALAEMONIDAE Rafinesque, 1815

Paraclimenaeus michaeli sp. nov.

(figs. 1-6)

Climeniperaeus spec. — Anker & De Grave, 2016: 409, fig. 77.

Material.— Indonesia: MZB Cru 4603: ovigerous female holotype, pocl. 2.5 mm; SW Sulawesi, Spermonde Archipelago, Barang Lompo; 22.ix.2000; depth 7 m; in *Melophlus sarassinorum*; collected by A. Jansen Ms19/20900/BA/AJ/4 (ex RMNH.CRUS.D.50211). Paratypes — RMNH.CRUS.D.50208: 1 male, pocl. 1.2 mm; SW Sulawesi, Spermonde Archipelago, Samalona; 19.viii.2000; scuba diving; 11 m depth; in *Melophlus sarasinorum* (det. N. de Voogd); collected by A. Janssen Ms7/190800/SA/AJ/6. RMNH.CRUS.D.50209: 2 males, pocl. 1.2 and 1.6 mm; SW Sulawesi, Spermonde Archipelago, Kudingareng Keke; 29.viii.2000; scuba diving; 14 m depth; in *Melophlus sarasinorum* (det. N. de Voogd); collected by A. Janssen Ms11/290800/KK/AJ/5. RMNH.CRUS.D.50210: 1 male, pocl. 1.3 mm; SW Sulawesi, Spermonde Archipelago, Samalona; 05.ix.2000; scuba diving; 8 m depth; in *Melophlus sarasinorum* (det. N. de Voogd); collected by A. Janssen Ms12/050900/SA/AJ/6. RMNH.CRUS.D.50212: 1 ovigerous female, pocl. 2.0 mm; SW Sulawesi, Spermonde Archipelago, Kudingareng Keke; 26.ix.2000; scuba diving; 16 m depth; in *Melophlus sarasinorum* (det. N. de Voogd); collected by A. Janssen Ms26/26900/KK/AJ/6. RMNH.CRUS.D.50213: 1 ovigerous female, pocl. 1.6 mm; SW Sulawesi, Spermonde Archipelago, Samalona; 03.x.2000; scuba diving; 12 m depth; in *Melophlus sarasinorum* (det. N. de Voogd); collected by A. Janssen Ms28/03100/SA/AJ/4. RMNH.CRUS.D.50214: 1 male, pocl. 1.3 mm; SW Sulawesi, Spermonde Archipelago, Barung Lompo; 06.x.2000; scuba diving; 7 m depth; in *Melophlus sarasinorum* (det. N. de Voogd); collected by A. Janssen Ms32/06100/BA/AJ/4. RMNH.CRUS.D.50215: 5 females or juveniles, pocl. 1.0-1.2 mm; SW Sulawesi, Spermonde Archipelago, Samalona; 03.x.2000; scuba diving; 7 m depth; in *Melophlus sarasinorum* (det. N. de Voogd); collected by A. Janssen Ms30/03100/SA/AJ/5. RMNH.CRUS.D.50216: 1 ovigerous female, pocl. 1.9 mm; SW Sulawesi, Spermonde Archipelago, Samalona; 03.x.2000; scuba diving; 7 m depth; in *Melophlus sarasinorum* (det. N. de Voogd); collected by A. Janssen Ms30/03100/SA/AJ/4. RMNH.CRUS.D.50217: 1 female or juvenile, pocl. 1.4 mm; SW Sulawesi, Spermonde Archipelago, Barung Lompo; 06.x.2000; scuba diving; 16 m depth; in *Melophlus sarasinorum* (det. N. de Voogd); collected by A. Janssen Ms31/06100/BA/AJ/6. RMNH.CRUS.D.50218: 1 ovigerous female, pocl. 2.3 mm; SW Sulawesi, Spermonde Archipelago, Barung Lompo; 06.x.2000; scuba diving; 7 m

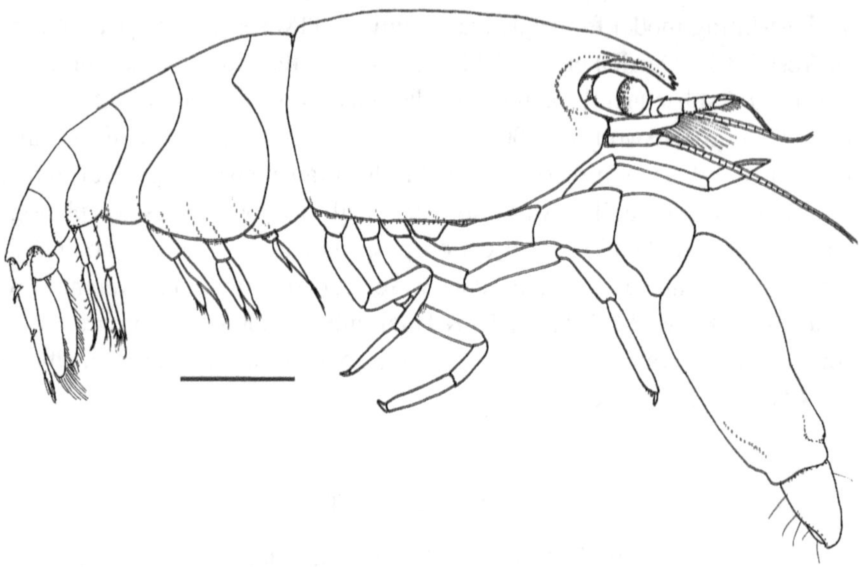

Fig. 1. *Paraclimenaeus michaeli* sp. nov., female holotype, MZB Cru 4603. Scale = 1 mm.

depth; in *Melophlus sarasinorum* (det. N. de Voogd); collected by A. Janssen Ms32/06100/BA/AJ/5. *Singapore* — ZRC 2014.0926: 1 female; sta. SB146, Strait of Singapore, W Pulau Hantu; 5-7 m depth; coral rubble brushing; leg. S. De Grave et al., 01 June 2013 (SIN-294).

Description.— Body (fig. 1) small, subcylindrical. Rostrum (fig. 2A, B) well developed, almost reaching distal margin of antennular peduncle, strongly curved downwards; tip acute with two (seldom one) acute dorsal teeth subdistally; broad lateral rostral carinae tapering distally to about 3/4 of rostrum; inferior orbital angle poorly developed; large acute antennal spine, slightly curved downwards, not exceeding antennal basicerite. Carapace smooth; no epigastric spine; orbit poorly developed dorsally, lower margin extending in antennal spine; branchiostegite anteriorly produced and round; posterior margin of carapace truncate.

Abdomen (fig. 1) smooth, pleura of first two segments rounded, and of third to fifth segments angular posteroventrally; posterolateral angle of sixth segment well developed and acutely produced, posteroventral angle triangularly produced; sixth segment about 1.25 times longer than fifth. Telson (fig. 2C) about 1.8 times length of sixth abdominal segment, about 2.2 times longer than its proximal width, about twice as broad as distal width; lateral margins straight; acute median process on posterior margin (fig. 2D); two pairs of acute dorsal spines, at about 0.2 and 0.6 of telson length from anterior margin; dorsal spines are about 1/6 of telson length, with most proximal pair between centre and lateral margins, and distal pair more submarginal; three pairs of acute posterior spines, of which lateral pair smallest, about 1/12 of telson length, intermediate pair about 0.2 times telson length, and setulose submedian pair almost as long as intermediate pair; two pairs of simple

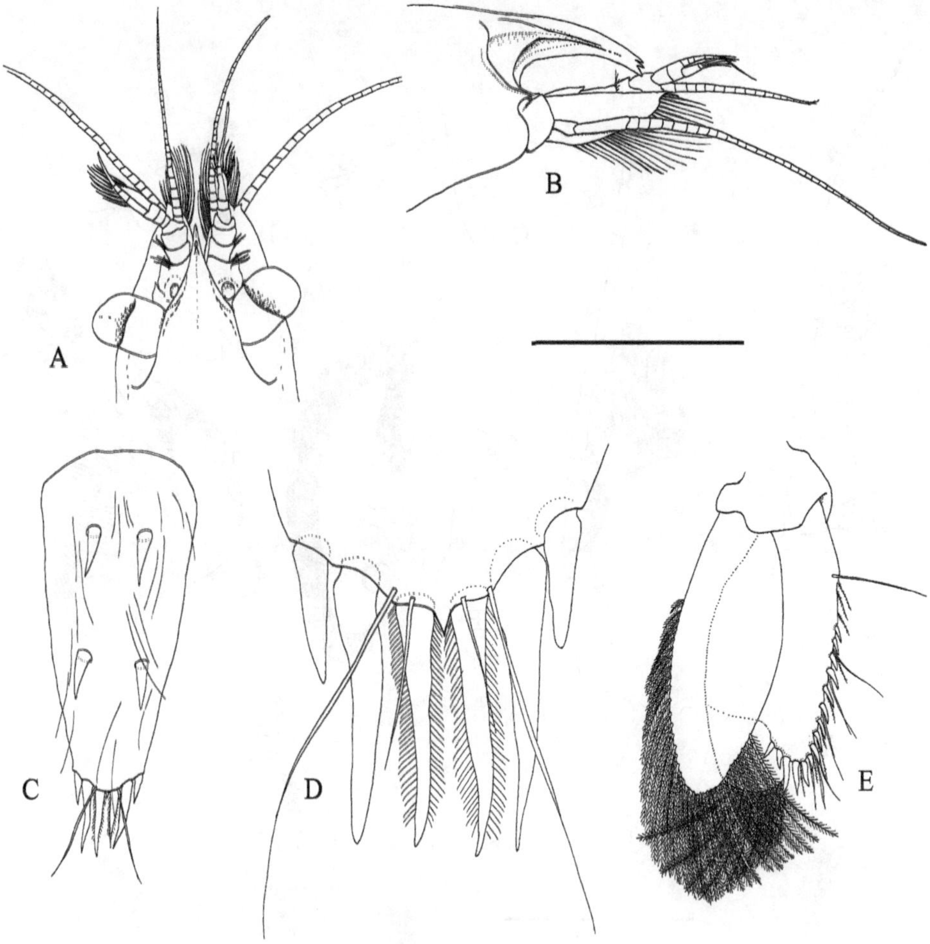

Fig. 2. *Paraclimenaeus michaeli* sp. nov., female holotype, MZB Cru 4603. A, anterior appendages, dorsal view; B, idem, lateral view; C, telson, dorsal view; D, idem, tip; E, uropod. Scale; A, B = 1 mm; C, E = 0.5 mm; D = 0.16 mm.

slender setae present, one pair submedially, shorter than submedian spines, and one pair between submedian and intermediate spines, extending well beyond posterior spines.

Eyestalk (fig. 2A) about as broad as long, with cornea slightly broader than stalk.

Antennula (fig. 3A) with peduncle and flagella well developed; stylocerite reaching almost to half of basal segment, broad, short, with acute distal tip and few simple setae laterally; basal segment with acute distolateral process; intermediate segment twice as broad as long; distal segment twice as long as intermediate segment, and about as broad as long; lower flagellum long, slender, with about 16

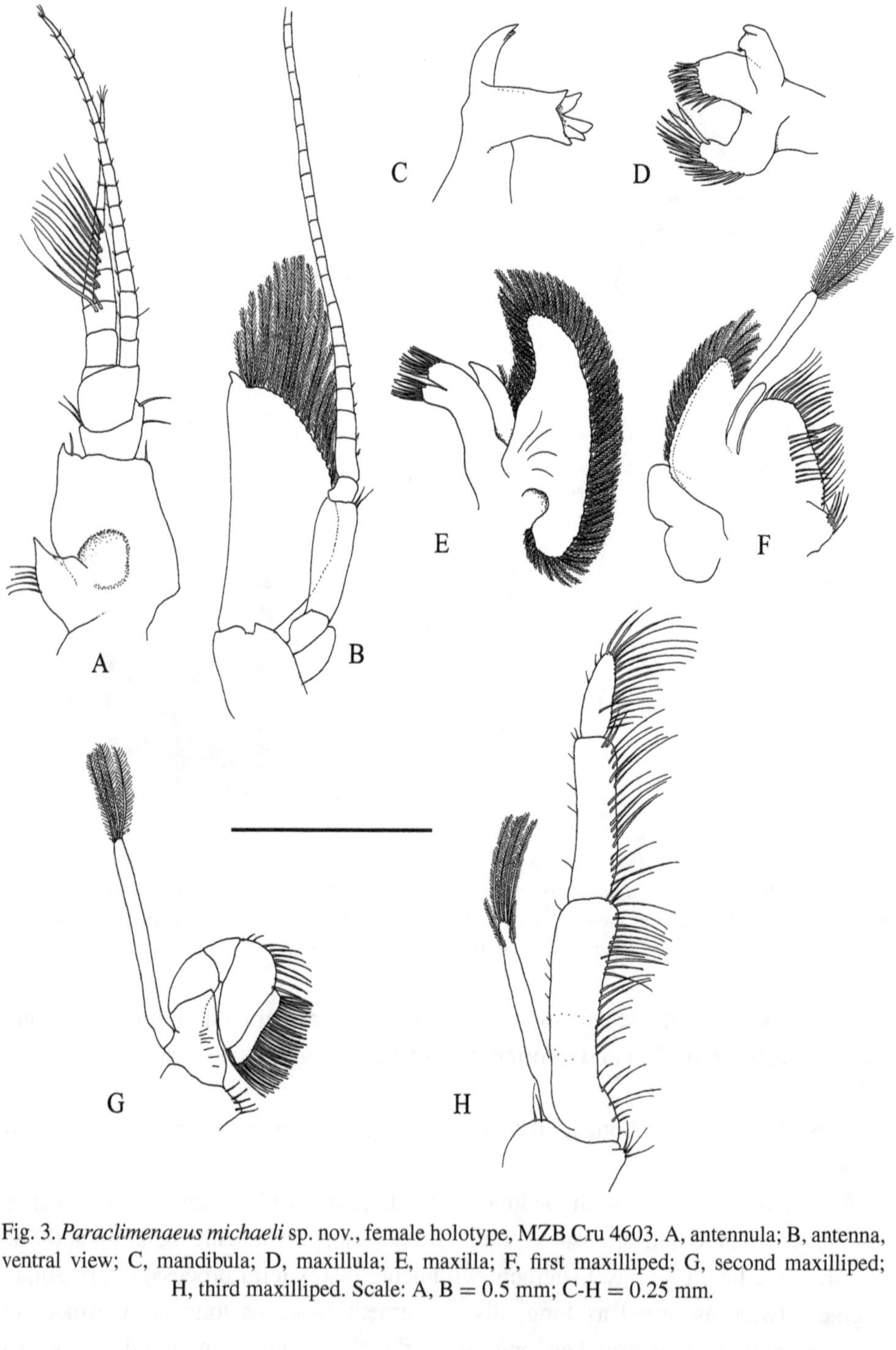

Fig. 3. *Paraclimenaeus michaeli* sp. nov., female holotype, MZB Cru 4603. A, antennula; B, antenna, ventral view; C, mandibula; D, maxillula; E, maxilla; F, first maxilliped; G, second maxilliped; H, third maxilliped. Scale: A, B = 0.5 mm; C-H = 0.25 mm.

segments; upper flagellum shorter, biramous, with proximal three segments fused; short free ramus consisting of 2 segments, with about 7 pairs of aesthetascs; longer free ramus long and slender, about twice as long as shorter ramus; distal margin of peduncle slightly overreaching tip of rostrum.

Antenna (fig. 3B) with basicerite, ischiocerite and merocerite normal; carpocerite short, slender, reaching to half lamella of scaphocerite, about three times longer than central width; flagellum slender, about as long as postorbital carapace length; scaphocerite with lamella about 2.3 times longer than central width, convex medially and anteriorly, ending in large acute hood-shaped distolateral tooth slightly bent laterally.

Mandible (fig. 3C) with incisor process slender with three distal acute teeth, central tooth smaller than outer teeth; molar process robust, with five blunt marginal distal teeth.

Maxillula (fig. 3D) with upper and lower lacinia well developed; upper lacinia with two rows of about ten short stout spines distally, truncate; lower lacinia slender with many simple and serrate setae and about three more robust serrate setae distally, bent medially with rounded distal margin; palp bilobed, larger lobe with small proximal tubercle.

Right maxilla (fig. 3E) with basal endite well developed and bilobed with lobes of about equal size, both with many stout setae disto-medially. Left maxilla with basal endite well developed but only one lobe, unclear whether second lobe lost or absent; coxal endite reduced; scaphognathite large, narrow, about three times longer than broad; posterior lobe narrow, round, about 1.5 times longer than its anterior width; anterior lobe about 1.5 times longer than its basal width, slightly tapering anteriorly; palp slender, tapering, shorter than lobe(s) of basal endite, with about four plumose setae on proximo-lateral margin.

First maxilliped (fig. 3F) with coxal endite triangular, with some simple setae, almost completely fused with basal endite; basal endite short, robust, with many simple setae on medial margin and few submarginally; exopod well developed, with four long plumose setae distally; caridean lobe small, narrow, with many plumose setae; epipod bilobed, large and heart shaped; palp simple, slender, non-setose.

Second maxilliped (fig. 3G) with well-developed slender exopod with about four long plumose setae distally; endopod with dactylar segment narrow, 4.5 times longer than broad, densely fringed with setae medially; propodal segment with distomedial margin round, bearing long simple setae; carpal segment small, about twice as broad as long; meral segment about twice as long as broad; basal and ischial segment fused, with few short simple setae submedially; coxal segment with medial margin broadly triangular, bearing about five simple setae.

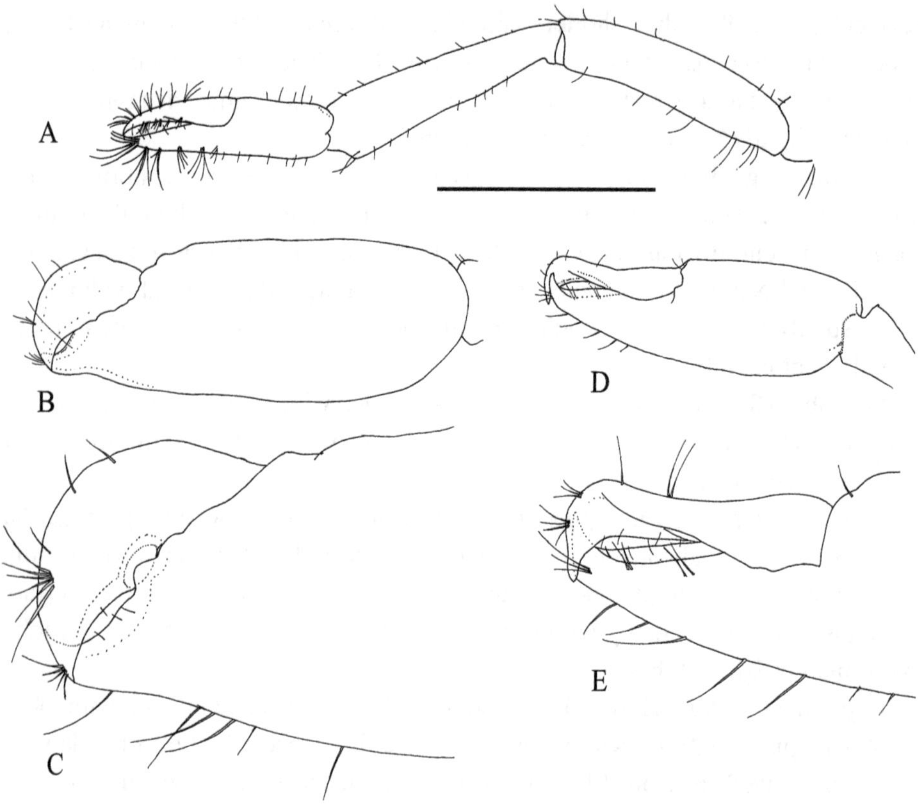

Fig. 4. *Paraclimenaeus michaeli* sp. nov., female holotype, MZB Cru 4603. A, first pereiopod; B, major second chela; C, idem, fingers; D, minor second chela; E, idem, fingers. Scale: A, C, E = 0.5 mm; B, D = 0.25 mm.

Endopod of third maxilliped (fig. 3H) slender, coxa reduced, with four short simple setae; lateral plate rounded, with about few short setae; ischiomerus completely fused to basis, with many long serrulose setae medially, and few very short setae laterally; penultimate segment long, slender, about four times longer than wide, and 2/3 of ischiomeral length, with many serrulose setae on medial margin; ultimate segment short, tapering distally, with many long setae along medial and disto-lateral border, about 1/3 of ischiomeral length, three times longer than wide; exopod well developed, slightly shorter than meral segment, with about four long plumose setae distally, and two shorter plumose setae subdistally.

First pereiopods (fig. 4A) exceeding carpocerite by chela and half of carpus; chela about four times longer than deep, subcylindrical; fingers subequal to palm length, with cutting edges entire, and acute tips; groups of setae present on outer margins of fingers and cutting edges; tip of fixed finger completely covered with setae; carpus about 1.3 times length of chela, about 4.5 times longer than distal width; carpus tapering proximally, with few simple short setae on margins; merus cylindrical, of about equal length as carpus, about five times longer than distal width.

Second pereiopods unequal, dissimilar. Major second pereiopod chela (fig. 4B, C) stout, subcylindrical, about 1.1 times postorbital carapace length; dactylus broad, strongly curved over fixed finger; about 0.25 of chela length; distal cutting edges entire; proximally two opposing fossae present at proximal third of cutting edge; few groups of simple setae distally on dactylus, and few single setae on distoventral margin of fixed finger; carpus stout, about 1/4 of length of chela, tapering proximally, about as long as distal width; merus short and broad, as long as carpus, less broad, slightly tapering proximally; ischium slender, about third length of chela, 2.5 times longer than distal width. Chela of minor second pereiopod (fig. 4D, E) tapering distally, with dactylus 0.45 of chela length, long and slender, about four times longer than its proximal width; dactylus twisted halfway, flattening distally, with acute tip; fixed finger broad, about twice as long as proximal width, slightly twisted, ending in acute tip; cutting edges entire, with few simple setae on dactylus and fixed finger.

Ambulatory pereiopods long and slender; dactylus of third pereiopod (fig. 5A, B) about 3 times its proximal width; unguis slender, only feebly curved, about 1/3 of dactylar length; corpus with two simple setae at base of unguis; large distal accessory tooth and 5 smaller forward directed teeth on flexor margin; dactylus about 1/7 of propodal length; propodus about 9 times longer than broad, about 1.5 times longer than carpus, and with distoventral pair and series of 12 serrate spines along ventral margin; carpus about 5 times longer than its proximal width, with feebly acute dorso-distal process; merus about as long as propodus, 3.5 times longer than its proximal width.

Fourth (fig. 5C, D) and fifth (fig. 5E, F) pereiopods similar to third. Propodus of fourth with series of 8 serrate spines along ventral margin, fifth with series of 7 serrate spines along ventral margin and with 3 oblique series of long serrulate setae distoventrally.

First pleopod of female (fig. 6A) with endopod about 5.5 times longer than proximal width, tapering distally, with few plumose setae laterally; exopod slender, tapering distally, about six times longer than proximal width, about 1.7 times endopod length, with many plumose setae disto-laterally; basipodite with long plumose setae disto-laterally and some disto-medially.

Second pleopod of female (fig. 6B) with endopod and exopod of similar length and size, both fringed with long plumose setae medio-distally, both about four times longer than maximal width; basipodite as long as exopod and endopod, slender, about 3.3 times longer than wide, with few long plumose setae laterally;

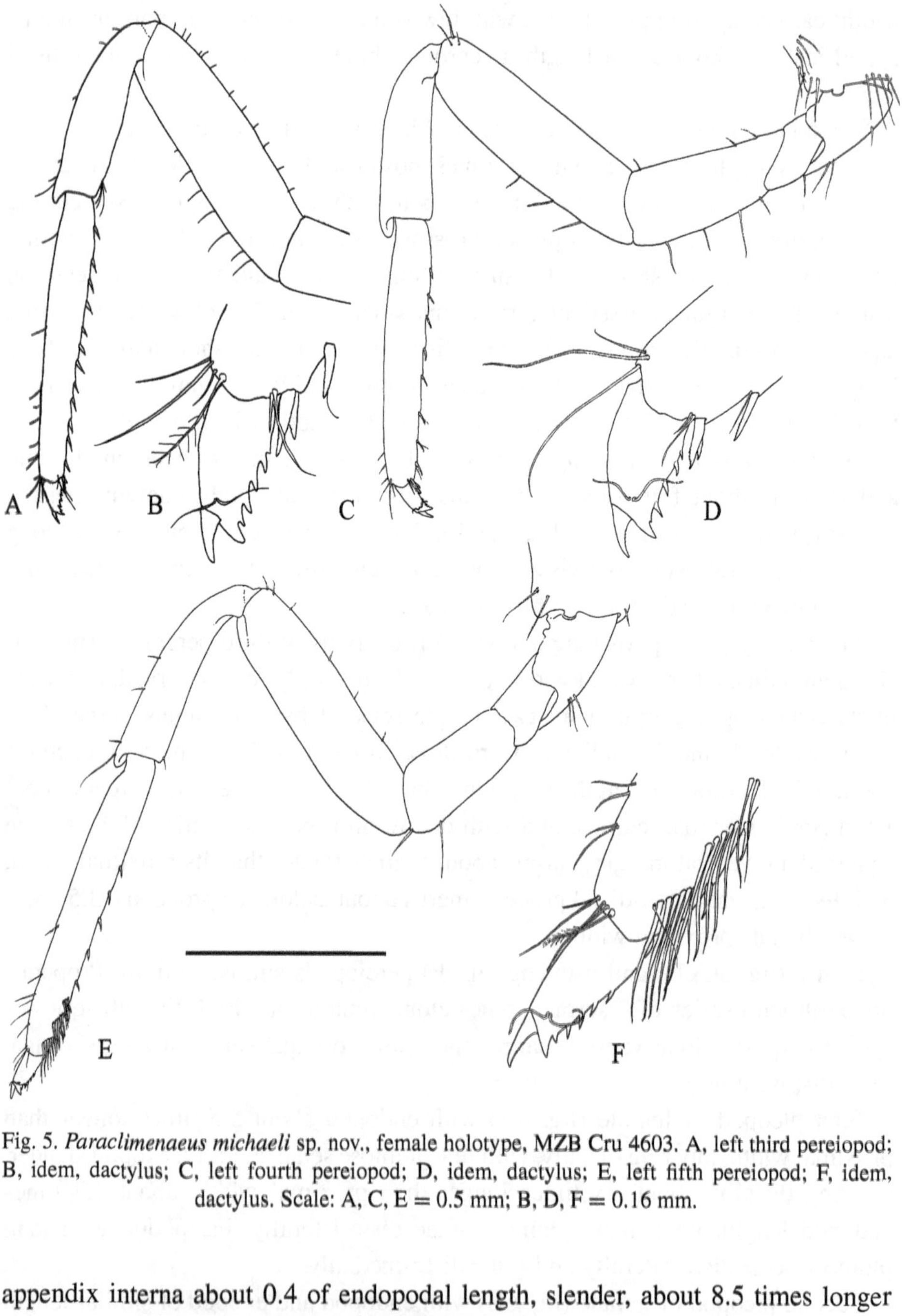

Fig. 5. *Paraclimenaeus michaeli* sp. nov., female holotype, MZB Cru 4603. A, left third pereiopod; B, idem, dactylus; C, left fourth pereiopod; D, idem, dactylus; E, left fifth pereiopod; F, idem, dactylus. Scale: A, C, E = 0.5 mm; B, D, F = 0.16 mm.

appendix interna about 0.4 of endopodal length, slender, about 8.5 times longer than wide, with few plumose setae laterally.

Endopod of male first pleopod (fig. 6C) slender, with about 5 short simple setae along median margin.

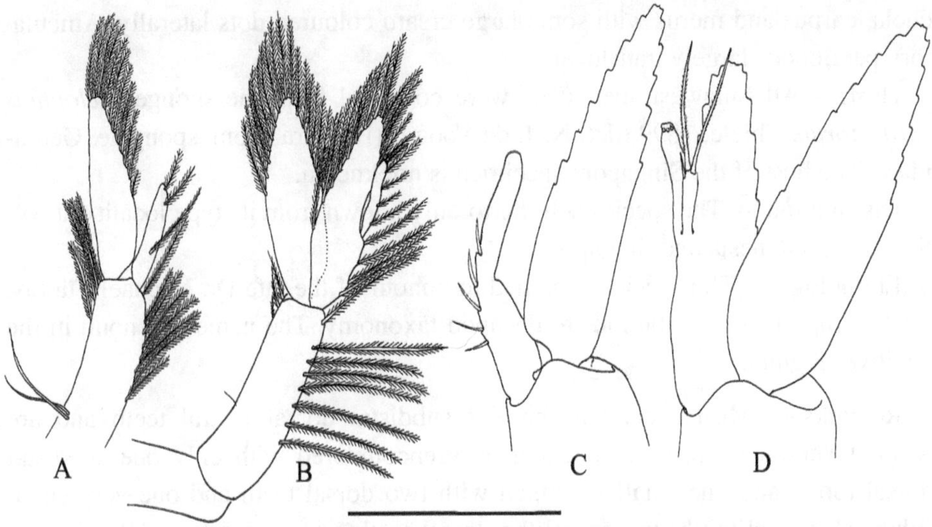

Fig. 6. *Paraclimenaeus michaeli* sp. nov., female holotype, MZB Cru 4603. A, first pleopod; B, second pleopod. Male paratype, pocl. 1.6 mm, RMNH.CRUS.D.50209. C, left first pleopod; D, left second pleopod. Scale: 0.25 mm.

Endopod of male second pleopod (fig. 6D) with appendix masculina slightly shorter than appendix interna, with 3-4 long, finely serrate setae distally and subdistally. Distal setae about twice as long as appendix masculina.

Uropod (fig. 2E) with protopodite unarmed; exopod broad, about 2.2 times longer than broad, lateral margin convex proximally, more straight distally; diaeresis produced distally, round; lateral margin of exopod with few simple and few plumose setae, with about 15 strong spines along margin and diaeresis, of which distalmost largest, decreasing in size proximo-laterally; distal lamina of exopod rounded, fringed with numerous long plumose setae; endopod about 2.7 times longer than wide, margins fringed disto-medially and disto-laterally with long plumose setae.

Eggs.— Ovigerous females carry between 20-30 eggs with a diameter of about 0.5 mm.

Size.— Largest ovigerous female has a pocl of 2.5 mm; largest male has a pocl of 1.6 mm.

Coloration (after Anker & De Grave, 2016, fig. 77).— Body generally translucent with a reddish tinge. Carapace with several large cream coloured spots anterodorsally. Abdominal segments with transverse line of small cream coloured dots at anterior dorsal margin. Larger cream coloured dots at midline of segments. Distal margin or uropods and telson cream coloured. Anterior appendages translucent, as well as first pereiopods. Second pereiopods with fingers of chela white, chela, carpus and merus with some large cream coloured dots laterally. Ambulatory pereiopods largely translucent.

Host.— All Sulawesi specimens were collected from the sponge *Melophlus sarasinorum* Thiele, 1899 (det. N. J. de Voogd) (Porifera, Demospongiae, Geodiidae). The host of the Singapore specimen is not known.

Distribution.— The species is hitherto only known from its type locality at SW Sulawesi, Indonesia and Singapore.

Etymology.— The species is named in honour of the late Dr. Michael Türkay, for his important contributions to decapod taxonomy. The name is a noun in the genitive singular.

Remarks.— Most specimens have 2 subdistal dorsal rostral teeth and are without ventral teeth. One specimen was encountered with only one subdistal dorsal tooth, and one small specimen with two dorsal teeth and one very small subdistal ventral tooth. In general the dorsal teeth are more subapically situated than in the smaller specimens in which the dorsal teeth are more widely spaced. Most specimens had about 15 spines along the distolateral margin of the exopod of the uropod. The lowest number counted was 12, the highest 18.

The species is closely related to *P. spinicauda*. However, it can be easily separated from *P. spinicauda* by 1) the number of dorsal rostral teeth (at most 2 in the new species whereas *P. spinicauda* has 4-6 dorsal teeth); and 2) the downcurved rostrum in the new species whereas it is straight in *P. spinicauda*.

MOLECULAR ANALYSIS

A phylogenetic tree (fig. 7) based on 16S sequences using ML and BI methods is here presented. The tree indicates the monophyly of *Paraclimenaeus* species with *P. michaeli* and *P. spinicauda* being sister species. The relationships among the three genera (*Paraclimenaeus*, *Climeniperaeus* and *Apopontonia*) is not resolved and remains a polytomy. The group of these genera is monophyletic in this reconstruction.

Several closely related species used in the present molecular analyses, i.e. *Paraclimenaeus michaeli*, *P. spinicauda*, and *Climeniperaeus orbitospinatus*, were collected from the same host sponge *Melophlus sarasinorum*. This raises the question which mechanisms leading to sympatric speciation in this group of genera are involved as host-switching events can be ruled out here (Horka et al., 2016). A comprehensive phylogenetic analysis involving further sponge-associated genera with more detailed information on host specificity could lead to a better understanding of these shrimp-host speciation patterns.

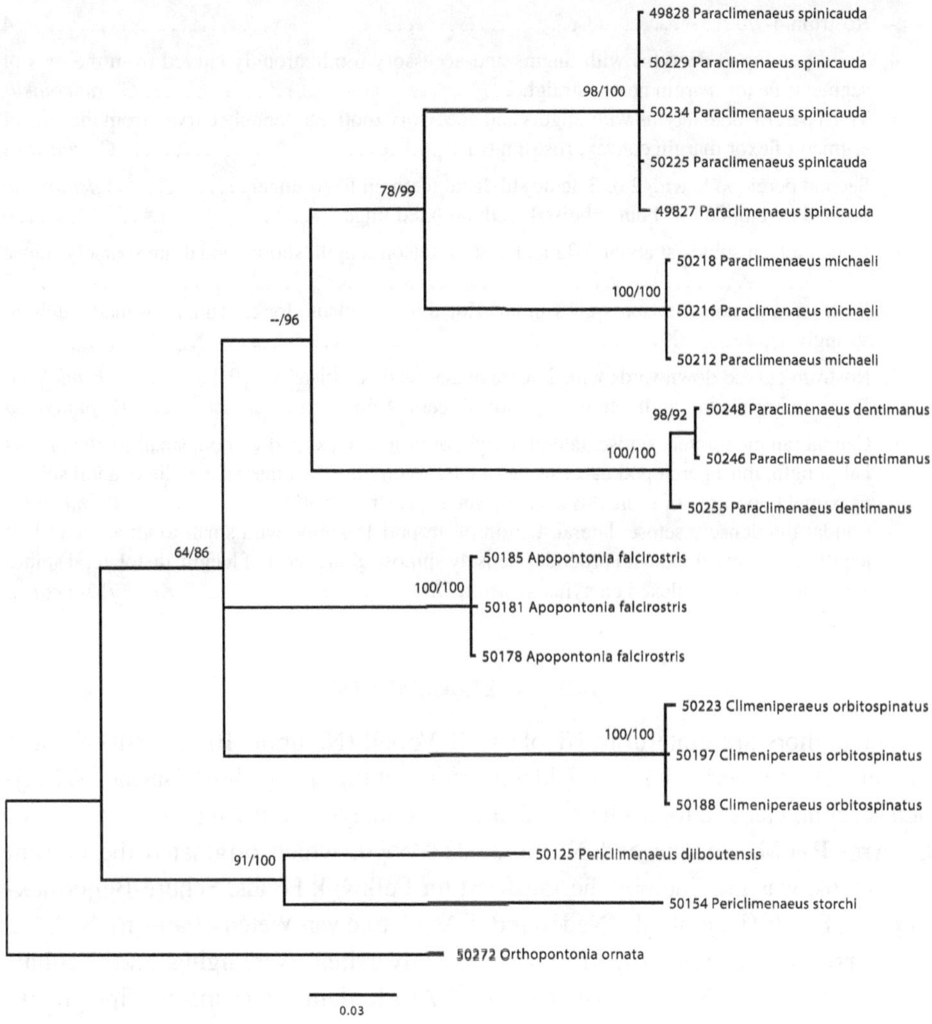

Fig. 7. Bayesian Inference (BI) tree with mean branch lengths of the 16S dataset. Confidence values at the nodes were obtained from 1000 bootstrap replicates of ML trees (first number) and Bayesian Inference (BI) (second number). Only values above 50% are indicated. Scale bar indicates genetic distance. Numbers in front of names refer to RMNH.CRUS.D collection registration numbers of voucher specimens.

KEY TO THE SPECIES OF THE GENERA *APOPONTONIA*, *CLIMENIPERAEUS* AND *PARACLIMENAEUS*

1. Second pereiopods similar, almost equal in size, without fossae .... *A. falcirostris*
– Second pereiopods dissimilar, unequal in size, two opposing fossae on major second pereiopod 2
2. Large protruding supraocular teeth present 3
– Large protruding supraocular teeth absent 5
3. Stout bifid rostrum with only one subdistal dorsal tooth .... *C. orbitospinatus*

– Rostrum 4-7/0 4

4. Third pereiopod dactylus with unguis and accessory tooth strongly curved from the axis of segment; flexor margin nearly straight . . . . *C. truncoideus*

– Third pereiopod dactylus with unguis and accessory tooth moderately curved from the axis of segment; flexor margin convex; rostrum 6/1 . . . . *C. truncatus*

5. Second pereiopods with 2 or 3 acute subdistal teeth on fixed finger . . . . *P. dentimana*

– Second pereiopods without subdistal teeth on fixed finger 6

6. Dorsal telson spines at about 0.33 and 0.66 of telson length, short; caudal fan sparsely setose 7

– Both pairs of dorsal telson spines in anterior third of telson, long; caudal fan moderately to strongly setose 8

7. Rostrum curved downwards with 2 acute dorsal teeth subdistally, 2/0-1 . . . . *P. michaeli*

– Rostrum horizontal, with 4 to 6 small dorsal teeth, 4-6/0-1 . . . . *P. spinicauda*

8. Caudal fan moderately setose, lateral margin of uropodal exopod without small tooth at about half length; third pereiopod dactylus flexing between long distomedial and distolateral spines, proximal two thirds of propodus sparsely setose; rostrum 5-9/0 . . . . *P. fimbriatus*

– Caudal fan densely setose, lateral margin of uropodal exopod with small tooth at about half length; third pereiopod with propodus densely spinose along most of length, distolateral spines short, not enclosing flexed dactylus; rostrum 7/0 . . . . *P. seticauda*

ACKNOWLEDGEMENTS

The authors are grateful to Nicole J. de Voogd (Naturalis Biodiversity Center, Leiden, The Netherlands) for the identification of the sponge host. Bastian T. Reijnen is acknowledged for his help with the DNA analyses. Funding for the fieldwork by Arni P.H.M. Janssen and Nicole J. de Voogd, which originated the present specimens, was provided by the Jan Joost ter Pelkwijk Fonds, Schure-Beijerinck-Popping Fonds (Koninklijke Nederlandse Academie van Wetenschappen), STUNT (University of Amsterdam), Amsterdamse Universiteits Vereniging, and Facultair Fonds, Faculty of Science (University of Amsterdam). The manuscript greatly improved from the reviews by Sammy De Grave and Zdenek Ďuriš.

REFERENCES

ANKER, A. & S. DE GRAVE, 2016. An updated and annotated checklist of marine and brackish caridean shrimps of Singapore (Crustacea, Decapoda). Raffles Bulletin of Zoology (Suppl.), **34**: 343-454.

BORRADAILE, L. A., 1915. Notes on carides. Annals and Magazine of Natural History, **15**: 205-213.

BRUCE, A. J., 1969. Preliminary descriptions of ten new species of the genus *Periclimenaeus* Borradaile, 1915 (Crustacea, Decapoda Natantia, Pontoniinae). Zoologische Mededelingen, Leiden, **44**: 159-175.

BRUCE, A. J., 1970. Further preliminary descriptions of new species of the genus *Periclimenaeus* Borradaile, 1915 (Crustacea, Decapoda Natantia, Pontoniinae). Zoologische Mededelingen, Leiden, **44**: 305-315.

BRUCE, A. J., 1976. Notes on some Indo-Pacific Pontoniinae, XXVII. *Apopontonia falcirostris* gen. nov., sp. nov., from Madagascar. Crustaceana, **31**: 301-311.

BRUCE, A. J., 1988. A re-description of *Periclimenaeus fimbriatus* Borradaile, 1915, with the designation of a new genus (Crustacea: Decapoda: Palaemonidae). Zoological Journal of the Linnean Society, London, **94**: 219-232.

BRUCE, A. J., 1989. A report on some coral reef shrimps from the Philippine Islands. Asian Marine Biology, **6**: 173-192.

BRUCE, A. J., 1996. Crustacea Decapoda: palaemonid shrimps from the Indo-West Pacific region, mainly from New Caledonia. In: A. CROSNIER (ed.), Résultats des campagnes MUSORSTOM, 15. Mémoires du Muséum National D'Histoire Naturelle **168**: 197-267. (Paris).

BRUCE, A. J., 2008. Palaemonoid shrimps from the Australian north west shelf. Zootaxa, **1815**: 1-24.

BRUCE, A. J., 2009. Notes on some Indo-Pacific Pontoniinae, XLVII. Re-evaluation of the genera *Apopontonia* Bruce, 1976, *Paraclimenaeus* Bruce, 1988 and *Climeniperaeus* Bruce, 1996. Crustaceana, **82**: 493-504.

CHACE JR., F. A. & A. J. BRUCE, 1993. The caridean shrimps (Crustacea: Decapoda) of the albatross Philippine expedition 1907-1910, part 6: superfamily Palaemonoidea. Smithsonian Contributions to Zoology, **543**: i-vii, 1-252.

DARRIBA, D., G. L. TABOADA, R. DOALLO & D. POSADA, 2012. jModelTest 2: more models, new heuristic and parallel computing. Nature Methods, **9**: 772.

DE GRAVE, S. & C. H. J. M. FRANSEN, 2011. Carideorum catalogus. The recent species of the dendrobranchiate, stenopodidean, procarididean and caridean shrimps (Crustacea: Decapoda). Zoologische Mededelingen, Leiden, **85**: 195-589.

HORKÁ, I., S. DE GRAVE, C. H. J. M. FRANSEN, A. PETRUSEK & Z. ĎURIŠ, 2016. Multiple host switching events shape the evolution of symbiotic palaemonid shrimps (Crustacea: Decapoda). Scientific Reports, **6**: 26486.

LARSSON, A., 2014. AliView: a fast and lightweight alignment viewer and editor for large data sets. Bioinformatics, **30**: 3276-3278.

MITSUHASHI, M. & T.-Y. CHAN, 2008. Pontoniine shrimps of the genus *Apopontonia* Bruce, 1976 (Crustacea: Decapoda: Palaemonidae) from the Philippines Panglao 2004 expedition, with description of one new species. The Raffles Bulletin of Zoology (Suppl.), **19**: 27-38.

PALUMBI, S., A. MARTIN & S. ROMANO, 1991. The simple fool's guide to PCR. (Department of Zoology and Kewalo Marine Laboratory, University of Hawaii, Honolulu, HI).

RATHBUN, M. J., 1906. The Brachyura and Macrura of the Hawaiian islands. Bulletin of the United States Fish Commission, **23**(3): 827-930.

RONQUIST, F. & J. P. HUELSENBECK, 2003. MRBAYES 3: Bayesian phylogenetic inference under mixed models. Bioinformatics, **19**: 1572-1574.

SWOFFORD, D. L., 2002. PAUP*. Phylogenetic analysis using parsimony (*and other methods). Version 4. (Sinauer Associates, Sunderland, MA).

First received 31 August 2016.
Final version accepted 7 February 2017.

[When citing this volume, please refer to Crustaceana 90 (2017) 771-1288]

Michael Türkay Memorial Issue

A SECOND REPORT ON MATERIAL FROM DR. MORTENSEN'S COLLECTION OF THALASSINIDEA AND CALLIANASSIDEA (DECAPODA) IN THE ZOOLOGICAL MUSEUM, COPENHAGEN

BY

KATSUSHI SAKAI[1])
Shikoku University, Tokushima, Japan

ABSTRACT

Material collected during the Pacific Expedition (1913-1916) under the leadership of Dr. Th. Mortensen was studied. As a result of the examination, *Jaxea novaezealandiae* Wear & Yaldwyn, 1966 (Laomediidae Borradaile, 1903) and *Eucalliax kensleyi* Dworschak, 2005 (Callianassidae Dana, 1852, sensu Sakai & Sawada, 2006) are recorded, and *Tuerkaygourretia galathea* gen. et sp. nov. (Gourretiidae Sakai, 1999) and *Garyia sachikoae* sp. nov. (Thomassiniidae De Saint Laurent, 1979) are newly described. The family Gourretiidae, reviewed on this occasion, is safely considered to comprise eight genera, including five new ones, *Heterogourretia* gen. nov., *Ivorygourretia* gen. nov., *Plantesgourretia* gen. nov., *Ruiyuliugourretia* gen. nov. and *Tuerkaygourretia* gen. nov.

ZUSAMMENFASSUNG

In dieser Untersuchung von Material, das während der Pazifischen Expedition (1913-1916) in einer Serie von Expeditionen (1889-1930) unter der Fahrtleitung von Dr. Th. Mortensen gesammelt wurde, sind *Jaxea novaezealandiae* Wear & Yaldwyn, 1966 in der Familie Laomediidae Borradaile, 1903 und *Eucalliax kensleyi* Dworschak, 2005 in der Familie Callianassidae Dana, 1852 (sensu Sakai & Sawada, 2006) bestätigt. Überdies sind *Tuerkaygourretia galathea* gen. et sp. nov. in der Familie Gourretiidae Sakai, 1999 und *Garyia sachikoae* sp. nov. in der Familie Thomassiniidae De Saint Laurent, 1979 bestimmt worden. Die Familie Gourretiidae, die bei dieser Gelegenheit überprüft worden ist, umfaßt mit Sicherheit acht Gattungen, einschließlich fünf neuer Gattungen, *Heterogourretia* gen. nov., *Ivorygourretia* gen. nov., *Plantesgourretia* gen. nov., *Ruiyuliugourretia* gen. nov. und *Tuerkaygourretia* gen. nov.

INTRODUCTION

The present work reports on the examination of material from a collection of Thalassinidea and Callianassidea, made during the Danish Pacific Expedition

[1]) e-mail: ksakai@mf.pikara.ne.jp

(1913-1916) under the leadership of Dr. Th. Mortensen, and deposited in the Zoological Museum, Copenhagen. This paper could be a sequel to the earlier paper by the present author (Sakai, 1992), that reported on the Axiidae collection of that museum.

New records of *Jaxea novaezealandiae* Wear & Yaldwyn, 1966 in the family Laomediidae and *Eucalliax kensleyi* Dworschak, 2005 in the family Callianassidae are reported. In addition, descriptions are presented of a new species of the genus *Garyia*, *Garyia sachikoae* sp. nov. in the family Thomassiniidae De Saint Laurent, 1979, and another new species under a new genus, *Tuerkaygourretia galathea* gen. et sp. nov. in the family Gourretiidae Sakai, 1999. The establishment of this new genus necessitated a careful review of the Gourretiidae, resulting in a reclassification of the genera included. It is here proposed that the Gourretiidae comprise eight genera, five of which are newly diagnosed: *Heterogourretia* gen. nov. based on *Gourretia aungtonyae* Sakai, 2002; *Ivorygourretia* gen. nov. based on *Gourretia barracuda* Le Loeuff & Intès, 1974; *Plantesgourretia* gen. nov. based on *Gourretia nosybeensis* Sakai, 2004; *Ruiyuliugourretia* gen. nov. based on *Gourretia sinica* Liu & Liu, 2010; and *Tuerkaygourretia* gen. nov. based on *T. galathea* sp. nov. This reclassification of the Gourretiid genera is based mainly on marked morphological differences detected in the male Plp1-2, which are safely considered to be of generic significance. These generic characters were already applied to the revision of the systematics of the Superfamilies Axioidea and Callianassoidea (Sakai, 2011). The present author's concept of classification differs from that applied by Hyžný & Klompmaker (2015: 403). The differences between the two systems are discussed in the remarks under the Gourretiidae.

Abbreviations used.— BMNH, Natural History Museum, London, U.K.; MBM, Marine Biological Museum, Chinese Academy of Sciences, China; RMNH, Naturalis Biodiversity Center, Leiden, the Netherlands; SMF, Forschungsinstitut Senckenberg, Frankfurt am Main, Germany; ZMUC, Zoological Museum, University of Copenhagen.

A1-2, antennae 1-2 [= antennules and antennae]; P1, 2, pereiopods 1, 2; Plp1-2, pleopods 1-2; TL/CL, total length/carapax length.

SYSTEMATICS

Infraorder **THALASSINIDEA** Latreille, 1831

Thalassinidea, Sakai & Sawada, 2006: 1355; Sakai &Türkay, 2014: 134.

Superfamily THALASSINOIDEA Latreille, 1831
Family LAOMEDIIDAE Borradaile, 1903

Laomediidae Borradaile, 1903: 540; Wear & Yaldwyn, 1966: 2; Sakai & Sawada, 2006: 1357.

Genera included.— *Laomedia* De Haan, 1849, *Jaxea* Nardo, 1847, *Saintlaurentiella* Paiva, Tavares & Silva-Neto, 2010, and *Naushonia* Kingsley, 1897.

Genus **Jaxea** Nardo, 1847

Jaxea Nardo, 1847: 3-4; Borradaile, 1903: 541; De Man, 1928: 16 (list).

Diagnosis.— Eyestalks reduced, hidden under rostrum, and invisible in dorsal view; pigmentation of cornea reduced. A1-2 peduncles elongate and relatively slender. Scaphocerite rudimentary but distinct. Maxilla 2 with 5-9 elongate whips. Carapace densely covered with very short setae. Linea thalassinica distinct. P1 subequal, relatively heavy but elongated, and chelate distally. P2 subchelate distally. Male Plp1 absent. Female Plp1 uniramous, slender, and two-segmented. Plp2-5 biramous, in male and female similar.

Type species.— *Jaxea nocturna* Nardo, 1847, original designation. The gender of the generic name, *Jaxea*, is feminine.

Species included.— *Jaxea nocturna* Nardo, 1847 (= *Calliaxis adriatica* Heller, 1863, junior synonym), and *J. novaezealandiae* Wear & Yaldwyn, 1966.

Jaxea novaezealandiae Wear & Yaldwyn, 1966
(fig. 1)

Jaxea novaezealandiae Wear & Yaldwyn, 1966: 6-10, figs. 1-2, table 1; Pervesler & Dworschak, 1985, table 3.

Material examined.— ZMUC-CRU 8341, 1 female (TL/CL 18.0/6.7 mm, lacking A1 on right side, A2, and P2-5), off Tiri-Tiri, Auckland, New Zealand, 27 m [15 fathoms], mud, dredge, 28.xii.1914, leg. Th. Mortensen aboard "Hinemoa".

Comparative material: *Jaxea nocturna*, SMF 18342, 11 males, 6 females, damaged, eastern Mediterranean, near Israel, 32°02.38′N 34°35.05′E, 95-103 m, 26.i.1987, leg. F.S. "Meteor"; BMNH 1962.vii.5.10, 1 male (TL/CL 48.0/17.1 mm), 1 mile south, Nady Isle, Ayrshire, Firth of Clyde, 40 m muddy sand, 30.x.1957, leg. R. Pike; BMNH 1991.2537-2538 (1999.7-2538), 1 male (TL/CL 30.0/11.1 mm), 1 ovig. female (TL/CL 38.0/11.8 mm), English Channel, Stn. 152, 50°17.8′N 4°16.9′W, 42 m, northeast Atlantic Ocean, det. R. Ingle, 1998; RMNH.CRUS.D.13544, 1 male (TL/CL 13.0/5.3 mm), Stn. 6: Mediterranean Sea, Israel, Haifa Bay, 29.ix.1956, leg. E. Gottlieb; RMNH.CRUS.D.13548, 1 male (TL/CL 26.0/9.3 mm), 1 female (TL/CL 26.0/9.1 mm), Mediterranean Sea, Israel, Tel Aviv, 20.viii.1949, leg. A. Wirszubski; RMNH.CRUS.D.23619, 1 male (TL/CL 38.0/12.8 mm), Stn. 1566: Spain, Ria de Arosa, NW Spain, 1.4 km SE Punta, 16.vii.1964, leg. Excursion RMNH "Jusepha Cambeilo"; RMNH.CRUS.D.13552, 1 male (TL/CL 18.0/6.2 mm), Mediterranean Sea, Israel, Gaza, 29.viii.1947, leg. A. Wirszubski; RMNH.CRUS.D.15177, 1 male

(TL/CL 59.0/19.0 mm), Croatia, Val Saline, N. of Rovinj, leg. 5.ix.1932; RMNH.CRUS.D.13550, 1 male (TL/CL 34.0/11.2 mm), Mediterranean Sea, Israel, Gaza, 21.vii.1947, leg. A. Wirszubski.

Description.— Rostrum lacking distal part. Eyestalks short, located under rostrum (fig. 1B). A1 peduncle elongate, with distal segment about as long as flagella (fig. 1A); A2 missing. Anterolateral margin of carapace with small spine above origin of linea thalassinica, followed by minute spinules (fig. 1A, B). Maxilla 2 scaphognathite with 9 posterior whips. Mxp3 (fig. 1C) ischium armed mesially with crest-like row of 12-14 strong teeth. P1 subequal in length, and chelate distally (fig. 1D, E); chela plus carpus subequal in length to carapace plus rostrum. In larger cheliped (fig. 1D) chela elongate, about 3.0 times as long as wide; fingers 1.2 times as long as palm; dactylus lacking distal part. In smaller cheliped (fig. 1E) chela 3.4 times as long as wide; fingers about 1.5 times as long as palm. P2-5 missing. Abdominal somite 2 (fig. 1F) slightly wider than each of abdominal somites 3-6; abdominal somites 2-5 sparsely setose on lateral surface, convex and minutely denticulate ventrally. Female Plp1 uniramous, slender and two-segmented; Plp2-5 biramous, endopods and exopods leaf-like. Telson (fig. 1G) longer than wide and rectangular; uropodal endopod and exopod each with distal flap and longitudinal median sulcus.

Remarks.— The female specimen (TL 18.0 mm) is heavily damaged, but can be considered almost identical in morphology to the male paratype (CL 11.0 mm) figured by Wear & Yaldwyn (1966, figs. 1, 2). This is a second record of *J. novaezealandiae*. Many comparative specimens of *J. nocturna* were also examined, and it is confirmed that in larger males from Croatia (RMNH.CRUS.D.15177) and Israel (RMNH.CRUS.D.13550), the abdominal somites are minutely denticulate ventrally as in *J. novaezealandiae*, and Plp1 is absent in males. In females, Plp1 is slender and two-segmented (ovig. female, BMNH 1991.2537-2538, TL/CL 38.0/11.8 mm, from the English Channel), but in a smaller female (RMNH.CRUS.D.13548, TL/CL 26.0/9.1 mm, from Tel Aviv, Israel), it is not yet observed, but it would supposedly develop in accordance with growth to maturity.

Type locality.— Shelly Bay, New Zealand.

Distribution.— New Zealand.

Infraorder **CALLIANASSIDEA** Dana, 1852 [new status, in Sakai & Sawada, 2006]

Superfamily CALLIANASSOIDEA Dana, 1852

Callianassoidea, Sakai & Sawada, 2006: 1357.

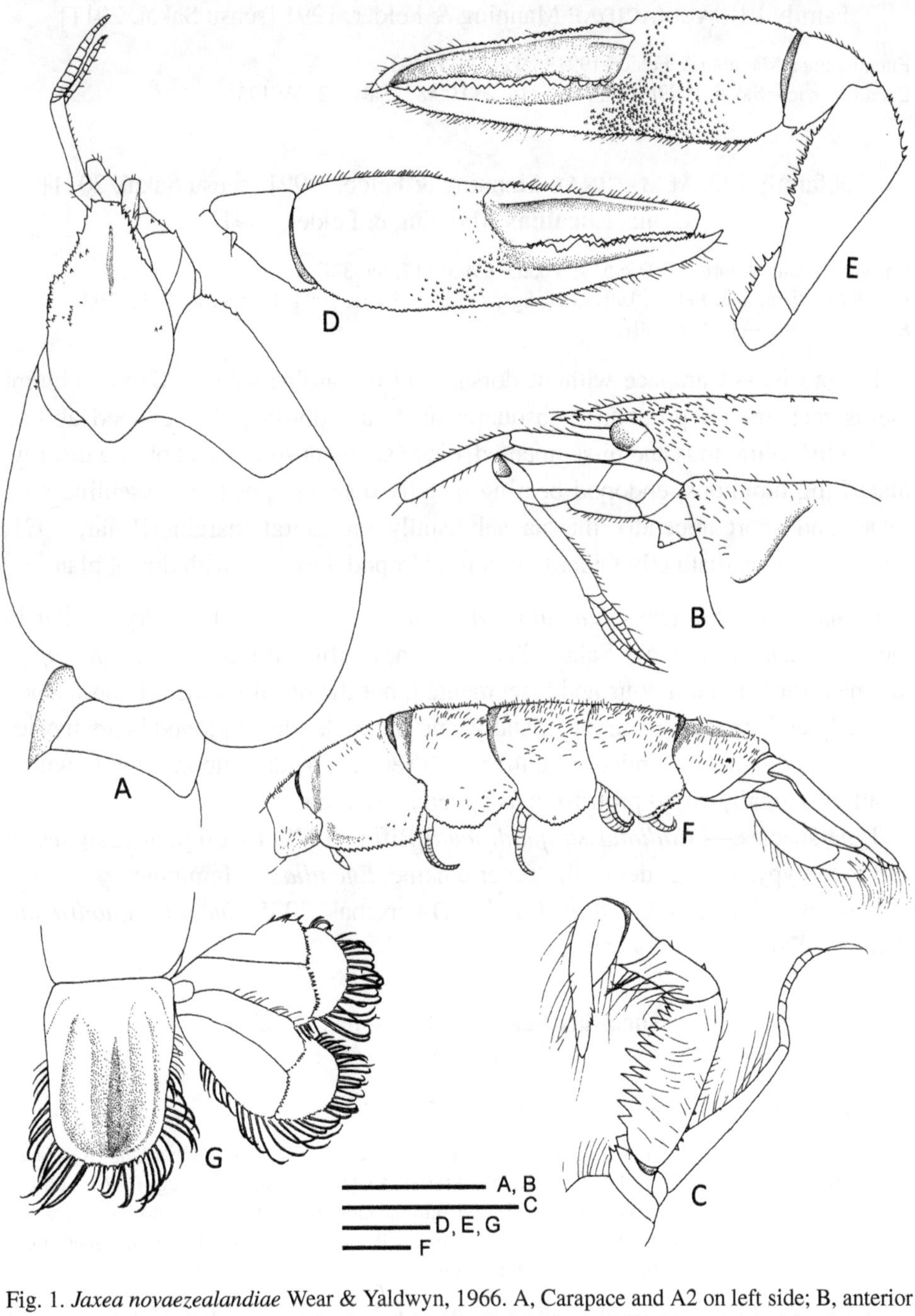

Fig. 1. *Jaxea novaezealandiae* Wear & Yaldwyn, 1966. A, Carapace and A2 on left side; B, anterior part of carapace, A1 peduncle and A2 on left side; C, Mxp3 on right side, mesial view; D, carpus and chela of P1 on right side, lateral view; E, P1 on left side, lateral view; F, abdominal somites and uropods on left side, lateral view; G, telson and uropods on right side. A-G, female (TL/CL 18.0/6.7 mm), off Tiri-Tiri, Auckland, New Zealand, 27 m. Scale bar A-G, 1 mm.

Family EUCALLIACIDAE Manning & Felder, 1991 [sensu Sakai, 2011]

Eucalliacidae Manning & Felder, 1991; Sakai, 2011: 491.
Eucalliacinae – Sakai, 1999b:108; Ngoc-Ho, 2003: 487; Sakai, 2005: 195.

Subfamily EUCALLIACINAE Manning & Felder, 1991 [sensu Sakai, 2011]
Genus **Eucalliax** Manning & Felder, 1991

Calliax De Saint Laurent, 1973: 514; Sakai, 1999b: 117, fig. 30C.
Eucalliax Manning & Felder, 1991: 781, figs. 7, 15a-e; Poore, 1994: 102; Sakai, 2011: 502.
Eucalliax s. str. — Felder, 2001: 440.

Diagnosis.— Carapace without dorsal oval or cardiac sulcus. Mxp3 ischium-merus rectangular; propodus subsquare; dactylus subtriangular; exopod absent. Male Plp1 uniramous and bi-segmented, distal segment simple and obtuse distally; male Plp2 biramous, endopod bearing distally distinct appendix masculina with setae, and short appendix interna subdistally on lateral margin (Biffar, 1971, fig. 2m). Telson distinctly wider than long. Uropodal exopod with dorsal plate.

Remarks.— The genus *Eucalliax* Manning & Felder, 1991 is very similar to the genus *Eucalliaxiopsis* Sakai, 2011, but they differ in that in *Eucalliaxiopsis* the male Plp1 is uniramous and bisegmented, but the distal segment is broadened medially and strongly incurvate distally, and the male Plp2 endopod bears mesio-medially a distinct appendix masculina with setae, provided subproximally with a small, proximally-fused protrusion as appendix interna.

Type species.— *Callianassa quadracuta* Biffar, 1970, by original designation and monotypy. The gender of the generic name, *Eucalliax*, is feminine.

Species included.— *Eucalliax kensleyi* Dworschak, 2005; *Eucalliax quadracuta* (Biffar, 1970).

Eucalliax kensleyi Dworschak, 2005
(fig. 2)

Eucalliax kensleyi Dworschak, 2005: 205, figs. 1-4, 5a, b.

Material examined.— ZMUC-CRU-8352, 1 female (TL/CL 34.0/9.4 mm), Ghardaqa (Red Sea), dug up from sand at low tide, ix.1937, leg. Dr. Th. Mortensen. [Journal] 6.v.1938, leg. Th. Mortensen; ZMUC-CRU 204, 1 ovig. female (TL/CL 35.0/8.1 mm, lacking P2 on right side, and P3-4), Danish Kei Island Expedition, Riemenbai, Ambon Bay, Ambon Island, Kei Islands, Indonesia, sand, tidal zone, hand-collected, 27.ii.1922, leg. Th. Mortensen.

Diagnosis.— Rostrum (fig. 2A) small and acutely triangular, lacking rostral carina. Carapace without dorsal oval, but with cervical groove located at posterior 1/3 of carapace, and cardiac sulcus. Mxp3 (fig. 2C) ischium-merus elliptic; propodus subovate; dactylus roundly triangular; exopod absent. P1 in female

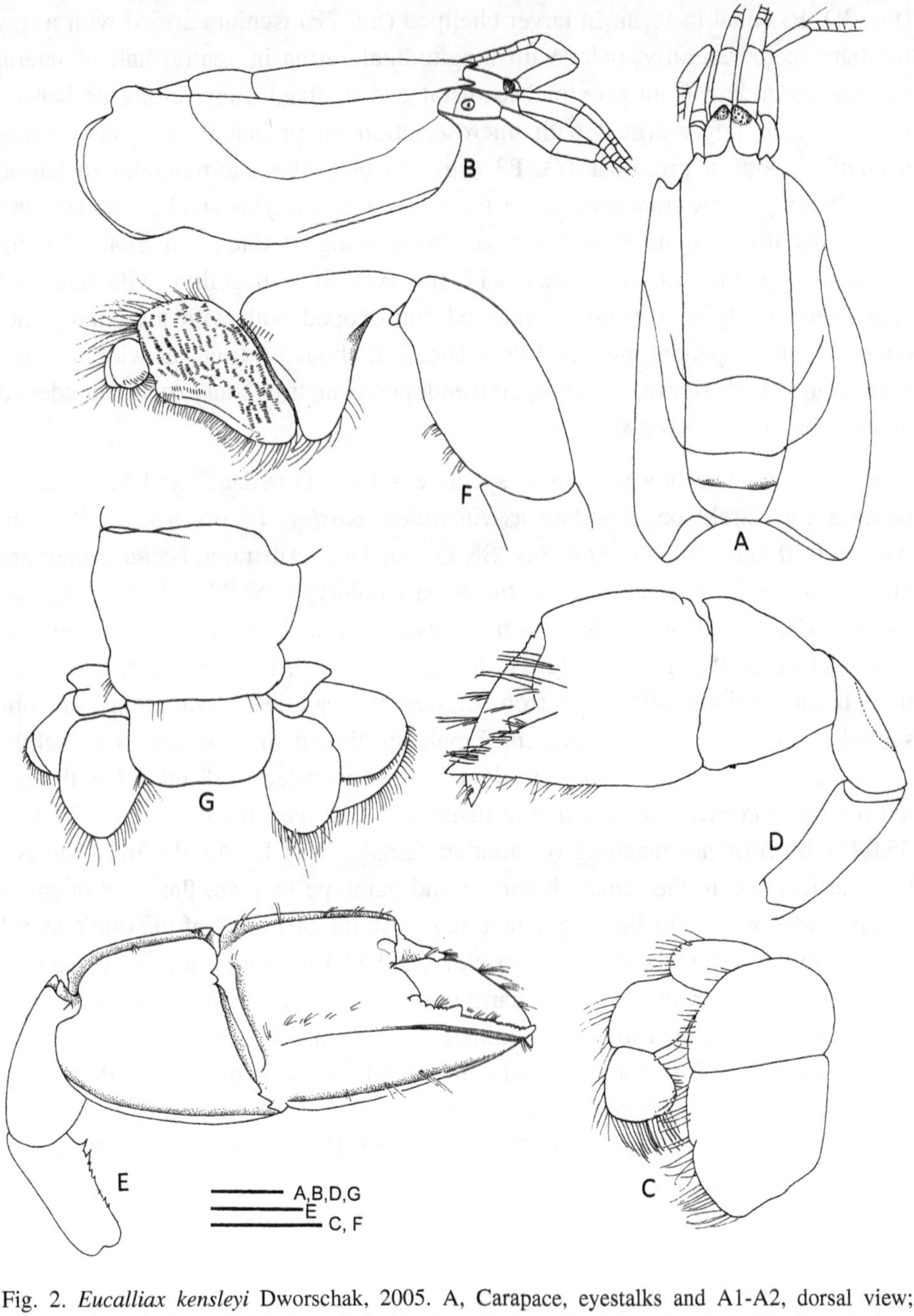

Fig. 2. *Eucalliax kensleyi* Dworschak, 2005. A, Carapace, eyestalks and A1-A2, dorsal view; B, same, lateral view; C, Mxp3 on left side, lateral view; D, cheliped on left side, lateral view; E, cheliped on right side, lateral view; F, P3 on left side, lateral view; G, abdominal somite 6, telson and uropods, dorsal view. A-C, E-G, female (TL/CL 34.0/9.4 mm), Chardaqa (Red Sea); D, ovig. female (TL/CL 35.0/8.1 mm), Riemenbai, Ambon Bay, Ambon Island, Kei Islands, Indonesia. Scale bar A-G, 1 mm.

(fig. 2D, E) equal in form. In larger cheliped (fig. 2E) ischium armed with a row of sharp teeth ventrally; palm with longitudinal carina in ventral half of lateral surface, extending from proximal to distal end of fixed finger along prehensile margin; fixed finger armed with microserration on prehensile margin, bearing triangular tooth at proximal 1/3. P3 (fig. 2F) propodus subtriangular in lateral view, bearing many transverse short rows of setae; dactylus small, rounded, and obtuse distally. Female Plp1 uniramous, consisting of thick proximal segment and deflexing thick distal segment with incurved thick flagellum with marginal setae; female Plp2-5 biramous, endopod and exopod both with marginal setae, endopod with appendix interna. Telson (fig. 2G) about 1.8 times as wide as long, with straight posterior margin; uropodal endopod oval; uropodal exopod broadened distally, bearing dorsal plate.

Remarks.— The present female specimens from Ghardaqa, Red Sea and Indonesia can safely be identified as *Eucalliax kensleyi* Dworschak, 2005 from Aqaba, Red Sea, and Moreton Bay, SE Queensland, Australia, because they are almost the same in morphology as the female holotype (NHMW 16779) and paratype (NHMW 16780) figured, in the eyestalks, Mxp3, P1, P3, and the tail-fan (Dworschak, 2005, fig. 1b, f, fig. 2g, h, fig. 3a, b, c, d, f, i, fig. 4a, b, c, d, e, f), though some minor differences from *E. kensley*, regarded as variations, are observed as follows: (1) in the present female specimens the eyestalks are slightly convex dorsally and egg-shaped in dorsal view, provided in distal 1/3 with dotted region as cornea and calcified at distal end, overreaching (in female, TL/CL 35.0/8.1 mm) or not reaching (in another female, TL/CL 34.0/9.4 mm) the A1 basal article (vs. in the female holotype and paratype they are flattened dorsally and provided with a distinct pigmented region in the distal 1/3 of the dorsolateral surface, overreaching the A1 basal article); (2) the P1 ischium is armed with a row of sharp teeth ventrally (vs. it is unarmed or very minutely spinulose ventrally); (3) the P1 palm bears a long longitudinal carina in the ventral half of the lateral surface, extending from the proximal to the distal end along the prehensile margin (vs. it bears a short longitudinal furrow along the prehensile margin).

Type locality.— Alona Beach, Panglao Island, Bohol, Philippines, 09°32.9′N 123°46.6′E, intertidal.

Distribution.— Bohol, Philippines; Ambon Island, Kei Islands, Indonesia; Chardaqa (Red Sea).

Family GOURRETIIDAE Sakai, 1999

Gourretiinae Sakai, 1999: 95; Ngoc-Ho, 2003: 498.
Gourretiidae — Sakai, 2004: 556; 2005: 217; 2011: 507; Sakai & Sawada, 2006: 1357.
Dawsoniinae — Sakai, 2006: 1276.

Diagnosis.— Rostrum small and unarmed on lateral margins, bearing no rostral carina. Carapace with linea thalassinica, but without dorsal oval; cardiac sulcus incomplete or absent; vertical row of setae rarely present on branchial region. Eyestalks flattened and contiguous. A2 scaphocerite present as small process. Maxilla 2 scaphognathite without posterior whip. Mxp3 ischium-merus rectangular; merus with mesial spine; propodus rectangular; dactylus digitiform; exopod present or absent. P1 chelate distally, unequal in size and dissimilar in shape. In larger cheliped merus with hook ventroproximally; palm rectangular; fingers elongate. In smaller cheliped merus usually with hook ventroproximally (not in *Gourretia lahouensis* Le Loeuff & Intès, 1974); fingers elongate. P2 chelate distally. P3 propodus subpediform. P4 simple. P5 chelate. Male Plp1 present or absent; when present, uniramous and two-, three-, or four-segmented, and distal segment hooked, subchelate, or chelate distally; male Plp2 biramous and slenderly foliaceous, endopod bearing appendices masculina and interna, or either appendix masculina or appendix interna. Abdominal somite 6 with or without lateral projections. Telson subtriangular or subsquare. Uropodal exopod bilobulate, or broadly rounded, bearing no dorsal plate; lateral notch present or absent.

Type genus.— *Gourretia* De Saint Laurent, 1973.

Genera included.— *Gourretia* De Saint Laurent, 1973; *Heterogourretia* gen. nov.; *Ivorygourretia* gen. nov.; *Laurentgourretia* Sakai, 2004; *Ruiyuliugourretia* gen. nov.; *Paracalliax* De Saint Laurent, 1979; *Plantesgourretia* gen. nov.; *Tuerkaygourretia* gen. nov.

Remarks.— The present review of the Gourretiidae, hitherto including five genera, became required by the results of the examination of additional specimens. This detailed study has clearly revealed that the male Plp1-2 show some distinct morphological differences as generic characters, which has necessitated the genera hitherto included in this family to be reconsidered and reclassified by focusing mainly on the major characters of the male Plp1-2. As a result of this review, eight genera, including five new ones, are safely determined as gourretiid genera, and are here presented in table I.

Male reproductive morphology has now been widely used to help define relationships in decapods, especially the male first and second gonopods in brachyuran crabs (e.g., Bott, 1970; Brandis, 2002; Sakai & Türkay, 2013), the male petasma of penaeid prawns (e.g., *Parapenaeopsis*, see Sakai & Shinomiya, 2011), and the male Plp1-2 have been extensively relied on to re-evaluate the genera of the superfamilies Axioidea Huxley, 1879, and Callianassoidea Dana, 1852 (Sakai, 2011, 2013). The morphology of the copulatory structures is shaped by evolutionary selection favouring efficient functioning during sperm transfer, and thus is less likely to be impacted by environmental factors. Thus, structural change

TABLE I
Comparison of morphological features among genera of Gourretiidae

		Mxp3 exopod	Abdominal somite 6	Uropod exopod	Male Plp1-2
1	*Gourretia* De Saint Laurent, 1973 Type sp.: *Gourretia denticulata* (Lutze, 1937)	Present or absent	With or without lateral projections	With or without lateral notch	Plp1 present and chelate or subchelate distally; Plp2 endopod bearing mesiosubdistally appendix masculina with small appendix interna mesially
2	*Heterogourretia* gen. nov. Type sp.: *Gourretia aungtonyae* Sakai, 2002	Present	Without lateral projections	Bilobulate	Plp1 absent; Plp2 endopod bearing distally segmented appendix interna; no appendix masculina
3	*Ivorygourretia* gen. nov. Type sp.: *Gourretia barracuda* Le Loeuff & Intès, 1974	Present	Without lateral projections	Without lateral notch	Plp1 present and hooked distally; Plp2 endopod bearing distally proximally-fused appendix masculina with proximally-fused appendix interna mesially
4	*Laurentgourretia* Sakai, 2004 Type sp.: *Laurentgourretia rhopalommata* Sakai, 2004	Absent	Without lateral projections	With lateral notch	Unknown
5	*Paracalliax* De Saint Laurent, 1979 Type sp.: *Paracalliax bollorei* De Saint Laurent, 1979	Present	Without lateral projections	Without lateral notch	Unknown
6	*Plantesgourretia* gen. nov. Type sp.: *Gourretia nosybeensis* Sakai, 2004	Present	Without lateral projections	Without lateral notch	Male Plp1 present and chelate distally; Plp2 biramous; exopod shorter than endopod; endopod bilobed distally, inner lobe bearing distally appendix masculina with setae, but no appendix interna
7	*Ruiyuliugourretia* gen. nov. Type sp.: *Gourretia sinica* Liu & Liu, 2010	Present	Without lateral projections	Without lateral notch	Plp1 present and hooked distally; Plp2 endopod bearing mesiosubdistally appendix masculina, but no appendix interna
8	*Tuerkaygourretia* gen. nov. Type sp.: *Tuerkaygourretia galathea* sp. nov.	Present	Without lateral projections	Bilobulate	Plp1 present and chelate distally; Plp2 exopod shaped as elongate oval, bearing marginal setae in distal half, and shorter than endopod; endopod consisting of two longitudinal sections, external section oval and larger than exopod, and mesial section bearing distally articulated appendix masculina with distal setae, with proximally-fused appendix interna in parallel with each other

tends to be relatively conservative, much slower than in non-sexual characters, and closely related species tend to have similar copulatory structures. This makes these organs significantly important for determining taxon relationships especially at the generic level. De Saint Laurent (1979: 80) also showed the validity of significant differences in the male Plp2, influencing even the classification at the species level, as in *Gourretia barracuda* Le Loeuff & Intès, 1974, *Gourretia* sp., and *G. serrata* De Saint Laurent & Le Loeuff, 1979 (= *G. denticulata* Lutze, 1937). The generic classification recognized here is significantly different from the recent views expressed by Hyžný & Klompmaker (2015), based on palaeontological studies. These authors disputed against Sakai's views (Hyžný & Klompmaker, 2015: 401, 404) as follows: "Sakai and colleagues erected an array of new genera, largely based on male pleopods (Sakai, 2011; Sakai & Türkay, 2014), but their diagnoses are not supplemented with information on chelipeds so that their recognition in the fossil record becomes much more difficult".

Neontological concepts and classifications are obviously based on a full complement of characters shown by the living animals, and therefore can be expected to be far more firmly founded than palaeontological systems based on relatively few, and inconsistently preserved, characters, such as chelipeds. Palaeontologists, therefore, must work with a limited dataset, which is why Hyžný & Klompmaker (2015) were forced to place great importance on cheliped characters (especially distal elements) for genus-level classification. In the opinion of the present author, however, chelipeds are too varied to be useful at the generic level. Sakai (1969) even asserted that cheliped morphology was often too variable even at the species level, to be used to classify Japanese species of *Callianassa* (now in *Trypaea*).

Kensley & Williams (1990) noted that "The taxonomic position of the present fossil specimens has been established by comparison with extant forms", but given the limited morphological characters typically preserved on fossil specimens, it is often impossible to reliably refer this material to extant taxa. Indeed, Schweitzer et al. (2012: 176) decided: "To avoid needless confusion and additional biogeographic chaos, it seems most prudent to erect a new genus for the new material from Río Turbio, instead of referring the material to an extant genus", though Hyžný & Klompmaker (2015: 422) criticized their view, remarking "It is questionable whether such approach can lead to a better understanding of the ghost shrimp fossil record. Careful comparison to extant taxa may resolve the affinities of fossil taxa better than erecting new genera without sound reasoning as has been shown by Hyžný and others". It seems that there has not yet appeared until now a systematic classification of extant callianassid species based mainly or only on cheliped characters as generic ones. This means that it would be prudent for palaeontologists not to try to refer to neontological taxa for their decision of genera by comparison, since there exist no reliable taxa based mainly on cheliped characters in neontological studies, that would be appropriate to compare their fossils with. It is evident from the above description that neontological generic concepts based primarily on the morphological differences of the male Plp1-2, cannot be easily related to genera established by palaeontologists, based only or mainly on characters on fossil chelipeds, and the two systems are largely incompatible for trying to understand evolutionary relationships.

Regarding DNA analysis, by which Tudge et al. (2000) showed that the two families Ctenochelidae and Callianassidae are both unambiguously monophyletic, it is still unknown how the patterns observed in the base-arrangements reflect the patterns found in the morphological characters, and this fact makes it impossible to decide species, genera, and higher level classifications based primarily or only on DNA analysis, since DNA analysis is basically dependent on the existing morphological classification (Sakai, 2013a: 1110).

Genus **Gourretia** De Saint Laurent, 1973

Gourretia De Saint Laurent, 1973: 514; Manning & Felder, 1991: 785; López de la Rosa et al., 1998: 394, fig. 1.
Dawsonius Manning & Felder, 1991: 785, figs. 4, 16.
Paragourretia Sakai, 2004: 568.

Diagnosis.— Mxp3 exopod present or not. Uropodal exopod with or without lateral notch. Abdominal somite 6 with or without lateral projections. Male Plp1 uniramous and bi- or tri-segmented, and chelate or subchelate distally; male Plp2 endopod bearing mesiosubdistally appendix masculina provided mesially with small appendix interna (Sakai & Türkay, 2005, fig. 3A).

Remarks.— The genus *Dawsonius* Manning & Felder, 1991 [sensu Sakai, 2005], based on *Callianassa latispina* Dawson, 1967 as the type species, is here synonymized with *Gourretia* De Saint-Laurent, 1973, because the male Plp1 is uniramous, bi-segmented and subchelate distally, and the male Plp2 endopod bears mesiodistally an appendix masculina with a small appendix interna mesially (Biffar, 1991, figs. 11i, 12e). Similarly *Paragourretia* Sakai, 2004, established for *Gourretia phuketensis* Sakai, 2002, by regarding then other characters than male Plp 1-2 as generic characters, is also here synonymized with *Gourretia*, because the male Plp1 is uniramous, bi-segmented and chelate distally (Sakai, 2004: 573), and the male Plp2 endopod bears mesiosubdistally an appendix masculina provided mesially with a small appendix interna.

Gourretia barracuda Le Loeuff & Intès, 1974, has long been admitted to be a member of *Gourretia*, but the present review has revealed that this species has a unique morphological type of the male Plp1-2 as follows: the male Plp1 is uniramous and bi-segmented, but hooked distally, and the male Plp2 endopod bears distally a distinct proximally-fused appendix masculina with a proximally-fused appendix interna (Le Loeuff & Intès, 1974, fig. 6t; De Saint Laurent & Le Loeuff, 1979, fig. 24a). These Plp1-2 characters are clearly different from those of the other *Gourretia* species, as well as those of species included in all other genera in the Gourretiidae (see table I), so *Gourretia barracuda* is here reclassified under a new genus, *Ivorygourretia* gen. nov. as *I. barracuda* (Le Loeuff & Intès, 1974).

The following three species, *Gourretia biffari* Blanco Rambla & Liñero Arana, 1994; *G. crosnieri* Ngoc-Ho, 1991; *Gourretia manihinae* Sakai, 1984 were described based on a female specimen as the holotype, so the male Plp1-2 are unknown, and *G. zarenkovi* (Sakai, 2010) was described on an incomplete male specimen without male Plp2, and the male Plp1 is uniramous, tri-segmented and chelate distally, so these four species remain provisionally as members of *Gourretia*, until exact identification becomes possible by examining male specimens of these species to be collected.

Type species.— *Callianassa denticulata* Lutze, 1937, a junior synonym of the preoccupied *Callianassa subterranea* var. *minor* Gourret, 1887. [*Gourretia minor* (Gourret, 1887); *Callianassa subterranea* var. *minor* Gourret, 1887.]

Not *Gourretia minor*, Le Loeuff & Intes, 1974: 26, fig. 4a-k. (= *Gourretia* sp. De Saint Laurent & Le Loeuff, 1979 = *Gourretia loeuffintesi* sp. nov.).

Species included.— *Gourretia biffari* BlancoRambla & Liñero Arana, 1994; *G. coolibah* Poore & Griffin, 1979; *G. crosnieri* Ngoc-Ho, 1991; *G. denticulata* Lutze, 1937 (= *Callianassa stebbingi*, Gottlieb, 1953; *Gourretia minor*, Manning & Števčić, 1982: 296; *G. serrata* De Saint Laurent & Le Loeuff, 1979; *G. laevidactyla* Liu & Liu, 2010; *G. loeuffintesi* Sakai, 2011); *G. lahouensis* Le Loeuff & Intès, 1974; *G. laresi* Blanco Rambla & Liñero Arana, 1994; *G. latispina* (Dawson, 1967); *G. manihinae* Sakai, 1984; *G. phuketensis* Sakai, 2002; *G. zarenkovi* (Sakai, 2010).

Genus **Heterogourretia** gen. nov.

Diagnosis.— Mxp3 exopod present. Abdominal somite 6 without lateral projections. Male Plp1 absent; male Plp2 endopod provided with segmented appendix interna distally, but no appendix masculina. Uropodal exopod bilobulate (Sakai, 2002, figs. 6H, 8C, D).

Remarks.— The genus *Paragourretia*, established by Sakai (2004) by other characters than those of male Plp1-2, regarded then as generic characters, is here synonymized with *Gourretia*, because the species included in *Paragourretia* have the same morphological type of the male Plp1-2 as in *Gourretia*, except *Paragourretia aungtonyae* (Sakai, 2002). This last-mentioned species has turned out by further examination and consideration, to possess morphological characters of the male Plp1-2 different from those of other members of *Gourretia*. Namely in *P. aungtonyae* the male Plp1 is absent and the male Plp2 endopod is provided distally with a segmented appendix interna, but there is no appendix masculina (see table I). These generic differences necessitate *P. aungtonyae* not to be included in *Gourretia*, and to be reclassified under a new genus, *Heterogourretia*, as *H. aungtonyae*.

Type species.— *Gourretia aungtonyae* Sakai, 2002.

Species included.— *Heterogourretia aungtonyae* (Sakai, 2002).

Etymology.— The new generic name "*Heterogourretia*" is derived from the Greek, "hetero", meaning different, with the existing generic name "*gourretia*" added. The gender is feminine.

Genus **Ivorygourretia** gen. nov.

Diagnosis.— Mxp3 exopod present. Abdominal somite 6 without lateral projections. Male Plp1 bi-segmented and hooked distally; male Plp2 endopod provided distally with stout proximally-fused appendix masculina with proximally-fused appendix interna mesially. Uropodal exopod without lateral notch.

Remarks.— In *Gourretia barracuda* Le Loeuff & Intès, 1974, the male Plp1 is hooked distally and the male Plp2 endopod bears distally a stout proximally-fused appendix masculina with a proximally-fused appendix interna mesially (Le Loeuff & Intès, 1974, fig. 6t; De Saint Laurent & Le Lœuff, 1979, fig. 24a). These are safely considered to be of generic significance (see table I), and therefore *G. barracuda* is here designated as the type species of a new genus, *Ivorygourretia*.

Type species.— *Gourretia barracuda* Le Loeuff & Intès, 1974.

Species included.— *Ivorygourretia barracuda* (Le Loeuff & Intès, 1974).

Etymology.— The generic name "*Ivorygourretia*" is derived from the type locality, Ivory Coast, where the holotype of *Gourretia barracuda* was collected, in combination with the existing generic name "*gourretia*". The gender is feminine.

Genus **Laurentgourretia** Sakai, 2004

Laurentgourretia Sakai, 2004: 557, figs. 1-3.

Diagnosis.— Mxp3 exopod absent. Abdominal somite 6 without lateral projections. Male Plp1-2 unknown. Uropodal exopod with lateral notch.

Remarks.— *Laurentgourretia rhopalommata* was described based on a single female as the holotype, so the male Plp1-2 are unknown. Thus the validity of this genus will need to be re-evaluated by examining male specimens to be collected.

Type species.— *Laurentgourretia rhopalommata* Sakai, 2004.

Species included.— *Laurentgourretia rhopalommata* Sakai, 2004.

Genus **Paracalliax** De Saint Laurent, 1979

Paracalliax De Saint Laurent, 1979: 1396; De Saint Laurent & Le Lœuff, 1979: 86, fig. 26a-I, 27a-c; 28a-h.

Diagnosis.— Mxp3 exopod present. Abdominal somite 6 with rounded lateral projections. Male Plp1-2 unknown. Uropodal exopod without lateral notch.

Remarks.— *Paracalliax* is the same in the characters of the Mxp3, abdominal somite 6, and uropodal exopod as *Ivorygourretia* gen. nov., *Plantesgourretia* gen. nov., and *Ruiyuliugourretia* gen. nov., and therefore might be synonymized with one of the three new genera, but not at present with any of them, because in *Paracalliax* the male Plp1-2 are unknown on account of the female holotype. Hence, *Paracalliax* remains provisionally as it is at the present generic status, until exact identification is made possible by examining male specimens of *Paracalliax bollorei*, yet to be collected.

Type species.— *Paracalliax bollorei* De Saint Laurent, 1979.

Species included.— *Paracalliax bollorei* De Saint Laurent, 1979.

Genus **Plantesgourretia** gen. nov.

Diagnosis.— Mxp3 exopod present. Abdominal somite 6 without lateral projections. Male Plp1 bi-segmented and chelate distally. Plp2 biramous; exopod shorter than endopod; endopod bilobed distally, and inner lobe bearing distally appendix masculina with setae, but no appendix interna (corrected from description of male Plp2 by Sakai, 2004, fig. 6D). Uropodal exopod without lateral notch.

Remarks.— *Gourretia nosybeensis* Sakai, 2004, was initially considered a member of *Gourretia*, but it should now be excluded from *Gourretia* and reclassified in a new genus *Plantesgourretia*, because significant differences can be observed in the male Plp1-2 of *G. nosybeensis* (see table I).

Type species.— *Gourretia nosybeensis* Sakai, 2004.

Species included.— *Plantesgourretia nosybeensis* (Sakai, 2004).

Etymology.— "*Plantesgourretia*" is derived from the name of Dr. Raphael Plante, who collected the holotype of the type species in Madagascar in 1965-70, in combination with the existing generic name "*gourretia*". The gender is feminine.

Genus **Ruiyuliugourretia** gen. nov.

Diagnosis.— Mxp3 exopod present. Abdominal somite 6 without lateral projections. Male Plp1 four-segmented and hooked distally (fig. 3A); male Plp2 foliaceous, endopod bearing mesiosubdistally appendix masculina with setae, but no appendix interna (fig. 3B). Uropodal exopod without lateral notch.

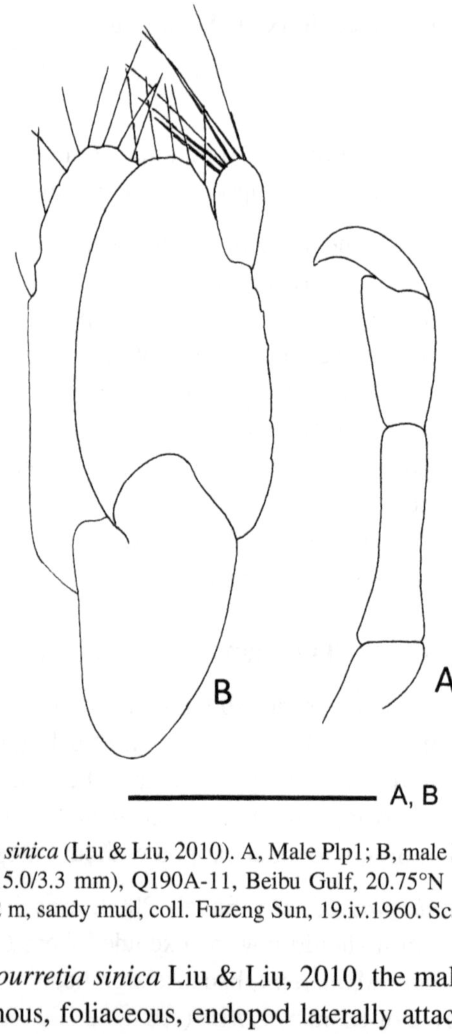

Fig. 3. *Ruiyuliugourretia sinica* (Liu & Liu, 2010). A, Male Plp1; B, male Plp2. A-B, MBM-210336, paratype male (TL/CL 15.0/3.3 mm), Q190A-11, Beibu Gulf, 20.75°N 107.5°E, off shore China, South China Sea, 32 m, sandy mud, coll. Fuzeng Sun, 19.iv.1960. Scale bar, A-B, 0.5 mm.

Remarks.— In *Gourretia sinica* Liu & Liu, 2010, the male Plp2 was described as "Pleopod 2 biramous, foliaceous, endopod laterally attached a wide appendix interna and distally without appendix masculina" (Liu & Liu, 2010: 754, fig. 9B). However, judging from their figure, "wide appendix interna" can rather be considered an appendix masculina, because it bears distal setae typical of an appendix masculina, but no patch of hooklets that would be expected to be present on an appendix interna. It has been confirmed by the present examination of the male paratype of *Gourretia sinica* (MBM-210336, paratype male (TL/CL 15.0/3.3 mm), Q190A-11, Beibu Gulf, 20.75°N 107.5°E, offshore, China, South China Sea, 32 m, sandy mud, 19.iv.1960, leg. Fuzeng Sun) that the Plp1 is four-segmented and hooked distally, and the Plp2 endopod bears mesiosubdistally an appendix masculina with setae, but no appendix interna (fig. 3B). This unique conformation of the male pleopods is considered generically important, and *G. sinica* is here designated as the type species of a new genus, *Ruiyuliugourretia.*

Type species.— *Gourretia sinica* Liu & Liu, 2010.

Species included.— *Ruiyuliugourretia sinica* (Liu & Liu, 2010).

Etymology.— "*Ruiyuliugourretia*" is dedicated to the late Dr. Ruiyu Liu of the Chinese Academy of Sciences, who was a leading figure in the Chinese Crustacean Study, in combination with the existing generic name "*gourretia*". The gender is feminine.

Genus **Tuerkaygourretia** gen. nov.

Diagnosis.— Rostrum acutely triangular. A1 short, but A2 elongate. Mxp3 with exopod. Male Plp1 uniramous and tri-segmented (fig. 5F) and chelate distally; male Plp2 biramous; exopod shaped as elongate oval and shorter than endopod, bearing marginal setae in distal half, endopod consisting of two longitudinal sections, lateral section shaped as elongate oval, larger than exopod, and mesial section bearing distally articulated appendix masculina with distal setae, with proximally-fused appendix interna in parallel with each other. Abdominal somite 6 without lateral projections. Telson subsquare. Uropodal exopod bilobulate.

Remarks.— *Tuerkaygourretia galathea* sp. nov., described below, is similar to the members of *Gourretia*, especially *G. crosnieri* Ngoc-Ho, 1991, but there exist several morphological differences between these two species as mentioned in the remarks under *Tuerkaygourretia galathea* sp. nov. Especially this new species possesses a very different type of male Plp1-2 as compared to any other species under any genera of the Gourretiidae (see table I). These unique morphological characteristics, which are considered differences at the generic level, suggest that it would hardly be possible to classify the new species in the genus *Gourretia*. That is why the new species is to be classified under a new genus established in the present paper, as *Tuerkaygourretia galathea* sp. nov.

Etymology.— The new generic name "*Tuerkaygourretia*" is dedicated to the late Prof. Dr. Michael Türkay, who devoted himself to crustacean taxonomy, greatly influencing many carcinologists, and strongly advocated the significance of genital organs such as male pleopods, petasma, or gonopods as important characters for understanding decapod phylogeny, and as the basis for generic classification. The gender is feminine.

Type species.— *Tuerkaygourretia galathea* sp. nov.

Species included.— *Tuerkaygourretia galathea* sp. nov.

Tuerkaygourretia galathea sp. nov.
(figs. 4, 5)

Material examined.— ZMUC-CRU-8367, holotype: female (TL/CL 26.0/7.0 mm), Galathea Stn. 290, off Tranquebar, SE India, 11°06′N 80°05′E, muddy sand, 94 m, Van Veen grab, 21.iv.1951.

ZMUC-CRU-8407, paratypes: 2 males (TL/CL 17.0/4.1 mm, 7.0/2.0 mm), same data as holotype; ZMUC-CRU-8431, 2 females (TL/CL 21.0/5.4, 10.0/2.9 mm), Galathea Stn. 304, India, Bay of Bengal, 20°37′N 87°33′E, muddy sand with shells, 50 m, Petersen grab, 26.iv.1951; ZMUC-CRU-8432, 1 male (TL/CL 21.0/4.4 mm), Galathea Stn. 357, 800 m S of Singapore [Tg. Gul.], muddy sand with coral and shell fragments, 10 m, Van Veen grab, 31.v.1951.

Description.— Small species. Rostrum acutely triangular and shorter than eyestalks (fig. 4A, B). A1 flagella short; A2 flagella elongate. Cervical groove located in posterior third of carapace including rostrum. Mxp3 with three-segmented exopod (fig. 5C). P1 unequal in size and dissimilar in shape. In larger cheliped (fig. 4C) ischium about 3 times as long as wide, unarmed dorsally and slightly convex distodorsally, but minutely denticulate to distinct distal spine ventrally; merus about as long as ischium, about 1.6 times as long as wide, and slightly arcuate and unarmed dorsally, but armed subproximally with sharp tooth followed distally by minute denticulation ventrally; carpus shaped as semicircle with rounded ventral margin, 0.7 times as long as wide and 0.5 times as long as merus; chela 4.0 times length of carpus; palm twice as long as carpus, about 1.2 times as long as wide and smooth dorsally and ventrally, distal margin unarmed, directed obliquely downward in ventral half to fixed finger; fixed finger 0.7 times length of palm, unarmed and smoothly curved on prehensile margin; dactylus slightly shorter than palm, curved downward distally and largely concave on distal half of unarmed prehensile margin. Smaller cheliped (fig. 4E) slenderly elongate and less massive than larger cheliped; ischium slender, slightly convex distodorsally, and unarmed on dorsal and ventral margins, though armed with 2 strong teeth midventrally in male (fig. 4D); merus about 0.8 times as long as ischium, twice as long as wide, shaped as oval, and unarmed dorsally, bearing sharp subproximal tooth ventrally; carpus as long as wide, 0.7 times length of merus; palm elongate, gradually convergent distally, 2.0 times as long as wide, and rounded ventroproximally; fixed finger 0.7 times length of palm and smooth on prehensile margin; dactylus smooth on prehensile margin, though armed with triangular denticles directed proximally on distal 2/3 of prehensile margin in male (fig. 4D). P2 chelate. P3 (fig. 5E) simple distally; ischium as long as wide, protruding ventrodistally; merus 3.0 times as long as wide; carpus 0.6 times length of merus; propodus subsquare, 1.2 times length of carpus, and 1.6 times as long as wide, bearing setae marginally and on lateral surface, though oval in smaller male specimen (fig. 5D); dactylus slenderly triangular, 0.4 times as long as propodus. P4 simple distally, propodus subquadrate with setae laterally. P5 subchelate distally, propodus and dactylus setose laterally. Female Plp1 simple and elongate (fig. 5G). Abdominal somite 6 without lateral projections. Uropodal endopod oval, smaller than exopod, bearing distinct longitudinal median carina; uropodal exopod bilobulate (fig. 4F). Telson shaped as stout inverted triangle with rounded posterior margin.

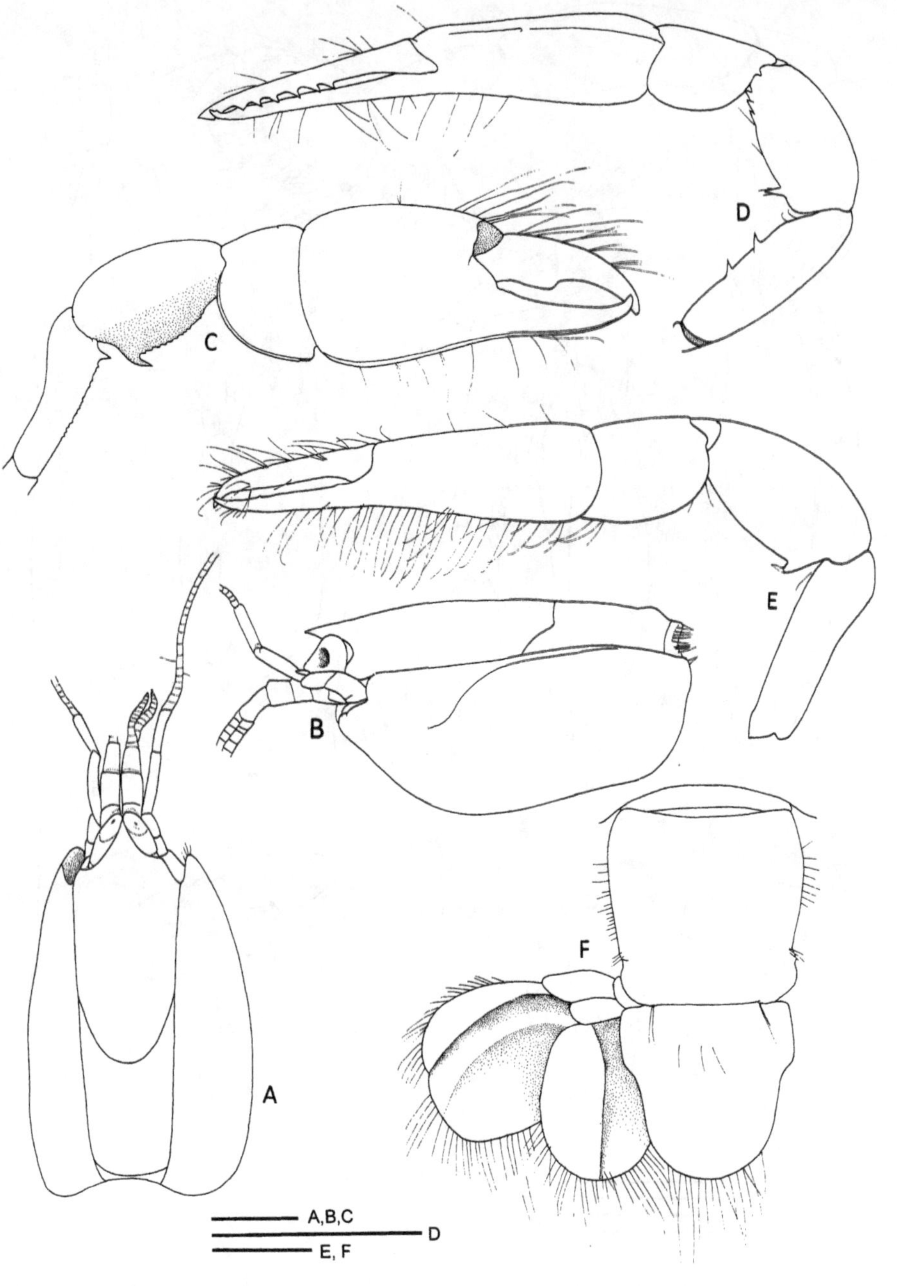

Fig. 4. *Tuerkaygourretia galathea* sp. nov. A, Carapace, eyestalks and A1-A2, dorsal view; B, same, lateral view; C, larger cheliped on right side, lateral view; D, smaller cheliped on left side, lateral view; E, smaller cheliped on left side, lateral view; F, abdominal somite 6, telson and uropods on left side, dorsal view. A, B, C, E, F, female (TL/CL 26.0/7.0 mm), Galathea Stn. 290, off Tranquebar, SE India, 11°06′N 80°05′E; D, male (TL/CL 33.0/7.7 mm), Thai-Danish Expedition, Stn. 1160. Scale bar, A-F, 1 mm.

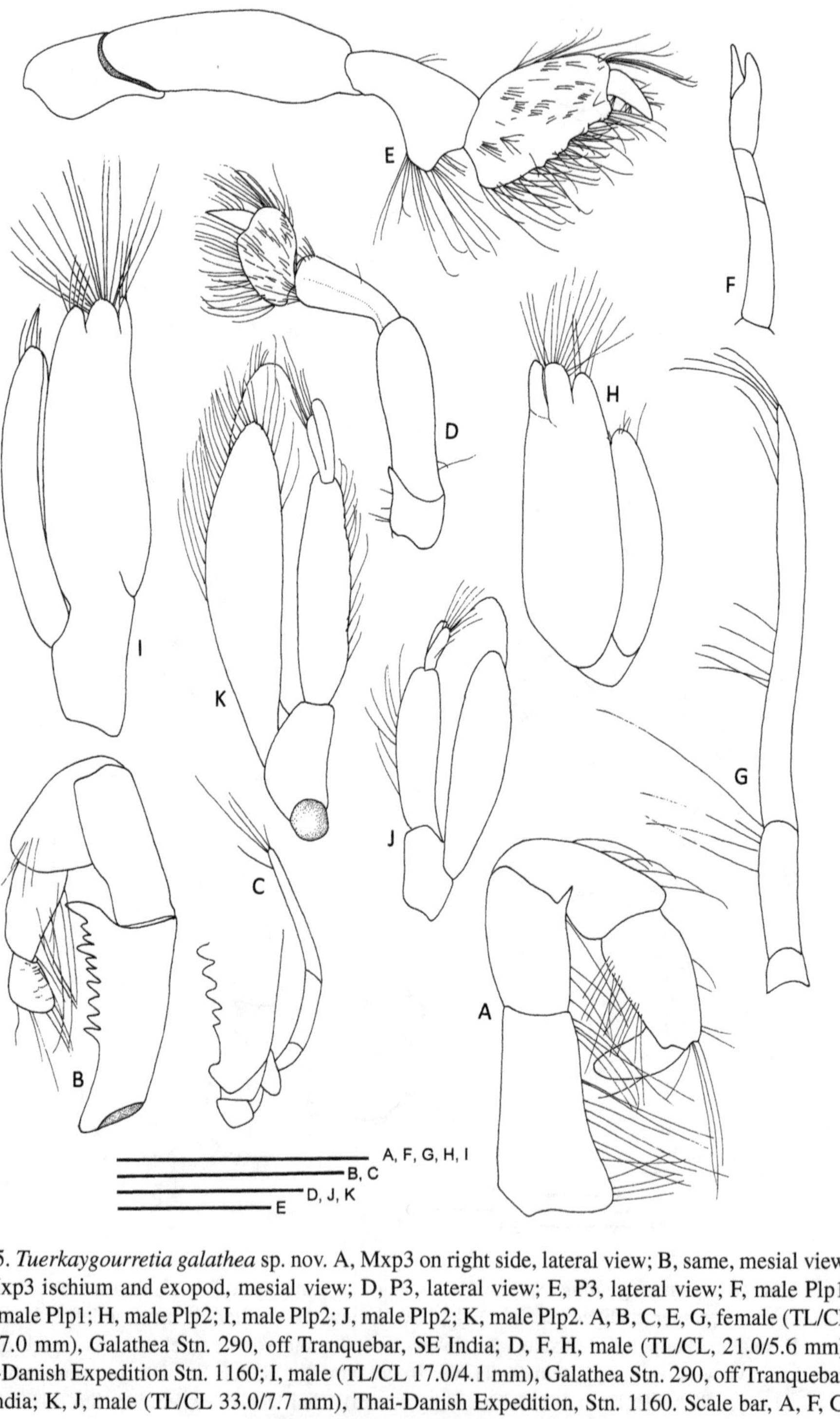

Fig. 5. *Tuerkaygourretia galathea* sp. nov. A, Mxp3 on right side, lateral view; B, same, mesial view; C, Mxp3 ischium and exopod, mesial view; D, P3, lateral view; E, P3, lateral view; F, male Plp1; G, female Plp1; H, male Plp2; I, male Plp2; J, male Plp2; K, male Plp2. A, B, C, E, G, female (TL/CL 26.0/7.0 mm), Galathea Stn. 290, off Tranquebar, SE India; D, F, H, male (TL/CL, 21.0/5.6 mm), Thai-Danish Expedition Stn. 1160; I, male (TL/CL 17.0/4.1 mm), Galathea Stn. 290, off Tranquebar, SE India; K, J, male (TL/CL 33.0/7.7 mm), Thai-Danish Expedition, Stn. 1160. Scale bar, A, F, G, H, I, 2 mm; B, C, D, E, J, K, 1 mm.

In male specimens, Plp1 three-segmented and chelate distally (fig. 5F); Plp2 in larger male biramous, exopod shaped as elongate oval and shorter and narrower than endopod, bearing marginal setae in distal half; endopod consisting of two longitudinal sections, external section shaped as large elongate oval, larger than exopod, and mesial section narrow, bearing distally articulated appendix masculina with distal setae, and proximally-fused appendix interna in parallel with each other (fig. 5J, K), but Plp2 endopod in smaller specimens slightly narrowing and trilobed distally, bearing marginal setae (fig. 5H, I).

Remarks.— The present specimens look similar to *Gourretia crosnieri* Ngoc-Ho, 1991 from off Tranquebar, but some morphological differences are detected between the present female specimen as the holotype and the female holotype of *G. crosnieri* as follows: (1) in the present female holotype, the A1 peduncle slightly overreaches the A2 penultimate segment (fig. 4A) (vs. in *G. crosnieri*, the A1 peduncle reaches the middle of the A2 penultimate segment); (2) the ventral margin of the ischium in the larger cheliped bears a distal spine (vs. the ventral margin of the ischium in the larger cheliped is unarmed); (3) the Mxp3 merus bears a distomesial spine (vs. the Mxp3 merus bears no distomesial spine); (4) the Mxp3 exopod is three-segmented (vs. the Mxp3 exopod is unsegmented). Furthermore the present male specimens possess a unique morphology of male Plp1-2, differentiated at the generic level from the *Gourretia* species (see table I). It would be suggested from their morphological features described in the above, that the present specimens are to be determined as a new species, which could not be classified under any genera of the family Gourretiidae. This means that this new species is to be classified under a new genus as a member of the Gourretiidae.

Distribution.— Off Tranquebar, India (type locality); W. of Subitana, Port Moresby, Papua New Guinea; West Malay Peninsula; Singapore Island.

Etymology.— The new species name "*galathea*" is derived from the famous Danish Galathea Expedition. It thus is a noun in apposition to the generic name.

Family THOMASSINIIDAE De Saint Laurent, 1979

Thomassiniidae De Saint Laurent, 1979: 1396; De Saint Laurent & Le Lœuff, 1979: 95; Sakai, 1992a: 18; Poore, 1997: 398.

Remarks.— The family Thomassiniidae was synonymized with Callianideidae by Poore (2015b). However, they are clearly different from each other, because in Thomassiniidae the male Plp2 endopod and exopod bear no filaments, while in Callianideidae they bear filaments.

Genus **Garyia** Sakai, 2011

Garyia Sakai, 2011: 530.

Diagnosis.— Rostrum obsolete. Cervical groove present. Linea thalassinica complete. Eyestalks contiguous, cornea present apically. A2 scaphocerite absent. P1 subequal. P3 propodus rectangular. P5 simple. Male Plp1 absent; male Plp2 biramous, endopod bearing mesially appendix masculina with distal setae, provided mesially with appendix interna; male Plps3-5 endopods with appendix interna. Telson wider than long and trapezoid. Uropodal exopod ovate, lacking dorsal plate. Setal rows present on carapace and uropodal endopod. (After Sakai, 2011: 530.)

Remarks.— The genus *Garyia* (Sakai, 2011: 530) was established based on *Thomassinia moorea* Poore, 1997, excluded from the members of the genus *Thomassinia* De Saint Laurent, 1979. The reasons are that in *Thomassinia moorea* the linea thalassinica is complete, the male Plp1 is absent, and the male Plp2 endopod bears an appendix masculina with an appendix interna; whereas in the type species of the genus *Thomassinia*, *T. gebioides* Poore, 1997, the linea thalassinica is incomplete, running only on the anterior three-fourths of the carapace, the male Plp1 is absent, but the male Plp2 endopod bears distally fused appendices, i.e., masculina and interna. As is evident from the above description, these two species are quite different from each other at the generic level, because of the morphological difference of the male Plp2. Hence, the genus *Garyia* was validly established. In the present paper, a new species, *G. sachikoae* sp. nov., from the collection of the Galathea Expedition is added. *Garyia arno* (Poore, 1997), which was erroneously included in the genus *Garyia* (Sakai, 2011: 531), is to remain in the genus *Thomassinia* as *T. arno* Poore, 1997.

Type species.— *Thomassinia moorea* Poore, 1997, by original designation (Sakai, 2011).

Species included.— *Garyia moorea* (Poore, 1997) and *G. sachikoae* sp. nov.

Garyia sachikoae sp. nov.
(fig. 6)

Material examined.— ZMUC-CRU-8443, holotype male (TL/CL 11.0/2.5 mm), Galathea Stn. 258, off Mombasa, 4°06′S 39°43′E, Kenya, 53 m, coralline gravel with mud and shells, 2.v.1951.

Description.— Rostrum small (fig. 6A). Eyestalks stout and bulged laterally, with dotted cornea in anterior part, overreaching A1 proximal segment. A1 peduncle slightly overreaching A2 peduncle. Carapace deep in anterior part; cervical groove located at posterior third and linea thalassinica complete (fig. 6A, B); submarginal vertical setal row present anteriorly in branchial region (fig. 6A). Mxp3 (fig. 6C) ischium and merus provided with thick setae on straight mesial margin.

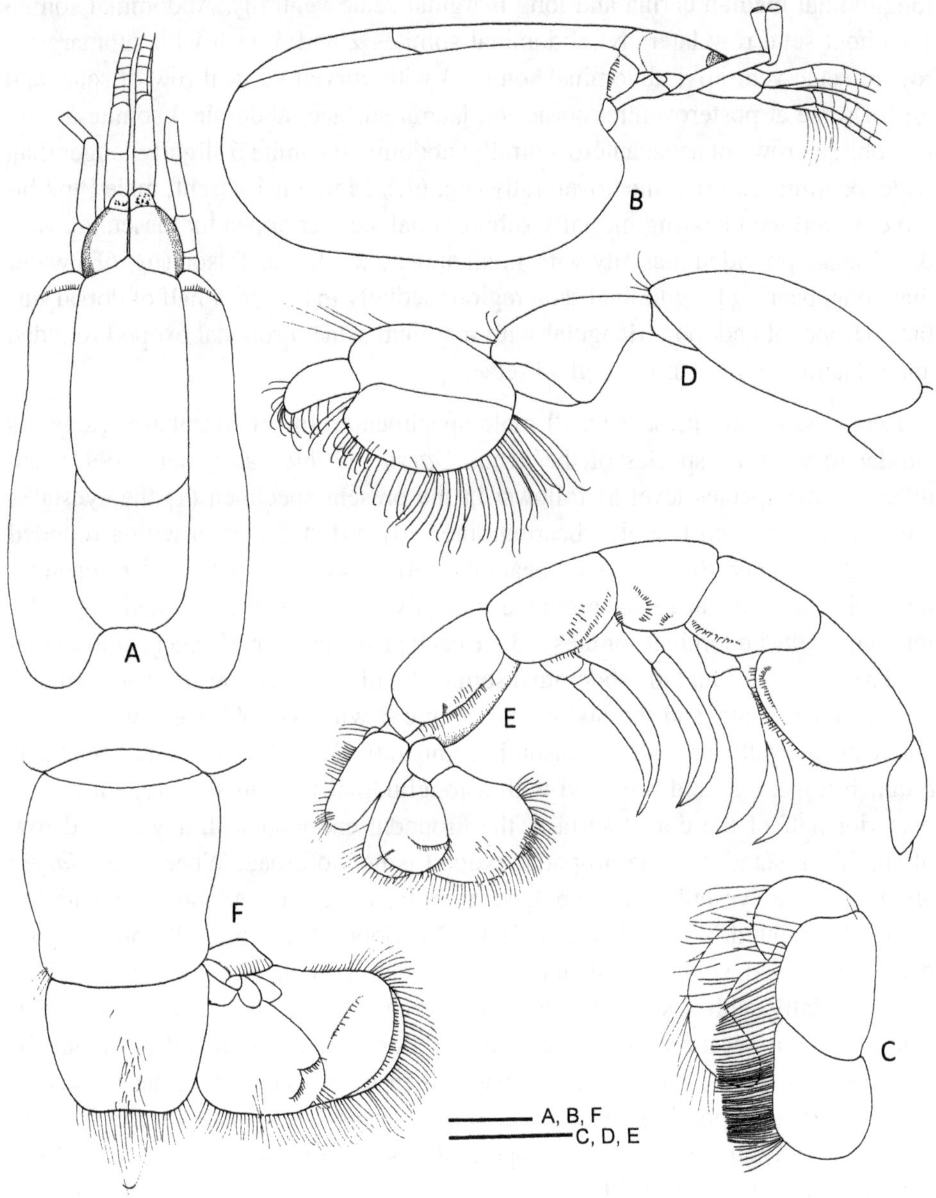

Fig. 6. *Garyia sachikoae* sp. nov. A, Carapace with eyestalks and A1-A2, dorsal view; B, same, lateral view; C, Mxp3 on left side, lateral view; D, P3 on left side, lateral view; E, abdominal somites, telson and uropods, lateral view; F, abdominal somite 6, telson and uropods on right side, dorsal view. A-F, holotype female (TL/CL 11.0/2.5 mm), Galathea Stn. 258, off Mombasa. Scale bar, A-E, 0.5 mm.

P1 missing. P2 chelate. P3 (fig. 6D) propodus rounded in lateral view, bearing longitudinal median carina and long marginal setae ventrally. Abdominal somite 1 without setal row laterally; abdominal somites 2 and 3 each with submarginal row of setae ventrally; abdominal somite 4 with curved vertical row of setae and tuft of setae at posteroventral corner on lateral surface; abdominal somite 5 with two oblique rows of setae anteroventrally; abdominal somite 6 slightly longer than wide, bearing setal row medioventrally (fig. 6E). Male Plp1 absent; male Plp2 biramous, endopod bearing mesially subproximal slender appendix masculina with distal setae, provided mesially with small appendix interna. Telson (fig. 6E) wider than long, bearing longitudinal setal region medially in posterior half of dorsal surface. Uropodal endopod triangular with marginal setae; uropodal exopod rounded anterolaterally with dense marginal setae.

Remarks.— The present small male specimen from off Mombasa, Kenya is similar to the type species of the genus *Garyia*, *G. moorea* (Poore, 1997), but differs at the species level as follows: in the present specimen (1) the eyestalks are stout and bulged laterally, bearing anteriorly a dotted cornea with a rounded apex, (2) the male Plp2 endopod bears mesially a subproximal slender appendix masculina with distal setae, provided mesially with a proximal small appendix interna, (3) the abdominal somites 2-3 are each provided with a longer submarginal setal row ventrally, and the abdominal somite 4 with a curved row of setae medially on the ventral half, and the abdominal somite 5 with two oblique rows of setae anterosubventrally, and (4) the telson is comparatively wider than long, shaped as a tallish trapezoid, and provided with a longitudinal median setal region on the posterior half of the dorsal surface, the uropodal endopod with a V-shaped row of spiniform setae, and the uropodal exopod is not so broad. Whereas in *Garyia moorea* (1) the eyestalks are not bulged laterally, and triangular anteriorly with the cornea pigmented black, (2) the male Plp2 endopod bears mesially an elongate distally-incurved subproximal appendix masculina with distal setae, provided mesiomedially with a slender appendix interna, (3) the abdominal somites 2-6 bear indistinctly scattered setae submarginally or on the ventral half of the lateral surfaces, and (4) the telson is much wider than long, shaped as a flat trapezoid, the uropodal endopod is provided with a short transverse row of spiniform setae posterolaterally, and the uropodal exopod is broadly rounded distally (Poore, 1997: 416, fig. 38A, H; Sakai, 2011: 531-532, fig. 71, fig. 72A, B, C). Hence, this male specimen is safely determined as a new species classified in *Garyia*, and named *Garyia sachikoae* sp. nov.

Type locality.— Mombasa, 4°06′S 39°43′E, Kenya, 53 m.

Distribution.— Only known from the type locality.

Etymology.— The new species name "*sachikoae*" is derived from the first name of the present author's wife, Mrs. Sakai, who has been cooperating with him over 50 years with sincere support and encouragement, especially under difficult situations around him. The name thus is a noun in the genitive singular.

ACKNOWLEDGEMENTS

The present author is much indebted to Dr. D. Eibye-Jacobsen, Zoological Museum, Natural History Museum of Denmark, Copenhagen, who provided me with the data on the "Galathea" Expedition (Wolff, T., 1964); Mrs. K. van Dorp of the, Naturalis Biodiversity Center in Leiden, Mrs. M. Lowe, Natural History Museum, Zoology, Cromwell Road, London, and Mr. A. Allspach and Mrs. B. Trautwein of the Senckenberg Forschungsinstitut und Naturmuseum, Frankfurt am Main, who sent me valuable material of *Jaxea nocturna* Nardo, 1847, for comparison with *J. novaezealandiae* Wear & Yaldwyn, 1966; Dr. X. Li, Institute of Oceanology, Chinese Academy of Sciences, Beijing, China, and Dr. W. Liu, Associate Professor, School of Ecological and Environmental Sciences, East China Normal University (ECNU), Shanghai, China, who kindly sent me the type specimen of *Gourretia sinica* Liu & Liu, 2010, and its photos; and Dr. Sammy De Grave, Oxford University Museum of Natural History, U.K., and Dr. Arthur Anker, Museu Paraense Emílio Goeldi, Belém, Brazil, who kindly sent me valuable references. Thanks are also extended to Dr. J. Carel von Vaupel Klein, formerly at Leiden University, and to Mr. Peter J. F. Davie, Queensland Museum, Brisbane, who critically read this manuscript for publication.

REFERENCES

BIFFAR, T. A., 1970. Three new species of callianassid shrimp (Decapoda, Thalassinidea) from the western Atlantic. Proc. Biol. Soc. Washington, **83**(3): 35-50, figs. 1-3.

BIFFAR, T. A., 1971. The genus *Callianassa* (Crustacea, Decapoda, Thalassinidea) in South Florida, with keys to the western Atlantic species. Bulletin of Marine Science, **21**(3): 637-715.

BLANCO RAMBLA, J. P. & I. LIÑERO ARANA, 1994. New records and new species of ghost shrimps (Crustacea: Thalassinidea) from Venezuela. Bulletin of Marine Science, **55**(1): 16-29.

BORRADAILE, L. A., 1903. On the classification of the Thalassinidea. Annals and Magazine of Natural History, **12**(7): 534-551, 638.

BOTT, R., 1970. Die Süßwasser-Krabben von Europa, Asien, Australien und ihre Stammesgeschichte. Eine Revision der Potamoidea und Parathelphusoidea (Crustacea, Decapoda). Abhandlungen der Senckenbergischen Naturforschenden Gesellschaft, **526**: 1-338, figs. 1-8, pls. 1-58, 1 map. (Frankfurt).

BRANDIS, D., 2002. On the taxonomic status and biogeography of the Isolapotamidae Bott, 1970 (Decapoda, Brachyura). J. Nat. Hist., **36**: 1291-1339, 22 figs.

DANA, J. D., 1852. Macroura. Conspectus Crustaceorum & Conspectus of the Crustacea of the Exploring Expedition under Capt. C. Wilkes, U.S.N. Proc. Acad. Nat. Sci., Philadelphia, **6**: 6-28.

DAWSON, C. E., 1967. *Callianaassa latispina* (Decapoda, Thalassinidea), a new mud shrimp from the northern Gulf of Mexico. Crustaceana, **13**(2): 190-196, fig. 1.

DE HAAN, W., 1833-1850. Crustacea. In: P. F. VON SIEBOLD (ed.), Fauna Japonica sive descriptio animalium, que in itinere per Japoniam, jussu et auspicies superiorum, qui Summum in India Batavia Imperium tenent, suscepto, annis 1823-1830 collegit, notis, observationibus et adumbrationibus illustravit: 1-243, pls. 1-55, A-Q. (Amsterdam).

DE MAN, J. G., 1928. The Thalassinidae and Callianassidae collected by the Siboga Expedition, with some remarks on the Laomedidae. The Decapoda of the Siboga-Expedition. Part 7. Siboga Expeditie mon., **39a** (6)(109): 1-187, pls. 1-10.

DE SAINT LAURENT, M., 1973. Sur la systématique et la phylogénie des Thalassinidea: definition des familles des Callianassidae et des Upogebiidae et diagnose de cinq genres nouveaux (Crustacea Decapoda). C. R. Acad. Sci. Paris, **277**: 513-516.

DE SAINT LAURENT, M., 1979. Sur la classification et la phylogénie des Thalassinidea: définition de la superfamille des Thomassiniinae et de la deux genres nouveaux (Crustacea Decapoda). C. R. Acad. Sci. Paris, **288**(18): 1395-1397.

DE SAINT LAURENT, M. & P. LE LŒUFF, 1979. Upogebiidae et Callianassidae. Crustacés Décapodes: Thalassinidea. 1. Ann. Inst. océanogr. Monaco, **55**(Suppl.): 29-101, figs. 1-28.

DWORSCHAK, P. C., 2005. A new species of *Eucalliax* Manning & Felder, 1991 (Decapoda: Callianassidae) from the Red Sea. Proc. Biol. Soc. Washington, **118**(1): 209-217.

GOTTLIEB, E., 1953. Decapod crustaceans in the collection of the Sea Fisheries Research Station, Caesarea, Israel. Bull. Res. Counc. Israel, **2**: 440-441.

HELLER, C., 1863. Die Crustaceen des südlichen Europa. Crustacea Podophthalmia. Mit einer Übersicht über die horizontale Verbreitung sämtlicher europäischer Arten: i-xi, 1-336, pls. 1-10. (Vienna).

HYŽNÝ, M. & A. A. KLOMPMAKER, 2015. Systematics, phylogeny and taphonomy of ghost shrimps (Decapoda): a perspective from the fossil record. Arthropod Systematics & Phylogeny, **73**(3): 401-437.

KENSLEY, B. & A. B. WILLIAMS, 1990. A new thalassinidean shrimp (Crustacea, Decapoda, Axiidae) from the Middle Eocene of South Carolina. J. Paleont, **64**(5): 798-802.

KINGSLEY, J. S., 1897. On a new genus and two new species of macrurous Crustacea. Bulletin of the Essex Institute, **27**: 95-99.

LATREILLE, P. A., 1831. Cours d'entomologie ou de l'histoire naturelle des Crustacés, des Arachnides, des Myriopodes et des Insectes. Exposition méthodique des ordres, des familles et des genres des trois premières classes. A l'usage des élèves de l'école du Muséum d'Histoire naturelle: i-xiii, 1-568, Atlas, 1-26, pls. 1-24. (Paris).

LE LOEUFF, P. & A. INTÈS, 1974. Les Thalassinidea (Crustacea, Decapoda) du golfe de Guinée: Systématique-écologie. Cahiers de l'Office de Recherches Scientifiques et Techniques Outre-Mer, (Océanographique), **12**(1): 17-69, figs. 1-22, tables 1-5.

LIU, W., R. LIU & J. Y. LIU, 2010. Two new species of the axiidean genus *Gourretia* de Saint Laurent, 1973 (Decapoda: Ctenochelidae) from the South China Sea. Journal of Crustacean Biology, **30**: 745-756.

LÓPEZ DE LA ROSA, I., J. E. GARCÍA RASO & A. RODRÍGUEZ MARTÍNI, 1998. First record of *Gourretia denticulata* (Lutze, 1937) (Crustacea, Decapoda, Thalassinidea) from the Atlantic coast of Spain. Sci. Mar., **62**(4): 393-395.

LUTZE, J., 1937. Eine neue *Callianassa*-Art aus der Adria. Note dell'Istituto Italo-Germanico di Biologia Marina di Rovigno d'Istria, **2**(1): 1-12, figs. 1-7, 1 map.

MANNING, R. B. & D. L. FELDER, 1991. Revision of the American Callianassidae (Crustacea: Decapoda: Thalassinidea). Proc. Biol. Soc. Wash., **104**(4): 764-792.

MANNING, R. B. & Z. ŠTEVČIĆ, 1982. Decapod fauna of the Piran Gulf. Quaderni Lab. Tecnol. Pesca Ancona, **3**(2-5): 285-304.

NARDO, G. D., 1847. Sinonimia moderna delle specie registrate nell'opera intitolata: Descrizione de Crostacei, de Testacei e de Pesci che abitano le lagune e golfo Veneto rappresentati in figure, a chiaro-scuro dall'abate Stefano Chriereghini Ven. Clodiense applicata per commissione governativa: i-xi, 1-127. (Venezia [= Venice]).

NGOC-HO, N., 1991. Sur quelques Callianassidae et Upogebiidae de Nouvelle-Calédonie (Crustacea, Thalassinidea). In: B. RICHER DE FORGES (ed.), Le benthos des fonds meubles des lagons de Nouvelle Calédonie, **1**: 281-311, figs. 1-11. (Paris).

NGOC-HO, N., 2003. European and Mediterranean Thalassinidea (Crustacea, Decapoda). Zoosystema, **25**(3): 439-555.

PAIVA, T. S., M. TAVARES & I. D. SILVA-NETO, 2010. Removing the homonymy between *Laurentiella* Dragesco & Njine, 1971 (Protista) and *Laurentiella* Le Loeuff & Intès, 1974 (Crustacea). Crustaceana, **83**(6): 767-768.

PERVESLER, P. & P. C. DWORSCHAK, 1985. Burrows of *Jaxea nocturna* Nardo in the Gulf of Trieste. Senckenbergiana marit., **17**(1/3): 33-53, 3 figs., 4 tabs., 4 pls.

POORE, G. C. B., 1994. A phylogeny of the families of Thalassinidea (Crustacea: Decapoda) with keys to families and genera. Mem. Mus. Victoria, **54**: 79-120, figs. 1-9.

POORE, G. C. B., 1997. A review of the thalassinidean families Callianideidae Kossmann, Micheleidae Sakai, and Thomassiniidae De Saint Laurent (Crustacea, Decapoda) with description of fifteen new species. Zoosystema, **19**(2-3): 345-420, figs. 1-38.

POORE, G. C. B. & D. J. G. GRIFFIN, 1979. The Thalassinidea (Crustacea: Decapoda) of Australia. Rec. Aust. Mus., **32**(6): 217-321, fig. 1-56.

SAKAI, K., 1969. Revision of Japanese callianassids based on the variations of larger cheliped in *Callianassa petalura* Stimpson and *C. japonica* Ortmann (Decapoda: Anomura). Publs Seto Mar. Biol. Lab., **27**(4): 209-252, pls. IX-XV, 8 text-figs.

SAKAI, K., 1984. Some thalassinideans (Decapoda: Crustacea) from Heron Is., Queensland, eastern Australia, and a new species of *Gourretia* from East Africa. The Beagle, Records of the Museums and Art Galleries of the Northern Territory, **1**: 95-108.

SAKAI, K., 1992. Axiid collections of the Zoological Museum, Copenhagen, with the description of one new genus and six new species (Axiidae, Thalassinidea, Crustacea). Zool. Scr., **21**(2): 157-180, figs. 1-18.

SAKAI, K., 1992a. The families Callianideidae and Thalassinidae, with the description of two new subfamilies, one new genus and two new species. Naturalists, Tokushima, **4**: 1-33, figs. 1- 11.

SAKAI, K., 1999. Description of *Ctenocheles balssi* Kishinouye, 1926, with comments on its systematic position and establishment of a new subfamily Gourretiinae (Decapoda, Callianassidae). Crustaceana, **72**(1): 85-97, figs. 1-3.

SAKAI, K., 1999b. Synopsis of the family Callianassidae, with keys to subfamilies, genera and species, and the description of new taxa (Crustacea: Decapoda; Thalassinidea). Zoologische Verhandelingen, **326**: 1-152.

SAKAI, K., 2002. Callianassidae (Decapoda, Thalassinidea) in the Andaman Sea, Thailand. In: N. L. BRUCE, M. BERGGREN & S. BUSSAWARIT (eds.), Proceedings of the International Workshop on the Biodiversity of Crustaceans of the Andaman Sea. Phuket mar. Biol. Center, (Spec. Publ.) **23**: 461-532.

SAKAI, K., 2004. Dr. R. Plante's collection of the families Callianassidae and Gourretiidae (Decapoda, Thalassinidea) from Madagascar, with the description of two new genera and one new species of the Gourretiidae Sakai, 1999 (new status) and two new species of the Callianassidae Dana, 1852. Crustaceana, **77**: 553-601, figs. 1-23.

SAKAI, K., 2005. Callianassoidea of the world (Decapoda, Thalassinidea). Crustaceana Monographs, **4**: i-vi, 1-285, figs. 1-44.

SAKAI, K., 2006. A new subfamily, Dawsoniinae in the Callianassoidea Dana, 1852 (Decapoda, Thalassinidea). Crustaceana, **79**(10): 1275-1278.

SAKAI, K., 2010. Callianassoidea from the Gulf of Tonkin and the Red Sea, in the Zoological Museum of Moscow University (Decapoda, Thalassinidea). Crustaceana, **83**(12): 1431-1467.

SAKAI, K., 2011. Axioidea of the world and a reconsideration of the Callianassoidea (Decapoda, Thalassinidea, Callianassida). Crustaceana Monographs, **13**: 1-616, 72 figs., 14 tabs.

SAKAI, K., 2013. A new genus *Kiictenocheloides* gen. nov., in the family Ctenocheloidae Sakai, 2011 (superfamily Callianassoidea Dana, 1852) (Decapoda, Pleocyemata). Crustaceana, **86**(13-14): 1689-1694.

SAKAI, K. & T. SAWADA, 2006. The taxa of the infraorders Astacidea, Thalassinidea, Palinura, and Anomura (Decapoda, Pleocyemata) classified by the form of the prepyloric ossicle. Crustaceana, **78**(11): 1353-1368, figs. 1-24 (colour).

SAKAI, K. & S. SHINOMIYA, 2011. Preliminary report on eight new genera formally attributed to *Parapenaeopsis* Alcock, 1901, sensu lato (Decapoda, Penaeidae). Crustaceana, **84**(4): 491-504.

SAKAI, K. & M. TÜRKAY, 2005. A redescription of *Callianassa denticulata* Lutze, 1937 with the designtion of a neotype (Thalassinidea, Gourretiidae). Crustaceana, **78**(3): 323-334.

SAKAI, K. & M. TÜRKAY, 2014. A review of the collections of the infraorders Thalassinidea Latreille, 1831 and Callianassidea Dana, 1852 (Decapoda, Pleocyemata) lodged in three German Museums, with revised keys to the genera and species. Crustaceana, **87**(2): 129-211.

SCHWEITZER, C. E., R. M. FELDMANN, S. CASADÍO & M. R. RAISING, 2012. Eocene decapod Crustacea (Thalassinidea and Brachyura) from Patagonia, Argentina. Annals of Carnegie Museum, **80**: 173-186.

TUDGE, C. C., G. C. B. POORE & R. LEMAITRE, 2000. Preliminary phylogenetic analysis of generic relationships within the Callianassidae and Ctenochelidae (Decapoda: Thalassinidea: Callianassoidea). Journ. Crust. Biol., **20**(Spec. No. 2): 129-149, figs. 1-4, appendices 1-3.

WEAR, R. G. & J. C. YALDWYN, 1966. Studies on thalassinid Crustacea (Decapoda, Macrura Reptantia) with a description of a new *Jaxea* from New Zealand and an account of its larval development. Zoology Publications from Victoria University of Wellington, **41**: 1-26, 26 figs.

WOLFF, T., 1964. The Galathea Expedition 1950-52. List of benthic stations from 0-400 metres, near-surface stations, and land stations. Vidensk. Medd., **127**: 195-258.

First received 6 January 2016.
Final version accepted 28 July 2016.

[When citing this volume, please refer to Crustaceana 90 (2017) 771-1288]

Michael Türkay Memorial Issue

A NEW WESTERN ATLANTIC SPECIES OF *COLLODES* STIMPSON (DECAPODA, BRACHYURA, INACHOIDIDAE)

BY

WILLIAM SANTANA[1,3]) and MARCOS TAVARES[2])

[1]) Universidade do Sagrado Coração — USC, Pró-Reitoria de Pesquisa e Pós-Graduação, Rua Irmã Arminda, 10-50, Jd. Brazil, 17011-160, Bauru, SP, Brazil

[2]) Museu de Zoologia, Universidade de São Paulo, Av. Nazaré 481, Ipiranga, 04263-000, São Paulo, SP, Brazil

ABSTRACT

A new species of spider crab, *Collodes tuerkayi* n. sp., is described and illustrated from the Western Atlantic (Puerto Rico and northern Brazil). The new species superficially resembles *C. inermis* A. Milne-Edwards, 1878, in the general appearance of the body, but is distinguished by the details of the carapace, thoracic sternum and antennal ornamentation and the proportion of the pereopods.

Key words. — Biodiversity, Brazil, Puerto Rico, Majoidea, spider crab

ZUSAMMENFASSUNG

Hier wird eine neue Seespinnen-Art, *Collodes tuerkayi* n. sp. aus dem Westatlantik (Puerto Rico und Nordbrasilien) beschrieben und illustriert. Oberflächlich betrachtet ähnelt die neue Art *C. inermis* A. Milne-Edwards, 1878 durch das generelle Erscheinungsbild ihres Körpers, kann aber durch Details des Carapax, des Brustbeins und der Ausschmückung der Antennen, sowie durch die Proportionen der Pereopoden unterschieden werden.

INTRODUCTION

The amphi-American endemic genus *Collodes* Stimpson, 1860 currently comprises 14 species inhabiting shallow waters down to 700 m. A reappraisal of specimens previously assigned to *Collodes inermis* A. Milne-Edwards, 1878 (e.g., Rathbun, 1901, 1925, 1933; Melo, 1996) revealed the existence of a new species of *Collodes* herein described and illustrated.

[3]) Corresponding author; e-mail: william_santana@yahoo.com.br

Specimens examined are deposited in the collections of the Museu de Zoologia, Universidade de São Paulo, São Paulo, SP, Brazil (MZUSP), Museum of Comparative Zoology, Harvard University, Cambridge, MA, U.S.A. (MCZ), and the National Museum of Natural History, Smithsonian Institution, Washington, DC, U.S.A. (USNM). Abbreviations used: cl, carapace length, taken along the dorsal midline from the base of the rostral sinus to the posterior margin of the carapace; cw, carapace width, taken at the level of its widest point; P2–P5, pereopods 2 to 5 (P1 is the cheliped); RV, research vessel; stn, station; WA, western Atlantic Ocean; EP, eastern Pacific Ocean.

SYSTEMATICS

Superfamily MAJOIDEA Samouelle, 1819
Family INACHOIDIDAE Dana, 1851
Collodes Stimpson, 1860

Type species.— *Collodes granosus* Stimpson, 1860 by monotypy. Gender masculine.

Included species.— *Collodes armatus* Rathbun, 1898; *C. gibbosus* (Bell, 1835) (formerly in *Microrhynchus*); *C. granosus* Stimpson, 1860; *C. inermis* A. Milne-Edwards, 1878; *C. leptocheles* Ratbhun, 1894; *C. levis* Ratbhun, 1901; *C. nudus* Stimpson, 1871; *C. obesus* A. Milne-Edwards, 1878; *C. robsonae* Garth, 1958; *C. robustus* Smith, 1883; *C. rostratus* A. Milne-Edwards, 1878; *C. tenuirostris* Rathbun, 1893; *C. trispinosus* Stimpson, 1871; *C. tumidus* Rathbun, 1898.

Collodes tuerkayi sp. nov.
(figs. 1A-B; 2A-B; 3A, C; 4A, C)

Collodes inermis (pro parte) — Coelho, 1971: 138, fig. 2; Melo, 1996: 198; 1998: 460 (not *Collodes inermis* A. Milne-Edwards, 1878).

Type material.— Puerto Rico, Johnson-Smithsonian Expedition, stn 16, 18°31′N 66°10′15″W, 03.ii.1933, M. J. Rathbun det., 70 m: 1 male paratype (USNM 67832). Johnson-Smithsonian Expedition, stn 10, 18°30′24″N 66°04′15″W, 02.ii.1933, 183 m: 1 male cl 7.8 mm, cw 6.4 mm, 2 juvenile males, 1 juvenile female paratypes (USNM 67835). Brazil, Amapá, "Almirante Saldanha", stn 1910A, 03°28′N 49°52′W, 5.v.1968, 70 m: female holotype, cl 10 mm, cw 8 mm (MZUSP 8822); Amapá, "Almirante Saldanha", stn 1926A, 04°44′N 51°33′W 06.v.1968: 1 male, 1 female paratypes (MZUSP 18741), 1 female paratype (MZUSP 6543).

Comparative material.— *Collodes inermis* A. Milne-Edwards, 1878: Colombia, RV "Oregon", stn 4906, 10°05′N 75°53′W, 28.v.1964, W. Santana det., 73-110 m, 1 male (USNM 1256360). Brazil, Bahia, RV "Hassler", 11°49′S 37°27′W, 18.i.1872, A. Milne-Edwards det., 22-31 m, male holotype, cl 7.4, cw 5.8 mm (MCZ 2949). Bahia, Parcel de Abrolhos, RV "Almirante Saldanha", stn 2050, 13.vii.1987, G. A. S. Melo and P. A. Coelho det., 1 male (MZUSP 8881). Espírito Santo, Projeto Rio Doce, stn RD-78, 19°33′08″S 39°37′04″W, 1973, sand and rubble bottom,

32 m: 1 female (MZUSP 9356). RV "Almirante Saldanha", stn 1378 [further details unavailable]: 1 female (MZUSP 8289). São Paulo, Caraguatatuba, Biota/Fapesp, fishing boat, stn SnC 13-16, from 23°21′02″S to 23°58′54″S and 44°11′42″W to 45°34′35″W, A. Fransozo coll., 2001, G. A. S. Melo det.: 1 male (MZUSP 14172). RV "W. Besnard", Projeto Integrado, stn 4949, 23°47′S 44°58.2′W, 27.vii.1986, G. A. S. Melo det., 50 m: 1 male (MZUSP 9280). *Collodes leptocheles* Ratbhun, 1894: United States of America, Florida, Gulf of Mexico, Panama City, RV "Albatross", stn 2400, 28°41′00″N 86°07′00″W, grey clay, 14.iii.1885, M. J. Rathbun det., 309 m: male holotype, cl 21.3, cw 17.8 mm, 3 males paratypes, 1 female paratype, 2 ovigerous females paratypes (USNM 9748). *Collodes trispinosus* Stimpson, 1871: United States of America, Georgia, off Sapelo Island, 30°52′N 80°01′15″W, 23.vi.1963, 503 m: 1 ovigerous female (USNM 156460). Florida, off Cape Canaveral, 28°31′18″N 80°12′42″W, H. Cortez col., 15.viii.1974, L. E. Scotto det., Trawl, 40 m: 1 ovigerous female (USNM 169797). Brazil, São Paulo, Caraguatatuba, Biota/Fapesp, Litoral norte, fishing boat, stn SnC 11, 12, 16, from 23°21′02″S to 23°58′54″S and 44°11′42″W to 45°34′35″W, A. Fransozo coll., 2001, G. A. S. Melo det.: 1 female (MZUSP 14176). RV "Prof. W. Besnard", Projeto Integrado, stn 4854, 23°47′S 44°58′W, heavy silt, 17.xii.1985, G. A. S. Melo det., 47 m: 1 male (MZUSP 11050).

Type locality.— Brazil, Amapá, 03°28′N 49°52′W, 70 m.

Description.— Carapace piriform, longer than wide (wider in females than in males), densely covered with strong tubercles, especially on cardiac, branchial, metabranchial and intestinal regions, evenly covered with hooked setae. Gastric, branchial, cardiac, intestinal regions without spines. Gastric, hepatic, branchial, cardiac, intestinal regions well delimited laterally by deep grooves. Protogastric, mesogastric, metagastric regions inconspicuously defined. Hepatic, gastric regions with few tubercles; branchial, cardiac, intestinal regions densely covered with subequal tubercles. Thoracic pleurites 5-8 gimnopleura, not fused to one another, usually densely covered with small granules, hooked setae.

Rostrum short, straight, slightly bifurcated. Supraorbital spine absent; orbital margin unarmed dorsally. Postorbital spine short, laterally directed, long hooked setae dorsally and ventrally.

Antennular fossae longitudinally ovate; anteroventral margin inermis. Interantennular septum longer than epistomial spine, compressed laterally, forming ventrally-directed keel.

Antenna (flagellum included) distinctly exceeding rostral length. First and second antennal articles fused to epistome. Second article protruding anterolaterally; ventrolateral surface a with strong longitudinal keel, armed with row of blunt, forward-directed teeth, increasing in size anteriorly; ventrolateral margin dentate, teeth acute. Third antennal article massive; fourth longest, cyclindrical; fifth article minute.

Epistome markedly wider than long. Epistomial spine separated by small gap from interantennular septum. Mouth field subrectangular. Pterygostomial region subtriangular, smooth near mouth frame, densely tubercular laterally, tubercles subequal. Subhepatic region strongly swollen, delimited from pterygostome by distinct slope, densely covered with subequal tubercles, several long setae.

Third maxillipeds almost completely covering buccal frame, ischia leaving distinct gap. Exopod long, nearly reaching distal margin of merus; dorsal face with sparse, small tubercles, irregularly distributed; lateral margin with strong lobe in proximal third. Ischium distinctly longer than broad, dorsal face with longitudinal, smooth, deep groove; lateral margin with distinct tubercles increasing in size distally; mesial margin slightly convex; crista dentata with small, rounded, irregular in size teeth. Merus slightly longer than half of ischium, dorsal surface granular near margins. Anterior margin deeply incised, anterolateral and anteromesial margins strongly expanded. Palp cylindrical, slightly overreaching ischiomeral suture. Carpus dorsolaterally sparsely covered with tubercles. Propodus granular, with long distomesial setae. Dactyl smooth, fringed with row of long setae.

Thoracic sternites II-IV broadly triangular in males, evenly and densely tubercular, with sparse hooked setae, small tubercles medially. Anterior half of fused sternites I-IV strongly sloping down in ventral view in males, sutures 2/3, 3/4 faint. Male sternites IV-VII densely covered with distinct tubercles, smooth in females.

Male and female chelipeds subcylindrical, homochelous, robust in males, slender in females. Dactylus, fixed finger approximately of same length as palm in males, females with fingers slightly longer than palm, covered with sparse setae. Cutting edges of fingers in males with calcareous teeth, subequal in distal half, leaving distinct proximal gap, with strong tooth in subproximal edge of dactyl. Distal two-thirds of cutting edges of fingers with subequal teeth in females. Male propodus conspicuously inflated, with sparse setae, small tubercles in dorsal margin, fewer tubercles in ventral margin. Propodus slender, smooth in females. Carpus setose, with several tubercles dorsally, inermis in females. Merus densely tubercular ventrally, dorsally, scattered tubercles in lateral and mesial faces, with long setae. Ischium sparsely tubercular, setose. Pereopods 2-5 similar, slender, cylindric. P2 longest, P3-P5 progressively decreasing in length. Dactyl, propodus, carpus, merus densely setose, densely covered with long setae interspersed with hooked setae. Dactylus broadly curved, tapering distally, spinose ventrally. Coxa granular ventrally.

Male abdominal somites 1-5 free, sixth fused to telson. Pleotelson subtriangular, rounded distally. Female abdominal somites 1-4 free, somites 5, 6, telson fused. Pleotelson markedly arched, transversally oval, densely tubercular. Male, female first abdominal somite with strong spine medially, scattered tubercles, somites 2-5 cambered medially. Male somites 2-3 densely tubercular; somites 4-6, telson with scattered tubercles. Female somites 2-4 densely tubercular laterally; somites 5-6, telson evenly, densely tubercular.

Distribution.— Known from Puerto Rico and Brazil (Amapá), between 70 m and 183 m.

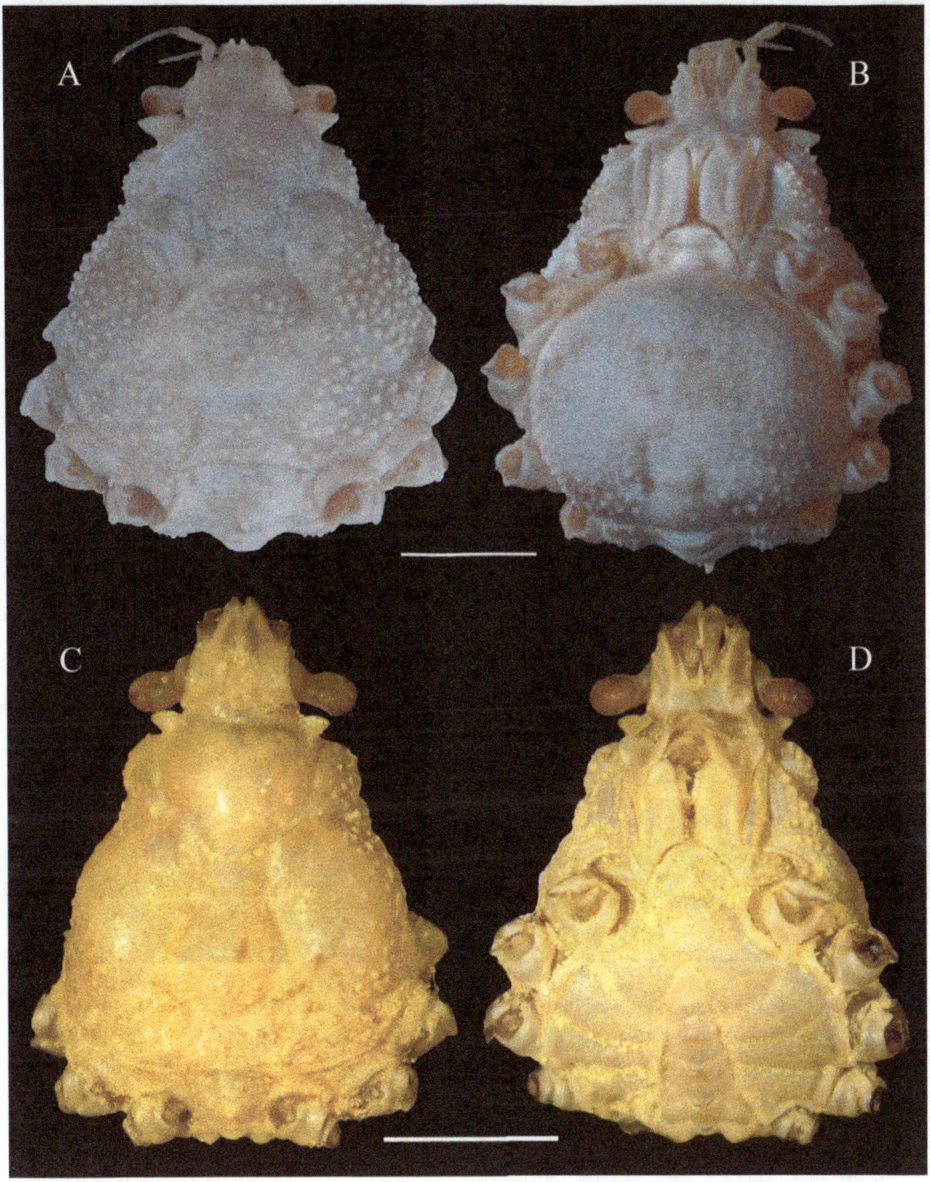

Fig. 1. A-B, *Collodes tuerkayi* sp. nov., female holotype, cl 10 mm, cw 8 mm (MZUSP 8822); C-D, *Collodes inermis* A. Milne-Edwards, 1878, male holotype, cl 7.4, cw 5.8 mm (MCZ 2949). A-B, cephalothorax dorsal view; C-D, cephalothorax ventral view. Scale bars = 3 mm.

Etymology.— The new species is named in honour of the late Professor Michael Türkay (Senckenberg Naturmuseum, Frankfurt am Main), in recognition for his outstanding contributions to the taxonomy of decapod crustaceans and dedication to marine biology.

Fig. 2. A-B, *Collodes tuerkayi* n. sp., male paratype, cl 7.8 mm, cw 6.48 mm (USNM 67835); C-D, *Collodes inermis* A. Milne-Edwards, 1878, male, cl 10.16, cw 7.76 mm (USNM 1256360). A, C, habitus, dorsal view; B, D, cephalothorax, ventral view. Pereopods removed electronically in B and D. Scale bars = 5 mm.

Remarks.— *Collodes tuerkayi* n. sp. superficially resembles *Collodes inermis* A. Milne-Edwards, 1878, in the mostly granular appearance of the carapace and in the thoracic sternum being tubercular, which lead previous authors to conclude that *C. inermis* represented only one species. However, the specimens from Puerto Rico (USNM 67832 and USNM 67835) and part of the specimens from Brazil attributed to *C. inermis* by Coelho (1971) and Melo (1996, 1998) actually belong to *C. tuerkayi* n. sp.

Collodes tuerkayi n. sp. differs from *C. inermis* A. Milne-Edwards, 1878 s. str. (as revealed by the examination of the comparative material listed above) by (i) the

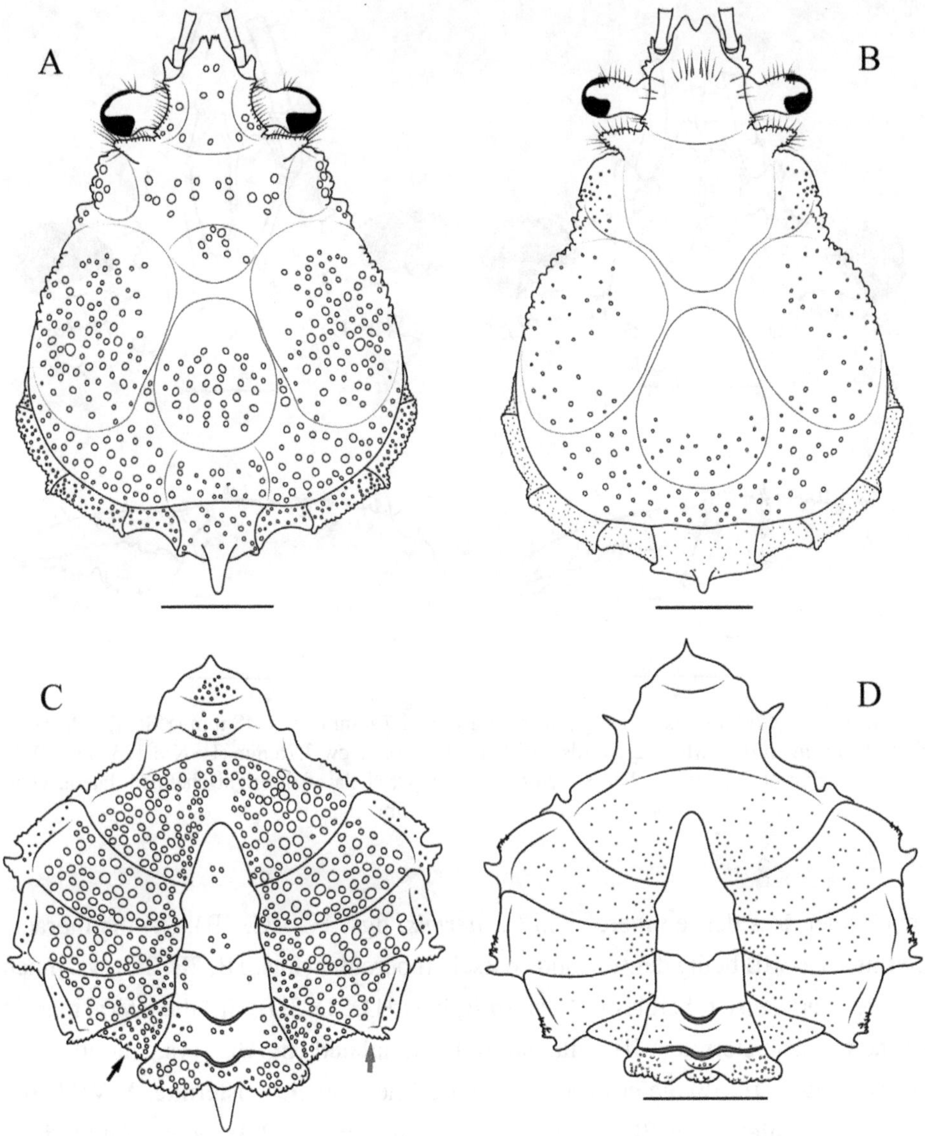

Fig. 3. A, C, *Collodes tuerkayi* n. sp., male paratype, cl 7.8 mm, cw 6.48 mm (USNM 67835); B, D, *Collodes inermis* A. Milne-Edwards, 1878, male, cl 10.16, cw 7.76 mm (USNM 1256360). A-B, cephalothorax dorsal view; C-D, cephalothorax ventral view. Note in C the lateral margin of the thoracic sternite VIII (black arrow) and the posterior margins of the thoracic episternites V-VII (grey arrow) being strongly crenulated. Scale bars = 2 mm.

carapace being markedly more tubercular, with the tubercles distinctly denser and stronger especially on the cardiac, branchial, metabranchial and intestinal regions (figs. 1A, 2A, 3A) (vs. cardiac, branchial, metabranchial and intestinal regions with

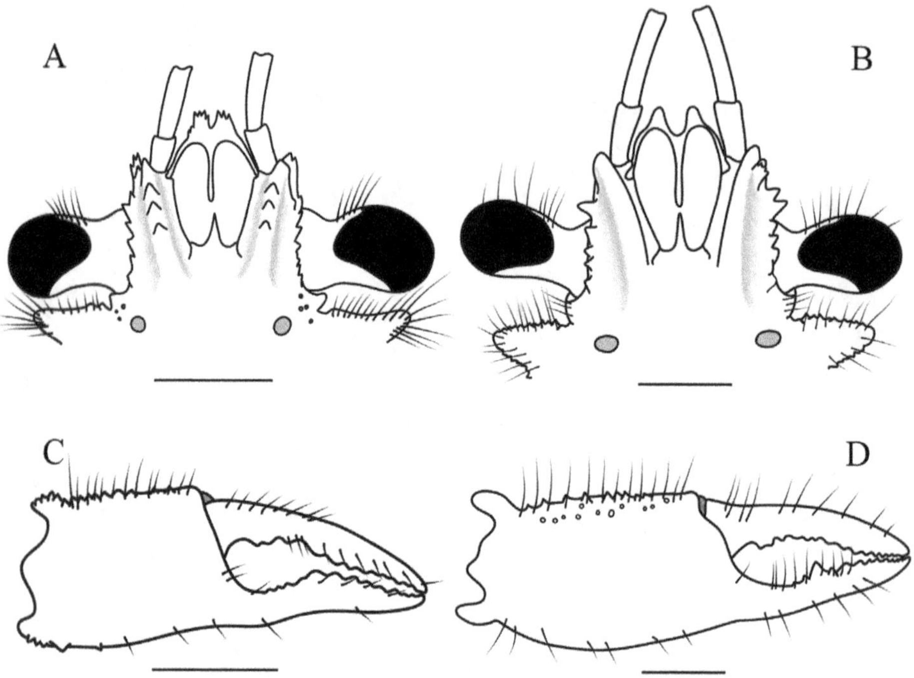

Fig. 4. A, C, *Collodes tuerkayi* n. sp., male paratype, cl 7.8 mm, cw 6.48 mm (USNM 67835); B, D, *Collodes inermis* A. Milne-Edwards, 1878, male, cl 10.16, cw 7.76 mm (USNM 1256360). A-B, anterior region of the cephalothorax, ventral view; C-D, chela of the right cheliped, lateral view. Scale bars = 1 mm.

much finer, less dense tubercles in *C. inermis*, figs. 1C, 2C, 3B); (ii) the thoracic sternites IV-VIII being densely and coarsely tubercular (figs. 1B, 2B, 3C) (vs. much smaller, less dense tubercles in *C. inermis*, figs. 1D, 2D, 3D); (iii) the lateral margin of the thoracic sternite VIII being strongly crenulated (fig. 3C) (vs. smooth in *C. inermis*, fig. 3D); (iv) the posterior margins of the thoracic episternites V-VII being strongly crenulated (fig. 3C) (vs. smooth in *C. inermis*, fig. 3D); (v) the longitudinal carina of the second antennal article provided with four strong, blunt teeth (fig. 4A) (vs. smooth in *C. inermis*, fig. 4B); (vi) the spine of the first abdominal somite being markedly stronger (figs. 1A, 2A-B, 3A) (vs. distinctly smaller in *C. inermis*, fig. 1C, 2C, 3B); (vii) the male abdominal somite V and pleotelson provided with medial tubercles (fig. 3C) (vs. devoid of tubercles in *C. inermis*, fig. 3D); (viii) the shorter P2-P5 (fig. 2A) (vs. longer in *C. inermis*, fig. 2C). The morphological characteristics of the first gonopod in *C. tuerkayi* n. sp. and *C. inermis* are as in the other species of the genus *Collodes*.

ACKNOWLEDGEMENTS

We gratefully thank Rafael Lemaitre (USNM) and Adam Baldinger (MCZ) for granting access to the collections under their responsibility and for providing working space. Thanks also to Karen Reed (USNM), Luciane Ferreira, Joana Darc and Maria José Coelho (MZUSP) for their kind assistance throughout the execution of this work. WS thanks FAPESP for financial support through grant no. 2013/01201-0. MT thanks CNPq (301806/2010-1) and PETROBRAS (4600224970) for funding studies on systematics and taxonomy of decapod crustaceans. Célio Magalhães (Instituto Nacional de Pesquisas da Amazônia, Manaus) is thanked for inviting us to contribute to the special issue in honour of the late Michael Türkay. This paper greatly benefited from comments of two anonymous reviewers and of Sammy De Grave (Oxford University Museum of Natural History, Oxford, U.K.).

REFERENCES

COELHO, P. A., 1971. Nota prévia sôbre os Majidae do norte e nordeste do Brasil. Arquivos do Museu Nacional, **54**: 137-143.

MELO, G. A. S., 1996. Manual de identificação dos Brachyura (caranguejos e siris) do litoral Brasileiro: 1-603. (Plêiade/FAPESP, São Paulo).

MELO, G. A. S., 1998. Malacostraca — Eucarida. Brachyura. Oxyrhyncha and Brachyrhyncha. In: P. S. YOUNG (ed.), Catalogue of Crustacea of Brazil. Série Livros, **6**: 455-515. (Museu Nacional, Rio de Janeiro).

MILNE-EDWARDS, A., 1878. Études sur les Crustacés Podophtalmaires de la région méxicaine. In: Mission Scientifique du Mexique et dans l'Amérique Centrale. Recherches zoologiques pour servir à l'Histoire de la faune de l'Amérique Centrale et du Mexique, **5**: 121-184. (Ministère de l'Instruction Publique, Paris).

RATHBUN, M. J., 1901. The Brachyura and Macrura of Porto Rico. Bulletin of the United States Fish Commission for 1900, **2**(2): 1-137.

RATHBUN, M. J., 1925. The spider crabs of America. Bulletin of the United States National Museum, **129**: 1-613.

RATHBUN, M. J., 1933. Brachyuran crabs of Porto Rico and the Virgin Islands. Scientific survey of Porto Rico and the Virgin Islands. Annals of the New York Academy of Sciences, **15**: 1-121.

First received 29 September 2016.
Final version accepted 2 December 2016.

[When citing this volume, please refer to Crustaceana 90 (2017) 771-1288]

Michael Türkay Memorial Issue

PERISESARMA TUERKAYI, A NEW SPECIES OF MANGROVE CRAB FROM VIETNAM (DECAPODA, BRACHYURA, SESARMIDAE), WITH AN ASSESSMENT OF ITS PHYLOGENETIC RELATIONSHIPS

BY

ADNAN SHAHDADI[1,2,4]), PETER J. F. DAVIE[3]) and CHRISTOPH D. SCHUBART[1])

[1]) Zoologie & Evolutionsbiologie, Universität Regensburg, 93040 Regensburg, Germany

[2]) Department of Biology, Faculty of Science, University of Hormozgan, Bandarabbas, Iran

[3]) Queensland Museum, P.O. Box 3300, South Brisbane, QLD 4101, Australia

ABSTRACT

A new species of *Perisesarma*, *P. tuerkayi*, is described from mangroves of Tan Thoi Island, southern Vietnam. Morphologically, the new species differs most significantly from congeners by the tuberculation pattern of the chelar dactylus, its unique G1 morphology, an unusually large maximum body size, and relatively short and broad ambulatory legs. Genetically, *P. tuerkayi* n. sp. is markedly divergent from other congeneric species, both in mitochondrial and nuclear DNA. It is the fifth species of *Perisesarma* reported from Vietnam.

ZUSAMMENFASSUNG

Wir beschreiben eine neue Art von *Perisesarma*, *P. tuerkayi*, aus den Mangroven der Insel Tan Thoi in Süd-Vietnam. Morphologisch unterscheidet sich die Art von Gattungsgenossen am deutlichsten durch das Tuberkelmuster auf den Dactyli der Scheren, die einzigartige Form der G1, einer überdurchschnittlichen Körpergröße und relativ kurzen und breiten Laufbeine. Genetisch lässt sich *P. tuerkayi* n. sp. zweifelsfrei von anderen Arten der Gattung unterscheiden, sowohl in mitochondrialer wie nukleärer DNA. Es handelt sich hierbei um die fünfte Art der Gattung aus Vietnam.

INTRODUCTION

Brachyuran crabs of the family Sesarmidae Dana, 1851 are among the most important faunal components of world mangroves (Davie, 1994; Cannicci et al., 2008; Ragionieri et al., 2009). The taxonomy of the constituent genera of this family has been unstable and has undergone several rearrangements (Serène & Soh, 1970; Holthuis, 1977; Davie, 2010). Ng et al. (2008) have provided the

[4]) Corresponding author; e-mail: adnan1361@gmail.com

most recent widely accepted compilation of sesarmid genera and species. There are, however, a number of genera that are still likely to be polyphyletic. The composition of *Perisesarma* is under particular scrutiny (see Shahdadi & Schubart, 2015), and ongoing studies are likely to cause some significant changes to the generic definition and delimitation. The current concept of *Perisesarma* contains 24 species distributed across the Indo-West Pacific region and along the Atlantic coast of southern West Africa (Shahdadi & Schubart, 2015).

The mangrove crab fauna of Vietnam is still relatively poorly studied and, until now, only four species of *Perisesarma* have been recorded, viz., *P. bidens* (De Haan, 1835), *P. eumolpe* (De Man, 1895), *P. haswelli* (De Man, 1887), and *P. maipoense* (Soh, 1978) (see Kosuge et al., 1997; Nhuong, 2003; Ng et al., 2010; Diele et al., 2013). However, considering the fact that sesarmid mangrove crabs have marine planktonic larvae (Guerao et al., 2004) and assuming connectivity of mangrove forests of Vietnam with adjacent areas (Giri et al., 2011), it seems likely that more targeted collecting will reveal additional species.

Here we describe a new species of *Perisesarma* from the Mekong River Estuary, south of Ho Chi Minh City in southern Vietnam and present the phylogenetic position of the newly described species among other related species based on their genetic relationship.

MATERIAL AND METHODS

Measurements (in mm) given in the text and in table I are of carapace width (cw) followed by carapace length (cl). Abbreviations: G1, male first gonopod; NHM, Natural History Museum, London, U.K.; QM, Queensland Museum, Brisbane, Queensland, Australia; RMNH, Naturalis Biodiversity Center, Leiden, The Netherlands; SMF, Forschungsinstitut und Museum Senckenberg, Frankfurt am Main, Germany; ZRC, Zoological Reference Collection of the Lee Kong Chian Natural History Museum, National University of Singapore, Singapore.

Molecular analyses

Genetic analyses were undertaken for species of *Perisesarma*, so as to make comparisons between *P. tuerkayi* n. sp. with its most morphologically similar relatives, other species from neighbouring areas in South East Asia, and the type species of the genus, *P. dussumieri*. For completeness, we included previously published sequences, this way covering 20 of the 24 species currently included in *Perisesarma* (table I).

Genomic DNA was isolated from leg muscle tissue using a modified Puregene method (Gentra Systems, Minneapolis, MN, U.S.A.) and a Mollusc DNA

TABLE I

Material examined for genetic comparisons in this study, with size (cw × cl, in mm), sex (M = male, F = female), locality, museum catalogue number and GenBank accession numbers for Cox1, 16S and NaK genes (see main text for museum abbreviations)

Species	Size, Sex	Locality	Catalogue No.	Cox1	16S	NaK
P. bengalense Davie, 2003	14.77 × 12.70, M	Thailand, Phuket	SMF 49919	KX400901	KX423809	
P. bidens (De Haan, 1835)		South China Sea		KM605220*		
P. bidens		Japan, Wakayama			AJ621183*	
P. brevicristatum (Campbell, 1967)	16.37 × 13.39, M	Australia, Queensland, Murray River	QM-W8270	KX400906	KX423800	
P. cricotum Rahayu & Davie, 2002	14.34 × 12.16, F	Indonesia, Irian Jaya, Kamora	ZRC2016.0522	KX400897	KX423796	KX394811
P. darwinense (Campbell, 1967)	12.24 × 10.24, M	Australia, Northern Territory, Darwin Island	SMF 49921	KX400904	KX423798	
P. dussumieri (H. Milne Edwards, 1853)	25.21 × 21.81, M	Thailand, Phuket	ZRC2016.0525	KX400916	KX423814	
P. eumolpe (De Man, 1895)	16.83 × 13.92, M	China, Hainan, Wen Chang	SMF 49922	KX400891		KX394816
P. eumolpe		Singapore, Lim Chu Kang mangrove			AJ621184*	
P. foresti Rahayu & Davie, 2002	14.25 × 11.85, F	Indonesia, Irian Jaya, Ajkwa	ZRC2016.0523	KX431204	KX423794	

TABLE I
(Continued)

Species	Size, Sex	Locality	Catalogue No.	Cox1	16S	NaK
P. guttatum (A. Milne-Edwards, 1869)		East Africa		KX374922*		
P. guttatum		Kenya, Gazi Bay			AJ621185*	
P. holthuisi Davie, 2010	19.16 × 16.16, M	Western Australia, Ashburton	QM-W28880, Paratype	KX400907	KX423806	KX394827
P. indiarum (Tweedie, 1940)	28.96 × 24.76, M	Indonesia, Ambon	RMNH. CRUS. D.141, Lectotype	KX761164	KX761171	
P. lanchesteri (Tweedie, 1936)	24.76 × 18.78, F	Singapore, Simpang, Mak Wai River	ZRC1967.11.8.3	KX761168	KX761174	
P. lividum (A. Milne-Edwards, 1869)	17.76 × 15.19, M	New Caledonia, Nanmea mangroves	QM-W24243	KX400893	KX423802	KX394812
P. longicristatum (Campbell, 1967)	18.50 × 14.94, M	Australia, Queensland, Port Alma	QM-W2464, Paratype	KY198240	KY198245	KY198249
P. maipoense (Soh, 1978)	25.27 × 19.57, M	Vietnam, mouth of the Red River	ZRC2009, 0800	KX400931		
P. messa (Campbell, 1967)	19.36 × 16.74, M	Australia, Queensland, Brisbane	ZRC 1999.0650	KX431205	KX423795	KX394835
P. onychophorum (De Man, 1895)	23.93 × 19.29, M	Singapore, Mandai mangroves	ZRC2000.1490	KX400913	KX423812	

TABLE I
(Continued)

Species	Size, Sex	Locality	Catalogue No.	Cox1	16S	NaK
P. samawati Gillikin & Schubart, 2004		Kenya, Mida Creek			AJ621186*	
P. semperi (Bürger, 1893)	11.72 × 9.58, M	Indonesia, Irian Jaya Ajkwa	QM-W27532	KX400910	KX423804	
P. tuerkayi n. sp.	28.23 × 24.00, M	Vietnam, Tan Thoi Island, Cua Tieu River	QM-W28348 Holotype	KY198241	KY198246	KY198248
P. tuerkayi n. sp.	25.27 × 21.20, F	Vietnam, Tan Thoi Island, Cua Tieu River	QM-W27002 Paratype	KY198242	KY198247	
P. tuerkayi n. sp.	23.46 × 20.30, M	Vietnam, Tan Thoi Island, Cua Tieu River	QM-W27002 Paratype	KY198243		
P. tuerkayi n. sp.	23.24 × 20.46, F	Vietnam, Tan Thoi Island, Cua Tieu River	QM-W27002 Paratype	KY198244		

* Previously published sequences from GenBank (https://www.ncbi.nlm.nih.gov/).

TABLE II
Primers used in the present study

Gene	Primer	Sequence ($5' \rightarrow 3'$)	Reference
Cox1	COL6	TYTCHACAAAYCATAAAGAYATYGG	Schubart (2009)
	COH6	TADACTTCDGGRTGDCCAAARAAYCA	Schubart & Huber (2006)
16S rRNA	16L29	YGCCTGTTTATCAAAAACAT	Schubart et al. (2001) (as 16L2)
	16H11	AGATAGAAACCRACCTGG	Schubart (2009)
NaK	NaK for-b2 (new name)	ATGACAGTCGCYCAYATGTGGTT	Tsang et al. (2008) (modified from NaK for-b)
	NaK rev3	GGAGGRTCAATCATRGACAT	Tsang et al. (2014)

kit (D3373-02, Omega Biotek, Norcross, GA, U.S.A.) using manufacturers' protocols. Three genes, including the mitochondrial protein-coding gene cytochrome oxidase subunit 1 (Cox1), the mitochondrial gene encoding the rRNA of the large 16S ribosomal subunit (16S), and, for a subset of species, the nuclear protein-coding gene sodium-potassium ATPase alpha-subunit (NaK), were partially amplified. Nuclear DNA was only included for those species with closer phylogenetic relationship to the new species according to our Cox1 and 16S phylogenies. Polymerase chain reactions (PCR) were carried out with different primer compositions (i.e., COL6/COH6; 16L29/16H11; NaK for-b2/NaK rev3, see table II) and the following profile: initial heating 4 min at 94°C; 40 cycles with 45 s at 95°C for denaturing, 60 s at 48°C (Cox1), 50°C (16S) and 58°C (NaK) for annealing, 60 s at 72°C for extension; and a final extension step with 5 min at 72°C.

The primers NaK for-b2 (forward) and NaK rev3 (reverse) amplify a segment of about 604 bp (including the primer regions) of the NaK gene, the decapod-specific primer combination COL6/COH6 amplifies a segment of 709 bp of Cox1 (658 without primers), corresponding to the generally used barcoding region (Hebert et al., 2003), and the combination 16L29/16H11 amplifies a segment of approximately 584 bp of aligned 16S rDNA (table II). PCR products were outsourced for sequencing to Macrogen Europe. Sequences were proofread using Chromas Lite (v. 2.1.1) (Technelysium, Brisbane, QLD, Australia), primer regions were removed, and the remaining sequences aligned automatically with ClustalW (Thompson et al., 1994) implemented in BioEdit 7.0.5 (Hall, 1999). For construction of a phylogenetic dendrogram, the sequences of Cox1 and 16S genes were concatenated to create a single alignment. This concatenated mtDNA alignment and the NaK dataset were converted with FaBox (Villesen, 2007) to Nexus files for phylogenetic analyses. The best evolutionary model describing our data was determined with the aid of jModelTest (v. 2.1.4) (Guindon & Gascuel,

2003; Darriba et al., 2012) and selected with the Akaike information criterion (AIC) (Posada & Buckley, 2004). Because of uncertainties in the phylogenetic relationships among species of *Perisesarma* and its sister genus *Parasesarma* (see Schubart et al., 2006), we chose not to force any of the constituent species of these two genera into the outgroup and rather used an unrooted dendrogram for presenting the mtDNA phylogenetic relationships. This dendrogram was reconstructed with a Maximum Likelihood (ML) algorithm using the software raxmlGUI (v. 1.3) (Silvestro & Michalak, 2012). A maximum parsimony network (Tempelton et al., 1992) for the NaK gene was constructed via PopART (http://popart.otago.ac.nz). The Maximum Likelihood (ML) dendrogram was obtained with 2000 bootstrap pseudoreplicates. Pairwise genetic distances (Kimura 2-parameter, K2P) based on the Cox1 sequences among the new species and possibly related congeners were calculated with the software MEGA version 5.2.2 (Tamura et al., 2011). New sequences were submitted to GenBank (NCBI), and accession numbers are given in table I.

RESULTS

TAXONOMY

Family SESARMIDAE Dana, 1851

Perisesarma De Man, 1895

Perisesarma tuerkayi n. sp.

(figs. 1, 2, 3, 4, 7A and 8A)

Material examined.— Holotype male (28.2 × 24.0 mm) (QM-W28348), Vietnam, Tan Thoi Island, Cua Tieu River (Mekong River Estuary), ca. 70 km south of Ho Chi Minh City, 10°17′N 106°42′E, coll. Nguyen van Xuan, Feb. 2002. Paratypes: 1 male (23.5 × 20.3 mm), 2 females (25.3 × 21.2 mm; 23.2 × 20.5 mm) (QM-W27002), same data as for holotype.

Comparative material.— *P. eumolpe* (De Man, 1895): 1 male (16.8 × 13.9 mm) (SMF 49922), China, Hainan, Wen Chang. *P. maipoense* (Soh, 1978): 1 male (27.7 × 21.0 mm) holotype (NHM1976.106), Hong Kong, Mai Po marshes; 1 male (25.2 × 19.6 mm) (ZRC2009, 0800), Vietnam, mouth of the Red River.

Diagnosis.— Medium-sized sesarmid crab (maximum size of studied specimens 28.2 × 24.0 mm); carapace rectangular, slightly broader than long; front moderately deflexed, with broad median margination, postfrontal lobes prominent, median lobes slightly broader than lateral ones, separated by deep median furrow; dorsal carapace regions discernible, anterolateral margin with sharp exorbital angle and well developed epibranchial tooth. Chelipeds homochelous, large; upper surface of palm with 2 transverse pectinated crests, dorsal surface of dactylus with 22-24 low, but distinct, transversely broad tubercles. Ambulatory legs relatively short (ca. 1.5 times as long as carapace width) and broad (merus of third pair ca.

1.9 as long as wide. Male pleon triangular; telson slightly longer than basal width. G1 relatively stout, straight, apical corneous process apically truncated, bent at angle of about 35° to vertical axis, arched in cross section, aperture terminal.

Description.— Medium-sized sesarmid crab (maximum size of studied specimens 28.2 × 24.0 mm), carapace subrectangular, slightly broader than long, greatest width between exorbital angles (cw/cl 1.13-1.19, $N = 4$), body relatively vaulted (body height/cw ca. 0.57). Carapace surface smooth, shining, punctate, with numerous short, transverse to slightly oblique crests edged with rows of short setae, sometimes forming low tufts. Front ca. 0.53 times carapace width, moderately deflexed, with broad, relatively deep, median concavity. Post-frontal lobes prominent, median lobes slightly broader than lateral ones, separated by deep furrow. Dorsal carapace regions moderately well indicated; gastric region most strongly demarcated; cardiac region separated from intestinal region, lateral branchial ridges prominent, upper orbital border smooth, lower orbital border finely granulate, anterolateral margin with sharp exorbital angle, well developed epibranchial tooth (carapace width between epibranchial teeth equal to, or only slightly less than, width between exorbital angles), no indication of second epibranchial tooth, lateral margin slightly concave, edged with row of short setae. Eyes pigmented, cornea slightly wider than eyestalk (figs. 1A, 2A, C).

Chelipeds homochelous.— Chela large (palm length ca. 0.77 × carapace width), robust (length ca. 1.74 × width) (fig. 1). Merus with granulate dorsal border and distinct subdistal spine; ventral border granulate; anterior border granulate, with distinct subdistal spine; inner face smooth with 2 longitudinal rows of setae, ventral row more prominent, continuous, with longer setae, dorsal row interrupted, setae extending to dorsal border. Carpus with inner angle not produced; inner margin granular; outer surface striated, granular. Upper surface of palm with 2 transverse pectinated crests (fig. 2D); distal (primary) crest composed of 18-21 tall, broad teeth (18 and 21 on opposite claws of holotype) (fig. 2G); secondary crest well developed, with 13-16 teeth (13 and 16 on holotype); both crests terminate at inner end in short swollen, tubercular ridge and several blunt granules, 1 or 2 row(s) of coarse granules (some with chitinous cap) proximal to second crest. Upper margin of palm with strong tubercles (some with chitinous tip). Outer surface of palm without setae, coarsely granular except for smooth, punctate fixed finger, lacking any indication of median longitudinal ridge (fig. 2E); inner surface of palm coarsely granular except area facing carpus (fig. 2F); ventral border of chela almost straight, length of cutting margin of fixed finger ca. 0.35 times length of entire propodus. Dactylus straight, stout, ca. 0.58 times propodus length; dorsal surface bearing 22-24 low, but distinct, transversely broadened tubercles, distinct to tip, each tubercle has series of fine longitudinal lines that give a finely wrinkled

Fig. 1. *Perisesarma tuerkayi* n. sp., holotype, male (28.2 × 24.0 mm) (QL-W28348). A, Overall dorsal view; B, ventral view. Scale bar = 1 cm.

appearance, tubercles evenly rounded, lacking transverse sulcus (figs. 2D, E, 8A). Row of about 11-17 rounded tubercles on proximal two-thirds of inner edge of dorsal surface. Fingers with chitinous tips, adult males with or without a narrow gape when fingers closed. Cutting edge of both fingers with a series of variably sized teeth (fig. 2E, F).

Ambulatory legs relatively short, broad; third pair longest (figs. 1A, B, 4A, B), total length ca. 1.5 times carapace width, merus with anterior margin crenulated, ca. 1.9 times as long as wide, carpus twice as long as wide, propodus ca. 2.5 times as long as wide, dactylus ca. 0.6 times length of propodus; propodus of first ambulatory leg relatively wide (ca. 1.9 times as long as wide) and both margins covered with dense long setae (fig. 2H).

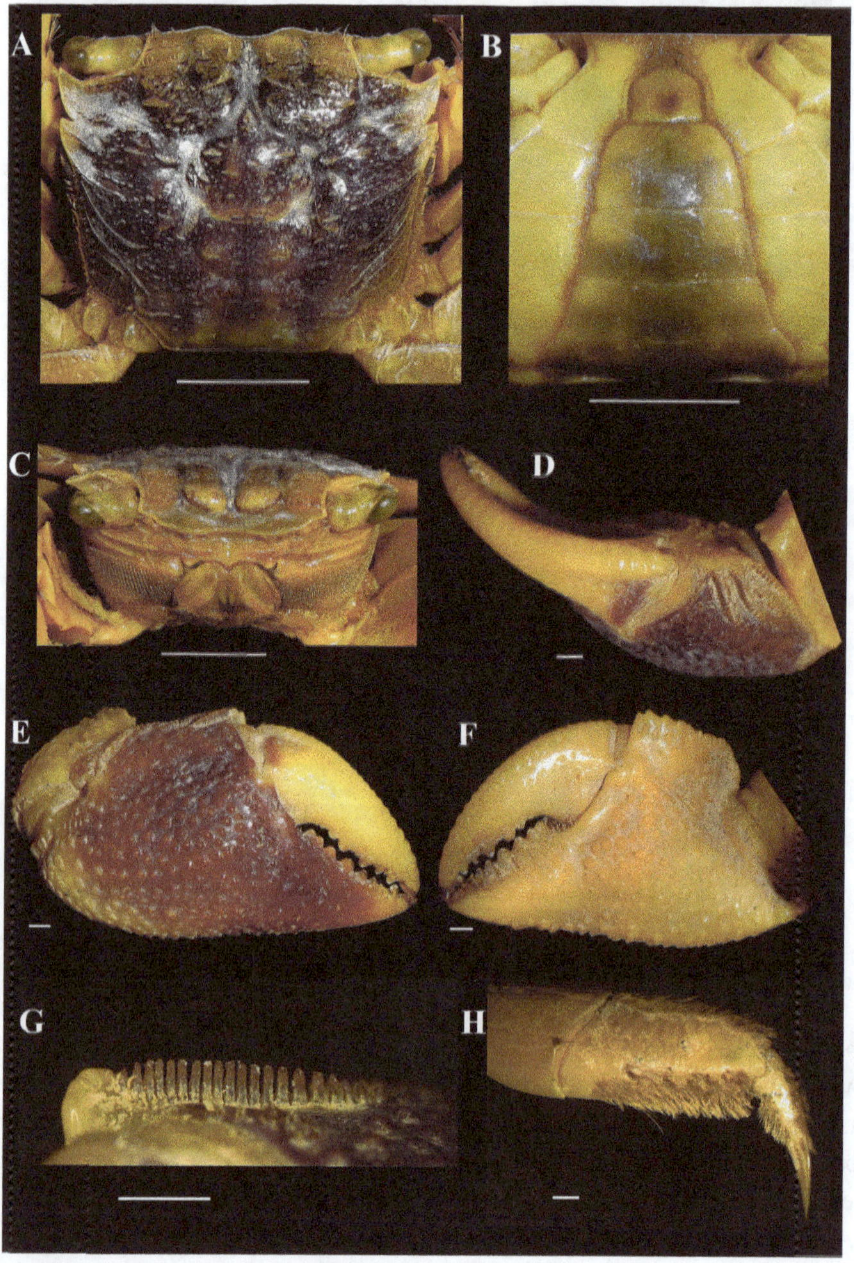

Fig. 2. *Perisesarma tuerkayi* n. sp., holotype male (28.2 × 24.0 mm) (QL-W28348). A, Carapace dorsal view; B, pleon; C, frontal view of carapace; D, dorsal view of left chela; E, outer view of right chela; F, inner view of right chela; G, frontal view of primary pectinated crests; H, inner side of propodus and dactylus of second pereiopod. Scale bars: A-C = 1 cm; D-H = 1 mm.

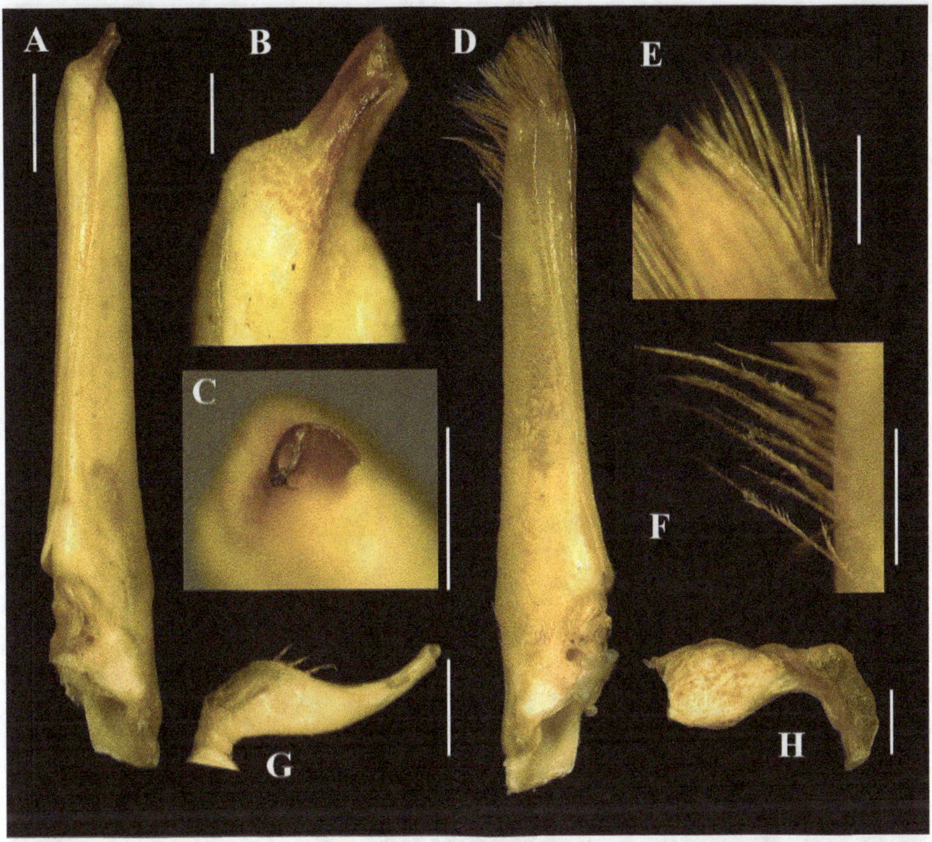

Fig. 3. *Perisesarma tuerkayi* n. sp., holotype male (28.2 × 24.0 mm) (QM-W28348). A, Dorsal view of denuded right G1; B, apical process of right G1; C, cross view of apical corneous process and aperture; D, left G1 with setae; E, simple setae on apical process of left G1; F, plumose setae on apical process of left G1; G, second gonopod; H, penis. Scale bars: A, D = 1 mm; B, C, E-H = 0.5 mm.

Male pleon triangular, with telson slightly longer than basal width (length ca. 1.1 times width), slightly shorter than somite 6 (ca. 0.9 times length of somite 6); somite 6 longer than others, ca. 1.8 times wider than long; somite 5 trapezoidal, width ca. 2.41 times length; somite 4 also trapezoidal, width ca. 3.16 times length; somite 3 widest, laterally convex; somite 2 medially longer than lateral edges (figs. 1B, 2B).

G1 relatively stout, straight (fig. 3A, D); stem triangular with blunt angles in cross section; apical corneous process truncated, bent at an angle of about 35° to vertical axis (fig. 3B), arched in cross section, aperture terminal (fig. 3C). Few scattered short setae along most of gonopod, apical end covered by longer setae, almost completely obscuring corneous tip (fig. 3D); two kinds of setae observed, long simple setae (fig. 3E), and some plumose setae (fig. 3F) restricted to outer

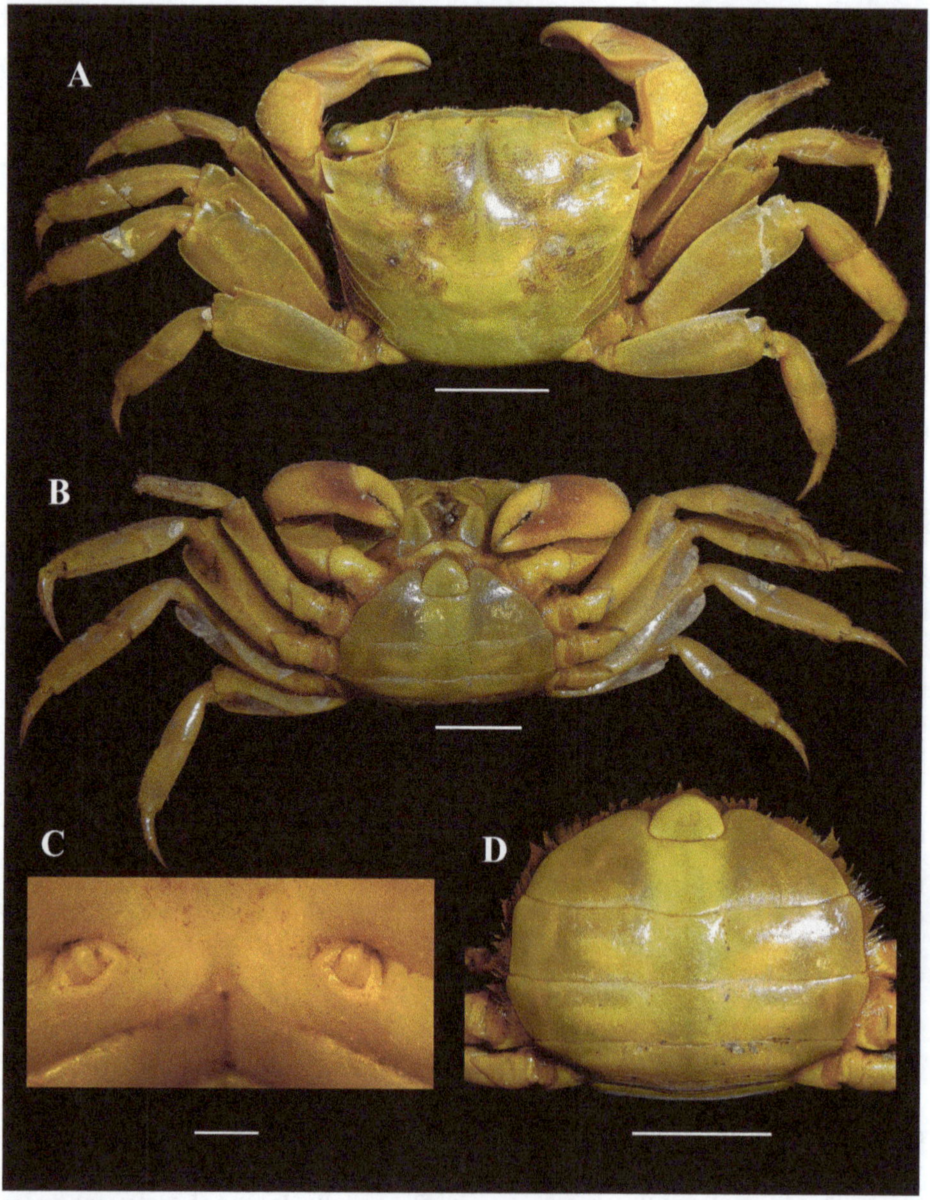

Fig. 4. *Perisesarma tuerkayi* n. sp., paratype female (25.27 × 21.20 mm) (QM-W27002). A, Overall dorsal view; B, ventral view; C, vulva; D, pleon. Scale bars: A, B, D = 10 mm; C = 1 mm.

side of subapical part. G2 short, curved, width of tip about one-third of width of base (fig. 3G). Penis short, stout, curved (fig. 3H).

Female with smaller chelipeds (fig. 4A, B), ratio of palm length to carapace width ca. 0.63; distal dactylar pectinate crest well developed, prominent as in male, but proximal crest reduced to row of granules with short chitinous tip. Pleon broad, evenly rounded, broadest at somite 4, fringed with long setae (fig. 4D). In adult female specimens, pleon touching coxae of ambulatory legs, telson ca. 1.2 times wider than long, inserted into somite 6 more than half length (fig. 4B, D). Vulva in depression on anterior edge of sternite 5, somewhat covered by posterior margin of sternite 4; operculum in inner part (fig. 4C).

Colour.— The colour in life was not recorded, but the preserved material still shows that adult males appear to have a characteristic chela colour pattern, with the outer and upper surfaces of the male palm and fixed finger dark reddish brown, while the movable finger is a pale yellowish cream (except for a narrow proximal reddish patch on the upper face adjacent to the articulation) (fig. 2D, E).

Etymology.— Warmly dedicated to the memory of Prof. Dr. Michael Türkay, a decapod crustacean specialist who made a significant contribution to our taxonomic understanding of Brachyura.

Habitat and distribution.— Full ecological information is lacking for this species, but Davie & Nguyen (2003), in describing the new varunid species *Metaplax gocongensis*, reported that it cohabited with both a species of *Parasesarma* and a species of *Perisesarma.* The latter is the species here described as *Perisesarma tuerkayi* n. sp. Therefore we know that *P. tuerkayi* n. sp. can at least tolerate the same conditions as *Metaplax gocongensis*, viz., the intertidal silty-mud zone of the banks of rivers, or ricefields subjected to tidal influence, where salinity ranges from 0-15 ppt. During the rainy season, the water may be completely fresh for up to six months, while in the dry months salinity will reach 10-15 ppt during high tides. As an estimate of relative abundances, a shrimp pond belonging to the My Loi community (Vam Co River system, Tiền Giang Province) was drained for harvesting, and all available burrows in the four dike walls surrounding the pond were excavated. The dike walls covered a total area of 2815 m^2. Of the collected 142 crab individuals, 16 were *Perisesarma tuerkayi* n. sp., nine were unidentified *Parasesarma* species, and the remaining 117 belonged to *Metaplax gocongensis* (Davie & Nguyen, 2003). *Perisesarma tuerkayi* n. sp. is only known with certainty from the Vam Co River system, ca. 40 km south of Ho Chi Minh City, Tiền Giang Province, and the Cua Tieu River (Mekong River Estuary), ca. 70 km south of Ho Chi Minh City, Vietnam. It can be expected to be more widespread through the region.

Molecular phylogeny

The cropped alignments of the concatenated mitochondrial dataset (Cox1 and 16S) and the nuclear NaK gene consist of 1181 and 501 bp, respectively, after removal of the primer sequences and adjacent regions. The alignments of the

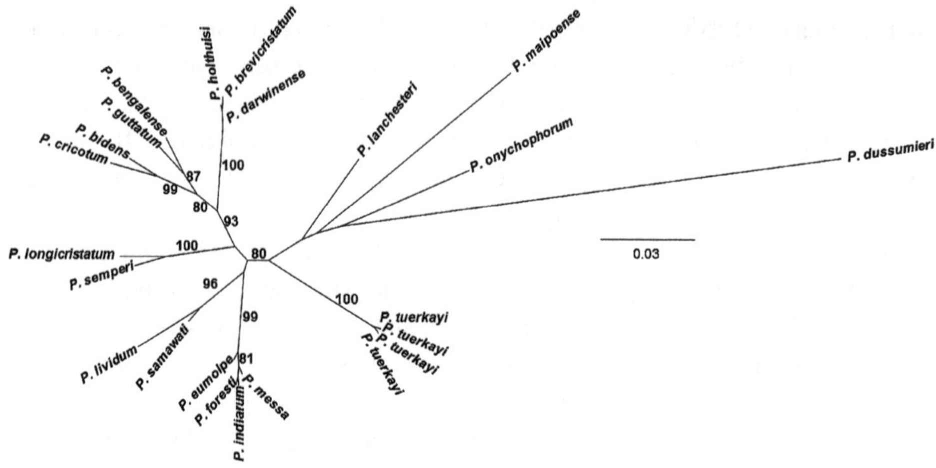

Fig. 5. Phylogenetic consensus dendrogram constructed with Maximum Likelihood (ML) (using the software raxmlGUI) of selected species of *Perisesarma*, including *P. tuerkayi* n. sp. from a concatenated dataset of Cox1 and 16S mitochondrial genes. Only bootstrap values higher than 50% are shown in the dendrogram.

protein coding genes contained no stop codons, which would have indicated that the gene is not translated to a functional protein, and thus the presence of a pseudogene (see Schubart, 2009). The best evolutionary model explaining the data was the General Time Reversible model plus Gamma (GTR+G, Rodriguez et al., 1990).

A phylogenetic dendrogram from the concatenated alignment (Cox1 and 16S genes) was constructed with Maximum Likelihood (ML) (fig. 5). The result shows that *P. tuerkay* n. sp. forms its own distinct clade and is well separated from other species-groupings (fig. 5).

In terms of Cox1 genetic distances, *P. tuerkayi* n. sp. is again clearly distant from other related species, the closest being *P. lanchesteri* (K2P = 0.069), *P. foresti* and *P. eumolpe* (both with K2P = 0.071) (table III). This result is also reflected in the nuclear NaK gene network (fig. 6) which confirms that *P. tuerkayi* n. sp. is separated from other potentially related species *P. eumolpe*, *P. holthuisi* and *P. lividum*, by at least five mutational steps.

The dendrogram shows that the type species, *Perisesarma dussumieri* holds a relatively distant position from the other species of the genus. Also, *P. maipoense*, *P. onychophorum* and *P. lanchesteri* hold isolated positions and do not group together with other species examined. In contrast, our analyses reveal some highly supported monophyletic clades among species of *Perisesarma*: *P. eumolpe*, *P. foresti*, *P. indiarum* and *P. messa* are genetically very close to each other and cluster together. The same is true for *P. longicristatum* and *P. semperi*. Another monophyletic clade is the one containing *P. bengalense*, *P. guttatum*, *P. bidens*,

TABLE III

Genetic K2P distances for the Cox1 gene among examined species of *Perisesarma*, including *P. tuerkayi* n. sp.

		1	2	3	4	5	6	7	8	9	10	11	12	13	14	15	16	17	18
1	*P. tuerkayi* n. sp.																		
2	*P. lanchesteri*	0.069																	
3	*P. foresti*	0.071	0.075																
4	*P. eumolpe*	0.071	0.069	0.015															
5	*P. indiarum*	0.073	0.077	0.005	0.013														
6	*P. lividum*	0.075	0.073	0.065	0.059	0.067													
7	*P. longicristatum*	0.075	0.083	0.066	0.064	0.068	0.082												
8	*P. messa*	0.075	0.075	0.010	0.012	0.008	0.063	0.070											
9	*P. brevicristatum*	0.078	0.081	0.080	0.070	0.082	0.079	0.069	0.076										
10	*P. darwinense*	0.078	0.079	0.076	0.068	0.080	0.078	0.074	0.074	0.012									
11	*P. semperi*	0.080	0.079	0.072	0.070	0.074	0.084	0.031	0.076	0.066	0.070								
12	*P. holthuisi*	0.082	0.079	0.080	0.072	0.085	0.082	0.074	0.079	0.008	0.007	0.070							
13	*P. cricotum*	0.084	0.081	0.068	0.060	0.073	0.078	0.087	0.066	0.069	0.055	0.083	0.063						
14	*P. bidens*	0.088	0.083	0.078	0.074	0.083	0.078	0.088	0.082	0.078	0.063	0.089	0.072	0.037					
15	*P. guttatum*	0.093	0.088	0.077	0.074	0.083	0.076	0.078	0.085	0.076	0.064	0.079	0.072	0.052	0.050				
16	*P. bengalense*	0.093	0.088	0.077	0.066	0.079	0.068	0.085	0.077	0.078	0.068	0.088	0.076	0.050	0.056	0.031			
17	*P. onychophurum*	0.102	0.081	0.107	0.098	0.111	0.095	0.096	0.109	0.096	0.091	0.098	0.091	0.111	0.104	0.100	0.112		
18	*P. maipoense*	0.110	0.089	0.106	0.101	0.104	0.112	0.122	0.099	0.113	0.111	0.117	0.111	0.134	0.127	0.129	0.125	0.099	
19	*P. dussumieri*	0.127	0.121	0.146	0.134	0.149	0.139	0.144	0.146	0.137	0.127	0.139	0.132	0.142	0.142	0.152	0.152	0.135	0.149

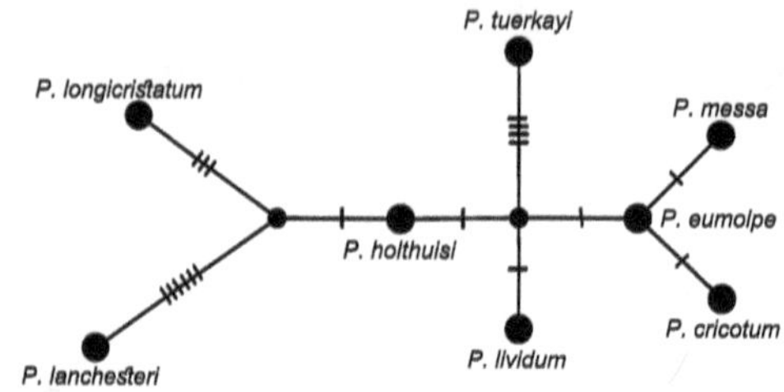

Fig. 6. Maximum parsimony network for NaK genotypes for a subset of species of *Perisesarma*, constructed with PopART.

P. cricotum, *P. brevicristatum*, *P. darwinense* and *P. holthuisi*. Of these, the three species *P. brevicristatum*, *P. darwinense* and *P. holthuisi* appear to be especially closely related. The same can be observed for *P. bidens*/*P. cricotum* and *P. bengalense*/*P. guttatum*. The dendrogram also shows sister species relationship of *P. lividum* and *P. samawati* (fig. 5).

DISCUSSION

According to traditional generic/subgeneric definitions (De Man, 1895; Campbell, 1967), the new species belongs to *Perisesarma* by having two transverse pectinated crests on the upper face of the palm (used as a plectrum for stridulatory sound production, see Davie et al., 2015), and a distinct anterolateral epibranchial tooth. *Perisesarma* is very close morphologically to *Parasesarma*, with the only significant diagnostic character between them being the presence (*Perisesarma*) or absence (*Parasesarma*) of a well-defined epibranchial tooth. However, the importance of this tooth is now considered of doubtful value (see Abele, 1975; von Hagen, 1978; Guerao et al., 2004) and its phylogenetic usefulness needs to be re-examined. It seems likely that a number of species of *Parasesarma* may be much more closely related to species of *Perisesarma* than previously thought (Schubart et al., 2006), and potential differences between these two genera will have to be re-evaluated.

Of the four species of *Perisesarma* previously recorded from Vietnam, the identity of *P. bidens* (originally described from Japan) and *P. haswelli* (originally described from Sullivan Island, Mergui Archipelago, Myanmar [= Burma]) need to be confirmed. Also *P. maipoense* has been previously reported from Vietnamese mangroves (Ng et al., 2010), but the two species are easily separated. *P. tuerkayi* n. sp. has a relatively straight cheliped dactylus, with more than 20 distinct dorsal tubercles, while *P. maipoense* has a curved dactylus with 5-8 flattened and indistinct tubercles (compare figs. 2D-F and 8A of present study with plate 1d in Soh, 1978 and figs. 1-3 in Ng et al., 2010). Also the G1 morphologies of these

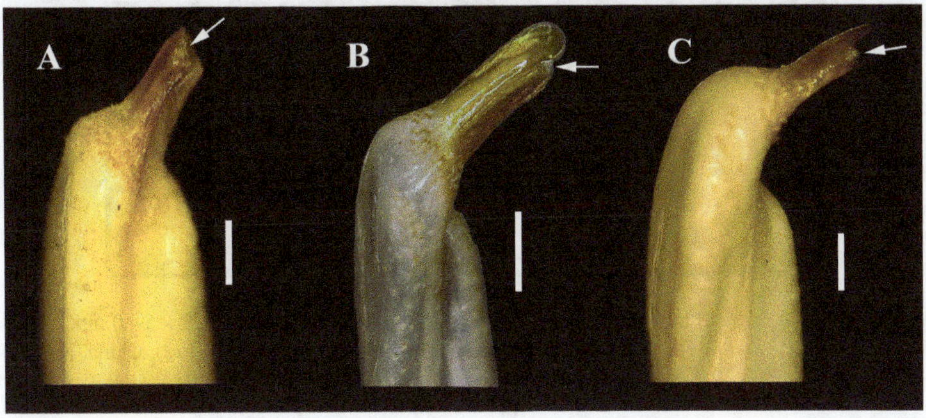

Fig. 7. Comparison of the apical processes of G1 (dorsal view) in selected species of *Perisesarma*. A, *P. tuerkayi* n. sp., holotype, male (28.2 × 24.0 mm) (QM- W27002); B, *P. eumolpe* (De Man, 1895), male (16.83 × 13.92 mm) (SMF 49922), China, Hainan, Wen Chang; C, *P. maipoense* (Soh, 1978) male (25.27 × 19.57 mm) (ZRC 2009.0800), Vietnam, mouth of the Red River; the arrows indicate position of aperture. Scale bar = 0.25 mm.

two species show clear differences: the apical corneous part in *P. maipoense* is narrower, longer, bent at an angle of about 62° to the vertical axis, and has a subterminal aperture; whereas in *P. tuerkayi* n. sp. the apical portion is stouter, shorter, less bent (about 35° to vertical), and the aperture is terminal (fig. 7A, C).

With regards to other species of *Perisesarma*, only *P. eumolpe* (also previously reported from Vietnam (Diele et al., 2013)) has a similar large number of dactylar tubercles (19-26) (Davie, 2010). But this species is different from *P. tuerkayi* in tubercle morphology and pattern, bearing obvious transverse sulci across the middle of each tubercle (fig. 8). Also the G1 of *P. eumolpe* has a long corneous process with a prominent rounded tip and a subterminal aperture; whereas the G1 of *P. turkayi* n. sp. has a shorter, less bent, somewhat truncated corneous process with a terminal aperture (fig. 7A, B).

The phylogenetic dendrogram based on mtDNA (Cox1 & 16S) and the NaK genotype network (figs. 5, 6) also show that *P. tuerkayi* n. sp. is clearly distinct from known congeners. The K2P distances also confirm significant genetic isolation of the new species from other *Perisesarma* (table III).

Our present phylogenetic results confirm that in most cases genetic relationships are in agreement with morphological similarities and also patterns of geographic distributions. For example, the two morphologically similar species (former subspecies), *P. longicristatum* and *P. semperi* turn out to be genetic sister species. This is also the case for *P. foresti* and *P. indiarum*. The three indigenous Australian species, *P. brevicristatum*, *P. darwinense* and *P. holthuisi* are geographically allopatric (east coast, north coast and west coast of Australia, respectively),

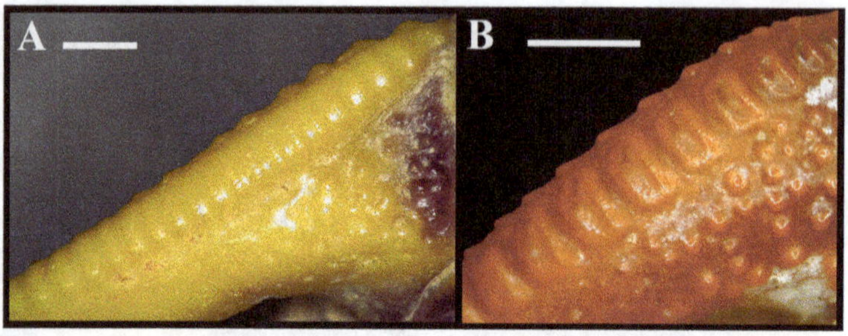

Fig. 8. Proximal dactylar tubercles in A, *P. tuerkayi* n. sp., holotype, male (28.2 × 24.0 mm) (QM-W27002); B, *P. eumolpe* (De Man, 1895), male (16.83 × 13.92 mm) (SMF 49922), China, Hainan, Wen Chang. Scale bar = 1 mm.

but show a close genetic association, suggesting evolutionary separation from a common ancestor isolated on that land mass. In contrast, *P. eumolpe* groups tightly with three species (*P. foresti*, *P. indiarum* and *P. messa*) that are morphologically strikingly different. Also two geographically distant species, the East African *P. samawati* and the western Pacific *P. lividum* show a close phylogenetic affinity. Further studies are needed to understand the relationships between biogeographic patterns, speciation mechanisms, and morphological divergences.

ACKNOWLEDGEMENTS

We are indebted to many institutions and colleagues who facilitated specimens, allowed access to samples or helped in the field. Among those our special thanks go to Peter K. L. Ng and Lee Bee Yan from the National University of Singapore (Singapore), to Miranda Lowe and Paul Clark from the Natural History Museum in London (U.K.), to Charles Fransen and Karen van Dorp from the Naturalis Biodiversity Center in Leiden (Netherlands), to Moritz Sonnewald and Bianca Trautwein from the Senckenberg Museum Frankfurt (Germany). We also greatly thank Dwi Listyo Rahayu, Katharina Wyschetzki, and Katharina Gryksa for continuing help. The first author's visits to the Natural History Museum in London (U.K.) (GB-TAF) and the Naturalis Biodiversity Center in Leiden (Netherlands) (NL-TAF) was supported by the SYNTHESYS Project (http://www.synthesys.info/), financed by the European Community, Research Infrastructure Action under the FP7 "Capacities". His visit to the Queensland Museum in Brisbane received support from the international PhD program at the University of Regensburg (iPUR). Both institutions are greatly acknowledged. Adnan Shahdadi is also being sponsored by the Iranian Ministry of Science, Research and Technology during his PhD studies at the University of Regensburg. The manuscript furthermore benefitted from comments of three anonymous reviewers.

REFERENCES

ABELE, L. G., 1975. The identity of *Sesarma hanseni* Rathbun, 1897, a supposedly West Indian species, with *S. dehaani* H. Milne Edwards, 1837 from the West Pacific. Crustaceana, **28**: 48-52.

CAMPBELL, B. M., 1967. The Australian Sesarminae (Crustacea: Brachyura): five species of *Sesarma* (*Chiromantes*). Memoirs of the Queensland Museum, **15**: 1-19.

CANNICCI, S., D. BURROWS, S. FRATINI, S. Y. LEE, T. J. SMITH III, J. OFFENBERG & F. DAHDOUH-GUEBAS, 2008. Faunistic impact on vegetation structure and ecosystem function in mangrove forests: a review. Aquatic Botany, **89**: 186-200.

DANA, J. D., 1851. Crustacea Grapsoidea (Cyclometopa, Edwardsii). Conspectus Crustaceorum quae in Orbis Terrarum circumnavigatione, Carolo Wilkes e Classe Reipublicae Foederatae Duce, lexit et descripsit J.D. Dana. Proceedings of the Academy of Natural Sciences of Philadelphia, **5**: 247-254.

DARRIBA, D., G. L. TABOADA, R. DOALLO & D. POSADA, 2012. jModelTest 2: more models, new heuristics and parallel computing. Nature Methods, **9**: 772.

DAVIE, P. J. F., 1994. Variations in diversity of mangrove crabs in tropical Australia. Memoirs of the Queensland Museum, **36**: 55-58.

DAVIE, P. J. F., 2010. A new species of *Perisesarma* (Crustacea: Brachyura: Sesarmidae) from Western Australia. In: C. H. J. M. FRANSEN, S. DE GRAVE & P. K. L. NG (eds.), Studies on Malacostraca: Lipke Bijdeley Holthuis memorial volume. Crustaceana Monographs, **14**: 195-207. (Brill, Leiden).

DAVIE, P. J. F., D. GUINOT & P. K. L. NG, 2015. Anatomy and functional morphology of Brachyura. In: P. CASTRO, P. J. F. DAVIE, D. GUINOT, F. R. SCHRAM & J. C. VON VAUPEL KLEIN (eds.), Decapoda: Brachyura (part 1). Treatise on zoology — anatomy, taxonomy, biology. The Crustacea, **9C-I**: 11-163. (Brill, Leiden).

DAVIE, P. J. F. & V. X. NGUYEN, 2003. A new species of *Metaplax* (Crustacea: Brachyura: Varunidae: Cyclograpsinae) from Vietnam. Raffles Bulletin of Zoology, **51**: 379-386.

DE MAN, J. G., 1895. Bericht uber die von Herrn Schiffscapitän Storm zur Atjeh, an den westlichen Küsten von Malakka, Borneo und Celebes sowie in der Java-See gesammelten Decapoden und Stomatopoden. Zoologische Jahrbücher, Abtheilung für Systematik, Geographie und Biologie der Thiere, **8**: 485-609.

DIELE, K., D. M. TRAN NGOC, S. J. GEIST, F. W. MEYER, Q. H. PHAM, U. SAINT-PAUL, T. TRAN & U. BERGER, 2013. Impact of typhoon disturbance on the diversity of key ecosystem engineers in a monoculture mangrove forest plantation, Can Gio Biosphere Reserve, Vietnam. Global and Planetary Change, **110**: 236-248.

GIRI, C., E. OCHIENG, L. L. TIESZEN, Z. ZHU, A. SINGH, T. LOVELAND, J. MASEK & N. DUKE, 2011. Status and distribution of mangrove forests of the world using Earth observation satellite. Global Ecology and Biogeography, **20**: 154-159.

GUERAO, G., K. ANGER, U. NETTELMANN & C. D. SCHUBART, 2004. Complete larval and early juvenile development of the mangrove crab *Perisesarma fasciatum* (Crustacea: Brachyura: Sesarmidae) from Singapore, with a larval comparison of *Parasesarma* and *Perisesarma*. Journal of Plankton Research, **26**: 1389-1408.

GUINDON, S. & O. GASCUEL, 2003. A simple, fast and accurate method to estimate large phylogenies by maximum-likelihood. Systematic Biology, **52**: 696-704.

HALL, T. A., 1999. BioEdit: a user-friendly biological sequence alignment editor and analysis program for Windows 95/98/NT. Nucleic Acids Symposium Series, **41**: 95-98.

HEBERT, P. D. N., A. CYWINSKA, S. L. BALL & J. R. DE WAARD, 2003. Biological identifications through DNA barcodes. Proceedings of the Royal Society, B, **270**: 313-322.

HOLTHUIS, L. B., 1977. The Grapsidae, Gecarcinidae and Palicidae (Crustacea: Decapoda: Brachyura) of the Red Sea. Israel Journal of Zoology, **26**: 141-192.

KOSUGE, T., K. WADA & P. D. TRONG, 1997. Crab distribution in the Cam River estuary, Haiphong, northern Vietnam. National Workshop on the Relationship between Mangrove Rehabilitation and Coastal Aquaculture in Vietnam: 178-184.

NG, P. K. L., D. GUINOT & P. J. F. DAVIE, 2008. Systema Brachyurorum, part I. An annotated checklist of extant brachyuran crabs of the world. Raffles Bulletin of Zoology, **17**(Suppl.): 1-286.

NG, P. K. L., H. NGOC KHAC & D. L. RAHAYU, 2010. On the taxonomy and ecology of the mangrove crab *Perisesarma maipoense* (Soh, 1978) (Crustacea: Decapoda: Brachyura: Sesarmiidae) from Vietnam. Raffles Bulletin of Zoology, **58**: 239-243.

NHUONG, D. V., 2003. Preliminary data on Brachyura in the Cangio mangrove area, Ho Chi Minh city. National Centre for Natural Science and Technology of Vietnam, Journal of Biology, **25**(1): 1-10. [In Vietnamese, with English summary.]

POSADA, D. & T. R. BUCKLEY, 2004. Model selection and model averaging in phylogenetics: advantages of Akaike information criterion and Bayesian approaches over likelihood ratio tests. Systematic Biology, **53**: 793-808.

RAGIONIERI, L., S. FRATINI, M. VANNINI & C. D. SCHUBART, 2009. Phylogenetic and morphometric differentiation reveal geographic radiation and pseudo-cryptic speciation in a mangrove crab from the Indo-West Pacific. Molecular Phylogenetics and Evolution, **52**: 825-834.

RODRIGUEZ, F. J., J. L. OLIVER, A. MARIN & J. R. MEDINA, 1990. The general stochastic model of nucleotide substitution. Journal of Theoretical Biology, **142**: 485-501.

SCHUBART, C. D., 2009. Mitochondrial DNA and decapod phylogenies: the importance of pseudogenes and primer optimization. In: J. W. MARTIN, K. A. CRANDALL & D. L. FELDER (eds.), Decapod crustacean phylogenetics. Crustacean Issues, **18**: 47-65.

SCHUBART, C. D., S. CANNICCI, M. VANNINI & S. FRATINI, 2006. Molecular phylogeny of grapsoid crabs (Decapoda, Brachyura) and allies based on two mitochondrial genes and a proposal for refraining from current superfamily classification. Journal of Zoological Systematics and Evolutionary Research, **44**: 193-199.

SCHUBART, C. D., J. A. CUESTA & A. RODRÍGUEZ, 2001. Molecular phylogeny of the crab genus *Brachynotus* (Brachyura: Varunidae) based on the 16S rRNA gene. Hydrobiologia, **449**: 41-46.

SCHUBART, C. D. & M. G. J. HUBER, 2006. Genetic comparison of German populations of the stone crayfish, *Austropotamobius torrentium* (Crustacea: Astacidae). Bulletin Francais de la Pêche et de la Pisculture, **380-381**: 1019-1028.

SERÈNE, R. & C. L. SOH, 1970. New Indo-Pacific genera allied to *Sesarma* Say, 1817 (Brachyura, Decapoda, Crustacea). Treubia, **27**: 387-416.

SHAHDADI, A. & C. D. SCHUBART, 2015. Evaluating the consistency and taxonomic importance of cheliped and other morphological characters that potentially allow identification of species of the genus *Perisesarma* De Man, 1895 (Brachyura, Sesarmidae). Crustaceana, **88**: 1079-1095.

SILVESTRO, D. & I. MICHALAK, 2012. RAxML GUI: a graphical front-end for RAML. Organisms Diversity and Evolution, **12**: 335-337.

TAMURA, K., D. PETERSON, N. PETERSON, G. STECHER, M. NEI & S. KUMAR, 2011. MEGA5: molecular evolutionary genetics analysis using maximum likelihood, evolutionary distance, and maximum parsimony methods. Molecular Biology and Evolution, **28**: 2731-2739.

TEMPLETON, A. R., K. A. CRANDALL & C. F. SING, 1992. A cladistic analysis of phenotypic associations with haplotypes inferred from restriction endonuclease mapping and DNA sequence data. III. Cladogram estimation. Genetics, **132**: 619-633.

THOMPSON, J. D., D. G. HIGGINS & T. J. GIBSON, 1994. CLUSTAL W: improving the sensitivity of progressive multiples sequence alignments through sequence weighting, position-specific gap penalties and weight matrix choice. Nucleic Acids Research, **22**: 4673-4680.

TSANG, L. M., K. Y. MA, S. T. AHYONG, T.-Y. CHAN & K. H. CHU, 2008. Phylogeny of Decapoda using two nuclear protein-coding genes: origin and evolution of the Reptantia. Molecular Phylogenetics and Evolution, **48**: 359-368.

TSANG, L. M., C. D. SCHUBART, S. T. AHYONG, J. C. Y. LAI, E. Y. C. AU, T.-Y. CHAN, P. K. L. NG & K. H. CHU, 2014. Evolutionary history of true crabs (Crustacea: Decapoda: Brachyura) and the origin of freshwater crabs. Molecular Biology and Evolution, **31**: 1173-1187.

VILLESEN, P., 2007. FaBox: an online toolbox for FASTA sequences. Molecular Ecology Notes, **7**: 965-968.

VON HAGEN, H. O., 1978. The systematic position of *Sesarma* (*Sesarma*) *rectum* Randall and a new definition of the subgenus *Sesarma* (Brachyura, Grapsidae). Crustaceana, **34**: 45.

ADDENDUM

Since the present paper was submitted, a separate study reviewing the status of the genera *Perisesarma* and *Parasesarma*, using both morphological and genetic analyses, has been accepted for publication (Shahdadi & Schubart, 2017). This separate work argues that most of the former species of *Perisesarma* are transferred to the genus *Parasesarma*. Due to the uncertainty of precise publication dates, it is possible that the work revising the genus *Perisesarma* may appear slightly before the present paper. In any case, the present new species will become *Parasesarma tuerkayi* (Shahdadi, Davie & Schubart, 2017).

SHAHDADI, A. & C. D. SCHUBART, 2017. Taxonomic review of *Perisesarma* (Decapoda: Brachyura: Sesarmidae) and closely related genera based on morphology and molecular phylogenetics: new classification, two new genera and the questionable phylogenetic value of the epibranchial tooth. Zoological Journal of the Linnean Society, London, **179**. DOI:10.1093/zoolinnean/zlx032.

First received 11 November 2016.
Final version accepted 15 February 2017.

[When citing this volume, please refer to Crustaceana 90 (2017) 771-1288]

Michael Türkay Memorial Issue

REPRODUCTIVE PERIOD AND RECRUITMENT OF PENAEOIDEA SHRIMP ON THE SOUTHEASTERN BRAZILIAN COAST: IMPLICATIONS FOR THE CLOSED SEASON

BY

SABRINA M. SIMÕES[1]), GISELE S. HECKLER and ROGERIO C. COSTA
Laboratório de Biologia de Camarões Marinhos e de Água Doce (LABCAM), Universidade Estadual Paulista, Av. Eng. Luiz Edmundo Carrijo Coube 14-01, 17033-360 Bauru, SP, Brazil

ABSTRACT

The aim of this study was to assess whether all Penaeoidea shrimps captured by trawling are protected by the current closed season. We investigated the reproductive period and juvenile recruitment of Penaeoidea shrimp species captured in the bay-estuary complex of Santos and São Vicente from May 2008 to April 2010. Samples were collected monthly using a shrimp boat at four sites in the estuary and four sites in the bay; environmental factors were monitored. The reproductive condition of females was determined by macroscopic observation of gonads. Juvenile and adult males were classified by observing the endopod forming the petasma; the reproductive condition was determined by the presence of spermatophore in the terminal ampoule. A dendrogram showed little similarity in the reproductive periods of the species sampled in this study. The greatest similarity in the recruitment period was in the white shrimp, *Litopenaeus schmitti*, and the pink shrimps *Farfantepenaeus paulensis* and *F. brasiliensis*, species that are protected by the current closed season. Pink shrimps are not fully protected, however, as they are captured indiscriminately in the estuary, to be used as fishing bait. Considering the number of individuals of reproductive females and juveniles, *Xiphopenaeus kroyeri* was the most abundant in the marine environment, and *L. schmitti*, *F. paulensis* and *F. brasiliensis* were the most abundant in the estuarine environment. The adequacy of the closed season is discussed, including issues such as the implementation of projects focusing on fishermen and the sustainable use of bait; the closure of fishing from October to December in coastal areas with depths below 20 m, and the application of a regional closed season instead of the current law applied in several states.

Key words. — *Artemesia*, environmental factors, *Farfantepenaeus*, juveniles, *Litopenaeus*, *Pleoticus*, *Rimapenaeus*, *Sicyonia*, trawling, *Xiphopenaeus*

ZUSAMMENFASSUNG

Als Ziel dieser Studie galt es herauszufinden, ob alle Garnelen der Penaeoidea, welche durch die Schleppnetzfischerei gefangen werden, durch das derzeitige saisonale Fangverbot geschützt sind. Wir untersuchten die Fortpflanzungszeit und den jungen Nachwuchs von Garnelenarten der

[1]) Corresponding author; e-mail: sabrinamsimoes@gmail.com

Penaeoidea, welche im Bucht-Ästuar-Komplex von Santos und São Vicente zwischen Mai 2008 und April 2010 gefangen wurden. Die Proben wurden monatlich mit einem Krabbenkutter jeweils an vier Positionen im Ästuar und in der Bucht genommen, während auch die Umweltfaktoren überprüft wurden. Der Fortpflanzungszustand der Weibchen wurde durch makroskopische Untersuchung der Gonaden bestimmt. Juvenile und adulte Männchen wurden durch Untersuchung des Endopoden (Petasma) klassifiziert; ihr Fortpflanzungszustand wurde anhand des Vorhandenseins von Spermatophoren in der Endampulle bestimmt. Ein Baumdiagramm zeigte leichte Ähnlichkeiten der Fortpflanzungszeiten der Arten, welche in dieser Studie untersucht wurden. Die größte Übereinstimmung ihrer Fortpflanzungszeiten zeigten die weiße Garnele *Lithopenaeus schmitti* und die rosafarbenen Garnelen *Farfantepenaeus paulensis* und *F. brasiliensis*, alle geschützt durch das derzeitige Fangverbot. Rosafarbene Garnelen sind jedoch nicht voll geschützt, da sie willkürlich im Ästuar als Fischköder gefangen werden. Bezüglich der Individuenzahl fortpflanzungsfähiger Weibchen und juveniler Tiere zeigte *Xiphopenaeus kroyeri* die größte Abundanz im Küstengebiet, während *L. schmitti*, *F. paulensis* und *F. brasiliensis* die größten Abundanzen im Mündungsgebiet aufwiesen. Die Angemessenheit des Fangverbots wird diskutiert, auch im Hinblick auf die Durchführung von Projekten, welche sich auf die Fischer und deren nachhaltige Nutzung von Ködern konzentrieren sollen. Diskutiert wird auch ein Fangverbot von Oktober bis Dezember in Küstengebieten mit einer Tiefe unter 20 m und die Anwendung eines saisonal-regionalen Fangverbotes anstelle des derzeitigen, in mehreren Staaten angewendeten Rechtes.

INTRODUCTION

Marine shrimp exploitation in Brazil contributes substantially to the national economy (Gillett, 2008), affecting much of the fishing sector in southeastern and southern Brazil. Brazil has the fourteenth-most productive shrimp fishery industry in the world; and 60% of the world's penaeid production comes from fishing, not from aquaculture (Gillett, 2008).

The southeastern region of Brazil is of great importance for fishery, responsible for 137 666 tonnes of landed fish products (IBAMA, 2007); 60% of all landed prawns come from the cities of Santos and Guarujá (Instituto de Pesca, 2013).

The seabob shrimp *Xiphopenaeus kroyeri* (Heller, 1862) and the pink shrimps *Farfantepenaeus brasiliensis* (Latreille, 1817) and *F. paulensis* (Pérez-Farfante, 1967) are three of the most captured species in Brazil, i.e., 26.7 and 17.9% of the production of marine crustaceans in Brazil (MPA, 2012), respectively. Another species with high commercial value caught on the Brazilian coast is the white shrimp *Litopenaeus schmitti* (Burkenroad, 1936), with 4077 tonnes captured in 2010 (MPA, 2012).

The decrease in pink shrimp production in southeastern and southern Brazil since the 1990s has forced the adoption of multi-species fisheries, including the capture of the stiletto shrimp *Artemesia longinaris* (Spence Bate, 1888) and the red shrimp *Pleoticus muelleri* (Spence Bate, 1888) (cf. Dias-Neto, 2011). Other Penaeoidea species captured on the coast are discarded due to lack of commercial interest because of their small size, such as *Sicyonia dorsalis* Kingsley, 1878, *S.*

typica Boeck, 1864 and *Rimapenaeus constrictus* (Stimpson, 1871) (Castilho et al., 2008a).

A closed season, the fisheries management measure most adopted in Brazil, prohibits fishing during specific periods of the year, in order to ensure replacement of the population (Dias-Neto, 2011). The closed shrimp season was created in 1983 by Brazilian legislation to reduce fishing pressure on pink shrimp juveniles, which were at their maximum capture level at that time (Franco et al., 2009). Since then the closed season has been changed several times to match to biological, economic and social changes in fisheries, and also due to pressure from the productive sector (Franco et al., 2009). Regulation of the closed season was last altered in 2008, when fishing of all shrimp species was banned in the southeastern and southern regions of Brazil from 1 March to 31 May (IBAMA, Normative instruction No. 189/September 23, 2008).

The main objective of this study is to determine whether the reproductive periods and/or juvenile recruitment of Penaeoidea shrimp (including those of no economic importance) are synchronized with the current closed season. We analysed the reproductive period and juvenile recruitment of species sampled in the bay-estuary complex of Santos and São Vicente, and compared environmental factors (temperature, salinity, phi, organic matter) to the abundance of juveniles and reproductive females.

MATERIAL AND METHODS

Sampling of marine shrimp

Sampling (license SISBIO/IBAMA No. 14653) was done monthly from May 2008 to April 2010 in the bay-estuary complex of Santos and São Vicente, on the central coast of São Paulo state. Collection was made using a shrimping boat equipped with trawl nets of 8 m mouth width, 10 m in length and 20 mm mesh size in the internodes, and with 18 mm of internode distance in the bag.

Four sites in the estuary were sampled and four in the bay, in areas with depths up to 15 m. Locations in the estuary were sampled from the mouth inwards to depths in which the vessel was able to move. In the bay, locations were defined according to a depth gradient (fig. 1).

The sampling effort was different in each environment, taking 10 min in the estuary (sample area = 5000 m^2) and 30 min in the bay (sample area = 15 000 m^2). The shorter time in the estuarine environment was due to large amounts of tangles in very muddy sediments, which hampered longer trawl time spent there.

All shrimps caught were sorted and stored in plastic bags with crushed ice, properly labelled and taken to the laboratory where each sample was analysed. The specimens were identified at species level according to Costa et al. (2003).

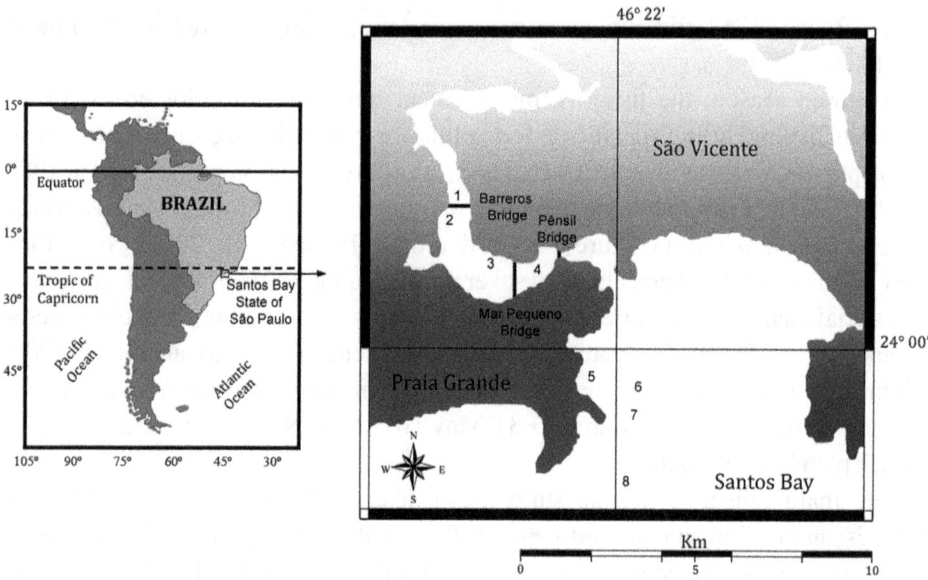

Fig. 1. Map of the study area, the bay-estuary complex of Santos and São Vicente, São Paulo state, southeastern Brazil, indicating the collection sites.

In this manuscript we follow the classification of penaeoid shrimps according to Pérez Farfante & Kensley (1997), because the phylogeny of *Penaeus* has not been fully elucidated in the recent molecular phylogeny study of Ma et al. (2011).

Sampling of environmental variables

Bottom and surface water samples were taken monthly at each collection site, using a Van Dorn bottle to determine the values of salinity, measured by a specific optical refractometer; and temperature, measured by a thermometer with 0.1°C accuracy. Depth was obtained monthly from an eco-probe at each sample site; geographical coordinates were determined using a global positioning system (GPS).

Sediment samples were collected monthly with a Van Veen grab, sampling an area of 0.06 m^2. The organic matter content was calculated by incineration and sediment texture was separated using the sieving method following the Wentworth scale (1922). Grain size categories were expressed on the phi scale ($= -\log_2$ grain diameter, mm) (Tucker, 1988): very coarse sand (> -1–0); coarse sand (>0–1); intermediate sand (>1–2); fine sand (>2–3); very fine sand (>3–4); and silt and clay ($\geqslant 4$). The procedures for the sediment texture and organic matter content analysis followed those of Costa et al. (2007).

Three main granulometric classes were defined, following Magliocca and Kutner (1965): Class A, corresponding to the percentage of medium sand (MS), coarse sand (CS), very coarse sand (VCS) and gravel (G); Class B, corresponding to the percentage of fine sand (FS) and very fine sand (VFS); Class C, corresponding to the percentage of silt + clay.

Data analysis

The total weight of each trawl was measured and a subsample of 400 g was taken. All individuals from the subsample were analysed according to their reproductive condition.

The reproductive condition of females was determined monthly by macroscopic observation of the reproductive system. It was defined as follows: immature or juvenile (IM), thin reproductive system of transparent colour (Castilho et al., 2007); rudimentary (RU), reproductive system ranging from white-translucent to light grey, occupying a small portion of the abdominal cavity (Pantone colour White/427C/7534); developing (D), a bigger reproductive system occupying the abdominal cavity and part of the carapace with colours ranging from light green to olive (Pantone colour 5803C/5773C/5763C); developed (DE), wider reproductive system occupying space to the end of the abdomen (Pantone colour 5757C/5753C/5743C) (Campos et al., 2009). Females were considered reproductive when they exhibited DE + DE ovaries.

Males were considered immature or juvenile (IM) when the endopodites forming the petasma were disunited; when they presented a united petasma they were considered adults (Castro et al., 2005). When the terminal ampoule showed secretion for spermatophore formation, males were considered mature (MA), and rudimentary (RU) when empty. The absolute frequency of males and females in each demographic category (IM, RU, MA) was plotted monthly.

The reproductive period was analysed monthly according to the percentage of reproductive females (D + DE) in relation to the total number of adult females. The juvenile recruitment period was determined from the percentage of juveniles (male + female) in relation to the total of sampled individuals.

The similarity of reproductive females and juveniles during all sample periods was analysed independently using cluster analysis. The number of individuals of each species was transformed ($\log(x + 1)$) to minimize data variability. Past Version 1.48 software was used for this analysis, which calculates the inverse index of Bray-Curtis (1957), since this index does not favour the most abundant species or reject the rarest (Krebs, 1989). The results of this index range from 0 to 1, and the closer to 1 the more similar the results presented. The connection method used was the Unweighted Pair-Group Method Using Arithmetic Average (UPGMA), since this best reflects the affinities of the original matrix on the dendrogram (Krebs, 1989).

The correlation of juveniles and reproductive females with temperature, salinity, organic matter and phi was tested by a redundancy analysis (RDA) using the R software (R Development Core Team, 2013), where $\alpha = 0.05$ (Zar 1999). The results of the statistical analysis (RDA) were demonstrated in a two-dimensional graph.

RESULTS

Reproductive period

The reproductive females of the Penaeoidea species *Xiphopeneaus kroyeri*, *Litopenaeus schmitti*, *Artemesia longinaris*, *Pleioticus muelleri*, *Rimapeneaus constrictus* and *Sicyonia dorsalis* were sampled only in the marine environment (Bay). From a total of 5253 subsampled females of *X. kroyeri*, 784 were reproductive females, captured during the two years of sampling. In contrast, the percentage of reproductive females of the rest of the species represented only 11% of the total sample (fig. 2).

Reproductive females of *X. kroyeri* were observed in all the sampled months and two peaks of higher intensity were recorded in 2009 (bimodal model): the first from February to April 2009 and the second from September to November 2009. The second peak (September to November 2009) was similar to that for the other three species (*L. schmitti*, *P. muelleri* and *S. dorsalis*); however, reproductive females of the other species were not captured in all sampled months (fig. 2).

The dendrogram constructed from the abundance of reproductive females sampled in the Bay showed two main groups with low similarity: the first

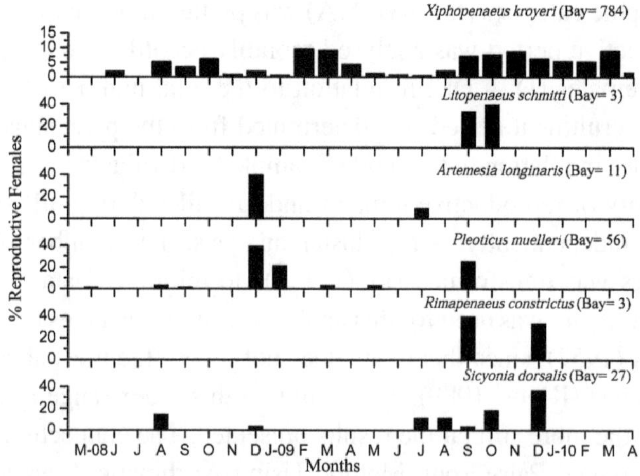

Fig. 2. Percentage of reproductive females captured in the Santos Bay, from May 2008 to April 2010. The total number of reproductive females is presented next to the name of each species.

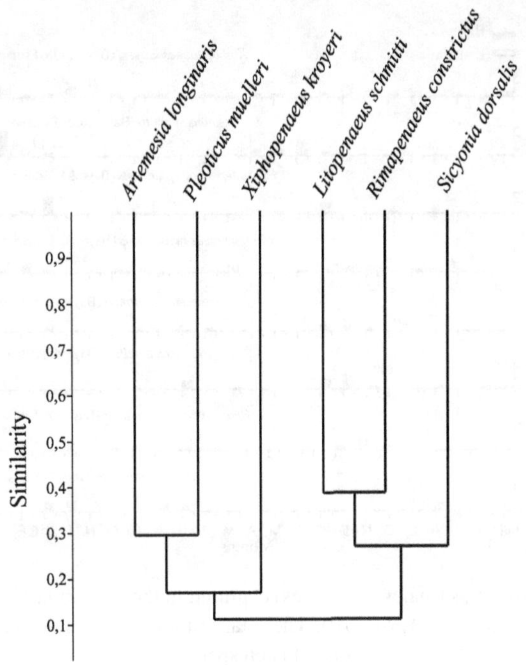

Fig. 3. Cluster analysis (UPGMA, Bray-Curtis), similarity between reproductive females collected in the Santos Bay, from May 2008 to April 2010.

comprised *A. longinaris*, *P. muelleri* and *X. kroyeri* (15% similarity) and the second *L. schmitti*, *R. constrictus* and *S. dorsalis* (25% similarity) (fig. 3).

Recruitment period

Juvenile recruitment of *X. kroyeri* occurred in the Bay during the two years of study, but the greatest intensity peaks were recorded from November to January in both years and in March and April 2010. The highest intensity peaks for *L. schmitti* were during January-March 2009 (Bay) and April and May 2009 in the estuarine environment. In 2010, the greatest abundance of juveniles was recorded from January to April 2010 (Estuary and Bay), and the last months synchronized with the recruitment of *X. kroyeri* (fig. 4). The cluster dendrogram for the Bay showed a similarity of 70% between *L. schmitti* and *X. kroyeri* (fig. 5).

The pink shrimps *F. paulensis* and *F. brasiliensis* recruited in similar months, from November 2008 to January 2009, and in January 2010 in the estuarine environment; and in December 2009 to April 2010 in the Bay and Estuarine environment (fig. 4). The cluster dendrogram showed high similarity between these shrimp species (>50%) (fig. 5). The main recruitments of both species were one to two months before that of *L. schmitti*, the species most abundant in the region.

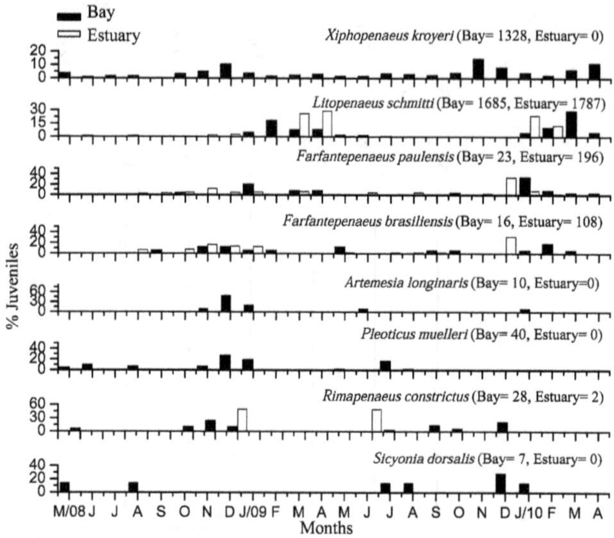

Fig. 4. Percentage of juveniles (males + females) captured in the bay-estuary complex of Santos and São Vicente, from May 2008 to April 2010. The total number of juveniles is presented next to the name of each species.

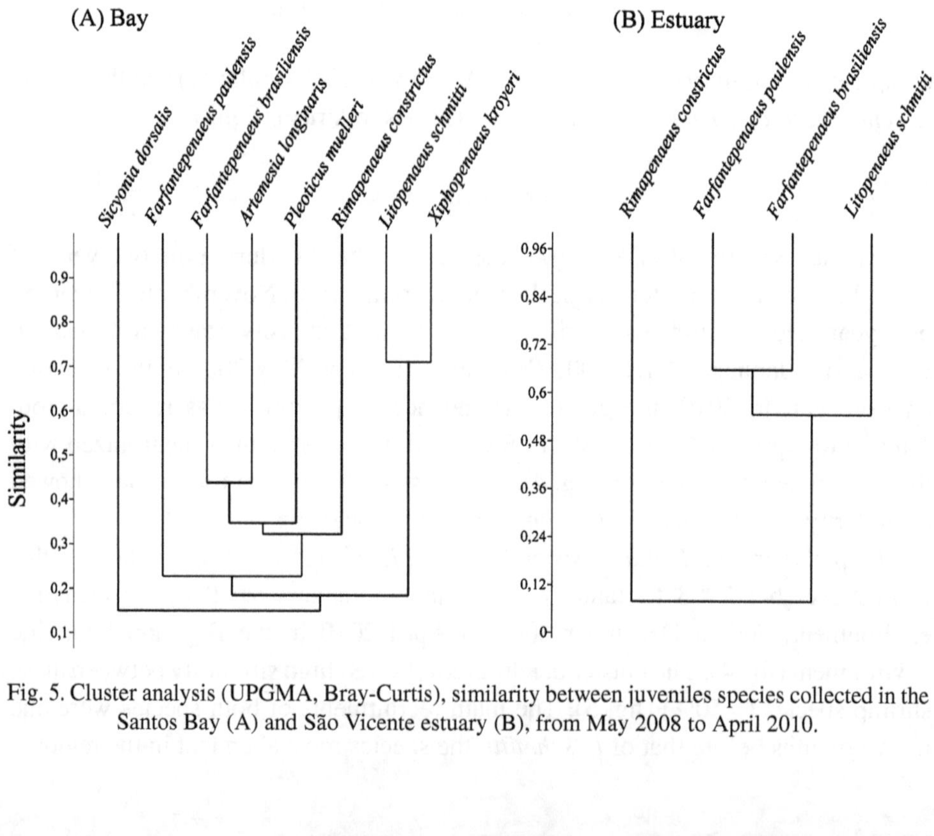

Fig. 5. Cluster analysis (UPGMA, Bray-Curtis), similarity between juveniles species collected in the Santos Bay (A) and São Vicente estuary (B), from May 2008 to April 2010.

Only a few juveniles of the species *A. longinaris*, *P. muelleri*, *R. constrictus* and *S. dorsalis* were captured in the study area; however, from October 2009 to January 2010 there was a greater abundance of *A. longinaris*, *R. constrictus* and *S. dorsalis* juveniles (fig. 4). The cluster dendrogram identified a similarity of 30% with *A. longinaris*, *P. muelleri* and *R. constrictus* (fig. 5).

Relationship of environmental factors with reproduction and recruitment

The lowest bottom temperatures in the estuary were recorded from November 2008 to January 2009, and the highest value was in March 2009. The lowest bottom salinities were recorded in January and March 2010, and the highest values were recorded in June 2009, October 2009 and March 2010 (fig. 6A).

In the Bay, the lowest bottom temperatures were recorded in November and December 2008, and the highest values were in March-April 2009 and February-March 2010. The highest bottom salinities were recorded in October and December 2009; the lowest values were recorded from July-August 2008 and March-April 2010 (fig. 6B).

A predominance of fine and very fine sand (Class B) was observed in all the sampled months, and phi values ranged slightly in variation (around 4). The highest percentages of organic matter were observed in December 2008, January 2009 and April 2010, but these values were constant in most months (fig. 7).

No statistically significant relationship was found between reproductive females and juveniles, and salinity, sediment texture or organic matter. Reproductive females of *X. kroyeri* and juveniles of *X. kroyeri* and *L. schmitti* were correlated positively only with temperature. The first axis of the bi-dimensional graph explained 98% of the total variance of the data, and the second explained only 1% (table I; fig. 8).

DISCUSSION

When considering the penaeoidean shrimp sampled in the present study, the Santos Bay proved to be an important spawning area for *Xiphopenaeus kroyeri*. The low frequency of reproductive females of the other shrimp species suggests that this area is not used for reproduction. There was thus no synchrony in the reproductive period of the penaeid shrimp captured in the complex bay-estuary of Santos and São Vicente. This can be confirmed by the dendrograms obtained, which generated clusters with low similarities among species.

The bottom temperature was crucial to the presence of reproductive females of *X. kroyeri*. Temperature has an influence on reproductive periodicity in penaeid shrimp and acts directly on gamete maturation (Sastry, 1983; Bauer & Lin, 1994;

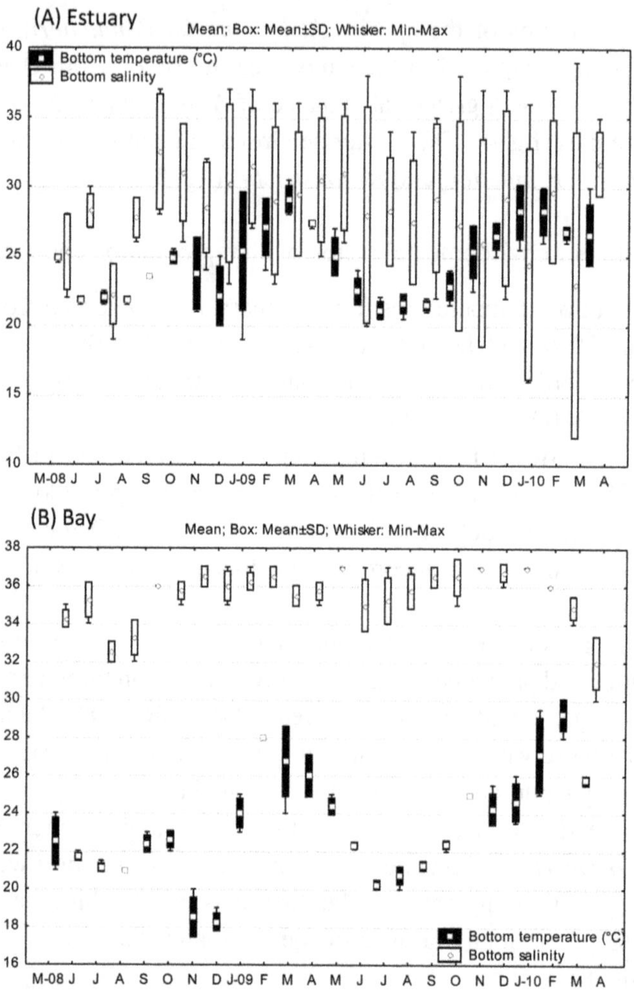

Fig. 6. The mean, standard deviation (SD) and minimum and maximum for bottom water temperature (°C) and salinity recorded from May 2008 to April 2010 in the São Vicente estuary (A) and Santos Bay (B).

Castilho et al., 2008b). Considering the abundance and constancy of reproductive females, the seabob shrimp *X. kroyeri* is extremely important in Santos Bay. This species is an important fishing resource in Brazil: it is the second most fished species in the study region, only sardines (*Sardinella brasiliensis* (Steindachner, 1879)) are fished more (Instituto de Pesca, 2013). In São Paulo state, the seabob shrimp is most captured in Santos and Guarujá, where about 1956.7 tonnes are landed per year (Instituto de Pesca, 2013).

The small number of reproductive females of *Farfantepenaeus paulensis*, *F. brasiliensis* and *L. schmitti* can be explained by the fact that adults spawn in

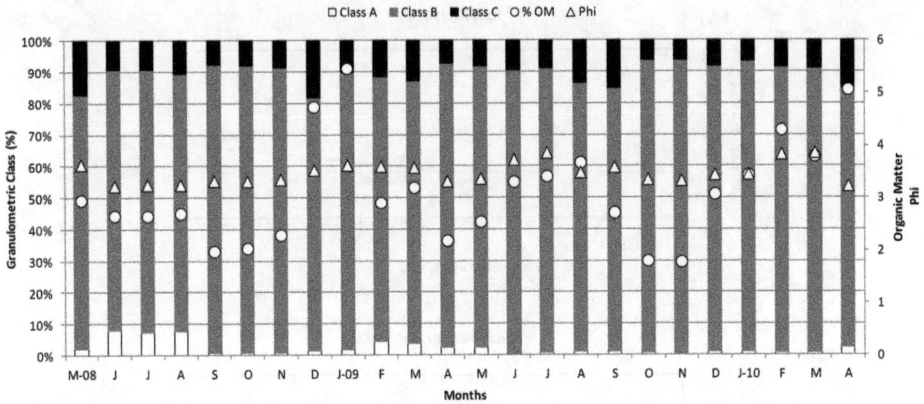

Fig. 7. Granulometric fractions, organic matter content (% OM) and the phi values (measures of central tendency of the sediment) collected in the bay-estuary complex of Santos and São Vicente, from May 2008 to April 2010. Class A: medium sand, coarse sand, very coarse sand and gravel; Class B: fine sand and very fine sand; Class C: silt + clay.

deeper locations (below 30 meters) (Zenger & Agnes, 1977). The low abundance of *Rimapeneaus constrictus* and *Sicyonia dorsalis* could possibly be to the sediment, predominantly composed of very fine sand, limiting their presence in the region, since these species prefer sediments predominantly composed of silt and clay (Costa & Fransozo, 2004; Costa et al., 2005). The Santos Bay is also very close to a large estuarine area, and these species prefer areas of higher salinities.

The inter-annual variation of recruits and reproductive females of *Artemesia longinaris* and *Pleioticus muelleri* observed in Santos Bay followed the same pattern as noted by Batista et al. (2011), in which abundance was associated

TABLE I

Results of redundancy analysis (RDA) of the reproductive females and juveniles abundances of Penaeoidea shrimps in relation to environmental variables that explained the most variance of the data collected in the bay-estuary complex of Santos and São Vicente from May 2008 to April 2010

		Proportion explained	Environmental factor	Explained variance	r^2	p
Reproductive females	RDA1	0.9925	Temperature	0.9853	0.4717	0.002*
	RDA2	0.0051	Salinity	0.9902	0.1754	0.098
			Organic Matter	−0.9999	0.0931	0.343
			Phi	−0.9937	0.0950	0.325
Juveniles	RDA1	0.9903	Temperature	0.9943	0.6332	0.001*
	RDA2	0.0086	Salinity	−0.9682	0.0776	0.428
			Organic Matter	0.8698	0.1007	0.309
			Phi	0.9847	0.0740	0.457

* Significant value ($p<0.05$).

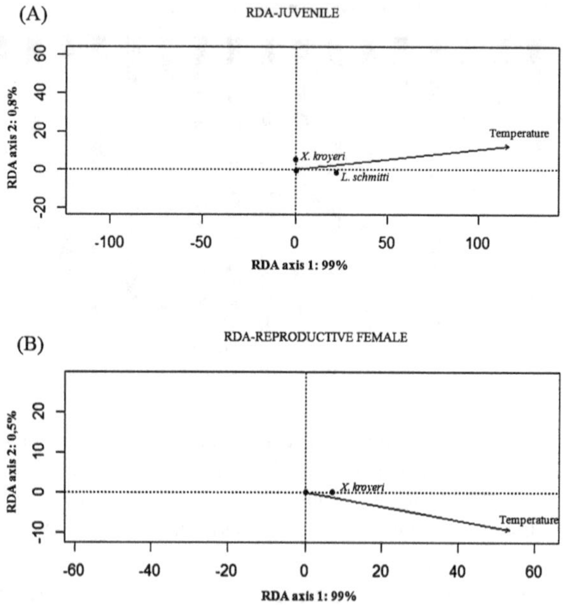

Fig. 8. Redundancy analysis (RDA) biplots illustrating the relationship between significant environmental variables and juveniles (A) and reproductive females (B) in the bay-estuary complex of Santos and São Vicente, from May 2008 to April 2010.

with low water temperature. The temperature decrease was probably caused by the intrusion of the South Atlantic Central Waters in spring and summer, which annually increases the abundance of these species. The recruitment period was synchronized in the estuarine environment between the white shrimp *L. schmitti* and pink shrimp *F. brasiliensis* and *F. paulensis*. Juveniles of these species are included in the current closed season; however, in 2009, in a period during which they would be protected by the closed season law, individuals did develop in the estuary but still continued to be fished indiscriminately with "gerival" nets, which are used manually with small boats without engines. This type of net provides a passive trawl during tidal movement (Mendonça, 2007).

The increase in abundance of juveniles of *L. schmitti*, together with the temperature increase, was also observed by Capparelli et al. (2012). These authors suggested that closely related species use temporal or spatial segregation to avoid competition, as for *L. schmitti* and *F. paulensis*. The main recruitment peak of *L. schmitti* occurred in the warmer months in Ubatuba, and for *F. paulensis* it occurred in the colder months.

An implementation of projects targeting the bait culture (juveniles) without removing them from the environment would help the maintenance of stocks and the conservation of these shrimps. Shrimp farming for live bait is an important tool for conserving natural stocks (Jensen et al., 2014). Studies are being developed to improve these methods with native species such as *F. brasiliensis* and *F. paulensis* (Fóes et al., 2011; Emerenciano et al., 2012; Jensen et al., 2014).

In the marine environment (Bay), recruitment was synchronized for *X. kroyeri* and *L. schmitti*, but only during the months open for fishing (October, November and December). In the Thermaikos Gulf (North Aegean Sea), regulations for the closed season are determined spatially: fishing is closed at different times and less than two nautical miles from the coast, which allows stock sustainability (Kevrekidis & Thessalou-Legaki, 2006). If fishing was closed in the Santos Bay during October-December in coastal areas with depths below 20 m, juvenile Penaeoidea shrimp and reproductive female *X. kroyeri* would be protected, since their reproductive peaks occur precisely in these months.

Xiphopenaeus kroyeri is protected by the current legislation on the north coast of São Paulo, but three months are not enough for the protection of juveniles. As the species takes four months to reach adulthood, Heckler et al. (2013) suggested the inclusion of February in the current closed season in Ubatuba, São Paulo. The implementation of a regional closed season instead of the same season being applied in several states as it is today is an issue of major concern. The closed season may be correct for other Brazilian regions, such as the coast of Rio de Janeiro, whose protection period is in accordance with the recruitment period of *X. kroyeri* (Fernandes et al., 2011), but not for all areas simultaneously.

We conclude that the most relevant species for suitable management plans in the bay-estuary complex of Santos and São Vicente are *X. kroyeri*, *L. schmitti* for the marine environment and *L. schmitti*, *F. paulensis* and *F. brasiliensis* for the estuarine environment. Other important issues were raised in this study and they contribute to future discussions of the adequacy of the closed season, such as sustainable exploitation of shrimp bait by aquaculture, the inclusion of a regional closed season and the prohibition of fishing from October to December in coastal areas with depths below 20 m.

ACKNOWLEDGEMENTS

We are grateful to the São Paulo Research Foundation (FAPESP) for providing financial support (Proc. Nos 08/54991-0, 08/53999-7, 09/53942-8, 10/50188-8 and 05/04707-5) and to CNPq (Grant 479541/2007-08 and Research Scholarships PQ 305919/2014-8). We are also thankful to the LABCAM coworkers (especially to Mateus Lopes), Dr. Fernando J. Zara (Unesp-Jaboticabal) and Dr. Alvaro Reigada (Unisanta), for their help during fieldwork. All sampling in this study has been conducted in compliance with current applicable state and federal laws.

This paper is dedicated to Dr. Michael Türkay for his contributions to the study of the decapod Crustacea.

REFERENCES

BATISTA, A. C., S. M. SIMÕES, M. LOPES & R. C. COSTA, 2011. Ecological distribution of the shrimp *Pleoticus muelleri* (Bate, 1888) and *Artemesia longinaris* Bate, 1888 (Decapoda, Penaeoidea) in the southeastern Brazilian littoral. Nauplius, **19**: 135-143.

BAUER, R. T. & J. LIN, 1994. Temporal patterns of reproduction and recruitment in populations of the penaeid shrimps *Trachypenaeus similis* (Smith) and *T. constrictus* (Stimpson) (Crustacea: Decapoda) from the north-central Gulf of Mexico. Journal of Experimental Marine Biology and Ecology, **182**: 205-222.

BRAY, J. R. & J. T. CURTIS, 1957. An ordination of the upland forest communities of southern Wisconsin. Ecological Monographs, **27**: 325-349.

CAMPOS, B. R. D., L. F. C. DUMONT & J. O. BRANCO, 2009. Ovarian development and length at first maturity of the sea-bob-shrimp *Xiphopenaeus kroyeri* (Heller) based on histological analysis. Nauplius, **17**: 9-12.

CAPPARELLI, M. V., P. KASTEN, A. L. CASTILHO & R. C. COSTA, 2012. Ecological distribution of the shrimp *Litopenaeus schmitti* (Burkenroad, 1936) (Decapoda, Penaeoidea) in Ubatuba Bay, São Paulo, Brazil. Invertebrate Reproduction & Development, **56**: 173-179.

CASTILHO, A. L., R. C. COSTA, A. FRANSOZO & E. E. BOSCHI, 2007. Reproductive pattern of the South American endemic shrimp *Artemesia longinaris* (Decapoda, Penaeidae), off São Paulo State, Brazil. Revista de Biología Tropical, **55**: 39-48.

CASTILHO, A. L., R. C. COSTA, V. FRANSOZO & M. L. NEGREIROS-FRANSOZO, 2008b. Reproductive and recruitment of the South American red shrimp, *Pleoticus muelleri* (Crustacea: Solenoceridae), from the southeastern coast of Brazil. Marine Biology Research, **4**: 361-368.

CASTILHO, A. L., M. R. PIE, A. FRANSOZO, A. P. PINHEIRO & R. C. COSTA, 2008a. The relationship between environmental variation and species abundance in shrimp community (Crustacea: Decapoda: Penaeoidea) in south-eastern Brazil. Journal of the Marine Biological Association of the United Kingdom, **88**: 119-123.

CASTRO, R. H., R. C. COSTA, A. FRANSOZO & F. L. M. MANTELATTO, 2005. Population structure of the seabob shrimp *Xiphopenaeus kroyeri* (Heller, 1862) (Crustacea: Penaeoidea) in the littoral of São Paulo, Brazil. Scientia Marina, **69**: 105-112.

COSTA, R. C. & A. FRANSOZO, 2004. Abundance and ecologic distribution of the shrimp *Rimapenaeus constrictus* (Crustacea: Penaeidae) in the northern coast of São Paulo State, Brazil. Journal of Natural History, **38**: 901-912.

COSTA, R. C., A. FRANSOZO, F. A. M. FREIRE & A. L. CASTILHO, 2007. Abundance and ecological distribution of the "sete-barbas" shrimp *Xiphopenaeus kroyeri* (Heller, 1862) (Decapoda: Penaeoidea) in three bays of the Ubatuba region, south-eastern Brazil. Gulf and Caribbean Research, **19**: 33-41.

COSTA, R. C., A. FRANSOZO, G. A. S. MELO & F. A. M. FREIRE, 2003. An illustrated key for Dendrobranchiata shrimps from the northern coast of São Paulo state, Brazil. Biota Neotropica, **3**: 1-12.

COSTA, R. C., A. FRANSOZO & M. L. NEGREIROS-FRANSOZO, 2005. Ecology of the rock shrimp *Sicyonia dorsalis* Kingsley, 1878 (Crustacea: Sicyoniidae) in a subtropical region of Brazil. Gulf and Caribbean Research, **17**: 49-56.

DIAS-NETO, J., 2011. Proposta de plano nacional de gestão para o uso sustentável de camarões marinhos do Brasil. (MMA/IBAMA, Brasília).

EMERENCIANO, M., E. L. BALLESTER, R. O. CAVALLI & W. WASIELESKY, 2012. Biofloc technology application as a food source in a limited water exchange nursery system for pink shrimp *Farfantepenaeus brasiliensis* (Latreille, 1817). Aquaculture Research, **43**: 447-457.

FERNANDES, L. P., A. C. SILVA, L. P. JARDIM, K. A. KEUNECKE & A. P. M. DI BENEDITTO, 2011. Growth and recruitment of the Atlantic seabob shrimp, *Xiphopenaeus kroyeri* (Heller, 1862) (Decapoda, Penaeidae), on the coast of Rio de Janeiro, southeastern Brazil. Crustaceana, **84**: 1465-1480.

FÓES, G. K., C. FRÓES, D. KRUMMERNAUER, L. POERSCH & W. WASIELESKY, 2011. Nursery of pink shrimp *Farfantepenaeus paulensis* in biofloc technology culture system: survival and growth at different stocking densities. Journal of Shellfish Research, **30**: 367-373.

FRANCO, A. C. N. P., R. SCHWARZ JR., N. PIERRI & G. D. SANTOS, 2009. Levantamento, sistematização e análise da legislação aplicada ao defeso da pesca de camarões para as regiões sudeste e sul do Brasil. Boletim do Instituto de Pesca, **35**: 687-699.

GILLETT, R., 2008. Global study of shrimp fisheries. FAO Fisheries Technical Paper No. 475. (FAO, Rome).

HECKLER, G. S., S. M. SIMÕES, A. P. F. SANTOS, A. FRANSOZO & R. C. COSTA, 2013. Population dynamics of the seabob shrimp *Xiphopenaeus kroyeri* (Dendrobranchiata, Penaeidae) in a south-eastern region of Brazil. African Journal of Marine Science, **35**: 17-24.

IBAMA (INSTITUTO BRASILEIRO DO MEIO AMBIENTE E DOS RECURSOS NATURAIS RENOVÁVEIS), 2007. Estatística da Pesca 2007, Brasil: Grandes Regiões e Unidades da Federação, Brasflia-DF, Dezembro 2007.

IBAMA, 2008. Instrução Normativa N° 189, de 23 de setembro de 2008. Reunião final com representações das regiões sudeste e sul, ocorrida em Itajaí/SC, no dia 21 de agosto de 2008; Processo IBAMA/SC n° 2026.001828/2005-35. 2008.

INSTITUTO DE PESCA, 2013. Informe da Produção Pesqueira Marinha e Estuarina do Estado de São Paulo — Dezembro de 2012. Instituto de Pesca. N° 26 — fev./2013. (São Paulo).

JENSEN, L., P. S. FURTADO, M. M. FUGIMURA, L. O. GARCIA, L. H. POERSCH, J. R. VERANI & W. WASIELESKY, 2014. The effect of stocking density on the transport of pink shrimp *Farfantepenaeus brasiliensis* (Crustacea: Decapoda), as live bait for sport fishing in Brazil. Latin American Journal of Aquatic Research, **42**: 204-212.

KEVREKIDIS, K. & M. THESSALOU-LEGALI, 2006. Catch rates, size structure and sex ratio of *Melicertus kerathurus* (Decapoda: Penaeidae) from an Aegean Sea trawl fishery. Fisheries Research, **80**: 270-279.

KREBS, C. L., 1999. Ecological Methodology (2nd ed.): 620 pp. (Benjamin Cummings, Menlo Park, California).

MA, K. Y., T. Y. CHAN & K. H. CHU, 2011. Refuting the six-genus classification of *Penaeus* s.l. (Dendrobranchiata, Penaeidae): a combined analysis of mitochondrial and nuclear genes. Zoologica Scripta, **40**: 498-508.

MAGLIOCCA, A. & A. S. KUTNER, 1965. Sedimentos do fundo da Enseada do Flamengo, Ubatuba, SP. Contribuições do Instituto Oceanográfico, **198**: 1-15.

MENDONÇA, J. T., 2007. Gestão dos recursos pesqueiros do Complexo Estuarino-lagunar de Cananéia, Iguape e Ilha Comprida, litoral sul de São Paulo, Brasil: 1-385. (Ph.D. Thesis, Universidade Federal de São Carlos).

MPA — MINISTÉRIO DA PESCA E AQUICULTURA, 2012. Boletim estatístico da pesca e aquicultura (Brasil 2010). (Ministério da Pesca e Aquicultura, Brasília).

PÉREZ FARFANTE, I. & B. KENSLEY, 1997. Penaeoid and sergestoid shrimps and prawns of the world. Keys and diagnoses for the families and genera. Mémoires du Muséum national d'Histoire naturelle, **175**: 1-233.

SASTRY, A. N., 1983. Ecological aspects of reproduction. In: F. J. VERNBERG & W. B. VERNBERG (eds.), Environmental adaptations. The biology of Crustacea, **8**: 179-270. (Academic Press, New York, NY).

TUCKER, M., 1988. Techniques in sedimentology. (Blackwell, Oxford).
WENTWORTH, C. K., 1922. A scale of grade and class terms for clastic sediments. The Journal of Geology, **30**: 377-392.
ZAR, J. H., 1999. Biostatistical analysis (4th ed.). (Prentice-Hall, Upper Saddle River, NJ).
ZENGER, H. H. & J. L. AGNES, 1977. Distribuição do camarão *Penaeus brasiliensis* e *Penaeus paulensis* ao longo da costa Sudeste e Sul do Brasil. Série Documentos Técnicos, **21**: 1-105.

First received 2 May 2016.
Final version accepted 20 February 2017.

[When citing this volume, please refer to Crustaceana 90 (2017) 771-1288]

Michael Türkay Memorial Issue

COMPOSITION OF THE EPIBENTHIC DECAPOD CRUSTACEAN MEGAFAUNA OF THE GERMAN EXCLUSIVE ECONOMIC ZONE: COMPARISON AND ANALYSIS OF PAST AND RECENT SURVEYS

BY

MORITZ SONNEWALD[1]) and MICHAEL TÜRKAY[†])
Senckenberg Research Institute and Natural History Museum, Marine Zoology — Crustacean Section, Senckenberganlage 25, 60325 Frankfurt am Main, Germany

ABSTRACT

This study comprises and compares epibenthic samples of the marine crustacean decapod megafauna, taken with a beam trawl at numerous stations in the German Exclusive Economic Zone of the North Sea in the years 1987, 1990, 2007 and 2013. All decapod crustacean species were recorded quantitatively. An overview of the decapod faunal composition with respect to changes in space, species and abundance with time is given. A time-dependent decrease of rare and often northern species at the research area is obvious, as a result of an increase of some dominant species concentrating in coastal areas. In 2013, a more uniform species composition seems to have developed. Decapod species and abundance depletions may be influenced by many factors, among them also climate change and fisheries.

Key words. — Beam trawl, Crustacea, Decapoda, species number decrease, Epibenthos, German Exclusive Economic Zone

ZUSAMMENFASSUNG

Diese Studie umfasst und vergleicht Epibenthosproben der Megafauna mariner, dekapoder Crustaceen, welche mit einer Baumkurre an zahlreichen Stationen in der Ausschließlichen Wirtschaftszone Deutschlands innerhalb der Nordsee in den Jahren 1987, 1990, 2007 und 2013 genommen wurden. Alle Arten dekapoder Crustaceen wurden quantitativ erfasst. Es wird ein Überblick über die Zusammensetzung der Dekapodenfauna im Hinblick auf Areal-, Arten- und Abundanzverschiebungen im Zeitverlauf gegeben. Eine zeitabhängige Abnahme seltener und oft nördlicher Arten im Untersuchungsgebiet, welche mit mit der Zunahme dominanter Arten im Küstenbereich einhergeht, ist offensichtlich. Im Jahre 2013 scheint sich eine einheitlichere Artenzusammensetzung entwickelt zu haben. Die Abnahme von Dekapodenarten und -abundanzen wird vermutlich von vielen Faktoren beeinflusst, darunter auch der Klimawandel und die Fischerei.

---

[1]) Corresponding author; e-mail: Moritz.Sonnewald@Senckenberg.de

[†]) Deceased.

INTRODUCTION

As the North Sea is one of the most intensely studied marine environments of the world (Wieking & Kröncke, 2003), the German Exclusive Economic Zone (EEZ) is often included in large-scale studies either with expansion over the whole North Sea, or at least covering extensive parts of this European shelf sea. Apart from studies focusing solely on the endobenthos, some studies combine both epifaunal and infaunal research in the North Sea. Künitzer (1990) compared the infaunal and epifaunal composition in the offshore central North Sea and at some German Bight stations, showing the central and northern North Sea to be two separate areas concerning the epifaunal composition but the infauna separating into four different groups. Reiss et al. (2010) reported related results with an even higher resolution. Concerning the epibenthic fauna, a number of studies covering the whole North Sea were conducted. For example, Frauenheim et al. (1989), Zühlke et al. (2001) and Jennings et al. (1999) revealed epibenthic sessile and mobile species distribution, as well as the typical division into a southern and a northern clade. Spatial variability of epifaunal communities in different areas of the North Sea was further analysed by Neumann et al. (2008), also showing that temperature is influencing epifaunal reproduction (Neumann & Kröncke, 2011). Epifaunal long-term studies at the Dogger Bank showed a community structure. Time-dependent changes in the epifaunal community structure around the beginning of the 21st century, related to changes in climate, have also been reported (Sonnewald & Türkay, 2011, 2012a, b).

Furthermore, local studies with a focus on the EEZ and the German Bight in particular have been conducted. A report by Rachor & Nehmer (2003) gives a detailed overview about the hydrographic and faunal regime of the North Sea with special attention to the EEZ. The temporal variability of the EEZ epifaunal communities on a large spatial scale was analysed by Neumann et al. (2009) on the basis of epibenthos samples from nine ICES (International Council for the Exploration of the Sea) IBTS (International Bottom Trawl Survey) cruises conducted between 1998 and 2008, revealing a community shift related to a cold winter in 1995/1996. Additionally, Neumann et al. (2012) analysed different habitats of the EEZ and linked them with the spatial distribution of infaunal, epifaunal and fish communities. Field works for his study were started by the late Prof. Michael Türkay, who always held the opinion that we only can understand the marine faunal composition and its changes through time by empiric observation, and not based on common opinions alone. Hence, this study contributes to a better understanding of the present biodiversity and long-term changes in the animal group of decapod crustaceans of the EEZ, showing and analysing the results of four monitoring cruises.

MATERIAL AND METHODS

Research area and stations

Monitoring cruises with a comparable set of stations were conducted in 1987 (DEB87), 1990 (D), 2007 (D2007) and 2013 (D2013) in the German Exclusive Economic Zone (fig. 1), on "RV Senckenberg". In 1987, the German Bight was sampled up to 55°N and in 1990 the northern EEZ with the so-called "duck's bill" was sampled. For the years 2007 and 2013, the sampling grid was a combination of the 1987 and 1990 stations, with some exceptions: stations D-9, D-26 to D-35, DEB87-14 and -10 were omitted, station DEB87-8 was moved (station 22 in 2007 and 2013), and stations D2007/2013-1 and -9 were added.

Sampling methods

Epibenthic samples were taken with a 2-m beam trawl with a design related to the one described by Jennings et al. (1999), having a stretched mesh size of 1 cm at the cod end, being towed for one nautical mile at a speed of two knots at each station. A 1 cm^2 mesh size sieve was used to isolate, identify and count the megafaunal decapod crustaceans. All decapod species were determined to species level (crabs were also sexed) and counted on board. Smaller animals (<1 cm^2) and undetermined species were preserved in a 5% formaldehyde/sea water solution for subsequent lab determination. A quantitative dataset was built based on the identified decapod species and their respective abundance per station. For each decapod species detected on the four EEZ cruises, the percentage of stations per cruise with a record of the respective species was calculated.

Environmental data were recorded using a CTD probe, measuring temperature and salinity of the water column at each station in 2007 and 2013.

Statistics

The abundance records of species of decapod crustaceans from 1987, 1990, 2007 and 2013 were used one by one for a similarity analysis after Bray & Curtis (1957). The results were depicted as a multidimensional scaling (MDS)-plot and the groups were depicted on the station maps when necessary. Moreover, the individual decapod abundances per sample of the years 1987 and 1990 were assigned to group one, while the abundances of the years 2007 and 2013 were assigned to group two. The dataset was square-root transformed and a resemblance matrix using the Bray-Curtis similarity coefficient was generated and depicted in an MDS-plot. The result was evaluated using an analysis of similarity (ANOSIM) and a similarity percentage analysis (SIMPER). In order to resolve if the decapod community composition is related to different habitats and to enhance

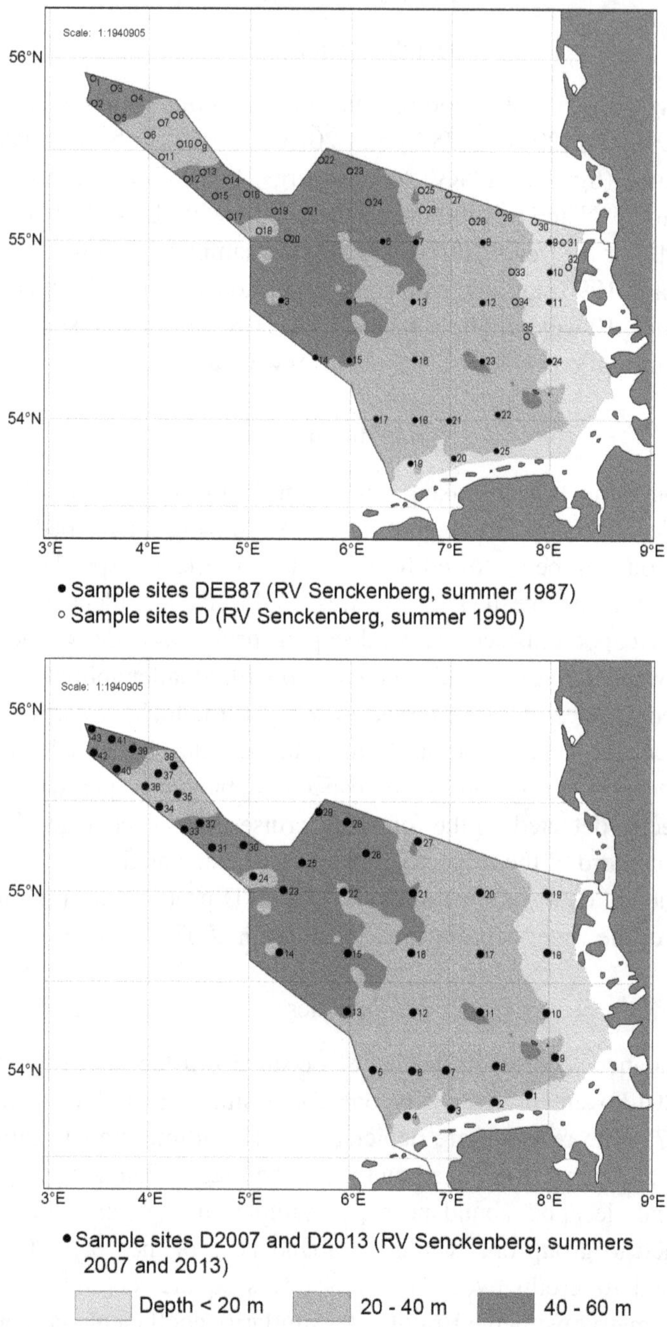

Fig. 1. Research area in the German Exclusive Economic Zone. Above, the stations of the 1987 and 1990 cruises are shown. The stations of the cruises 2007 and 2013 are depicted below.

the resolution of possible community changes with time, the data of 1987/90, 2007 and 2013 were analysed independently in the same way as described above, depicting the groups resolved by the MDS-plots to the station map if appropriate. All statistical analyses were done using the PRIMER V6 software (Clarke & Gorley, 2006).

RESULTS

Species records and time-dependent tendencies

A species list with the percentage of stations with positive records is presented in table I, sorted by percentage of occurrence, also depicting the decapod species numbers according to the respective cruise at the bottom line. These are showing a remarkable decrease in species richness from past to recent surveys, especially when considering that the 1987 and 1990 cruises can more or less be treated as one cruise with 36 identified decapod species, as they covered the northern and southern EEZ, respectively, which were sampled in combination starting from the 2007 cruise.

The percentages of positive records of decapod species in relation to the number of sampled stations also show a declining tendency from the past to recent surveys. This reduction is shown by the example of *Crangon allmanni* Kinahan, 1860 in fig. 2, where a loss of range and abundance with time is obvious.

Overall, the numbers of decapod species in 2007 (21; 0.48 species per station) and 2013 (15; 0.35 species per station) both are below the number of species in 1987/1990, which in sum is 36 (0.63 species per station). This tendency is a mixture of a decapod species decrease in the offshore areas and a slight increase at the coasts with time, which is shown in detail in fig. 3.

Temporal changes in species composition

The difference in species composition when comparing 1987/1990 to 2007/2013 was visualised in a MDS-plot (fig. 4), using the two previously assigned groups for temporal separation. Although an ANOSIM-test of the data showed a significance level of $p = 0.1$ and $R = 0.126$ (close to even distribution), the plot quite clearly shows a separation of both groups with a few overlaps. A SIMPER analysis of both groups identified the main contributors to this dissimilarity, namely *Pagurus bernhardus* (Linnaeus, 1758), *Liocarcinus holsatus* (Fabricius, 1798), *Crangon allmanni*, *C. crangon* (Linnaeus, 1758) and *Corystes cassivelaunus* (Pennant, 1777), as shown in table II. Except *C. allmanni* (see fig. 2), all above-mentioned species show an increase in abundance when comparing the times before and after the millennium, together with a general depletion of the most rare species (table I).

TABLE I

List of decapod crustacean species from the EEZ cruises

	DEB1987 (22 stations)		D1990 (35 stations)		D2007 (43 stations)		D2013 (43 stations)	
	∑	%	∑	%	∑	%	∑	%
Pagurus bernhardus (Linnaeus, 1758)	22	100%	35	100%	43	100%	35	81%
Liocarcinus holsatus (Fabricius, 1798)	17	77%	31	89%	37	86%	25	58%
Corystes cassivelaunus (Pennant, 1777)	13	59%	20	57%	28	65%	21	49%
Crangon allmanni Kinahan, 1857	21	95%	9	26%	11	26%	7	16%
Macropodia rostrata (Linnaeus, 1761)	8	36%	3	9%	11	26%	11	26%
Crangon crangon (Linnaeus, 1758)	9	41%	4	11%	11	26%	9	21%
Hyas coarctatus Leach, 1815	3	14%	14	40%	4	9%	2	5%
Liocarcinus depurator (Linnaeus, 1758)	1	5%	2	6%	14	33%	6	14%
Cancer pagurus Linnaeus, 1758	2	9%	2	6%	10	23%	6	14%
Philocheras bispinosus (Hailstone, 1835)	2	9%	15	43%	0	0%	0	0%
Philocheras trispinosus (Hailstone, 1835)	0	0%	7	20%	3	7%	0	0%
Carcinus maenas (Linnaeus, 1758)	0	0%	1	3%	1	2%	8	19%
Pandalus montagui Leach, 1814	2	9%	3	9%	2	5%	3	7%
Galathea intermedia Lilljeborg, 1851	0	0%	7	20%	0	0%	0	0%
Nephrops norvegicus (Linnaeus, 1758)	0	0%	2	6%	3	7%	1	2%
Liocarcinus pusillus (Leach, 1815)	0	0%	5	14%	0	0%	0	0%
Ebalia cranchii Leach, 1817	2	9%	4	11%	1	2%	0	0%
Anapagurus laevis (Bell, 1845)	0	0%	4	11%	1	2%	0	0%
Pagurus pubescens Krøyer, 1838	0	0%	4	11%	1	2%	0	0%
Pisidia longicornis (Linnaeus, 1767)	2	9%	4	11%	0	0%	0	0%
Ebalia tumefacta (Montagu, 1808)	0	0%	2	6%	1	2%	1	2%
Liocarcinus navigator (Herbst, 1794)	1	5%	0	0%	1	2%	2	5%
Goneplax rhomboides (Linnaeus, 1758)	0	0%	0	0%	3	7%	0	0%
Hyas araneus (Linnaeus, 1758)	3	14%	2	6%	0	0%	0	0%
Processa nouveli holthuisi Al-Adhub & Williamson, 1975	2	9%	2	6%	0	0%	0	0%
Pandalina brevirostris (Rathke, 1843)	1	5%	2	6%	0	0%	0	0%

TABLE I
(Continued)

	DEB1987 (22 stations)		D1990 (35 stations)		D2007 (43 stations)		D2013 (43 stations)	
	$\sum$	%	$\sum$	%	$\sum$	%	$\sum$	%
Callianassa subterranea (Montagu, 1808)	2	9%	0	0%	0	0%	0	0%
Processa modica modica Williamson & Rochanaburanon, 1979	1	5%	1	3%	0	0%	0	0%
Macropodia tenuirostris (Leach, 1814)	0	0%	2	6%	0	0%	0	0%
Pontophilus spinosus (Leach, 1816)	0	0%	2	6%	0	0%	0	0%
Liocarcinus marmoreus (Leach, 1814)	0	0%	0	0%	2	5%	0	0%
Pinnotheres pisum (Linnaeus, 1767)	0	0%	0	0%	1	2%	1	2%
Pilumnus hirtellus (Linnaeus, 1761)	1	5%	0	0%	0	0%	0	0%
Eualus pusiolus (Krøyer, 1841)	0	0%	1	3%	0	0%	0	0%
Galathea nexa Embleton, 1834	0	0%	1	3%	0	0%	0	0%
Hippolyte varians Leach, 1814	0	0%	1	3%	0	0%	0	0%
Inachus dorsettensis (Pennant, 1777)	0	0%	1	3%	0	0%	0	0%
Spirontocaris liljeborgii (Danielssen, 1859)	0	0%	1	3%	0	0%	0	0%
Upogebia deltaura (Leach, 1815)	0	0%	1	3%	0	0%	0	0%
Number of species per cruise	**20**		**33**		**21**		**15**	

$\sum$, sum of stations with positive records of the respective species per cruise; %, percentage of stations with positive records in relation to sum of stations per cruise.

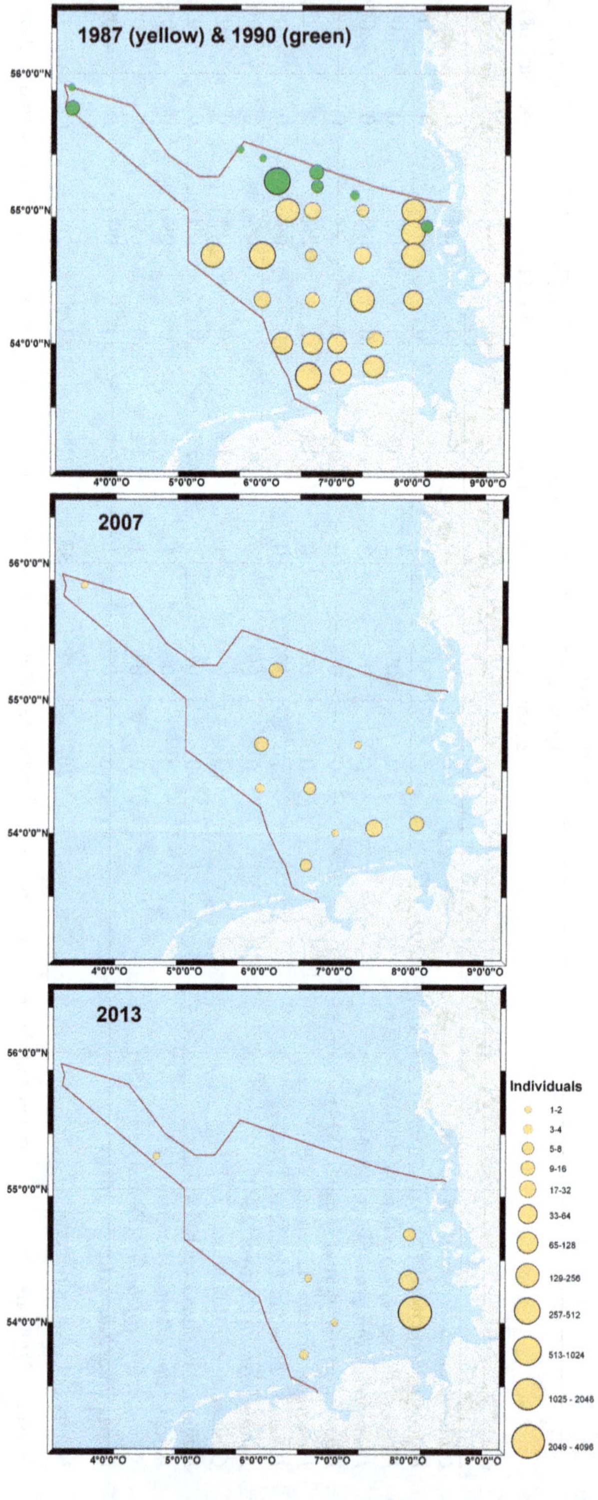
1987 (yellow) & 1990 (green)
2007
2013
56°0'0"N
55°0'0"N
54°0'0"N
4°0'0"O
5°0'0"O
6°0'0"O
7°0'0"O
8°0'0"O
9°0'0"O
Individuals
1-2
3-4
5-8
9-16
17-32
33-64
65-128
129-256
257-512
513-1024
1025 - 2048
2049 - 4096

Comparing the results of 1987/1990 and 2007, the decapod species composition can mainly be assigned to two groups: one group in the coastal and shallow area, and another in the more offshore and comparatively deeper areas, as shown by the transfer of the according groups resolved by Bray-Curtis similarity analyses to the station map (figs. 5 and 6). Surprisingly, the results of 2013, which were achieved by the same data treatment (fig. 7), showed no clear separation of groups. Hence, the decapod fauna must have become more uniform between 2007 and 2013.

DISCUSSION

Decapod species composition at the EEZ

The decapod species composition of the EEZ (table I), with 39 species detected from 1987 to 2013, is quite comparable with the composition of decapod crustaceans in the Helgoland Trench (HTR) (Schwinn et al., 2014). In this biodiversity hotspot south of the island of Helgoland, 37 decapod species were detected from 1977 to 2013. Except the species *Anapagurus laevis* (Bell, 1846), *Pagurus pubescens* Krøyer, 1838, *Goneplax rhomboids* (Linnaeus, 1758), *Macropodia tenuirostris* (Leach, 1814), *Pontophilus spinosus* (Leach, 1816), *Liocarcinus marmoreus* (Leach, 1814), *Inachus dorsettensis* (Pennant, 1777) and *Spirontocaris lilljeborgii* (Danielssen, 1859), all species detected at the EEZ were also recorded from the HTR. Regarding the EEZ, species unique for the HTR are *Palaemon varians* Leach, 1813 and *Palaemon serratus* (Pennant, 1777), which are nevertheless quite common in the estuaries of the southern North Sea (Schwinn et al., 2014). The hermit crab *Pagurus cuanensis* Bell, 1846 also seems to be only present in the HTR with rare occurrences, but not in the EEZ, since it is only known from the western part of the North Sea including the Dogger Bank (Sonnewald & Türkay, 2011) beside the HTR. *Galathea squamifera* Leach, 1814, known from the HTR and from rocky habitats around Helgoland (Schwinn et al., 2014), is not present at the EEZ and at the open North Sea but at the west coast of the UK (GBIF, 2014a). *Galathea dispersa* Bate, 1859 is known from the northwestern North Sea and the HTR (Schwinn et al., 2014), but not from the EEZ. Nevertheless, there is one record of *G. dispersa* from the Dutch sector of the North Sea (Adema et al., 1982). The velvet swimming crab *Necora puber* (Linnaeus, 1767), also recorded from the HTR, is very uncommon in the EEZ (Schwinn et al., 2014), but is known

---

Fig. 2. Abundance (individuals per station) of the pink shrimp *Crangon allmanni* Kinahan, 1860 in the research area in relation to sampling years. The years 1987 and 1990 were combined due to temporal proximity and different spatial coverage of sampling stations. Maps were generated with ArcGIS 10.1.

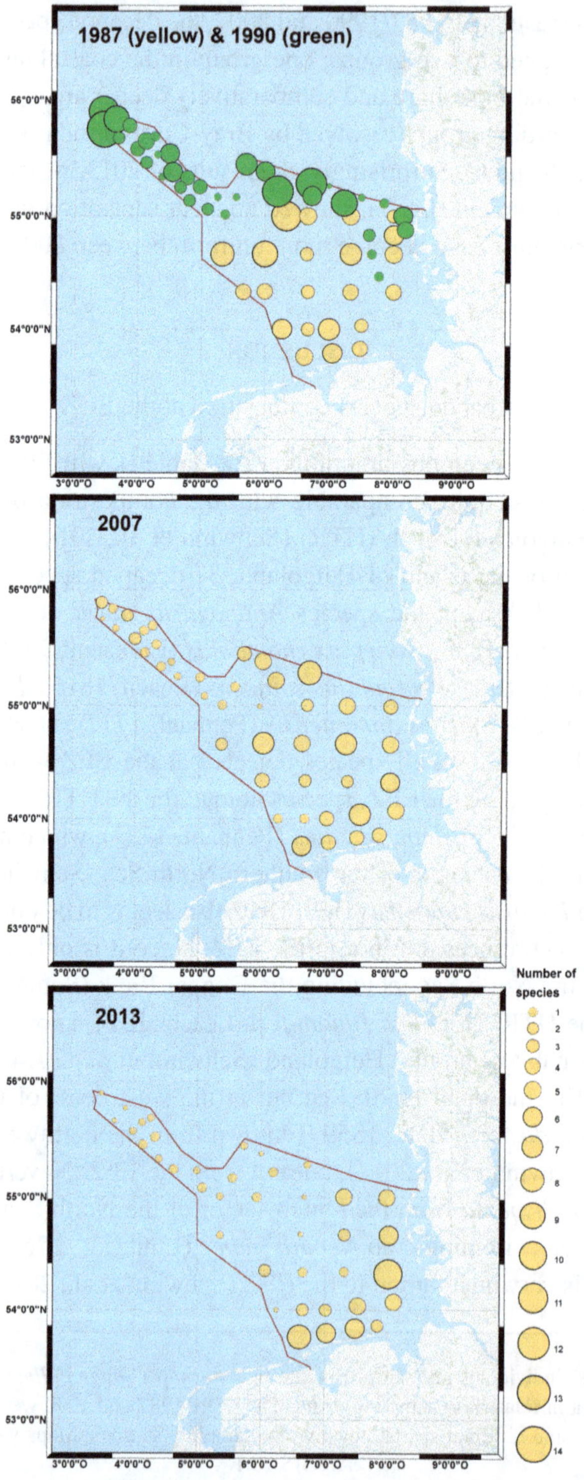
1987 (yellow) & 1990 (green)
2007
2013
56°0'0"N
55°0'0"N
54°0'0"N
53°0'0"N
3°0'0"O
4°0'0"O
5°0'0"O
6°0'0"O
7°0'0"O
8°0'0"O
9°0'0"O
Number of species
1
2
3
4
5
6
7
8
9
10
11
12
13
14

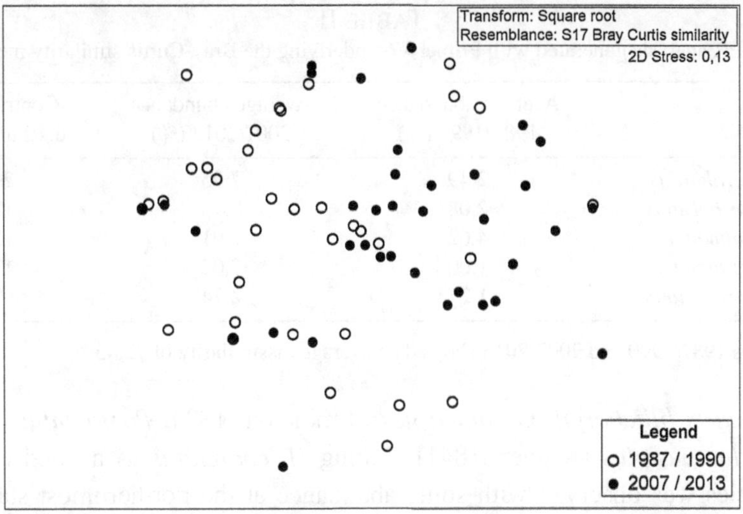

Fig. 4. MDS plot based on the Bray-Curtis similarity measure. Points and circles depict samples. Distances based on the 2nd root transformed abundance data of decapod crustaceans in each sample are shown.

from the Dogger Bank (Sonnewald & Türkay, 2011) and all British coasts (GBIF, 2014b).

The loss of species and possible reasons

The observed tendency of a loss of (mainly rare) species with time (table I) has to be taken as a loss of specialists that cannot find the proper habitat conditions anymore, which according to Gibson (1994) mainly are food, predators, temperature, salinity, oxygen, geological habitat structure, water depth and hydrodynamics. The loss of food is difficult to observe because many details of marine food webs would have to be taken into account, but the immediate coherence of foodwebs and fisheries (Pauly et al., 1998) should not be repelled. The pressure of predators may have risen, as the abundance of the omnivores *Pagurus bernhardus*, *Liocarcinus holsatus*, *C. crangon* and *Corystes cassivelaunus* has increased. Although mainly observed for various fish species (Perry et al., 2005), a change of their distributional range with increasing minimal, maximal and mean temperatures, especially at the time of our observations (Wiltshire & Manly, 2004), could also be the case for some of the decapods with a more northern distributional range in general. This can most obvious be the case for species like *Hyas coarctatus* Leach, 1816,

---

Fig. 3. Species numbers (decapods only) per station within the research area in relation to sampling years. The years 1987 and 1990 were put together due to temporal proximity and different spatial coverage of sampling stations. Maps were generated with ArcGIS 10.1.

TABLE II
SIMPER analysis generated with Primer V6 underlying the Bray-Curtis similarity measure

Species	Average abundance 1987/1990 (%)	Average abundance 2007/2013 (%)	Contribution to dissimilarity (%)
Pagurus bernhardus	2.49	7.95	21.10
Liocarcinus holsatus	2.08	8.18	18.29
Crangon allmanni	4.62	1.91	15.04
Crangon crangon	1.00	7.02	12.46
Corystes cassivelaunus	1.27	2.74	9.68

The groups 1987/1990 and 2007/2013 showed an average dissimilarity of 72.45%.

Spirontocaris lilljeborgii, *Galathea nexa* Embleton, 1834, *Pontophilus spinosus* and *Eualus pusiolus* (Krøyer, 1841). Taking *H. coarctatus* as a good example, this species was observed with some abundance at the northernmost stations in 1990, and had nearly completely vanished from there in 2007 and 2013. Salinity as a factor driving these changes can all but be excluded. By our own CTD measurements, we observed a salinity gradient ranging from 31 to 34 ppt with higher salinities at the most offshore stations with comparable values for each station in every year. Changes in the habitat structure as a cause for faunal changes cannot be excluded, but due to the lack of sediment samples cannot be further investigated. Nonetheless, ground net fishery and wind park construction in more recent years has almost certainly also led to structural habitat changes. Water depth and hydrodynamic conditions are not investigated further and are considered as relatively stable throughout the research period.

Changes of decapod distribution and abundance

The declining tendency of *C. allmanni* (fig. 2) is considered to be at least partially linked to seasonality, as shown by Blahudka & Türkay (2002). The 1987 cruise was conducted in May, the 1990 and 2007 cruises in August, and the 2013 cruise in July. The occurrence of this shrimp species in July 2013 seems to be reduced in coastal areas, whereas its spatial extent ranged all over the German Bight in May/August of 1987/1990. Survey data from two North Sea cruises, conducted in summer 1986 and winter 1987 — also described in Frauenheim et al. (1989) — showed a reduction of occurrence of *C. allmanni* in the southern North Sea from June to August. The spot with the highest occurrence of this species in 2013 (2306 animals) was characterized by muddy sediment and an extraordinary faunal composition with very high abundances of the brittle stars *Ophiura ophiura* (Linnaeus, 1758) (approx. 104 000) and *Ophiura albida* Forbes, 1839 (about 70 000) — this was the only station where it was necessary to calculate

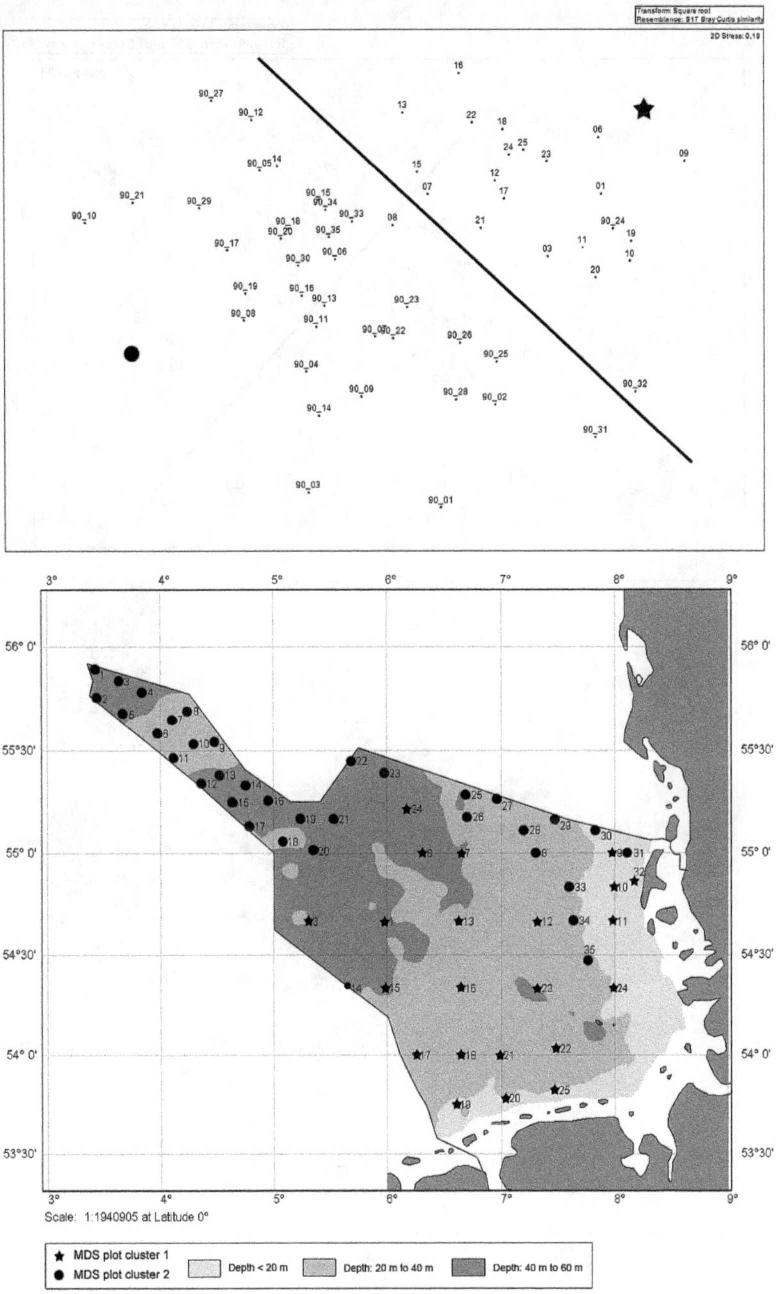

Fig. 5. MDS plot of 1987/90: square-root transformed decapod abundances analysed with Bray-Curtis similarity measure, assigning clusters 1 and 2 to the station map.

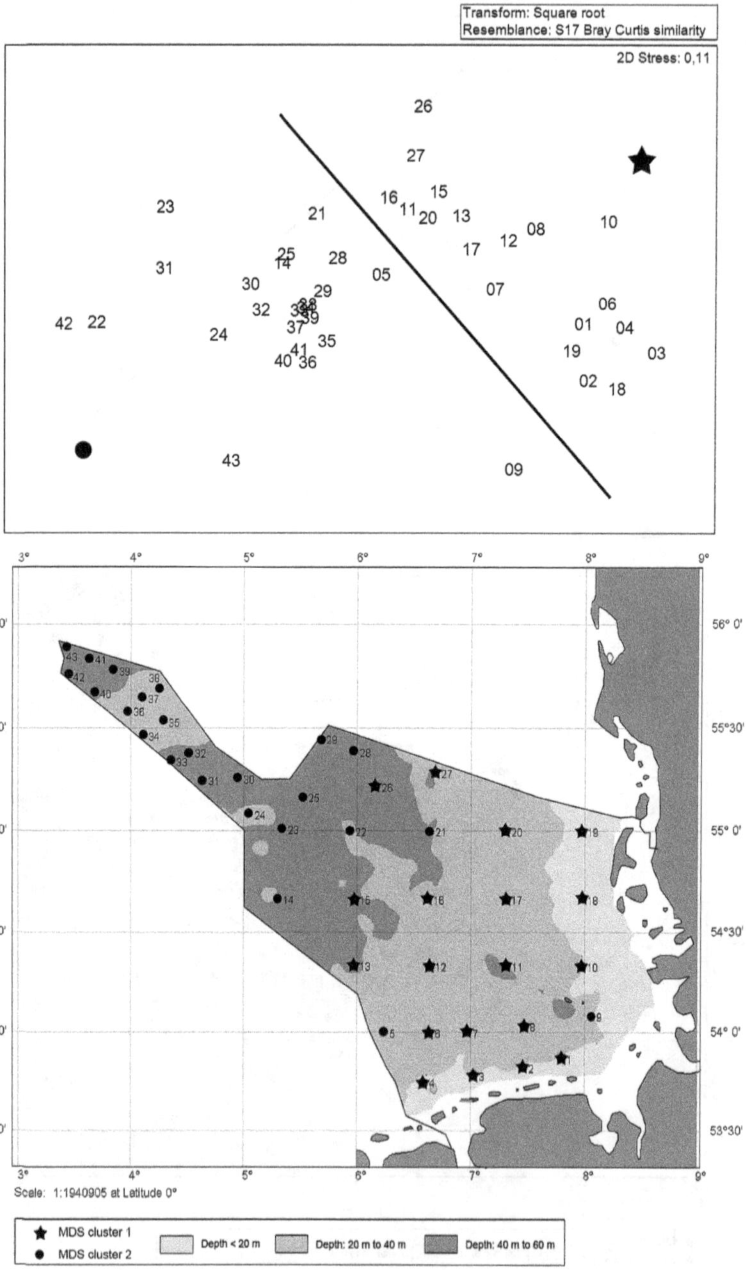

Fig. 6. MDS plot of 2007: square-root transformed decapod abundances analysed with Bray-Curtis similarity measure, assigning clusters 1 and 2 to the station map.

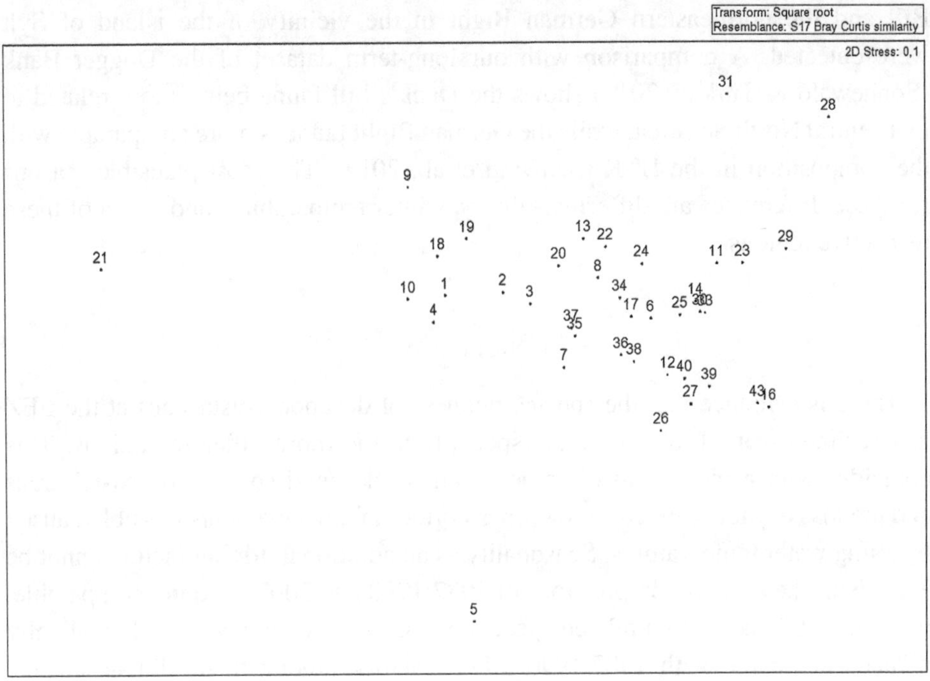

Fig. 7. MDS plot of 2013: square-root transformed decapod abundances analysed with Bray-Curtis similarity measure. No obvious clusters.

the abundances in 140 litres of sample volume by extrapolating the results of counting 10 litres due to loss of ship time.

According to fig. 3, a reduction in decapod species numbers with time is predominantly occurring in the more offshore areas, while an increase can be observed in the coastal zones. Decapod species with a relatively stable occurrence pattern are common and abundant species like *Pagurus bernhardus*, *Crangon crangon* and *Liocarcinus holsatus*, while others have declined predominantly at the offshore stations, as demonstrated above.

Faunal composition changes and spatial structure

The faunal composition of decapods from 1987/1990 to 2007/2013, shown in fig. 4, shows general dissimilarities between species abundances counted during the older and more recent time period. This mainly seems to be due to the development of a more uniform decapod species composition in 2013 (fig. 7), as the groupings of 1987/1990 and 2007 show similarities (figs. 5 and 7).

Figs. 5 and 6 clearly reveal a spatial structure of the decapod faunal composition. Generally, the faunal composition of the Duck's Bill and the German Bight seems to be different, while in 1990 similarities between the faunal assemblage of Duck's

Bill and the northeastern German Bight in the vicinity of the island of Sylt were detected. A comparison with our long-term dataset of the Dogger Bank (Sonnewald & Türkay, 2011) shows the Duck's bill fauna being more related to that central North Sea area, while the German Bight fauna is more comparable with the composition in the HTR (Schwinn et al., 2014). The most plausible reasons for these differences are differing salinity, winter temperatures and depth of these respective regions.

CONCLUSIONS

There is evidence that the species richness of decapod crustaceans at the EEZ has declined from 1987 to 2013, especially at the more offshore stations. This coincides with a movement of some dominant decapod species to coastal areas and the loss of more rare and sometimes northern faunal elements possibly caused by rising water temperatures. Seasonality as an additional driving factor cannot be excluded. The species compositions of 1987/1990 and 2007 are quite comparable, while in 2007 and 2013 a related species richness is shown, which is dramatically reduced compared with 1987/1990. Many factors, among them climate change and fisheries, have been considered to have caused these changes. Due to these findings, a continuation of regular monitoring programs at the German EEZ is necessary in order to detect changes by observation, in the sense of Prof. Michael Türkay.

ACKNOWLEDGEMENTS

We thank the crew (the old sea dogs as well as the newcomers) of the Research Vessel "Senckenberg", who always gave their best for the success of our cruises. Furthermore, the main author thanks the late Prof. Michael Türkay for all the support and teaching during his diploma- and PhD-thesis and for his support during the production of the present work, which happened to a great part only weeks before his decease.

REFERENCES

ADEMA, J. P. H. M., F. CREUTZBERG & G. J. VAN NOORT, 1982. Notes on the occurrence of some poorly known Decapoda (Crustacea) in the southern North Sea. Zool. Bijdr., **28**: 9-32.

BLAHUDKA, S. & M. TÜRKAY, 2002. A population study of the shrimp *Crangon allmanni* in the German Bight. Helgol. Mar. Res., **56**: 190-197.

BRAY, J. R. & J. T. CURTIS, 1957. An ordination of the upland forest communities of southern Wisconsin. Ecol. Monogr., **27**: 325.

CLARKE, K. R. & R. N. GORLEY, 2006. PRIMER v6: user manual/tutorial. (PRIMER-E, Plymouth).

FRAUENHEIM, K., V. NEUMANN, H. THIEL & M. TÜRKAY, 1989. The distribution of the larger epifauna during summer and winter in the North Sea and its suitability for environmental monitoring. Senckenb. marit., **20**: 101-118.

GBIF, 2014a. The Global Biodiversity Information Facility: backbone taxonomy, 2013-07-01. Available online at http://www.gbif.org/species/5178347 (accessed 4 April 2014).

GBIF, 2014b. The Global Biodiversity Information Facility: backbone taxonomy, 2013-07-01. Available online at http://www.gbif.org/species/4307449 (accessed 4 April 2014).

GIBSON, R. N., 1994. Impact of habitat quality and quantity on the recruitment of juvenile flatfishes. J. Sea Res., **32**: 191-206.

JENNINGS, S., J. LANCASTER, A. WOOLMER & J. GÖTTER, 1999. Distribution, diversity and abundance of epibenthic fauna in the North Sea. J. Mar. Biol. Ass. UK, **79**: 385-399.

KÜNITZER, A., 1990. The infauna and epifauna of the central North Sea. Meeresforsch., **33**: 23-37.

NEUMANN, H., S. EHRICH & I. KRÖNCKE, 2008. Spatial variability of epifaunal communities in the North Sea in relation to sampling effort. Helgol. Mar. Res., **62**: 215-225.

NEUMANN, H. & I. KRÖNCKE, 2011. The effect of temperature variability on ecological functioning of epifauna in the German Bight. Mar. Ecol., **32**: 49-57.

NEUMANN, H., H. REISS, S. EHRICH, A. SELL, K. PANTEN, M. KLOPPMANN, I. WILHELMS & I. KRÖNCKE, 2012. Benthos and demersal fish habitats in the German Exclusive Economic Zone (EEZ) of the North Sea. Helgol. Mar. Res., **67**: 445-459.

NEUMANN, H., H. REISS, S. RAKERS, S. EHRICH & I. KRÖNCKE, 2009. Temporal variability in southern North Sea epifauna communities after the cold winter of 1995/1996. ICES J. Mar. Sci., **66**: 2233-2243.

PAULY, D., V. CHRISTENSEN, J. DALSGAARD, R. FROESE & F. TORRES, 1998. Fishing down marine food webs. Science, **279**: 860-863.

PERRY, A. L., P. J. JOW, J. R. ELLIS & J. D. REYNOLDS, 2005. Climate change and distribution shifts in marine fishes. Science, **308**: 1912-1915.

RACHOR, E. & P. NEHMER, 2003. Erfassung und Bewertung ökologisch wertvoller Lebensräume in der Nordsee. Abschlussbericht für das F & E-Vorhaben FKZ 899 85 310 (Bundesamt für Naturschutz). (Alfred Wegener Institute for Polar and Marine Research, Bremerhaven).

REISS, H., S. DEGRAER, G. C. A. DUINEVELD, I. KRÖNCKE, J. ALDRIDGE, J. A. CRAEYMEERSCH, J. D. EGGLETON, H. HILLEWAERT, M. S. S. LAVALEYE, A. MOLL, T. POHLMANN, E. RACHOR, M. ROBERTSON, E. VANDEN BERGHE, G. VAN HOEY & H. L. REES, 2010. Spatial patterns of infauna, epifauna, and demersal fish communities in the North Sea. ICES J. Mar. Sci., **67**: 278-293.

SCHWINN, M., M. TÜRKAY & M. SONNEWALD, 2014. Decapod fauna of the Helgoland trench (Crustacea) a long-term study in a biodiversity hotspot. Mar. Biodivers., **44**: 491-517.

SONNEWALD, M. & M. TÜRKAY, 2011. The megaepifauna of the Dogger Bank (North Sea): species composition and faunal characteristics 1991–2008. Helgol. Mar. Res., **66**: 63-75.

SONNEWALD, M. & M. TÜRKAY, 2012a. Environmental influence on the bottom and near-bottom megafauna communities of the Dogger Bank: a long-term survey. Helgol. Mar. Res., **66**: 503-511.

SONNEWALD, M. & M. TÜRKAY, 2012b. Abundance analyses of mega-epibenthic species on the Dogger Bank (North Sea): diurnal rhythms and short-term effects caused by repeated trawling, observed at a permanent station. J. Sea Res., **73**: 1-6.

WIEKING, G. & I. KRÖNCKE, 2003. Macrofauna communities of the Dogger Bank (central North Sea) in the late 1990s: spatial distribution, species composition and trophic structure. Helgol. Mar. Res., **57**: 34-46.

WILTSHIRE, K. H. & B. F. J. MANLY, 2004. The warming trend at Helgoland Roads, North Sea: phytoplankton response. Helgol. Mar. Res., **58**: 269-273.

ZÜHLKE, R., J. ALVSVÅG, I. BOOIS, J. COTTER, S. EHRICH, A. FORD, H. HINZ, A. JARRE-TEICHMANN, S. JENNINGS, I. KRÖNCKE, J. LANCASTER, G. PIET & P. PRINCE, 2001. Epibenthic diversity in the North Sea. Senckenb. marit., **31**: 269-281.

First received 15 November 2016.
Final version accepted 21 December 2016.

[When citing this volume, please refer to Crustaceana 90 (2017) 771-1288]

Michael Türkay Memorial Issue

TWO NEW SPECIES OF *THALAMITA* LATREILLE, 1829 (DECAPODA, PORTUNIDAE)

BY

VASSILY A. SPIRIDONOV[1])

P. P. Shirshov Institute of Oceanology of the Russian Academy of Sciences, Nakhimovskiy Prospekt 36, Moscow 117997, Russia

ABSTRACT

Thalamita matzuzawai sp. n. is described from the subtidal zone of Shikoku, Japan. This species is most similar to *Thalamita picta* Stimpson, 1858 and *T. simillima* Crosnier, 2002. It also shows similarity to the group of *Thalamita* species with a supplementary tooth at the base of the first anterolateral tooth, thus representing an intermediate position between these two groups. *Thalamita williami* sp. n. is described from Brunei. It belongs to the *T. crenata* Rüppell, 1830 group and is most similar to *T. starobogatovi* Tien, 1969.

ZUSAMMENFASSUNG

Hier wird *Thalamita matzuzawai* sp. n. aus dem Sublittoral der Shikoku-Insel, Japan beschrieben. Diese Art ähnelt *Thalamita picta* Stimpson, 1858 und *T. simillima* Crosnier, 2002 aber auch einer Gruppe der *Thalamita*-Arten, welche ein zusätzliches Zähnchen an der Basis des ersten anterolateralen Zahnes haben. Somit befindet sich diese Art zwischen diesen Artengruppen. *Thalamita williami* sp. n. wurde von Brunei beschrieben. Sie gehört zu der *Thalamita crenata* Rüppell, 1830 — Artengruppe und zeigt die größte Änlichkeit mit *Thalamita starobogatovi* Tien, 1969.

INTRODUCTION

Thalamita Latreille, 1829 contains the greatest number of species among the swimming crabs (Portunidae) genera. According to the most recent inventory of the Brachyura, this genus includes about 90 recent species (Ng et al., 2008; De Grave et al., 2009). The size and the complexity of the genus have always posed problems for taxonomists working on *Thalamita*. Heinrich Balss (1924: 3) once wrote: “Die Arten dieser Gattung varieren sehr stärk, so das eine sichere

[1]) e-mail: vspiridonov@ocean.ru

Bestimmung oft unmöglich ist". In the last decades some *Thalamita* species were shown to be complexes of related taxa (Wee & Ng, 1995; Apel & Spiridonov, 1998; Crosnier, 2002) while several species names were made synonyms (Wee & Ng, 1995; Spiridonov & Neumann, 2008). Owing to the size of the genus, its complete revision is a tedious task that cannot be accomplished in a short time. In the process of the revisionary work, initiated about two decades ago, the author examined the collections of *Thalamita* in several European, American and Australian museums. During this study some *Thalamita* specimens were recognized to belong to yet undescribed species. They are described, illustrated and named in the present paper.

MATERIAL AND METHODS

The material used in the present study is deposited at the Muséum national d' Histoire naturelle, Paris, France (MNHN), Museum für Naturkunde, Berlin, Germany (ZMB), Naturalis Biodiversity Center, Leiden, the Netherlands, formerly Rijksmuseum van Natuurlijke Historie (RMNH), Natural History Museum of Stockholm University (NHMS), Forschungsinstitut und Naturmuseum Senckenberg, Frankfurt am Main, Germany (SMF), Zoological Institute of the Russian Academy of Sciences, St. Petersburg, Russia (ZIN RAN), Zoological Museum of the Oslo University, Oslo, Norway (ZMUO), Zoological Museum of the Moscow University, Moscow, Russia (ZMMU), Zoologisches Museum der Universität Göttingen (ZMG, currently deposited in SMF).

The set of characters used to describe and diagnose the species of *Thalamita* is based on the concept developed by Wee & Ng (1995), Apel & Spiridonov (1998) and Spiridonov & Neumann (2008). In addition to the terminology adopted by these authors the term "pleon" is used instead of "abdomen" and "pleomere" is preferred over "abdominal segment". The first anterolateral tooth is an accepted equivalent and used as a synonym to outer supraorbital lobe. This follows the tradition in the Portunidae studies and is necessary to provide consistency of anterolateral teeth counts in the contemporary literature and diagnoses in the classical works by Crosnier (1962) and Stephenson (1972) and earlier references herein. Terminology used to describe teeth pattern on chelae fingers is an English equivalent (Spiridonov et al., 2014) of the German terms proposed by Schäfer (1954). The following measurements were made using a caliper with accuracy of 0.1 mm: CL, carapace length along the midline; CW3, carapace width measured between 3rd anterolateral teeth; CW5, carapace width measured between 5th anterolateral teeth; FM, frontal margin width between mesial corners of inner supraorbital lobes; FSO, width of FM together with inner supraorbital lobes;

FO, total frontorbital width; PM, width of posterior margin of carapace. All measurements are in mm.

Other abbreviations used in the text and tables: coll., collector; CW, maximum carapace width; G 1, gonopod 1 (1st pleopod of male); G 2, gonopod 2 (2nd pleopod of male); ov, ovigerous; P 1-P 5, pereiopods of 1st to 5th pairs; *n*, number of specimens; SD, standard deviation; vs., versus.

SYSTEMATIC ACCOUNT

Order DECAPODA Latreille, 1802
Infraorder BRACHYURA Linnaeus, 1758
Section HETEROTREMATA Guinot, 1977
Family PORTUNIDAE Rafinesque, 1815
Subfamily THALAMITINAE Paul'son, 1875
Genus *Thalamita* Latreille, 1829

Thalamita matzuzawai sp. n.

(figs. 1A-B, 2A-F, 3A-C, 4)

Material.— Holotype: male (SMF 37113), Japan, Shikoku, Kochi Prefecture, Toza-Shimizu, of Tateishi town, about 32°48′N 133°E, 30 m, 02 June 2004, coll. unknown. Paratypes: 2 males, 2 females, 2 females ov (SMF 37114), same locality as holotype. 1 male (SMF 28906; identified as *Thalamita picta*), Japan, Shikoku, Kochi Prefecture, Toza-Shimizu, of Ashizuri, Ohama, from T. Sakai collection.

Comparative material.— *Thalamita picta* Stimpson, 1858: 3 males, 2 females, 1 female ov (SMF 21638), Kenya, S. Malindi, near Watamu, 0-0.5 m, dead corals overgrown with algae, 24.vii-07.ix.1989, coll. H-G. Müller; 3 males, 5 females (MNHN B 6237), New Caledonia, coll. Balansa; 2 males (MNL 476, identified as *Goniosoma lineatum* A. Milne-Edwards, 1861), French Polynesia, Tahiti, from the Museum Godeffroy collection, 1887. — *T. macropus* Montgomery, 1931: 1 male (Queensland Museum 03090), Australia, NIMBUS Cruise, St 24, about 92 m (51-52 fathoms). — *T. indistincta* Apel & Spiridonov, 1998, *T. loppenthini* Apel & Spiridonov, 1998, *T. macrodonta* Borradaile, 1903: see Apel & Spiridonov (1998).

Diagnosis.— Carapace 1.5-1.6 times as broad as long (in the type series, mean CW/CL = 1.54; SD = 0.059), dorsally finely granular and pilose; maximum width between fifth anterolateral teeth. All anterior carapace ridges present, frontals as small groups of granules, protogastrics sinuous, mesogastrics and metagastrics interrupted medially; posterior ridges less clearly expressed, cardiacs sinuous, and a pair of short mesobranchials. Front little shorter than posterior border (in the type series, mean FM/PM = 0.92; SD = 0.056), and consists of six lobes: rounded medians more protruding than others; submedians truncated, at least 1.5 times broader than medians; laterals smallest, subtriangular, separated with V-shaped incisions on both sides. Inner supraorbital lobe somewhat shorter than submedian frontal lobes but distinctly broader than medians and laterals, arcuate. Five anterolateral teeth; first tooth slightly broader than others; tiny (seen under magnification) knob may be present at its base. Second and 3rd teeth subequal, 4th smallest, 5th longest. Basal antennal segment nearly as long as major dimension of orbit, granular, with crest consisting of rounded granules fused at base. Chelipeds heterochelic, moderately granular. Upper surface of manus with 4 spines, besides the usual spine at carpus articulation, outer surface with 3 granular carinae, heavily granular between upper ones, lower and inner faces finely granular to smooth. P 2-P 4 with posterodistal corner of propodi rounded; dactyli relatively narrow. Propodi of P 5 with 5-8 spinules on posterior border. Pleomere tergum 4 of male with faint keel, 6th pleomere broader than long, with sinuous lateral borders. G 1 with basal sub-ovate lobe, neck evenly curved, tip flared and truncated; lateral face with row of spinules extending to distal fourth of neck, proximally with 5-6 sparsely set spinules, decreasing in size towards tip; 8-12 smaller subterminal spinules markedly decreasing in size distally and located on lateral, abdominal and sternal faces of tip. Female genital opening guttated, broad posteriorly, long axis subparallel to mesial margin of sternite, covered by a cuticular cup anterolaterally.

Description (based on holotype).— Carapace 1.48 times as broad as long, maximum width between fifth anterolateral teeth; dorsal surface nearly flattened, densely covered dorsally with fine granules and short pile; only margins smooth. All anterior carapace ridges finely granular: frontals as small groups of granules; protogastrics short, sinuous, widely separated, centered at level of lateral frontal teeth, and joining orbital groove; mesogastrics straight, separated, nearly equal in extension to frontal margin, not reaching level of lateral ends of protogastrics; metagastrics widely interrupted medially and straight. Epibranchial ridges slightly sinuous; cardiacs sinuous, uninterrupted, mesobranchials short, both latter ridges not clearly distinguishable (fig. 1A). A faint granular patch located at level of 2nd mesobranchial ridge. Front slightly convex, nearly as broad as posterior border, consisting of six lobes: rounded and more protruding medians; submedians truncated, more than 1.5 times broader than medians; laterals smallest, subtriangular, separated from submedians and inner supraorbital lobe with V-shaped incisions. Orbits occupying about 19% of frontorbital margin. Upper orbital margin with two very narrow and shallow notches, thus comprising of two lobes and outer supraborbital lobe (=1st anterolateral tooth) (figs. 1A, 2A). Inner supraorbital lobe somewhat shorter than submedian lobes, but distinctly broader than medians and laterals, arcuate. Inner suborbital lobe broadly triangular, protruding beyond inner supraorbital lobe with moderately granular margin, separated by a shallow V-shaped notch from 1st anterolateral tooth.

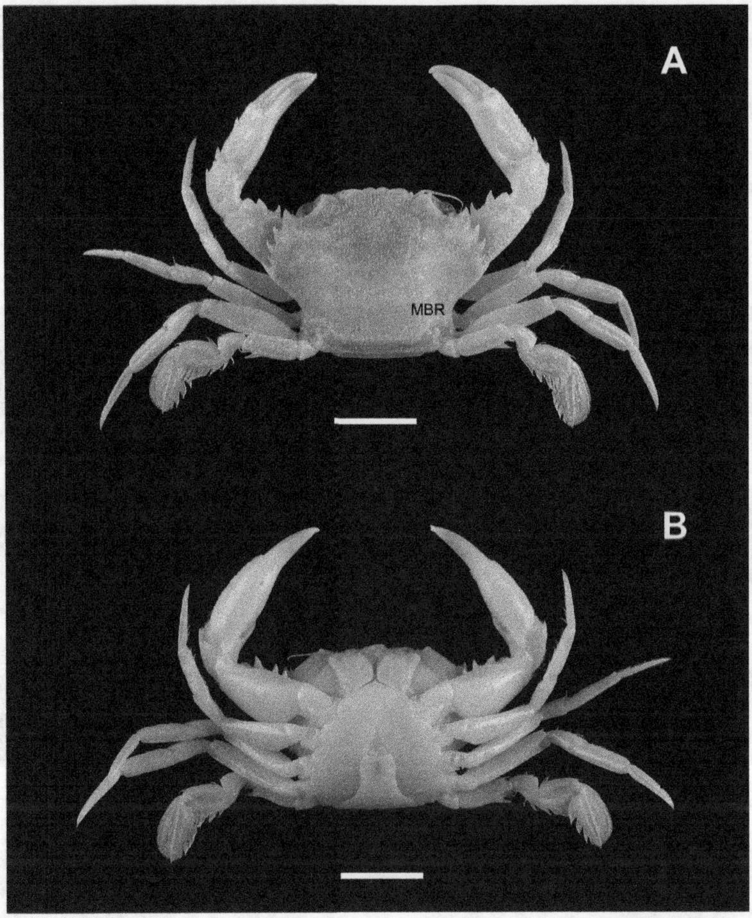

Fig. 1. *Thalamita matzuzawai* sp. n., holotype, male, CL 9.4 mm (SMF 37113). A, dorsal view; B, ventral view. MBR – mesobranchial ridge. Scale bars 5 mm.

Five anterolateral teeth, first slightly broader than others, its base with minute additional knob, seen only at optical magnification (fig. 2A). Other teeth slightly curved: 2nd and 3rd teeth subequal, 4th smallest, 5th longest. Tips of all 5 teeth dark-pigmented (fig. 1A).

Epistome with broadly triangular groove medially. Antennular fossae enlarged laterally, their posterior margins finely granular. Suborbital regions in general finely granular, with patches of larger granules located in lateral part and at contact area with basal antennal segment; from this patch, row of granules extending along border of pterygostomial region. Row of granules extending along suture between pterygostomial and suborbital region; pterygostomial and subhepatic regions finely granular; thoracic sternum smooth. All sutures between sternites recognisable, those between sterna and pereopod coxae obscure.

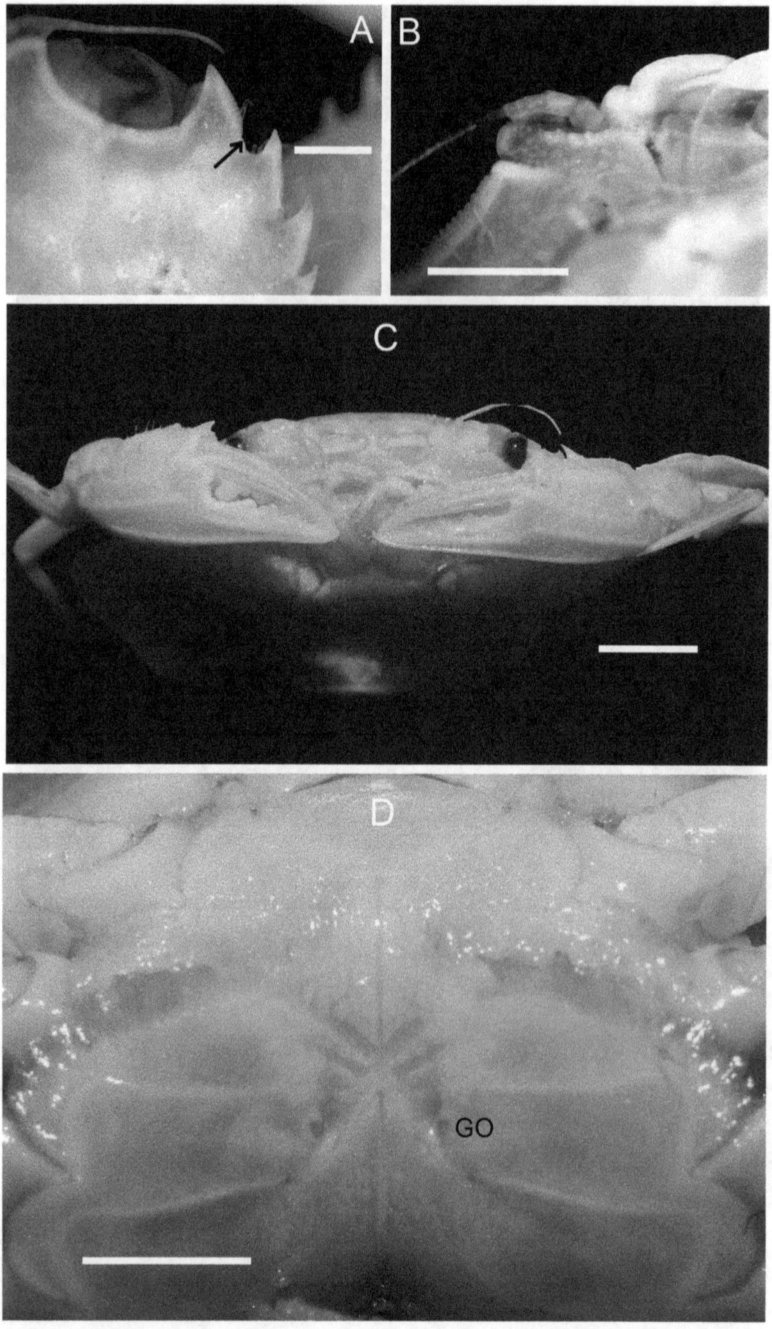

Fig. 2. *Thalamita matzuzawai* sp. n. A, holotype (SMF 37113), orbit and anterolateral margin; arrow points to a knob at the base of the 1st anterolateral tooth; B, same specimen as A, basal antennal segment; C, paratype, female ov (SMF 37114), frontal view; D, same specimen as C, thoracic sternal region and genital opening (GO). Scale bar: 2 mm (A, B), 5 mm (C, D).

Basal antennal segment nearly as long as width of orbit, granular, with crest bearing 6-7 rounded granules fused at base, scattered granules posterior to top of crest (fig. 2B).

Mouthparts not dissected; mesial lobe on endopodite of maxilliped 1 well developed. Maxilliped 3 with ischium of endopod and exopod microscopically granular; mesial margin of ischium with dense setae of increasing size anteriorly bordering; merus with somewhat stronger granules than those on ischium, located in irregular patches, covered with sparse setae on external face and along edges, mesial margin with usual brush of setae, anterior part of merus slightly convex, antero-external corner rounded and produced laterally; carpus with outer face bearing granules similar in size to those on merus; morphology and setation pattern of carpus, propodus and dactylus as typical for Portunidae.

Chelipeds with basis and ischium smooth. Merus with upper surface moderately granular, except proximal part touching pterygostomial region, lower surface finely granular; anterior margin of merus with 3 large spines, distal one being granular (fig. 1A, B), others smooth; large granular tubercle distally, near junction to carpus. Carpus with outer and upper face bearing 3 granular rows ending correspondingly in outermost, proximal median, and inner spines; additional short sinuous row of granules meeting usual spine at junction of carpus and manus; otherwise surface of carpus covered with different size granules. Left manus about as high as thick, right manus 1.4 times higher than thick; upper surface granular with most coarse granules along edges, with 4 spines besides of usual spine at carpus articulation; in larger cheliped distalmost outer spine reduced to tubercle. Outer surface with 3 granular carinae, heavily granular between upper ones, with few finer granules between median and lower carina. Lower and inner faces finely granular to smooth. Fingers shorter than palm, strong, with usual grooves, moderately curved. Right chela with moderately sized knobstick molariform tooth located proximally at dactylus, opposed to mucronate tuberous tooth at polex. Left chela with tuberous tooth opposing to molariform tooth at dactylus, following by another large flattened tuberous tooth. Distad to these proximal teeth, both dactylus and polex beset with sharp serial conical teeth, regularly alternating in size or trilobed teeth (fig. 2C).

Pereiopods 2-4 relatively robust, P 3 longest; proportions of segments similar in all legs. Meri about 3.3 times longer than broad, smooth, with dorsal face finely granular and sparsely haired, posterodistal corners angular. Carpi keeled and haired anteriorly along dorsal face. Propodi about 2.5 times longer than broad, keeled along dorsal face, with line of hairs anterior to keel, grooved along anterior and posterior faces, sparsely haired, posterodistal corners rounded. Dactyli narrower than propodi, about 9 times longer than broad, slightly curved, keeled along dorsal face, with line of hairs anterior to keel, grooved along anterior and posterior faces (fig. 2D).

Natatory leg (P 5) as usual for *Thalamita* morphology and setation pattern. Merus about 2 times as long as broad, with indistinct finely granulated groove on dorsal face and usual posterodistal spine; ventral face smooth. Dorsal face of propodus granular, posterior border with 5 main spines and 1-2 proximal spinules, beset with soft setae increasing in size proximally. Anterior margin of paddle-shaped dactylus beset with strong spinules in distal half, these spinules continuing to posterior margin up to short distance behind terminal spine; proximal half of anterior margin and most part of posterior margin bordered by softer setae (fig. 2E).

Pleonal terga smooth, terga 3-5 fused, suture between these fused terga and tergum 2 indistinct; smoothed keel along tergum 3 area. Pleomere 4 with obsolete keel, penultimate pleomere with lateral borders diverging up to last fourth, than converging again, broader than long (length : width ratio 0.67). Telson about as long as broad (fig. 1B).

G 1 with basal lobe sub-ovate, neck evenly curved, tip little broadened (compared to distalmost part of neck), flared and truncated. Membrane covering about one fourth of the aperture. Lateral face with row of spinules extending to distal fourth of neck: proximally with 5 sparsely set spinules, decreasing in size towards tip; 10 smaller subterminal spinules markedly decreasing in size distally (so that spinules in vicinity of tip recognizable only under strong magnification) located on lateral, abdominal and sternal faces of tip (fig. 3A-C).

Female characteristics.— Pleonal terga in ovigerous female with transverse keels along entire width of pleomeres 2 to 5; surface of terga smooth; pleomere 5 broadest; telson shaped like a rugby ball, half inserted in hollow on margin of pleomere 6. Female genital opening drop-shaped, with broad part posteriorly, long axis quasi-parallel to mesial margin of sternite, covered by cuticular cup antero-laterally (fig. 2D).

Size (CL × CW5) and measurements (mm).— Holotype: 9.4 × 13.9 (FM 4.2, FSO 5.8, FO 10.7, PM 4.3). Paratypes: males 8.0 × 12.0, 9.8 × 15.9, 15.5 × 23.2; females: 8.25 × 12.5, 10.3 × 16.7; ovigerous females 12.1 × 17.9, 14.65 × 23.9.

Coloration.— Carapace mostly red, hepatic, anterolateral, posterolateral regions marble-brownish; marble band extending and broadening from median frontal lobes to mesogastric region; pair of marble spots on both sides of metagastric region; a pair on mesogastric regions, and another pair of marble spots between metagastric and cardiac regions. Carapace ridges brownish red. Chelipeds red with brownish granules; anterior half of inner face of manus pink orange to opaque; proximal half of movable finger dark brown; distal halves of cheliped fingers light brown. Most of articles of P 2-P 5 orange-red with dark red punctations. Darker

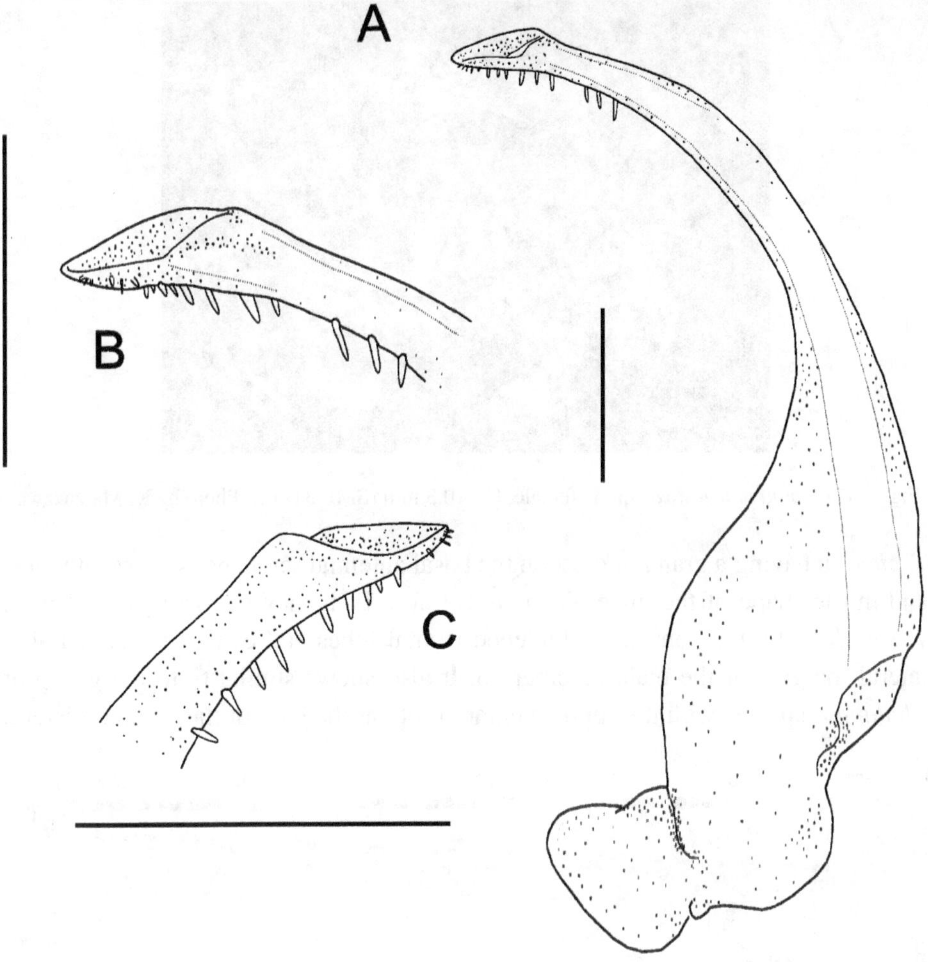

Fig. 3. *Thalamita matzuzawai* sp. n., holotype (SMF 37113). A, gonopod 1, abdominal view; B, tip of gonopod 1, pleonal view; C, tip of gonopod 1, sternal view. Scale bars 1 mm.

transverse bands on meri to propodi of P 2-P 4 and meri and carpi of P 5; dactyli of P 2-P 5 and propodus of natatory legs orange (fig. 4).

Etymology.— The species is named in honour of Mr. Keisuke Matzuzawa, a dedicated naturalist who made an extensive and a very valuable collection of Crustacea from the waters of Shikoku Island and supplemented most of specimens with photographs of live coloration.

Vertical range and habitats.— Subtidal, to 30 m.

Distribution.— So far, known from the Tosa Bay, Japan (fig. 5).

Affinities.— The present new species is most similar to *Thalamita picta* Stimpson, 1858 and *T. simillima* Crosnier, 2002. *T. matzuzawai* sp. n. differs from

Fig. 4. *Thalamita matzuzawai* sp. n., female, CL 10.3 mm (SMF 37114). Photo by K. Matzuzawa.

T. picta in having a granular crest on the basal antennal segment vs. a smooth crest and in the shape of the male pleomere 6 (table I). It may be distinguished from *T. simillima* by the convex vs. flattened frontal lobes and convex vs. subparallel lateral margins of the male pleomere 6. It also shows similarity to the group of *Thalamita* species with the supplementary tooth at the base of the 1st anterolateral

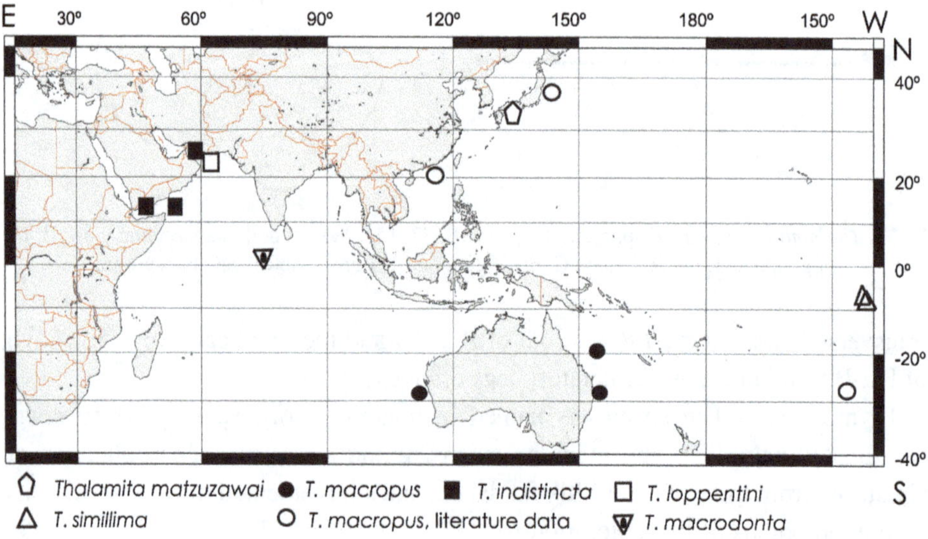

Fig. 5. Distribution of *Thalamita matzuzawai* sp. n. (present study), *T. macropus* Montgomery, 1931 (based on Montgomery, 1931; Sakai, 1976; Crosnier, 2002; Minemizu, 2002; Yang et al., 2012; present study), *T. indistincta* Apel & Spiridonov, 1998, *T. loppenthini* Apel & Spiridonov, 1998, *T. macropus* Montgomery, 1931 (based on Apel & Spiridonov, 1998), and *T. simillima* Crosnier, 2002 (based on Crosnier, 2002).

tooth: *T. indistincta* Apel & Spiridonov, 1998, *T. loppenthini* Apel & Spiridonov, 1998, *T. macrodonta* Borradaile, 1903, and *T. macropus* Montgomery, 1931. The general difference of *T. matzuzawai* sp. n. from that group of species is the absence of the supplementary anterolateral tooth. Other carapace and pleonal characters that can be used to distinguish between these species and *T. matzuzawai* are listed in table I.

The G 1 in these species also show some differences although they clearly belong to the same morphological type with a flared tip. The morphology of the tip in *T. matzuzawai* sp. n. differs from both *T. picta* and *T. simillima* by a row of spinules or bristles on the mesial face of the neck, which is present in the new species but is lacking in *T. picta* (see Wee & Ng, 1995, fig. 50E, F) and is poorly expressed in *T. simillima* (see Crosnier, 2002, fig. 28G, F). The G 1 of *T. matzuzawai* sp. n. differs from the gonopods of *T. macropus* (see Crosnier, 1975, fig. 4h; Yang et al., 2012, fig. 112f) and *T. loppenthini* (see Apel & Spiridonov, 1998, fig. 66c) by the absence of recognizable spinules on the lateral face of the G 1 neck. It also differs from *T. macropus* (see Crosnier, 1975, fig. 4h) by a sparser and less extending in proximal direction row of spinules on the G 1 mesial face.

Thalamita indistincta described from the Gulf of Aden shows a variable expression of the supplementary anterolateral tooth that may be lacking in some smaller specimens (see Apel & Spiridonov, 1998, fig. 58); its G 1 is also most similar to the G 1 of *T. matzuzawai* (see Apel & Spiridonov, 1998, fig. 57e). However, spines on the mesial face of the gonopod are relatively shorter and sparsely set in *T. matzuzawai* than in *T. indistincta*. *T. indistincta* is also a small sized species (CL < 6 mm), and most of specimens have supplementary teeth while *T. matzuzawai* sp. n. is a relatively large species lacking the supplementary tooth and reaching a size of 15.5×23.2 mm, much larger than *T. indistincta*.

Thalamita matzuzawai sp. n. thus holds an intermediate position between the species of *Thalamita*, with 6 anterolateral teeth (i.e., having a minute supplementary tooth at the base of the 1st anterolateral (exorbital tooth)) and *T. picta*, which indicates their close relationship.

The distribution range of the new species is within the range of *Thalamita picta*, a widespread Indo-Pacific species (Stephenson, 1972a; Wee & Ng, 1995) and *T. macropus*, which has been recorded in Japan (Sakai, 1976; Miyake, 1983; Minemizu, 2000), but does not overlap with the known distribution of *T. indistincta*, *T. loppenthini*, and *T. simillima* (fig. 5).

Remarks.— Colour illustrations (Sakai, 1976; Minemizu, 2000) indicate a generally similar pattern of coloration in *T. macropus* from the subtidal zone of Japan and the new species. Taking into account morphological differences between these species such similarity may seem surprising. This calls for a

TABLE I

Comparison of the carapace and pleonal characters between *Thalamita matzuzawai* sp. n. and morphologically similar species

Character	*Thalamita matzuzawai* sp.n	*Thalamita picta* Stimpson, 1858	*Thalamita simillima* Crosnier, 2002	*Thalamita macropus* Montgomery, 1931; *T. loppenthini* Apel & Spiridonov, 1998	*Thalamita indistincta* Apel & Spiridonov, 1998	*Thalamita macrodonta* Borradaile, 1903
Median frontal lobes	Markedly convex (figs. 1A, 2A, 3)	Markedly convex (Wee & Ng, 1995, fig. 50A)	Relatively flattened (Crosnier, 2002, fig. 28A)	Convex (Montgomery 1931, pl. 24, fig. 4; Apel & Spiridonov, 1998, fig. 65)	Convex (Apel & Spiridonov 1998, figs. 56, 58a-c)	Convex (Apel & Spiridonov, 1998, fig. 59a)
Supplementary anterolateral tooth at base of the first tooth	Absent but a microscopic knob may be present *	Absent*	Absent*	Present*	In most cases present*	Present*
Last anterolateral tooth	Longer than penultimate one*	Longer than penultimate one*	Same sized or shorter than penultimate one*	Longer than penultimate one*	Longer than penultimate one*	Longer than penultimate one which is minute*
Mesobranchial ridges	1 pair*	1 pair*	1 pair*	2 pairs*	1 pair*	1 pair, faint*
Basal segment of antenna	With a granular crest (fig. 2B)	With a smooth crest (Wee & Ng, 1995, fig. 50A)	With a granular crest (Crosnier, 2002, fig. 28B)	With a crest consisting of separate granules (Apel & Spiridonov, 1998, fig. 66a)	With a granular crest (Apel & Spiridonov, 1998, fig. 57a)	With a crest consisting of few separate granules (Apel & Spiridonov, 1998, fig. 59b)
Lateral margins of male pleomere 6	Convex (fig. 1B)	Parallel for most of length (Wee & Ng, 1995, fig. 50C)	Parallel for half length, then converging (Crosnier, 2002, fig. 28E)	Convex (Apel & Spiridonov, 1998, fig. 72)	Convex (Apel & Spiridonov, 1998, fig. 58f-h)	Converging (Apel & Spiridonov, 1998, fig. 59e)

* Same figures as for median frontal lobes for respective species.

more comprehensive description of morphological and colour variation along with reconstructions of molecular phylogeny within the entire group of species including *T. matzuzawai* sp. n., *T. macropus* from Australia and the northwestern Pacific, *T. loppenthini* and *T. indistincta*, *T. macrodonta*, and *T. picta*.

Thalamita williami sp. n.
(figs. 6A-B, 7A-B, 8A-E)

? *Thalamita danae*, form C — Stephenson, 1972b, figs. 6E-F, 7E-F only (not *T. danae* Stimpson, 1858).

Material.— Holotype, male (ZMB 12497, identified as *Thalamita stimpsoni* A. Milne-Edwards, 1861), Brunei (then British North Borneo), Borneo (Kalimantan) Island, Pagel, 24.viii.1907, coll. Volkner.

Comparative material of other species.— *Thalamita crenata* Rüppell, 1830: 1 male, 1 female (ZIN RAN 305), Natal, 1841, coll. Prof. Krauss; 2 males (ZIN RAN 79783), China, Hainan Is. Sanya, 24.xii.1958, dead corals; 1 male, 1 female (ZIN RAN 79756, identified as *Thalamita danae* Stimpson, 1859), Vietnam, without other data, coll. E.F. Gurjanova; 1 female ov (ZMMU Ma1894), Indonesia, Java, Chiligin, Stn J-262, 06.ix.1962, coll. O.B. Mokievsky; 1 male (ZMMU Ma 1886), Papua New Guinea, Port-Morsby, littoral, sandy beach near mangrove, 07.v.1958; 1 female (ZMH K-5721, identified as *Thalamita stimpsoni*), Papua New Guinea, Bismarck Archipelago, St. Mathias, east coast, Enaie, 26-27.viii.1908, coll. Dr. G. Duncler. — *Thalamita danae* Stimpson, 1859: 5 males (ZIN RAN 79767, identified as *Thalamita stimpsoni* by D.D. Tien), China, Hainan Is., littoral, 1958, coll. E.F. Gurjanova; 2 males, 1 female, 1 female ov (SMF 19734), Philippines, Bohol, N. Calibutan Is., Mahaba sand bank, 10°15′N 124°17′E, coll. A. Schreiber; 1 female, 1 juv Indonesia, Java, Madura Strait, Pasirputi, St. J-487p, littoral, 19.x.1962, coll. O.B. Mokievsky; 2 males, 1 female (SMF 3875), Indonesia, Molucca Islands, Ternate Is.; 1 male (ZMB 18027), New Guinea; 1 male (ZMB 4266, identified by A. Milne-Edwards as *Thalamita stimpsoni*), New Caledonia; 1 female (ZMB 5060, identified as *Thalamita picta*), Tonga Is., Letuka, «Gazelle» Expedition; 1 male (ZMB 13776), Admiralty Is., Park Is. 1909, coll. Schroede; 1 male (ZMH K-2415), Fidji, Viti Is.; 1 male (ZMUO, no catalogue number), Australia, Great Barrier Reef, High Peak Is., 21°57′S 150°42′E, coll. Vigeland. — *Thalamita foresti* Crosnier, 1962: 1 female ov (SMF 3877), Madagascar, coll. Ebeman; 1 male (ZMB 6013, identified as *Thalamita danae*), Mozambique, coll. Peters; 1 male, 1 female (ZMH K-K-2471), Zanzibar, coll. F. Schumann; 1 male (SMF 23860), Fidji, Suva East, reef, from corals, 31.i.1995, coll. M. Türkay; 1 male, 1 female (ZMH K-2422), Samoa, from the Godeffroy collection, M.G. 5793; 1 male (ZMB 5064, identified as *Thalamita danae*), Papua New Guinea, Bismarck Islands, New Ireland Is., «Gazelle» Expedition, coll. Casteret; 1 male (ZMH K-5732, identified as *Thalamita danae*), Papua New Guinea, Bismarck Islands, St. Mathias, east coast, Enaie, 07-24.viii.1908, coll. G. Duncler; 1 male (NHMS 14194, identified as *Thalamita danae* by H. Balss), Marshal Is., S. Bock Expedition, Jaluit, N of S entrance, 15.x.1917, coll. C. Hessle. 1 female (MNHN, no catalogue number), New Caledonia, Platier de Thio (St. Gabriel), 22.iv.1989; 1 female (MNHN uncatalogued), New Caledonia, southern coast, Platier de Goro, blocs, intertidal zone, 11.i.1993, coll. B. Richer de Forges. — *Thalamita holthuisi* Stephenson, 1975: 1 male (RMNH.CRUS.D. 29881, holotype), Indonesia, Snellius Expedition, Taliaboe, Soela, 18.iii.1930; 1 male, 1 female (RMNH.CRUS.D. 29883, paratypes), Indonesia, Snellius Expedition. Moluccas. Morotai, 3-10.vi.1930; 2 males (RMNH.CRUS.D.29885, paratypes), Indonesia, Kera, near Timor, 15-16.xi.1929; 3 males, 2 females (RMNH.CRUS.D. 29884, paratypes), Indonesia, exact locality unknown; 1 male (RMNH.CRUS.D. 29885), Indonesia, West Irian, Ouri Is., coll. W. Roosdorp. — *Thalamita spinicarpa* Wee & Ng, 1995: 2 males (ZMMU 5218) Labrador Beach, January 1996, coll.

P.K.L. Ng. — *Thalamita starobogatovi* Tien, 1969: 1 male (ZMB 12769, identified as *Thalamita stimpsoni*), Sri Lanka, Trincomali (Trimcomalee), 06.iii.1906, coll. H. Schroede; 1 male (ZIN RAN 58275, holotype), China, Hainan Island, Sin Tsung, middle of the inlet, from live corals, 18.iv.1958, coll. P.V. Ushakov; 1 male (ZIN RAN 58276, paratype), same locality as for holotype; 1 female ov (ZIN RAN 58277, paratype), China, Hainan Island, Sin Tsung, 23.iv.1958, coll. D.V. Naumov; 1 female ov (ZMMU Ma2450, identified as *Thalamita stimpsoni*), Gulf of Tonkin, Hongai, intertidal zone, 06.iii.1960, coll. N.A. Zarenkov; 1 male (ZMMU Ma4891), Vietnam, Haifon, pond near the Red River, coll. N.A. Zarenkov; 2 males, 1 female, 1 female ov (ZMG 864), Philippines. Bohol, Ubay, coll. C. Semper.

Diagnosis.— Carapace about 1.7 times as broad as long; maximum width between third anterolateral teeth pilose. All anterior carapace ridges present: meso- and metagastrics interrupted, no ridges behind epibranchials. Front markedly wider than posterior border; six subequal, closely set frontal lobes; inner suborbital lobe broad, arcuate, going subparallel to front distally. Five anterolateral teeth, fifth smaller than third, fourth tooth vestigial, attached to posterior margin of third tooth. Basal antennal segment markedly wider than orbit, with a granular crest and scattered granules. Upper surface of manus with four spines, granular band extending from spine near carpal articulation to distal outer spine. Outer surface with two rows of sharpened granules and lower carina, other surfaces smooth. Posterodistal corners of propodi of ambulatory legs rounded. Pleomere 4 of male with obsolete keel, pleomere 6 with markedly convex lateral margins. G 1 relatively stout, evenly curved; a clump of 3-4 subterminal conical spinules on mesial face just beneath lip followed by a short row of approximately 6 thinner and longer spinules. Row of spinules on mesial face starting right behind subdistal projection and continuing to approximatelly one half of neck. Female genital sternite and gonopore remain undescribed.

Description.— Carapace 1.7 times as broad as long; maximum width between third anterolateral teeth, not markedly convex, covered with very short pile except margins. All anterior carapace ridges moderately granular: frontals closely set, elevated; protogastrics arcuate, widely separated; mesogastric nearly straight with some interruption in midline, distally ending at about the same level as protogastrics; metagastrics with slight interruption, nearly straight; epibranchials slightly sinuous; cardiacs and mesobranchials absent (figs. 6A, 7A). Front markedly (1.23 times) wider than posterior border; six subequal, closely set lobes: medians at lower plane than others, with somewhat convex anterior margin; submedians truncated; laterals rounded, with oblique lateral margin. Inner supraorbital lobe broadly arcuate. Inner suborbital lobe broad, with inner part subparallel to inner supraorbital lobe.

Five anterolateral teeth; first broadest; second, third and fifth gradually reduced in size; fourth tooth vestigial, attached to posterior margin of third tooth. Subhepatic regions with elongated granular patch, otherwise smooth. Thoracic sternum smooth (figs. 6A, 7A).

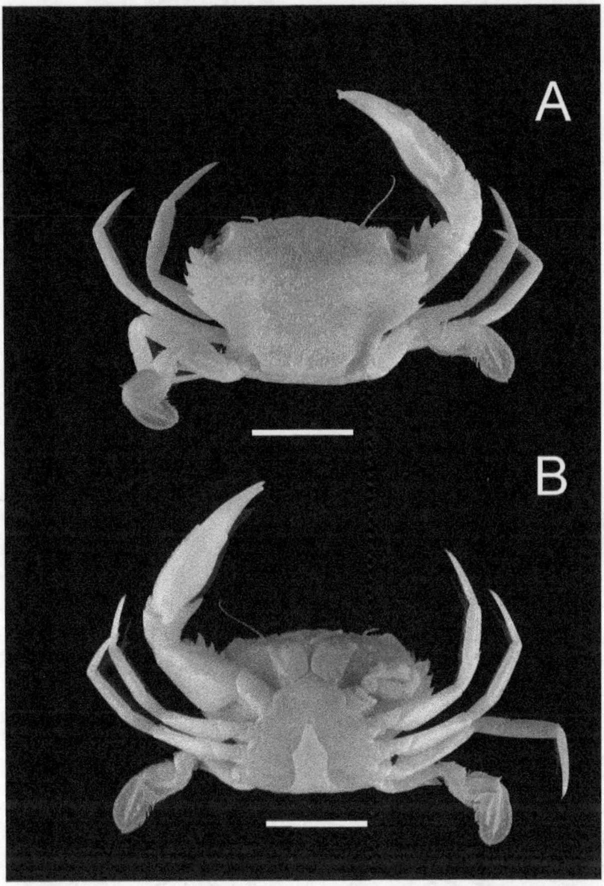

Fig. 6. *Thalamita williami*, holotype, male, CL 15.1 mm (ZMB 12497). A, dorsal view; B, ventral view. Scale bars 10 mm.

Basal antennal segment 1.4 times wider than orbit, with crest bearing 10-12 prominent rounded granules; scattered granules posterior to top of crest (fig. 7B).

Mouthparts not dissected. Maxilliped 1 with mesial lobe on endopodite. Maxilliped 3 smooth, anterior part of merus nearly straight, antero-external corner rounded and slightly produced laterally.

Only right cheliped present; merus with upper surface bearing smoothed granules, lower surface with longitudinal line of fine granules, anterior margin with three large spines increasing in size distally (fig. 6A, B), two proximal tubercles and one tubercle between two proximal teeth. Outermost and median spines of carpus short, depressed, inner spine sharp, curved; outer and upper face of carpus bearing 3 radial granular rows ending correspondingly in these spines; additional short, sinuous rows of granules between outer and median radial rows. Manus 1.48 times higher than thick; upper surface with 4 spines, granular band extending from

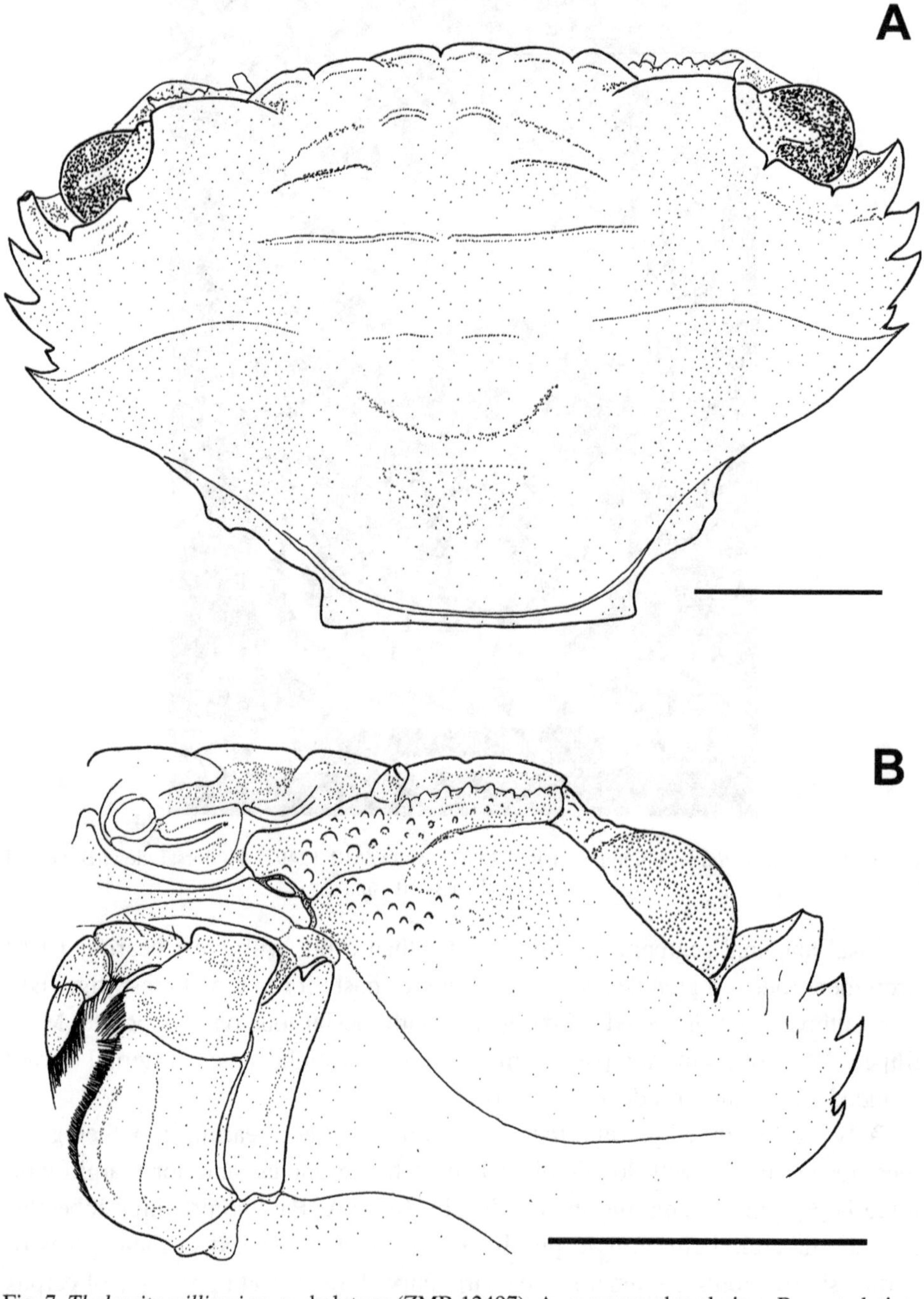

Fig. 7. *Thalamita williami* sp. n., holotype (ZMB 12497). A, carapace, dorsal view; B, ventral view of anterior part of cephalothorax, left side. Scale bars 5 mm.

spine near carpal articulation to distal outer spine. Outer surface with two rows of sharpened granules and lower carina, other surfaces smooth; fingers about as long as palm, strong, with usual grooves, moderately curved. There is no knobstick molariform tooth on right cheliped; serial conical teeth indistinctly bi- or trilobed, fused at bases (fig. 8A).

Pereiopods smooth, with posterodistal corners of meri sub-angular, those of propodi rounded; propodi keeled both anteriorly and posteriorly; dactyli much narrower than propodi, gently curved (fig. 8B). Merus of natatory leg (P 5) 2.12 times as long as broad with usual posterodistal spine; posterior border of propodus of natatory leg with 10-12 spinules, dactylus characteristic for *Thalamita*, broadening distally. Pleomere 4 with obsolete keel, lateral borders of pleomere 6 strongly convex. Telson about as long as broad (figs. 7B, 8E).

G 1 with quadrangular basal lobe, rather stout, evenly curved, tip with well-developed subdistal projection and spoon-shaped lip. Clump of 3-4 subterminal conical spinules on mesial face just beneath lip followed by short row of approximatelly 6 thinner and longer spinules. Row of spinules on mesial face beginning behind subdistal projection, continuing to about one half of neck (fig. 8C, D). G 2 about one fourth of G 1 length, slender, with nearly equal in size lateral and mesial lobes at tip.

Size (CL × CW5) and measurements (mm).— 15.1 × 25.2 (CW3 26.1, FM 10.5, FSO 15.3, FO 23.2, PM 8.3, manus length 17.6).

Coloration.— Live coloration unknown.

Etymology.— The species is named after the prominent American zoologist William Stimpson (1832-1872) who in particular made a significant contribution to the study of the genus *Thalamita*.

Vertical range and habitat.— No direct data on the holotype's habitat are available. The specimens identified by Stephenson (1972b) as *Thalamita danae* form C and presumably assigned to the new species (see Remarks) were collected from the coral reef, and at depth between 50 and 54 m on sand, coral and stones. *T. danae* sensu str. is, in turn, a littoral species (Wee & Ng, 1995; Spiridonov, 1999).

Distribution.— So far, the species is known with certainty only from Borneo (Kalimantan) Island. If it is conspecific with part of Stephenson's (1972b) material (see Remarks), the range of the new species includes Philippines and Java (fig. 9). According to the present record, the most morphologically similar species, *Thalamita starobogatovi*, occurs around the South-East Asia coast from Hainan Island to Sri Lanka, including Philippines (fig. 9).

Affinities.— The new species belongs to the group that includes *Thalamita crenata* Rüppell, 1830, *T. danae* Stimpson, 1859, *T. foresti* Crosnier, 1962; *T. holthuisi*

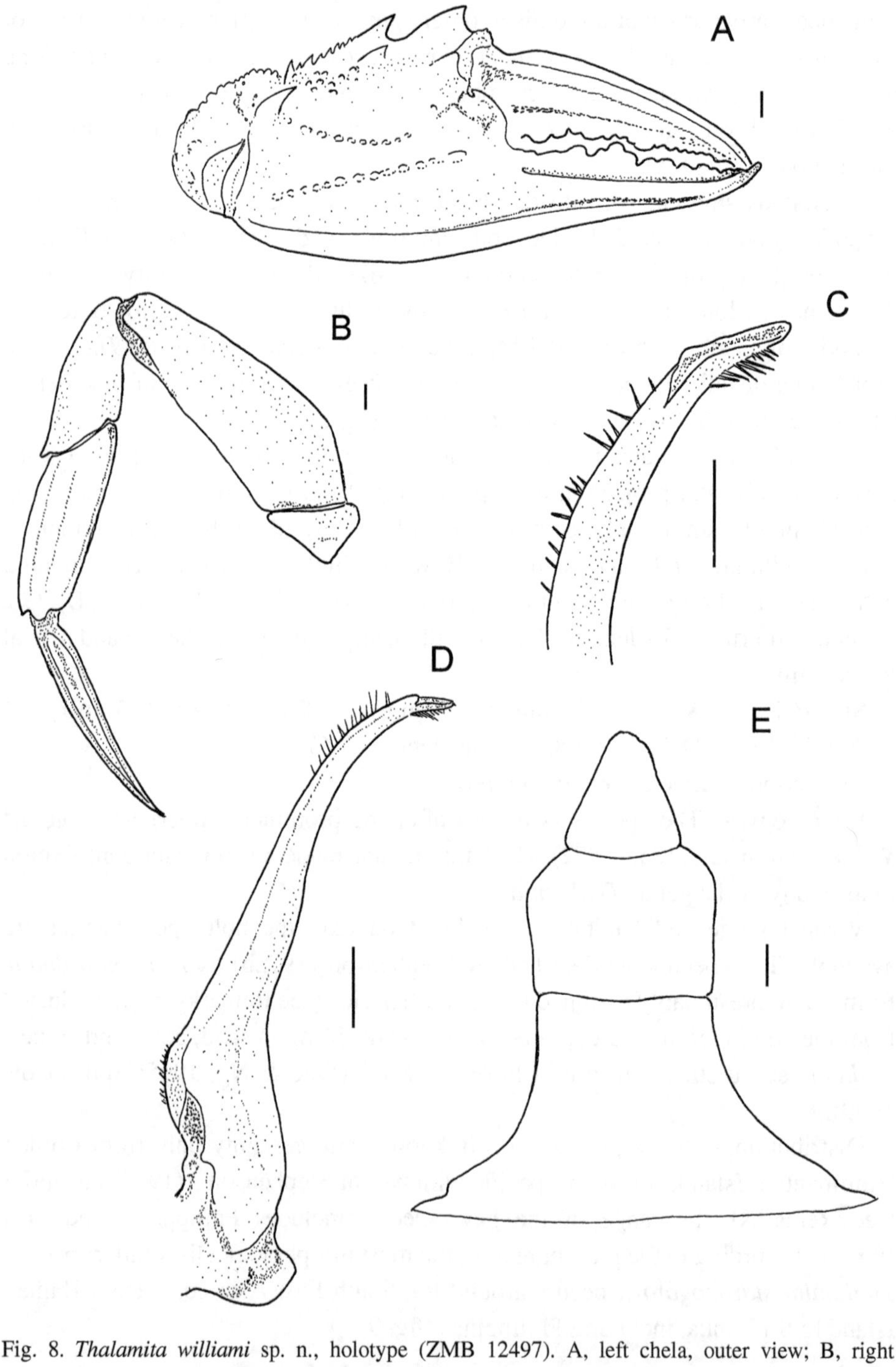

Fig. 8. *Thalamita williami* sp. n., holotype (ZMB 12497). A, left chela, outer view; B, right pereopod 3, ventral view; C, gonopod 1, tip, abdominal view; D, gonopod 1, abdominal view; E, male pleon (pleomeres 3-6 and telson). Scale bars 1 mm.

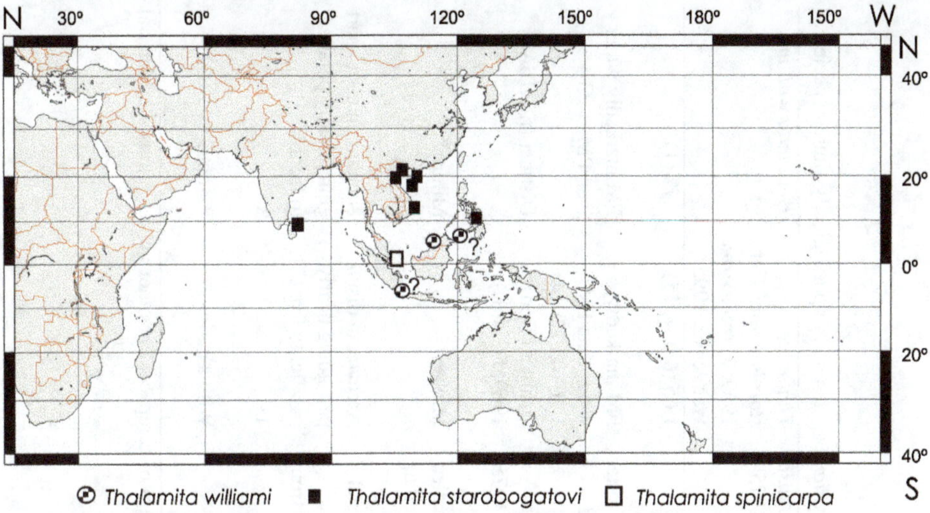

Fig. 9. Distribution of *Thalamita williami* sp. n. (present study and Stephenson, 1972b), *T. starobogatovi* Tien, 1969 (present study), and *T. spinicarpa* Wee & Ng, 1995 (Wee & Ng 1995).

Stephenson, 1975, and *T. starobogatovi* Tien, 1969 (table II). *T. spinicarpa* Wee & Ng, 1995 described from Singapore is very similar to the latter species, but larger; its status has to be discussed in a separate publication, but here the differences of *T. williami* from *T. starobogatovi* and *T. spinicarpa* taken together are discussed. With regard to the general appearance, *T. williami* is different from *T. crenata*, *T. danae*, *T. foresti*, and *T. holthuisi* by the greater CW/CL ratio and the front markedly broader than the posterior border. It is also different from all these species except *T. crenata* by the absence of a carina on the inner surface of manus (table II). It may be distinguished from these species by having the minute 4th anterolateral tooth vs. not markedly contrasting in size with other teeth; however, variation in this character needs to be analysed further. With regard to the carapace and cheliped morphology the present species is closest to *T. starobogatovi* and *T. spinicarpa* but differs from them by the absence of the additional spine on the upper surface of cheliped carpus.

The male of *T. williami* is distinguished from all species except *T. foresti* and *T. holthuisi* by the convex lateral margins of the 6th pleomere vs. gradually converging or subparallel. On the basis of males of these two species, *T. williami* clearly differs by a G 1 that is not very thin, with the neck bent as in *T. foresti* (see Apel & Spiridonov, 1998, fig. 55a). Considering further the G 1 morphology, the new species differs from *T. crenata* and *T. starobogatovi* in the presence of a clump of conical spinules just beneath the lip and a much denser row of spinules on the mesial face (fig. 8C). From *T. danae* (see Apel & Spiridonov, 1998, fig. 54a) it differs by a markedly separated tip with a more pronounced subdistal projection

TABLE II

Comparison of morphometric indexes and carapace and cheliped characters between *Thalamita williami* sp. n. and morphologically similar species

Character	*Thalamita crenata* Rüppell, 1830	*Thalamita danae* Stimpson, 1859	*Thalamita foresti* Crosnier, 1962	*Thalmita holthuisi* Stephenson, 1975	*Thalamita starobogatovi* Tien, 1969; *T. spinicarpa* Wee & Ng, 1995	*Thalamita williami* sp. n.
Mean CW/CL (SD; *n*)	1.52 (0.036; 13)	1.56 (0.031; 18)	1.51 (0.030; 9)	1.59 (0.040; 9)	1.65 (0.034; 15)	1.72 (1)
FM/PM (mean (SD; *n*))	FM slightly wider than PM; 1.10 (0.061; 14)	FM slightly wider than PM; 1.10 (0.124; 19)	FM slightly shorter than PM; 0.94 (0.030; 9)	FM slightly shorter than PM; 0.92 (0.030; 9)	FM markedly wider than PM; 1.26 (0.072; 14)	FM markedly wider than PM; 1.23
Frontal ridges	Faint (Wee & Ng, 1996, fig. 35A)	Distinct (Wee & Ng, 1996, fig. 42A-C)	Distinct (Crosnier, 1962, pl. XII (1))	Distinct (Stephenson, 1975, pl. 2)	Distinct (Tien, 1969, fig. 4(3))	Distinct (fig. 16A)
4th anterolateral tooth	As large or larger than 5th*	Variable but usually smaller than 5th*	Subequal to 3rd and 5th*	Somewhat smaller than 5th, shifted to 3rd*	Minute*	Minute*
Inner supra-orbital lobes	Arched, oblique to front*	Arcuate, in distal part going nearly parallel to front*	Arched, oblique to front*	Arcuate, in distal part going subparallel to front*	Arcuate, in distal part going nearly parallel to front*	Arcuate, in distal part going nearly parallel to front*
Spines on cheliped carpus	3	3	3	3	4	3
Inner surface of manus	Smooth	With a carina	With a carina	With a carina	With a carina	Smooth

SD, standard deviation; *n*, number of examined specimens; FM, frontal margin; PM, posterior margin of carapace. For other abbreviations see Material and Methods.

* Same figure as for frontal lobes for respective species.

and a better-developed lip (fig. 8C, D). *T. danae* also usually has a sparse row of short spinules on the mesial face vs. a dense row of relatively prominent spinules in *T. williami*.

Remarks.— Stephenson (1972b) recorded small specimens identified as *Thalamita danae*, which had the fourth anterolateral tooth minute and the fifth much smaller than the third one (his fig. 6E-F). The G 1 of these specimens is very similar to those of our new species (Stephenson, 1972b, fig. 7E-F) and most probably they belong to *T. williami*.

DISCUSSION

The current list of accepted species names for the genus *Thalamita* in the World Register of Marine Species (Davie & Türkay, 2017) is based on the check lists by Ng et al. (2008) and De Grave et al. (2009), and the latest regional monograph (Yang et al., 2012). It comprises of 91 names. Considering this and adding two species described in the present paper the species number becomes 93. The species inventory of *Thalamita* is, however, by far not completed. The description of a new species from Japan where the portunid fauna has been well studied, owing first of all to the lifetime effort of Tune Sakai (1939, 1976), indicates that even in these areas a discovery of new species is possible.

ACKNOWLEDGEMENTS

I am indebted to Prof. Michael Türkay who hosted me as an "annual" guest scientists in the Senckenberg Museum (SMF, Frankfurt am Main, Germany), attracted my attention to the remarkable collection of K. Matzuzawa (deposited in SMF) and repeatedly discussed this study. This is a very small tribute to Michael's memory, which will live with many of his colleagues and friends from different countries of the world. My gratitude extends to Angelika Brandt (ZMH), Oliver Coleman (ZMB), Marit Christiansen (ZMUO), Alain Crosnier (MNHN), Christer Erseus (NHMS), Charles Fransen (Naturalis Biodiversity Center), Victor Petryashov (ZIN RAN), for the loan of material and organization of working in the collections of their museums. I also thank Peter K. L. Ng (National University of Singapore, Singapore) for useful comments to the earlier version of the present paper, two anonymous reviewers of the present version for their suggestions to improve the text. I am grateful to Andreas Allspach, Kristin Arnold, Bianca Trautwein, Frank Mause and Sven Traenkner for technical support during my work in SMF, and Anna Zalota for proofreading and checking the English. The research visits to SMF and other German museums in the years 2000-2012 were supported by the guest grants of the Senckenberg Gesellschaft für Naturforschung, Frankfurt am Main, Germany.

REFERENCES

APEL, M. & V. SPIRIDONOV, 1998. Taxonomy and zoogeography of the portunid crabs (Crustacea: Decapoda: Brachyura: Portunidae) of the Arabian Gulf and the adjacent waters. Fauna of Arabia, **17**: 159-331, pls 1-12.

BALSS, H., 1924. Expedition S. M. Schiff "Pola" in das Rote Meer. Zoologische Ergebnisse. XXXIV. Decapoden des Roten Meeres III. Die Parthenopiden, Cyclo- und Catometopen. Denkschriften Akademie der Wissenschaft in Wien. Mathematisch-Naturwissenschaftlichen Klasse, **99**: 1-30, Taf. 1.

CROSNIER, A., 1975. Sur quelques Portunidae, Grapsidae et Ocypodidae (Crustacea Decapoda Brachyura) de Madagascar ou des îles avoisinantes, nouveaux, rares ou non encore signalés. Bulletin du Muséum national d'Histoire naturelle, 3e série, 304, Zoologie, **214**: 711-741, figs 1-10.

CROSNIER, A., 2002. Portunidae (Crustacea, Decapoda, Brachyura) de Polynésie française, principalement des îles Marquises. Zoosystema, **24**: 401-449, figs 1-28.

DAVIE, P. & M. TÜRKAY, 2017. *Thalamita* Latreille, 1829. In: World Register of Marine Species. Available online at http://www.marinespecies.org/aphia.php?p=taxdetails&id=106931 (accessed 30 April 2017).

DE GRAVE, S., D. PENTCHEFF, S. T. AHYONG, T.-Y. CHAN, K. A. CRANDALL, P. A. DWORSCHAK, D. L. FELDER, R. A. FELDMANN, C. H. J. M. FRANSEN, L. Y. D. GOULDING, R. LEMAITRE, M. Y. E. LOW, J. W. MARTIN, P. K. L. NG, C. E. SCHWEITZER, S. H. TAN, D. TSHUDY, R. WETZER & D. GUINOT, 2009. A classification of living and fossil genera of decapod crustaceans. Raffles Bulletin of Zoology, **21**(Suppl.): 1-109.

MINEMIZU, R., 2000. Marine decapod and stomatopod crustaceans mainly from Japan: 1-1344. (Bun-Ichi Sogo Shuppan, Tokyo).

MIYAKE, S., 1983. Japanese crustacean decapods and stomatopods in colour. In: Brachyura (crabs), **2**: 1-277. (Hoikusha, Osaka).

MONTGOMERY, S. K., 1931. Report on the Crustacea Brachyura of the Percy Sladen Trust Expedition to the Abrolhos Islands under the leadership of Prof. W. J. Dakin, in 1913 along with other crabs from the Western Australia. Zoological Journal Linnean Society, **37**: 405-465, pls 24–30.

SAKAI, T., 1939. Studies on the crabs of Japan. IV. Brachygnatha, Brachyrhyncha. Science Reports of the Tokyo Bunrika Daigaku (B), **3**(Suppl. 3): 365-741, pls. 42–111.

SAKAI, T., 1976. Crabs of Japan and the adjacent Seas. 3 volumes: xxix+773 pp., figs. 1-379; 16 pp., pls. 1-251; 461 pp., figs. 1-2. (Kodansha, Tokyo).

SCHÄFER, W., 1954. Form und Funktion der Brachyuren-Schere. Abhandlugen der Senckenbergischen Naturforschenden Geselschaft, **489**: 1-65.

SPIRIDONOV, V. A., 1999. Results of the Rumphius Biohistorical Expedition to Ambon (1990). Part. 8. Swimming crabs of Ambon (Crustacea: Decapoda: Portunidae). Zoologische Mededelingen, **73**(4): 63-97.

SPIRIDONOV, V. A., T. V. NERETINA & D. SCHEPETOV, 2014. Morphological characterization and molecular phylogeny of Portunoidea Rafinesque, 1815 (Crustacea Brachyura): implications for understanding evolution of swimming capacity and revision of the family-level classification. Zoologischer Anzeiger, **253**: 404-429.

SPIRIDONOV, V. & V. NEUMANN, 2008. Coral-inhabiting swimming crabs (Crustacea, Decapoda, Portunidae) of the Sudanese Red Sea. Organisms, Diversity and Evolution, **8**(3): 170e1-170e19.

STEPHENSON, W., 1972a. An annotated check-list and key to the Indo-West-Pacific swimming crabs (Crustacea: Decapoda: Portunidae). Bulletin of Royal Society New Zealand, **10**: 1-64.

STEPHENSON, W., 1972b. Portunid crabs from the Indo-West-Pacific and western America in the Zoological Museum, Copenhagen (Decapoda, Brachyura, Portunidae). Steenstrupia, **2**: 127-156.

STEPHENSON, W., 1975. Biological results of the Snellius Expedition, XXVI. The Portunidae (Decapoda, Brachyura) of the Snellius Expedition (Part II). Zoologische Mededeelingen, **49**(14): 173-206.

TIEN, D. D., 1969. New species and new subspecies of crabs Portunidae of Tonkin Gulf and the Hainan Island. Zoologicheskij Zhurnal, **48**: 505-511 (in Russian).

YANG, S., H. CHEN & A. DAI, 2012. Fauna Sinica. In: Invertebrata. Crustacea. Decapoda. Portunidae, **49**: 1-431, 14 pls. (Science Press, Beijing).

First received 2 October 2016.
Final version accepted 30 April 2017.

[When citing this volume, please refer to Crustaceana 90 (2017) 771-1288]

Michael Türkay Memorial Issue

BIOLOGY OF THE SHRIMP *OPLOPHORUS SPINOSUS* (BRULLÉ, 1839) (DECAPODA, OPLOPHORIDAE) IN THE CONTINENTAL SLOPE WATERS OF THE COAST OF NORTHWEST AFRICA

BY

S. A. SUDNIK[1])

Kaliningrad State Technical University, Department of Ichthyopathology and Hydrobiology, 1 Soviet Prospect, 236000 Kaliningrad, Russia

ABSTRACT

The size-sex structure and reproduction of the pelagic shrimp *Oplophorus spinosus* (Brullé, 1839) in the northwestern African continental slope is described herein. A total of 175 males and 53 females were caught by pelagic and bottom trawls at depths of 354-913 m in October-May 2003-2005. Total body length (TL) was 20-55.5 mm in males and 28.5-54 mm in females. Female TL at first maturity was 31 mm, the TL of ovigerous females ranged from 39.5-54 mm. Fecundity varied from 5 to 25 eggs with an outer diameter of 3.21 ± 0.19 mm at the beginning of embryogenesis and 3.14 ± 0.16 mm at half-time. *O. spinosus* has a continuous spawning period with a possible peak in spring. Its main reproductive traits are similar to those of this species from the Pacific and are comparable with the other species of this genus. In general, *Oplophorus* spp. are typical K-strategists with large-sized eggs and low fecundity.

ZUSAMMENFASSUNG

Hier wird das Größen- und Geschlechterverhältnis, sowie die Reproduktion der pelagischen Garnele *Oplophorus spinosus* (Brullé, 1839) am Nordwestafrikanischen Kontinentalhang beschrieben. Mit Grundschleppnetz und pelagischem Netz wurden von Oktober bis Mai 2003-2005 insgesamt 175 Männchen und 53 Weibchen in Tiefen von 354-913 m gefangen. Die Gesamtlänge (GL) der Männchen betrug 20-55.5 mm und die Weibchen maßen 28.5-54 mm. Die GL der Weibchen betrug 31 mm bei der 1. Geschlechtsreife, eiertragende Weibchen erreichten 39.5-54 mm. Die Fruchtbarkeit variierte zwischen 5 und 25 Eiern mit einem äußeren Radius von 3.21 ± 0.19 mm zu Beginn der Embryogenese und 3.14 ± 0.16 mm in der zeitlichen Mitte. *O. spinosus* laicht das ganze Jahr hindurch mit einem möglichen Höhepunkt im Frühling. Seine Hauptreproduktionsmerkmale gleichen denen von Vertretern derselben Art aus dem Pazifik und sind überdies vergleichbar mit denen der anderen Arten dieser Gattung. Allgemein sind *Oplophorus* spp. typische K-Strategen mit großen Eiern und einer niedrigen Fruchtbarkeit.

[1]) e-mail: lanasudnik@mail.ru

INTRODUCTION

Caridean shrimps of the genus *Oplophorus* H. Milne Edwards, 1837 include four oceanic species (De Grave & Fransen, 2011). A typical representative of this genus is *Oplophorus spinosus* (Brullé, 1839), with a cosmopolitan distribution range. It inhabits tropical and subtropical waters of the Atlantic, the Pacific and the Indian Ocean. It can be found at depths of 10-3050 m, as during daily vertical migrations at night it rises almost to the surface (Hendrickx & Estada-Navarrete, 1989; Burukovsky, 1994, 1996; Pequegnat & Wicksten, 2006). *O. spinosus*, being a micronektonic predator feeding on copepods, euphausiids, small shrimp, squid and fish, is abundant in the North-Atlantic subtropical convergence area (Burukovsky, 1996). On the other hand, it serves as a prey for some fish, squids, shrimps and dolphins, playing an important role in the pelagic trophic food web (Burukovsky, 2009). Despite its ecological importance, other biological aspects of *O. spinosus* have been poorly studied. Most of the data published on this species describe its morphology, taxonomy, geographic distribution, bathymetry, size composition and feeding habits. There is little information on the sizes of *O. spinosus* collected in southwestern Brazil waters (Cardoso & Young, 2005) and on its fecundity and egg size in the Pacific Ocean (Omori, 1974; Griffiths & Brandt, 1983), while there is no information available on the size structure and reproductive traits of this species from the eastern Atlantic Ocean. There are some data on body size, fecundity and egg sizes of other *Oplophorus* species, i.e., *O. gracilirostris* A. Milne-Edwards, 1881 and *O. typus* H. Milne Edwards, 1837 [in H. Milne Edwards, 1834-1840] (Omori, 1974). Only for *O. novaezeelandiae* (De Man, 1931) there is a relatively complete description of its size-age structure, migrations and reproductive traits (Burukovsky, 1994).

The aim of this work is to describe size structure and main reproductive traits of *O. spinosus* in the continental slope waters of the northwestern coast of Africa, and compare the results to the ones obtained for the other three *Oplophorus* species. The male individuals prevailed (sex ratio = 3.4 : 1), showing nearly no sexual dimorphism, with the females being larger in general. The largest animals, as well as freshly moulted shrimps (males : females = 2 : 1) were caught in May, while no juveniles were found in any of the months sampled. Spawning occurred in all observed months. The individuals of *O. spinosus* are supposed to only spawn once in a lifetime.

MATERIAL AND METHODS

The studied material was collected during the expeditions of RV "Atlant-NIRO" and RV "Atlantida" (AtlantNIRO, Kaliningrad, Russia) on the continental slope waters of northwestern Africa from 30°30′0″N 13°20′8″W to 17°29′0″N,

16°40′5″W at depths between 354 and 913 m (table 1, fig. 1). Samples were collected from 18 catches with a semi-pelagic trawl (RTTM-70: 4 hauls in October 2003, December 2004 and January 2005) and a bottom trawl (HAKE-4M: 14 hauls in May 2005), both with a 10 mm minimum mesh size at the cod end. Shrimp samples were preserved in 4% borax neutralized formaldehyde for further laboratory analyses. A total of 228 specimens were analysed: 175 males and 53 females including 24 non-ovigerous and 29 ovigerous ones (table I).

In order to study the reproductive traits, we used the methodology for biological analysis of shrimps after Burukovsky (1992) that had been adopted for pelagic shrimps (Sudnik & Falkenhaug, 2014). The analysis included:

- measurements of the total body length (TL) (table I, fig. 2);
- measurements of the carapace length (CL) (fig. 2);
- sex identification (by pleopods 2 based on appendix masculina; the structure of endopodite of pleopod 1 was the same in both sexes; table I);
- the phase of the moult cycle (on the condition of cuticle: moulting/inter-moulting), assigning female ovaries maturity stage (OMS) and the development stage of embryos (EDS) (table I);
- counting fecundity of eggs (= embryos) and measurement of their size (table I).

For fecundity estimation, all eggs from pleopods were counted directly but only females with embryos in EDS 1 were used for analysis, in order to avoid bias caused by egg loss during the incubation period as occurs in shrimps (Sudnik & Falkenhaug, 2014). In each ovigerous female with embryos in EDS 1-3 at least 10 eggs were randomly selected from each clutch and measured using an ocular microscope with an accuracy of 0.01 mm. Afterwards, the mean value was calculated. As eggs had an ellipsoid form, their volume was estimated by the formula: $V = 4/3 \times \pi \times r_1 \times (r_2)^2$, where r_1 is half the egg length and r_2 half the egg width (Oh & Hartnoll, 2004). Statistical analyses were carried out using Statistica 8.0: Multiple Regression Analysis (Orlov, 2004).

RESULTS

Sex ratio and size composition

The majority of shrimps studied was collected during spring (69%), and much less in autumn (16%) and winter (15%) (table I). In total, 175 males and 53 females of *Oplophorus spinosus* were caught; no juveniles were observed. The overall sex ratio was 3.4 : 1. Males prevailed in all samples at different seasons and localities (tables I and II).

TABLE I

Data on sampling and demographic characteristics of *Oplophorus spinosus* (Brullé, 1839) in northwestern African waters (1-4, pelagic trawl, horizon, mean depths; 5-18, bottom trawl)

No	Date	Coordinates	Depth/ horizon (m)	Number ♂♂/♀♀	TL (mm) (♂♂/♀♀)	Number of ♀♀ with ovaries at different OMS					Number of ♀♀, NOV/OV	Moulting (♂♂/♀♀)	Number of ♀♀ with embryos at EDS 1-3		
						II	III	IV	V	VI-II			EDS 1	EDS 2	EDS 3
1	23 October 2003	28°08′1N 13°15′8W	520-549/ 5-35-65	2/5	39, 41/ 35.5-45	1	–	–	1	3	2/3	–/–	3	–	–
2	27 October 2003	25°04′5N 16°15′8W	469-354/ 5-35-65	23/6	22.5-47/ 29.5-35	6	–	–	–	–	6/–	–/–	–	–	–
3	16.12.2004	23°15′6N 17°05′6W	583/ 5-35-64	24/5	32-47/ 31.5-38	2	3	–	–	–	5/–	–/–	–	–	–
4	13.01.2005	17°29′2N 16°40′5W	470/ 5-35-64	5/1	41-45/44	–	–	–	–	1	–/1	–/–	1	–	–
5	06 May 2005	30°01′5N 10°11′0W	808-867	3/1	37-53/40.5	1	–	–	–	–	1/–	–/1	–	–	–
6	08 May 2005	29°04′6N 12°09′2W	719-740	1/3	40/40.5-43	–	1	1	–	1	–/3	1/1	–	1	2
7	09 May 2005	28°30′4N 12°52′6W	477-490	–/1	–/47	–	–	–	–	1	–/1	–/–	1	–	–
8	09 May 2005	28°27′9N 13°08′4W	862-864	6/5	35.5-47.5/ 40.5-47	–	4	–	–	1	–/5	1/–	1	2	1
9	10 May 2005	28°03′4N 13°20′8W	628-700	6/2	39-51.5/ 33, 43	1	1	–	–	–	1/1	1/–	–	–	1
10	10 May 2005	28°00′5N 13°24′0W	852-913	2/1	31, 51/44	–	–	–	–	1	–/1	–/–	1	–	–

TABLE I
(Continued)

No	Date	Coordinates	Depth/ horizon (m)	Number ♂♂/♀♀	TL (mm) (♂♂/♀♀)	Number of ♀♀ with ovaries at different OMS					Number of ♀♀, NOV/OV	Moulting (♂♂/♀♀)	Number of ♀♀ with embryos at EDS 1-3		
						II	III	IV	V	VI-II			EDS 1	EDS 2	EDS 3
11	12 May 2005	26°59′6N 13°53′6W	521-523	1/2	55.5/ 42, 42.5	1	1	–	–	–	1/1	1/–	–	1	–
12	13 May 2005	26°27′9N 14°47′8W	826-855	3/–	44.5-49/–	–	–	–	–	–	–/–	–/–	–	–	–
13	15 May 2005	25°58′8N 15°37′7W	841-852	–/1	–/36	1	–	–	–	–	1/–	–/–	–	–	–
14	19 May 2005	21°01′0N 17°42′3W	726-743	–/2	–/33.5-43	1	–	–	–	1	1/1	–/–	1	–	–
15	21 May 2005	22°03′4N 17°32′2W	814-816	1/4	46/ 28.5-45.5	2	–	–	–	2	2/2	–/–	1	2	–
16	24 May 2005	23°33′4N 17°12′0W	810-816	2/–	42.5-43/–	–	–	–	–	–	–/–	–/–	–	–	–
17	27 May 2005	24°59′0N 16°26′1W	617-629	3/1	43.5-51/34.5	–	1	–	–	–	1/–	–/–	–	–	–
18*	06-27 May 2005	30°30′0N 13°20′8W	337-913	93/13	20-55/41-54	3	3	1	–	6	3/10	51/6	7	3	–
Total				228		19	14	2	1	17	24/29	55/8	16	9	4

NOV, non-ovigerous females; OV, ovigerous females; EDS, development stage of embryos (5 stages): 1, no visual signs of embryo development; 2, yolk occupies up to 70% of egg volume, embryo crescent-shaped; 3, yolk occupies up to 30%, curled embryo shape, tail reaching or even passing over the head; rudimental eye crescent-shaped, eye cornea partly pigmented; 4, well-developed eye (oval shaped, pigmented cornea), thoracic appendages developed; 5, complete formed larva with yolk droplet inside cephalothorax; free body appendages and free tail in embryos (females with embryos in EDS 4 and 5 were not found); OMS, female ovaries maturity stage (6 stages): I, II, immature ovaries; III, IV, maturing; V, mature; VI-II, immature in second time of post-spawned females (without vitellogenic oocytes).

* Aggregated sample of shrimps from different catches that were collected for student's zoological practical training, unfortunately without detail labels.

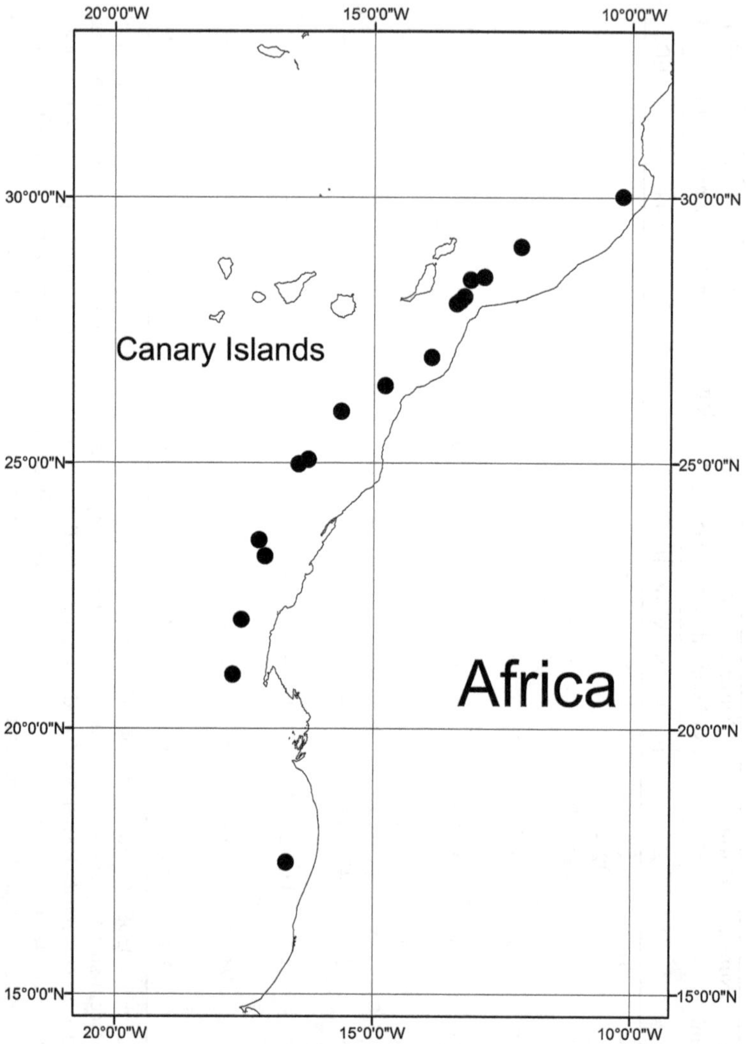

Fig. 1. Location of sampling stations of *Oplophorus spinosus* (Brullé, 1839) on the northwestern African continental slope.

In total, TL varied from 20.0 to 55.5 mm and CL ranged from 7.0 to 20.0 mm (males: TL 20.0-55.5 mm, CL 7.0-20.0 mm; females: TL 28.5-54.0 mm, CL 8.7-17.0 mm) (fig. 2). Most large males and females were observed in the May samples (table II).

Moulting shrimps

Moulting shrimps occurred in May only ($N = 63$), with male predomination (table I). The proportion of moulting shrimps was 46% among males and 22%

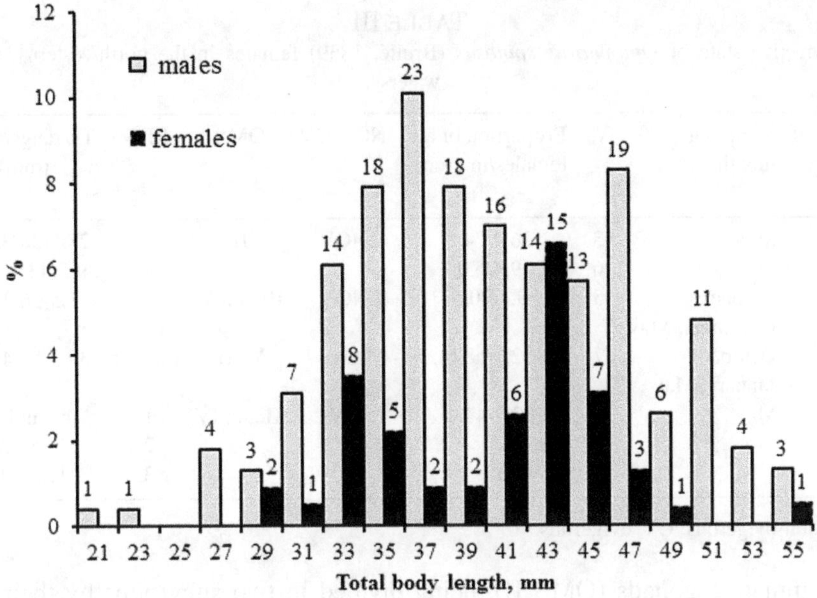

Fig. 2. The male and female size composition of *Oplophorus spinosus* (Brullé, 1839) in northwestern African waters. Numerals in columns indicate number of studied specimens in each size group.

TABLE II

Seasonal variability of *Oplophorus spinosus* (Brullé, 1839) sizes in northwestern African waters

Sex	May 2005		October 2003		December 2004-January 2005	
	N	TL (mm)	*N*	TL (mm)	*N*	TL (mm)
Male	121	20.0-55.5	25	22.5-47.0	29	32.0-47.0
Female	36	28.5-54.0	11	29.5-45.0	6	31.5-44.0

TL, total body length from the posterior margin of orbits to the median posterior edge of telson; resolution = 0.1 mm.

among females. Size of males ranged from 27 to 55.5 mm TL (mean 39 mm). Four moulting non-ovigerous females (40.5-43.5 mm TL) had immature ovaries at OMS II. Four other moulting females (40.5-49.5 mm) were ovigerous with eggs at EDS 1 and 2. Among them, 3 females had immature gonads at OMS VI-II (post-spawning) and 1 female at the beginning of maturation (OMS III).

Female reproductive traits

The studied females can be divided into four groups based on the following criteria: the OMS, presence or absence of eggs on pleopods, the ESD and body sizes (table III, fig. 3). These groups are the following: (a) non-ovigerous females

TABLE III
Reproductive state of *Oplophorus spinosus* (Brullé, 1839) females in the northwestern African waters

Group of females	Sampling month	*N*	Proportion of all females/in group (%)	NOV/OV	OMS	EDS	TL range/mode (mm)
a	May	13	26.4/74.0	NOV	II	–	28.5-38/32-34
		6	9.4/6.0			–	39.5-43.5/42-44
b	October, December, May	5	9.4/20.0	NOV	III and V	–	31.5-36/34-36
c	October, January, May	17	32.1/58.6	OV	VI-II	1 and 2	39.5-54/41-47
d	May	2	22.7/41.4	OV	III and IV	1	43 and 46
		6				2	40.5-45.5
		4				3	41.5-43.5

NOV, non-ovigerous; OV, ovigerous.

with immature gonads (OMS II), being divided in two subgroups by their TL; (b) non-ovigerous females with maturing and mature gonads (OMS III and V); (c) recently spawned females: ovigerous with embryos in EDS 1 and EDS 2 and immature gonads (OMS VI-II); (d) ovigerous females with embryos in EDS 1-3 and maturing ovaries (OMS III and IV).

In May samples, a total of 21 ovigerous females had embryos in EDS 1 and 2, while 4 specimens carried embryos in EDS 3 (i.e., at the midpoint of embryogenesis). In autumn and winter, only ovigerous females with embryos in EDS 1 were found (3 and 1, respectively). Generally, females with embryos at more advanced stages of their pleopods also had more developed gonads (table IV).

Egg size and fecundity

Eggs ranged in size from 3.0-3.57 × 1.98-2.40 mm (egg length × width) in females of 39.5-54 mm TL, carrying embryos in EDS 1 (2.94-3.30 × 1.84-2.26 mm) and in females of 41.5-43.5 mm TL, carrying embryos in EDS 3 (table V). Egg volume varied from 10.52 (EDS 1, 46 mm TL) to 5.07 mm^3 (EDS 2, 40.5 mm TL) (table V). The mean egg volume in females of 39.5-54 mm TL was 7.69 ± 1.38 mm^3. There was no correlation between TL of females and egg parameters: egg length ($r = 0.472$, $P = 0.065$), egg width ($r = 0.283$, $P = 0.288$) and egg volume ($r = 0.379$, $P = 0.148$).

Fecundity was counted in 29 females. Mean fecundity of the studied females of 39.5-54.0 mm TL was 14 ± 5 eggs, ranging from 5 (47 mm TL) to 25 eggs (54 mm TL) (fig. 4). There was no correlation between fecundity of females with embryos in EDS 1 and their TL ($r = 0.301$, $P = 0.257$).

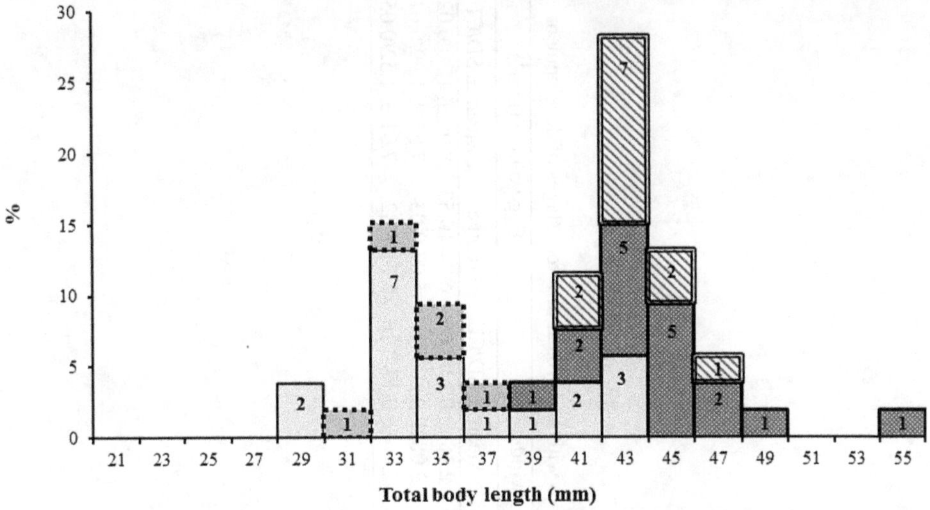

Fig. 3. Size composition of ovigerous (EDS 1-3) and non-ovigerous females of *Oplophorus spinosus* (Brullé, 1839) with ovaries in different maturity stages (OMS II—VI-II) (% of all studied females). Numbers in columns indicate number of specimens studied in each category.

TABLE IV

Correspondence between the OMS and EDS in *Oplophorus spinosus* (Brullé, 1839) ovigerous females in northwestern African waters

OMS	Number of ovigerous females		
	EDS 1	EDS 2	EDS 3
VI-II	14	3	–
III	2	5	3
IV	–	1	1

For OMS and EDS, see legend to table I.

DISCUSSION

No juvenile shrimp was discovered during the study at depths of 5-64 m, but mostly at 337-913 m (table I). They are usually supposed to inhabit the epipelagic layers between 50-200 m (Crosnier & Forest, 1973; Burukovsky, 1994) where only four trawls were carried out. Also these small shrimps possibly could slip through mesh of the trawl (10 mm).

TABLE V

Egg sizes, egg volume, fecundity and sizes of *Oplophorus spinosus* (Brullé, 1839) ovigerous females with embryos at different stages of development

EDS	N	TL (mm)		Fecundity (Range)	Egg length (mm)		Egg width (mm)		Egg volume (mm^3)	
		Range	Mean ± SD		Range	Mean ± SD/CI	Range	Mean ± SD/CI	Range	Mean ± SD/CI
1	16	39.5-54	44.8 ± 3.3/3.28	5-25	3.00-3.57	3.21 ± 0.19/0.09	1.98-2.40	2.18 ± 0.13/0.06	6.34-10.52	8.06 ± 1.30/0.02
2	9	40.5-45.5	42.7 ± 1.9/1.2	12-20	2.99-3.42	3.2 ± 0.15/1.22	1.8-2.27	2.08 ± 0.16/0.10	5.07-9.03	7.35 ± 1.39/0.04
3	4	41.5-43.5	42.8 ± 0.9/0.85	16-20	2.94-3.30	3.14 ± 0.16/0.85	1.84-2.26	2.06 ± 0.17/0.16	5.21-8.82	7.01 ± 1.52/0.05

CI, correspondence interval.

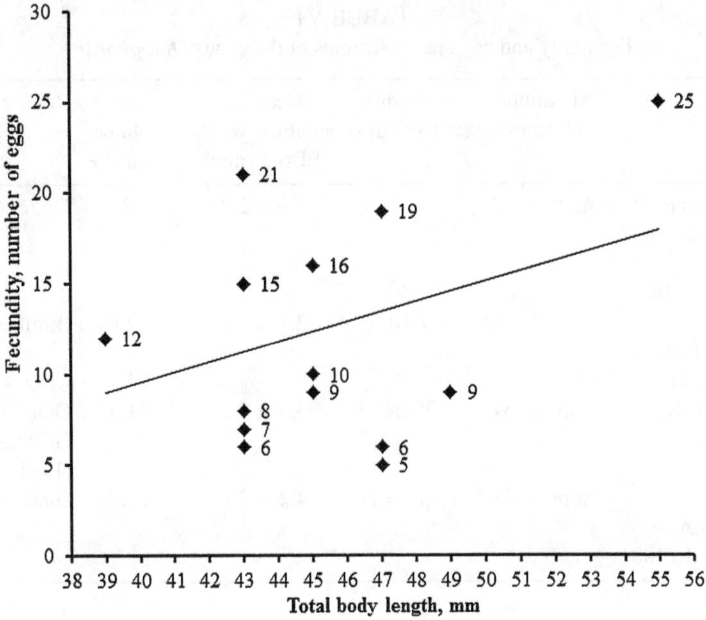

Fig. 4. Relationship between fecundity of *Oplophorus spinosus* (Brullé, 1839) females with embryos in EDS 1 and their TL. Numbers near symbols indicate fecundity value.

The sex ratio in all studied samples from different areas and seasons was stable, with a pronounced predominance of males (tables I and II). This situation probably is the result of a vertical stratification of specimens of different sexes, as well as a high female mortality after the larval release and a second spawning event (Nilssen, 1990). A similar sex ratio was observed in spring and summer in *Oplophorus novaezeelandiae* from Namibian waters: males have been predominant more than 3 times (Burukovsky, 1994). Sexual dimorphism in body size is nearly absent in *O. spinosus* as in *O. novaezeelandiae* (Burukovsky, 1994).

O. spinosus attains the largest TL compared to other congeners (table VI). The size structure of studied central-eastern Atlantic *O. spinosus* was similar to that of this species from the Pacific (Omori, 1974; Griffiths & Brandt, 1983). The maximal TL of *O. spinosus* collected in southwestern Brazil waters (up to 20 mm CL in males and to 24 mm CL in females) (Cardoso & Young, 2005) exceeded sizes of this species from the central-eastern Atlantic and from the Pacific (up to 17.5 mm CL) (Omori, 1974; Griffiths & Brandt, 1983).

Moulting *O. spinosus* were observed in May only, with males in absolute predominance. Among females, a total of 4 moulting non-ovigerous shrimps were at the beginning of their first maturing and soon must be mating, stimulating the development of the ovaries (Charniaux-Cotton, 1985). Four other females with recently laid eggs were in pre-spawning moulting, after which caridean females

TABLE VI
Fecundity and egg size in shrimps of the genus *Oplophorus*

Species	Maximum TL (mm)	Fecundity (No. of eggs)	Mean egg length × width, EDS 5 (mm)	Egg volume (mm^3)	References
Oplophorus typus H. Milne Edwards, 1837 [in H. Milne Edwards, 1834-1840]	Approx. 42	–	2.9 × 2.1	6.7	Omori (1974)
Oplophorus novaezeelandiae (De Man, 1931)	48	7-10	3.1 × 2.1	7.6	Burukovsky (1994)
Oplophorus spinosus (Brullé, 1839)	Approx. 57	25-36	3.5 × 2.5	11.4	Omori (1974); Griffiths & Brandt (1983); our data
Oplophorus gracilirostris A. Milne-Edwards, 1881	Approx. 48	16 ($n = 1$)	4.2 × 2.6	14.8	Omori (1974)

mate, spawn and begin the next cycle of ovarian maturing. The same relations of the moult and reproductive cycles in females was described for different groups of shrimps (Nelson, 1991; Burukovsky, 1994; Sudnik, 2007)

In the studied area spawning *O. spinosus* females were found in all sampled months. Therefore, spawning of *O. spinosus* in the waters of northwestern Africa could be continuous, possibly with a peak in spring. *O. novaezeelandiae* from Namibia waters had a spawning peak in summer (Burukovsky, 1994). Year-round spawning was described for a number of other tropical shrimps including lower meso- and bathypelagic ones (Omori, 1974).

On the basis of the data on the possible length of juveniles (up to 28 mm TL, based on data in *O. novaezeelandiae* by Burukovsky, 1994) and the size composition of *O. spinosus* females in the studied area (fig. 3), we can propose the following working hypothesis on female main reproductive traits in the waters of northwestern Africa: for the estimation of the TL at first maturity, the TL of maturing and mature non-ovigerous females at OMS III-V were taken into consideration, as shrimp females with developing gonads do not moult and do not grow (Burukovsky, 1994). Therefore, the TL at fist maturity varies between 31 and 36 mm on average. This estimate for *O. spinosus* is rather close to data for *O. novaezeelandiae* (25-44 mm; Burukovsky, 1994). Ovigerous females of *O. spinosus* ($N = 28$) of 39.5-49.5 mm TL might be extruding eggs for the first time, of which 16 females of the same size range had immature gonads (OMS VI-II) and carried recently extruded eggs (EDS 1). Twelve other ovigerous females, 40.5-46 mm TL with maturing ovaries at OMS III and IV, probably have been maturing for the second time. Only one, the largest recently spawned female of 54 mm TL with immature ovaries caught in May, could have spawned for a second time in her lifespan. These data can be indirect evidence for the high mortality levels of females during each pre-spawning moulting event and after larvae hatching, as a similar result was also found in *O. novaezeelandiae* (Burukovsky, 1994). This implies the low probability of a second spawning event of *O. spinosus* females and their short lifespan in general. The same was described for *O. novaezeelandiae* (Burukovsky, 1994).

In Brazilian waters, the maximal length of 6 ovigerous *O. spinosus* was up to 17 mm CL (Cardoso & Young, 2005), the same as that in this species from the central-eastern Atlantic (17 mm CL). However, 36 Brazilian non-ovigerous females with a size of up to 24 mm CL (Cardoso & Young, 2005) were considerably larger than those 24 ovigerous females from the area studied here (17 mm CL).

The average size of eggs at the final stage of development of *O. spinosus* from the Tasmanian Sea was 3.5 × 2.5 mm (Omori, 1974), which is close to our findings from the Atlantic (average 3.1 × 2.1 mm for eggs with embryos in the middle of embryogenesis). All four *Oplophorus* species have relatively large eggs compared to other shrimps, and the size range of their eggs is narrow (2.9-4.2 mm, table VI). These differences in egg size seem to be caused by differences in adult female body size between these species.

The larger egg volume in *O. spinosus* females with recently laid embryos at EDS 1 compared with that in females with more developed embryos at EDS 2 and 3 can be explained by the larger body sizes of females of the former group (table V). Additionally, the egg volume can decrease during incubation due to a decrease of egg organic content and reduced water filling (Krol et al., 1992).

The maximal value of fecundity in *O. spinosus* from the Atlantic Ocean by our data (25 eggs of 39.5-54 mm TL) was somewhat less than that of the same species from the Pacific (36 eggs of 43.7-50.2 mm TL; Griffiths & Brandt, 1983). But on the other hand, the largest female from the Pacific (57 mm TL) had 29 eggs (Omori, 1974). Fecundity of *Oplophorus* spp. is low and its maximum values known varied between 10 and 36 eggs (table VI).

The main reproductive traits, including spawning mode, egg size and fecundity, of *O. spinosus* are totally comparable with those in *O. novaezeelandiae*, *O. gracilirostris* and *O. typus* (Omori, 1974; Griffiths & Brandt, 1983; Burukovsky, 1994). In general, representatives of *Oplophorus* are typical K-strategists with large-sized eggs and low fecundities as a natural fit to their mostly meso- and bathypelagic mode of life.

ACKNOWLEDGEMENTS

I wish to thank A. N. Golub, A. B. Miheev and A. A. Fetisov for sampling of material in cruises, A. K. Zalota for the help in translation of the article into English, R. N. Burukovsky. V. V. Laptikhovsky and M. R. Yakushev for kind help with our work, and I am very grateful to Ch. M. Nigmatullin for critical manuscript reading and editing, and invaluable suggestions. Special thanks go to the anonymous reviewer and M. Thiel for important and useful remarks.

REFERENCES

BURUKOVSKY, R. N., 1992. Method of the biological analysis of some tropical and subtropical shrimps. In: B. G. IVANOV (ed.), Fishery-biological investigations of marine invertebrates: 77-84. (VNIRO, Moscow). [In Russian.]

BURUKOVSKY, R. N., 1994. On the biology of shrimp *Oplophorus novaezealandiae* De Man, 1931, from Namibian waters (Crustacea, Decapoda, Oplophoridae). Arthropoda Selecta, **3**: 3-12.

BURUKOVSKY, R. N., 1996. Shrimp fauna (Crustacea: Decapoda) on the north Atlantic subtropical convergence. Archipélago, **14A**: 7-15.

BURUKOVSKY, R. N., 2009. Feeding and trophic interactions of shrimps. (Kaliningrad State Technical University Publishing, Kaliningrad). [In Russian with English summary, content, figure and table legends.]

CARDOSO, I. & P. YOUNG, 2005. Deep-sea Oplophoridae (Crustacea Caridea) from southwestern Brazil. Zootaxa, **1031**: 1-76.

CHARNIAUX-COTTON, H., 1985. Vitellogenesis and its control in malacostracan crustaceans. American Zoology, **25**: 197-206.

CROSNIER, A. & J. FOREST, 1973. Les crevettes profondes de l'Atlantique oriental tropical. Faune Tropicale, **19**: 1-409. (ORSTOM, Paris).

DE GRAVE, S. & C. H. J. M. FRANSEN, 2011. Carideorum catalogus: the recent species of the dendrobranchiate, stenopodidean, procarididean and caridean shrimps (Crustacea, Decapoda). Zoologische Mededelingen, **85**: 195-588.

GRIFFITHS, F. B. & S. B. BRANDT, 1983. Distribution of mesopelagic decapod Crustacea in and around a warm-core eddy in the Tasman Sea. Marine Ecology Progress Series, **12**: 175-184.

HENDRICKX, M. E. & F. D. ESTRADA-NAVARRETE, 1989. A checklist of the species of pelagic shrimps (Penaeoidea and Caridea) from the eastern Pacific with notes on their geographic and depth distribution. CalCofi Reports, **30**: 104-121.

KROL, R. M., W. E. HAWKINS & R. M. OVERSTREET, 1992. Reproductive components. In: F. W. HARRISON & A. G. HUMES (eds.), Microscopic anatomy of invertebrates, Decapod Crustacea, **10**: 295-343. (Wiley-Liss, New York, NY).

NELSON, K., 1991. Scheduling of reproduction in relation to molting and growth in malacostracan crustaceans. In: A. WENNER & A. KURIS (eds.), Crustacean egg production. Crustacean Issues, **7**: 77-113. (A. A. Balkema, Rotterdam/Brookfield).

NILSSEN, E. M., 1990. Demography, fecundity, growth and mortality of the deep-water prawn *Pandalus borealis* from Isfjord, Spitsbergen (78°N). ICES 1990/Shell Paper No 109: 1-36.

OH, C.-W. & R. G. HARTNOLL, 2004. Reproductive biology of the common shrimp *Crangon crangon* (Decapoda: Crangonidae) in the central Irish Sea. Marine Biology, **144**: 303-316.

OMORI, M., 1974. The biology of pelagic shrimps in the ocean. Advances in Marine Biology, **12**: 233-324.

ORLOV, A. I., 2004. Applied statistics. (Examine, Moscow). [In Russian.]

PEQUEGNAT, L. H. & M. K. WICKSTEN, 2006. Oplophorid shrimps (Decapoda: Caridea: Oplophoridae) in the Gulf of Mexico and Caribbean Sea from the collections of the research vessels Alaminos, Oregon and Oregon II. Crustacean Research, **35**: 92-107.

SUDNIK, S. A., 2007. Oogenesis of shrimp *Glyphus marsupialis* Filhol 1884 (Decapoda, Pasiphaeidae). In: V. I. SOCOLOV (ed.), Marine commercial invertebrates and algae (biology and fishery). VNIRO Proceedings, **147**: 237-253. (VNIRO, Moscow). [In Russian with English abstract.]

SUDNIK, S. A. & T. FALKENHAUG, 2014. The method of biological analysis for caridean shrimps (Decapoda: Natantia: Caridea) with emphasis on pelagic shrimps. Science and society in condition of globalization: H34 International scientific-practical conference. (Ufa, 21-22 April 2014): RIO ICIPT: 7-11.

First received 30 April 2016.
Final version accepted 14 November 2016.

[When citing this volume, please refer to Crustaceana 90 (2017) 771-1288]

Michael Türkay Memorial Issue

TWO NEW SPECIES OF THE GENUS *ACTUMNUS* DANA, 1851 (DECAPODA, BRACHYURA, PILUMNIDAE) FROM THE RYUKYU ISLANDS, SOUTHWEST JAPAN

BY

MASATSUNE TAKEDA[1]) and HIRONORI KOMATSU
Department of Zoology, National Museum of Nature and Science, 4-1-1 Amakubo, Tsukuba, Ibaraki 305-0005, Japan

ABSTRACT

Two new species of the crab genus *Actumnus* Dana, 1851 (family Pilumnidae) are reported from the Ryukyu Islands, southwest Japan. *Actumnus tsurukaii* sp. nov. is similar to *A. squamosus* (De Haan, 1835), *A. setifer* (De Haan, 1835) and *A. forficigerus* (Stimpson, 1858) in possessing a poorly defined dorsal surface of the carapace with a coat of short setae, but differs in having relatively larger, fewer and more widely spaced conical granules or tubercles on the outer surfaces of the palms of both chelipeds. *Actumnus uformis* sp. nov. is similar to *A. dorsipes* Stimpson, 1858 and *A. granotuberosus* Garth & Kim, 1983 in having a prominently areolated dorsal surface of the carapace with a soft thick tomentum, but can be distinguished by the stronger areolation and the distinctly U-shaped protogastric region.

Key words. — Crustacea, Pilumnoidea, new species, Okinawa-jima Island, taxonomy

ZUSAMMENFASSUNG

Hier werden zwei neue Arten der Krabbengattung *Actumnus* Dana, 1851 (Familie Pilumnidae) von den Ryuku-Inseln, Südwest-Japan gemeldet. *Actumnus tsurukaii* sp. nov. ähnelt *A. squamosus* (De Haan, 1835), *A. setifer* (De Haan, 1835) und *A. forficigerus* (Stimpson, 1858) durch die schwach definierte Dorsaloberfläche seines Carapax mit einer Schicht kurzer Setae, unterscheidet sich jedoch durch relativ größere, weiter voneinander getrennte, konische Körnchen oder Tuberkel in geringerer Anzahl auf den äußeren Oberflächen der Palpen beider Chelipeden. *Actumnus uformis* sp. nov. ähnelt *A. dorsipes* Stimpson, 1858 und *A. granotuberosus* Garth & Kim, 1983 durch die markant gefelderte Dorsaloberfläche des Carapax mit einem weichen, dicken Flaum, kann jedoch leicht von dieser durch die stärker ausgeprägten Felder und durch die deutlich U-förmige Protogastralregion unterschieden werden.

[1]) Corresponding author; e-mail: takeda@kahaku.go.jp

INTRODUCTION

In the Department of Zoology of the National Museum of Nature and Science, Tsukuba, Japan, many crab specimens from subtidal waters of Nakagusuku Bay at the southeast coast of Okinawa-jima Island in the central Ryukyu Islands have been deposited. Dr. N. Shikatani, then a postgraduate student of the Ocean Research Institute, University of Tokyo, Tokyo, Japan, collected these crabs during the mid-1980s. Takeda & Shikatani (1900) reported on *Calappa* Weber, 1795 (family Calappidae) from the collection, but the other specimens remain unidentified. Recently, the authors started studies on the remaining specimens, and in this report two new species of the genus *Actumnus* Dana, 1851, of the family Pilumnidae Samouelle, 1819, are described.

Most of the known *Actumnus* species are usually subtidal or shallow water inhabitants, differing from the species of the genus *Pilumnus* Leach, 1816, mainly in having a less transverse carapace, with a strongly convex dorsal surface which is deeply or moderately areolated and usually covered with pearly granules, a distinctly sunken posterolateral surface of the carapace, and variously armed palms of the chelipeds with relatively short fingers. The two new species described in this paper are clearly belonging to *Actumnus* as defined here.

The type specimens are preserved in the Department of Zoology, National Museum of Nature and Science, Tsukuba (NSMT), and the Ryukyu University Museum, Fujukan, Okinawa (RUMF). The width and length of the carapace are abbreviated as cw and cl, respectively, with measurements in mm.

TAXONOMY

Family PILUMNIDAE Samouelle, 1819
Genus *Actumnus* Dana, 1851
Actumnus tsurukaii sp. nov.
(figs. 1, 2, 3A-C)

Material examined.— Chinen, Nakagusuku Bay, southeast coast of Okinawa-jima I., from refuse caught in gill-net for fish: 1♂ (cw 11.5 × cl 8.7 mm), holotype, NSMT-Cr 24599, 09-v-1985; 1♂ (cw 10.0 × cl 7.8 mm), 1♀ (cw 9.7 × cl 7.0 mm), paratypes, RUMF-ZC-4319, 09-v-1985; 1 juv. (cw 5.5 × cl 4.2 mm), non-type, RUMF-ZC-4322, 30-v-1985; 1♂ (cw 11.0 × cl 7.8 mm), paratype, NSMT-Cr 24600, 17-vii-1985; 2♂♂ (cw 10.5 × cl 7.9 mm, cw 6.4 × cl 5.6 mm), paratypes, RUMF-ZC-4320, 06-viii-1986; 1 ♂ (cw 10.4 × cl 7.9 mm), paratype, RUMF-ZC-4321, vi-1986.

Description of holotype (male, NSMT-Cr 24599).— Carapace (figs. 1A, 2A) narrow, width to length ratio 1.32, moderately convex longitudinally, similar to *A. squamosus* (De Haan, 1835) (fig. 6A, B), *A. setifer* (De Haan, 1835) (fig. 6C, D) and *A. forficigerus* (Stimpson, 1858) (fig. 6G, H). Dorsal surface of carapace

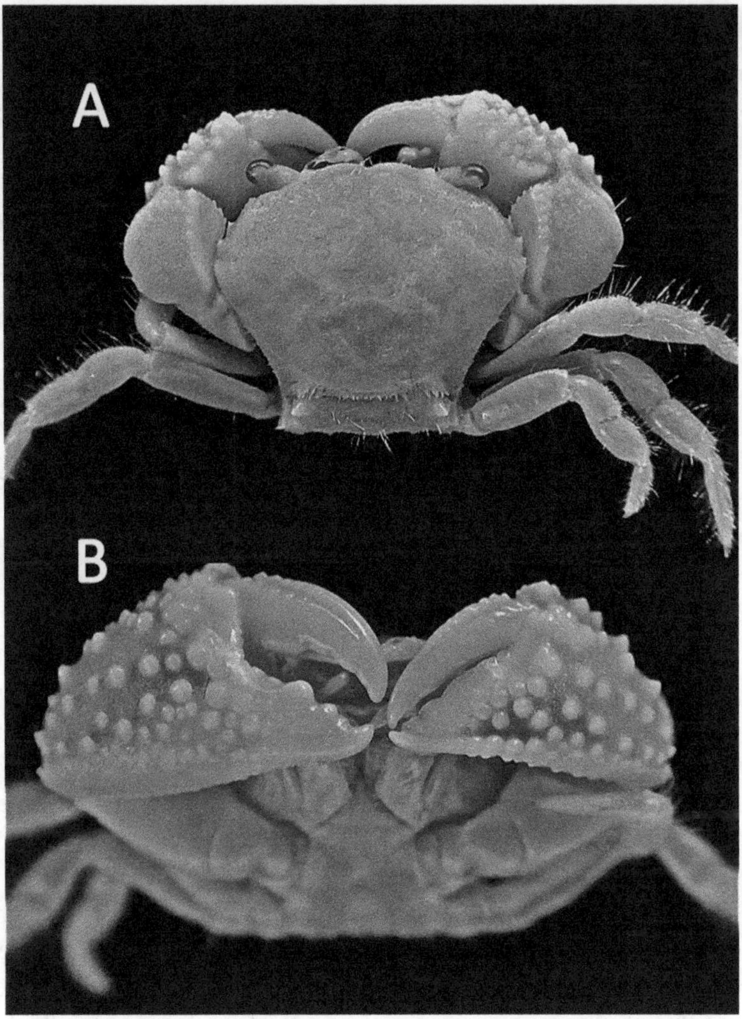

Fig. 1. *Actumnus tsurukaii* sp. nov., holotype, male (NSMT-Cr 24599; cw 11.5 × cl 8.7 mm). A, dorsal view; B, ventral view.

divided into regions separated by shallow furrows, covered with dense short setae and sparse long setae; protogastric, mesogastric, cardiac and branchial regions hardly distinguishable, without granules; posterolateral surface deeply sunken for reception of posterior ambulatory legs. Frontal margin (fig. 2A) directed downwards, subdivided into two convex lobes with a median shallow depression; no submarginal fringe of longish setae. Supraorbital margin shallow in dorsal view, merging laterally into carapace anterolateral margin by low external orbital angle.

Anterolateral margin (fig. 1A) with three subequal teeth; each tooth directed almost anteriorly rather than obliquely, sharp with a spinule at anterior end;

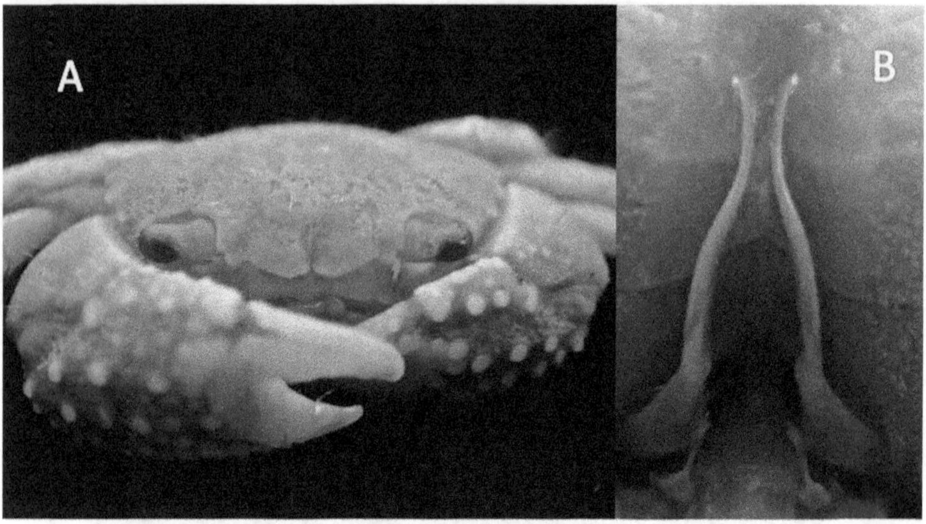

Fig. 2. *Actumnus tsurukaii* sp. nov., holotype, male (NSMT-Cr 24599; cw 11.5 × cl 8.7 mm). A, frontal view; B, first and second pleopods in ventral view.

distance between external orbital angle and tip of first tooth subequal to distance between tips of first and third (last) teeth; outer margins of anterior two teeth nearly straight; outer margin of third tooth weakly convex outwards, strongly convergent posteriorly as anterior part of carapace posterolateral margin, without discrimination from carapace posterolateral margin. Posterior margin of carapace hardly convex posteriorly, fringed with short setae, as wide as frontal margin between supraorbital angles of both sides.

Both chelipeds (figs. 1A, B, 2A) stout, unequal, right cheliped larger. Merus short, mostly hidden under carapace when viewed dorsally, with thickened upper margin. Carpus (fig. 1A) large, unarmed, covered with short and longish setae like on carapace dorsal surface; inner upper margin thickened, fringed with clusters of granules, unarmed at its distal end; inner surface truncated to fit with subhepatic surface together with inner surface of merus. Both palms (fig. 1B, 2A) different in size, similar in armature; outer surface with obtuse conical granules or tubercles, some of which are of relatively large size, interspaced with short stiff setae; granules well separated from each other, not adjacent; several granules on main part of outer surface roughly arranged into two longitudinal rows; lower margin of palm with fringe of close-set conical granules. Both of movable and immovable fingers in right (larger) chela markedly stout (fig. 1B); cutting edge of movable finger with a low molariform tooth basally, occupying one third length of cutting edge; immovable finger irregularly toothed on cutting edge, with prominent conical median tooth; movable finger stout, with curved distal half; tip of immovable finger curved towards tip of movable finger, leaving narrow gape between both fingers

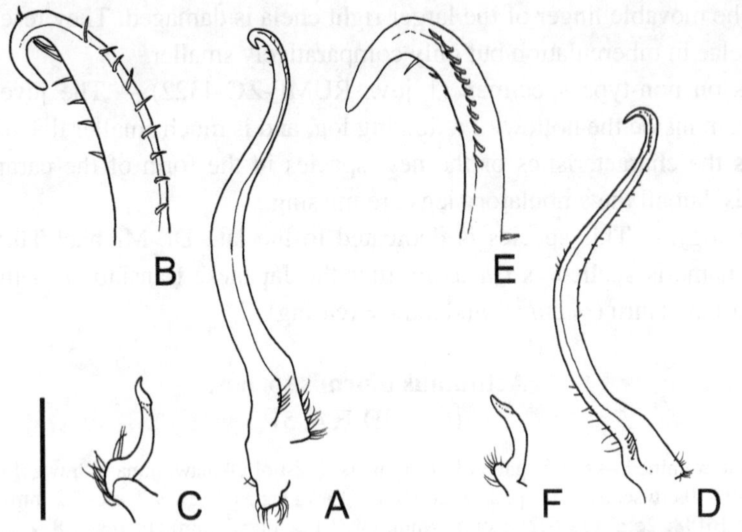

Fig. 3. Male pleopods of *Actumnus tsurukaii* sp. nov., holotype, male (NSMT-Cr 24599; cw 11.5 × cl 8.7 mm) (A-C), and *A. uformis* sp. nov., holotype, male (NSMT-Cr 16198; cw 9.7 × cl 7.2 mm) (D-F). A, D, right first pleopod in mesial view; B, E, distal part of the same; C, F, right second pleopod in mesial view. Scale = 1 mm for A, C, D, F; 0.25 mm for B, E.

when closed; left (smaller) chela similar to right chela, fingers relatively more slender; immovable finger with three or four lobate teeth on cutting edge; movable finger of left chela not curved, no teeth on cutting edge, leaving no space against immovable finger when closed.

Ambulatory legs (fig. 1A) comparatively slender, with stout dactyli, fringed with sparse longish setae of variable lengths.

Abdomen six-segmented with telson, covered with dense short setae and sparse long setae, like carapace dorsal surface; telson 1.3 times as long as sixth somite measured along median line, with sharply pointed terminal margin.

Shaft of first pleopod (figs. 2B, 3A, B) slender, convex ventrally along median one-third, with small beak-like structure strongly curved ventrally. Second pleopod (figs. 2B, 3C) very short, with weakly hooked tip.

Paratypes (1♂, NSMT-Cr 24600; 1♂, 1♀, RUMF-ZC-4319; 2♂♂, RUMF-ZC-4320).— Four male paratypes agree completely with the holotype male in the size, setation and poor areolation of the dorsal surface of the carapace, possession of three teeth on the anterolateral margin, and the dispersed conical granules or tubercles interspaced with dark brown tomentum on the palms of both chelipeds. In some specimens, the chelipeds and some ambulatory legs are detached or missing, but all the specimens possess the diagnostic features of the new species. The paratype female agrees with the males (holotype and four paratypes) except for sexual features and the cheliped size. Both chelipeds are detached, and the distal half of the movable finger of the larger right chela is damaged. They resemble the male chelae in tuberculation but only comparatively smaller.

Notes on non-type specimen (1 juv., RUMF-ZC-4322).— The juvenile was found from inside the hollow of a floating log, and is much smaller than the types. It shows the characteristics of the new species in the form of the carapace and chelipeds, but all the ambulatory legs are missing.

Etymology.— This species is dedicated to the late Dr. Michael Türkay. The specific name is spelled as *tsurukaii* after the Japanese translation of the family name Türkay: tsuru (= crane) and kai (= rearing).

Actumnus uformis sp. nov.
(figs. 3D-F, 4, 5)

Material examined.— Off Nagannu I., Kerama Is., west of Okinawa-jima I., trawl, TV *Toyoshio Maru* TY-04-10 cruise, 53 m deep, 22-v-2004, M. Osawa leg.: 1♂ (cw 9.7 × cl 7.2 mm), holotype, NSMT-Cr 16198; 2♂♂ (cw 7.2 × cl 5.5 mm, cw 7.1 × cl 5.3 mm), 1♀ (cw 7.8 × cl 5.6 mm), paratypes, NSMT-Cr 24601; 1♀ (cw 7.4 × cl 5.4 mm), with isopod parasite in left gill-chamber, 2 juv. (cw 5.2 × cl 3.5 mm, cw 4.5 × cl 3.0 mm), non-types, NSMT-Cr 24602. Chinen, Nakagusuku Bay, southeast coast of Okinawa-jima Island, in refuse caught with gill-net for fish: 1♀ (cw 13.2 × cl 8.3 mm), paratype, URMF-ZC-4324, 24-ii-1986; 2♀♀ (cw 12.1 × cl 8.5 mm, cw 9.7 × cl 7.2 mm), paratypes, URMF-ZC-4325, 06-viii-1986.

Description of holotype (male, NSMT-Cr 16198).— Carapace (fig. 4A) narrow, width to length ratio 1.35, convex in both directions, wholly covered with soft thick tomentum; dorsal surface distinctly divided into epigastric, protogastric, mesogastric, cardiac, hepatic and branchial regions, separated by deep furrows; epigastric region narrowly ovate, anteriorly convex in dorsal view; protogastric region (figs. 4A, 5A) prominent, occupying a quarter width of carapace, subdivided for its most part by wide median furrow or depression, forming distinct U-shape; on denudation, main surface of protogastric region shallowly sculptured to form U-shape, with sparse microscopic granules on anterolateral part; posterior part of mesogastric region subdivided into two by longitudinal median furrow; anterior prolongation of mesogastric region undivided, reaching to anterior margin of protogastric region; cardiac region prominent, transversely quadrate, not convex dorsally; hepatic region prominent, convex, clearly isolated from outer half of supraorbital margin; main surface of hepatic region with microscopic granules on anterior marginal part; branchial region inside of last anterolateral tooth of carapace oblique, mostly subdivided into two by oblique furrow; main surfaces of branchial regions with microscopic granules on anterior marginal parts; posterodorsal surface concave.

Frontal region (figs. 4A, 5A) in front of epigastric region flattened, moderately deflexed obliquely downwards, with nearly truncated free margin; median interruption deep, slit-like through tomentum, narrowly U-shaped after denudation. Orbital

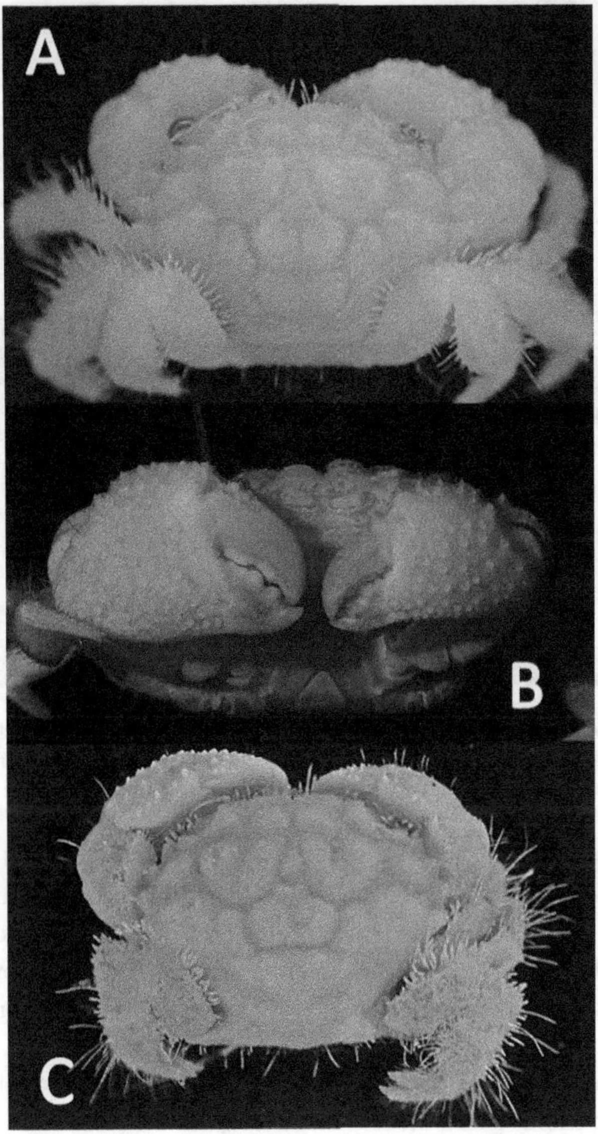

Fig. 4. *Actumnus uformis* sp. nov. A, holotype, male (NSMT-Cr 16198; cw 9.7 × cl 7.2 mm) in dorsal view; B, chelae of holotype in outer view; C, paratype, female (URMF-ZC-4325; cw 9.7 × cl 7.2 mm) in dorsal view.

region narrow, weakly raised along supraorbital margin. Orbit deep, wholly occupied by stout eyestalk with large cornea; external orbital angle formed on outer notch of supraorbital margin, directed forwards, tip sharp, with straight outer margin.

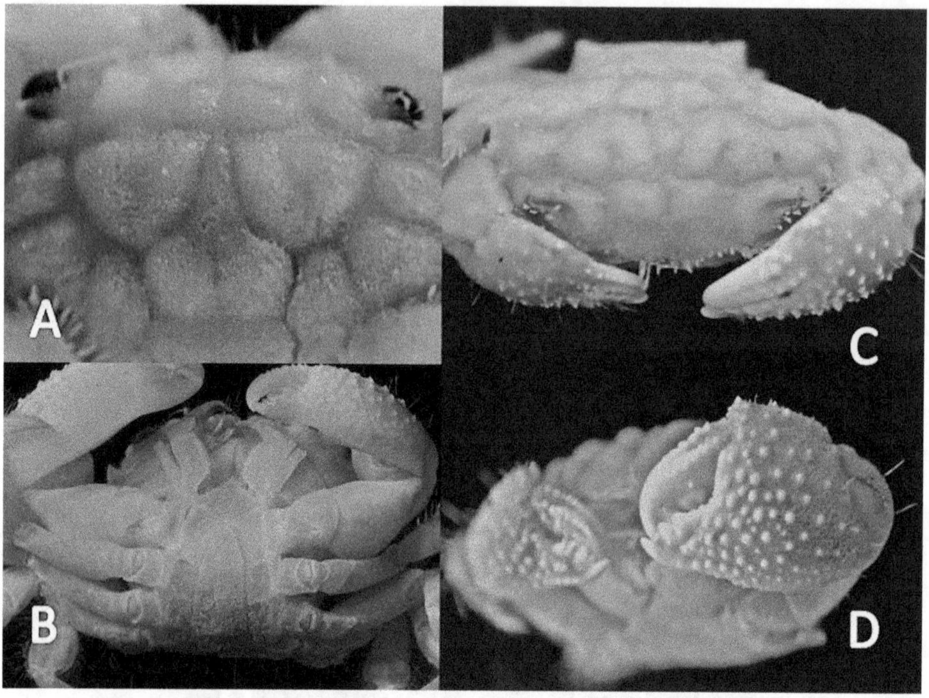

Fig. 5. *Actumnus uformis* sp. nov., holotype, male (NSMT-Cr 16198; cw 9.7 × cl 7.2 mm) (A, B), and paratype female (URMF-ZC-4325; cw 9.7 × cl 7.2 mm) (C, D). A, anterior half of carapace dorsal surface in dorsal view; B, ventral view; C, antero-dorsal view; D, chelae in outer view.

Anterolateral margin (fig. 4A) with three teeth behind external orbital angle; each tooth tipped with spinule, become smaller and sharper posteriorly; first tooth flattened, triangular in dorsal view, second tooth narrower, directed obliquely outwards, third tooth more strongly directed obliquely outwards; outer margin of last tooth not separated from posterolateral margin of carapace, directed towards posterolateral corner of carapace.

Both chelipeds (fig. 4B) unequal, stout, covered with thick tomentum, right cheliped larger. Merus short, hidden under carapace on dorsal view; carpus large, with furrow along distal margin lined with granules. Main surface of carpus (fig. 4A) covered with sparse conical granules, inner margin with conical or slightly sharper tubercle close to distal end. Both palms (fig. 4B) different in size but similar in setation and granulation; outer surface covered with very short setae mixed with sparse long setae; size of conical granules fairly constant; granulation relatively dense, granules well-spaced, not adjacent to each other; granules fewer on outer upper surface than on outer lower surface. Fingers (fig. 4B) stout, sharply toothed on cutting edges in both chelae; outer surface of immovable finger covered with conical granules in smaller chela, but nearly smooth in larger chela; movable finger with conical granules at basal part of longitudinal ridge of smaller chela, worn out in larger chela.

Ambulatory legs (fig. 4A) stout, covered with soft tomentum, fringed with short setae or setae of variable lengths.

Abdomen (fig. 5B) six-segmented with telson, covered with soft thick tomentum and sparse long setae; telson 1.3 times as long as sixth somite, with sharply pointed terminal margin.

First pleopod (fig. 3D, E) sinuous, with shaft moderately curved along subdistal part. Second pleopod (fig. 3F) short, stout, subacute at tip.

Paratypes (2♂♂, 1♀, NSMT-Cr 24601; 3♀♀, URMF-ZC-4324, 4325).— Two males and four females are designated as the paratypes. Two males and one female in NSMT are comparatively small but still show the diagnostic characters of the dorsal areolation of the carapace. In the smaller male, both chelipeds are detached and some ambulatory legs are missing. The female (URMF-ZC-4324) is the largest of all the type specimens. It agrees well with the holotype male in all the main characteristics, although each region of the dorsal surface of the carapace is more convex dorsally and distinctly isolated from the adjacent regions, and the U-shaped protogastric region, the most diagnostic character of this species, is more pronounced, and is directed somewhat obliquely towards the posterolateral margins. Both chelipeds are distinctly different in size, with stronger heterochely than in the holotype. Both palms are inflated, the right being larger, but the density and size of the tubercles on the outer surface of the palm of the cheliped are similar to those of the holotype. One of the paratype females (URMS-ZC-4325) is of good size and similar to the female discussed above in size and areolation of the dorsal surface of the carapace. Unfortunately, all the left ambulatory legs are missing. Another female with the same registered number is figured here (figs. 4C, 5C, D).

Notes on non-type specimens (1♀, 2 juv., NSMT-Cr 24602).— The female specimen (cw 7.4 × cl 5.4 mm) is parasitized probably by an isopod crustacean, with the left half of the carapace deformed, and in the right half, the deformity is not as prominent as on the left. Even though the two juvenile specimens are much smaller than the paratypes, the characteristic dorsal areolation of the carapace is still obvious. Unfortunately, most of their chelipeds and ambulatory legs are detached and impossible to separate out. As such, these specimens are excluded from the type series.

Etymology.— The specific name, *uformis*, alludes to the shape of the protogastric region of the carapace, the most obvious character which distinguishes the new species from the congeneric species.

DISCUSSION

A total of 27 Indo-West Pacific species were listed in the genus *Actumnus* Dana, 1851 by Ng et al. (2008), with the type species, *A. tomentosus* Dana, 1852, generally considered to be synonymous with *A. setifer* (De Haan, 1835). Most of the known species fit the generic definition relatively well, although Sakai (1983) stated that *Platypodia digitalis* Rathbun, 1907 might belong to a new distinct genus intermediate between *Platypodia* Bell, 1835, and *Actumnus*. Takeda & Miyake (1969) followed Bouvier (1915) and recorded the species from the Ryukyu Islands as a species of *Actumnus*, noting that *Platypodia digitalis* is without doubt a pilumnid. Its generic status, however, needs to be re-examined. The precise identities of several poorly known species will also need to be established after fresh revisions in the future: *A. obesus* Dana, 1852, *A. globulus* Heller, 1861, *A. elegans* De Man, 1888, *A. targionii* Cano, 1889, *A. amirantensis* Rathbun, 1911, *A. davaoensis* Ward, 1941, and *A. antelmei* Ward, 1942. The genus *Neoactumnus* Sakai, 1965, is also not clearly distinguished from *Actumnus*. According to the original description, *Neoactumnus* can be separated from *Actumnus* only by differences in the frontorbital margin. The absence of a notch of the frontal margin and presence of an interruption on the supraorbital margin, however, are doubtful generic characters. Ho et al. (2001) and Naderloo & Ng (2011) discussed this problem at length. Clearly, much work still needs to be done on the various taxonomic problems associated with *Actumnus*. The present two new species from the Ryukyu Islands are nevertheless clearly belonging to *Actumnus* s. str., affiliated with several well-known species, and the present paper will focus on their taxonomy.

The general appearance of the carapace, chelipeds and ambulatory legs of *A. tsurukaii* sp. nov. most closely allies it with *A. squamosus* (De Haan, 1835) (cf. Sakai, 1976). In both species, the carapace (figs. 1A, 2A, 6A) is moderately convex in both directions, with a thin coat of short stiff setae and sparse long soft setae. The dorsal surface of the carapace is divided into more flattened regions by shallow furrows in both species, with those on *A. squamosus* slightly wider and deeper than in *A. tsurukaii*. In both species, the anterolateral margin of the carapace is armed with three anteriorly directed teeth, the posterolateral margin is concave laterally and dorsally near the posterior end. The new species is distinctly different from *A. squamosus* in the tuberculation of the palms of both chelipeds. In both chelipeds of *A. squamosus*, the outer surfaces of the palm and immovable finger are densely covered with closely packed conical granules or tubercles (fig. 6B). In contrast, the outer surfaces of both palms and immovable fingers of *A. tsurukaii* are more sparsely covered with conical granules or tubercles of distinctly larger sizes and are more widely spaced (fig. 1B). *Actumnus squamosus* has two junior synonyms:

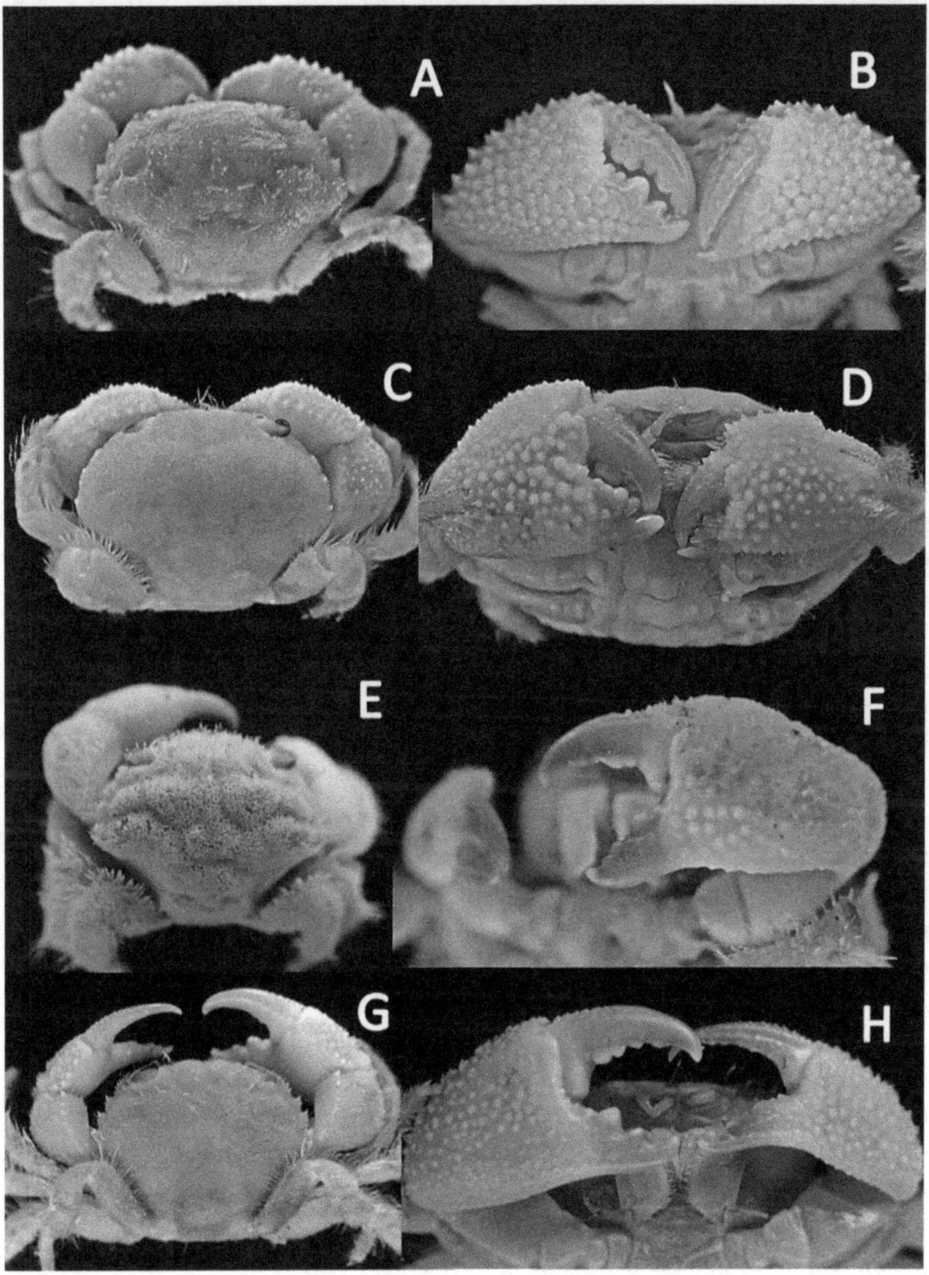

Fig. 6. A, B, *Actumnus squamosus* (De Haan), male (NSMT—Cr 8358; cw 12.8 × cl 10.0 mm) from the Ariake Sea, west of Kyushu; C, D, *A. setifer* (De Haan), female (URMF-ZC, unregistered; cw 17.3 × cl 12.5 mm) from Nakagusuku Bay, Okinawa-jima I., Ryukyu Is.; E, F, *A. dorsipes* Stimpson, male (NSMT-Cr 6383; cw 11.7 × cl 8.5 mm) from Kushimoto, Kii Penin.; G, H, *A. forficigerus* Stimpson, male (NSMT-Cr 13891; cw 12.0 × cl 9.0 mm) from Ariake Sea, west of Kyushu.

Pilumnus lapillimanus Stimpson, 1858, and *P. dehaanii* Miers, 1879 (cf. Ng et al., 2008). *Pilumnus lapillimanus* has the palm of the cheliped (Stimpson, 1907: pl. 8 fig. 7a) closely resembling what is here described for *A. squamosus* and thus quite different from *A. tsurukaii*. *Pilumnus dehaani* is poorly known and no figures were provided. Miers (1879: 32), however, described the outer surface of the palm possessing "numerous, crowded, unequal, conical, acute tubercles" which suggests an arrangement of tubercles similar to that of what is here figured for *A. squamosus*, i.e., packed closely together (fig. 6B).

Actumnus tsurukaii is superficially similar to *A. setifer* (De Haan, 1835) (fig. 6C). In *A. setifer*, however, the dorsal surface of the carapace is more strongly convex fore and aft, and is densely covered with a coat of short stiff setae, with the interregional furrows relatively narrower. In both species, the anterolateral margin of the carapace is armed with spine-tipped teeth as usual, but the tips do not protrude from the general contour of the carapace in *A. setifer* (fig. 6C). The outer surfaces of both palms of *A. setifer* are covered with conical or subconical granules of variable sizes (fig. 6D), but all the granules are relatively smaller and more densely and closely packed than those of *A. tsurukaii* (fig. 1B). The ambulatory legs are stout and fringed with long setae in *A. setifer* (fig. 6C), but rather slenderer and possess only sparse setae in *A. tsurukaii* (fig. 1A). *Actumnus setifer* has one junior synonym, *A. tomentosus* Dana, 1852 (cf. Ng et al., 2008), but this species is almost identical (Dana, 1855: pl. 14 fig. 2) to what is here figured for *A. setifer*, and as such, all the differences discussed above for *A. tsurukaii* also apply for it.

Actumnus tsurukaii can also be compared with *A. forficigerus* (Stimpson, 1858), which occurs in Japanese waters from Sagami Bay to the East China Sea, with a bathymetric range from 35 to 200 m (Takeda & Miyake, 1969). *Actumnus forficigerus* (fig. 6G) is more similar to *A. squamosus* (fig. 6A) than *A. setifer* (fig. 6C) in the general appearance of the carapace. Comparing *A. forficigerus* with *A. setifer*, the dorsal surface of the carapace of *A. forficigerus* is relatively less convex with the regions poorly defined and covered with velvety tomentum mixed with long silky setae (fig. 6C, G). In both *A. setifer* (fig. 6D) and *A. forficigerus* (fig. 6H), the outer surface of the palm is covered with relatively smaller granules or tubercles which are more closely packed than in *A. tsurukaii*. Furthermore, as mentioned by Takeda & Miyake (1969) and also confirmed in the present study, the larger granules on the outer surface of the palm are each tipped with a strongly procurved, horny spinule in *A. forficigerus*, a diagnostic feature of this species.

The second new species, *Actumnus uformis* sp. nov., is very distinct and readily distinguished from all the congeners by having a U-shaped protogastric region of the carapace. The general shape of the deeply areolated dorsal surface of the carapace and the thick coat of soft tomentum allies *A. uformis* with *A. dorsipes* Stimpson, 1858 (fig. 6E, F), which is known only from Japan, Guandong (China) and Hong Kong. The records of *A. dorsipes* are relatively few, but the species has been well characterized by Stimpson (1858, 1907), Sakai (1939, 1965, 1976), and Dai & Yang (1991), with the male first pleopod clearly figured (Takeda & Miyake, 1969; Dai & Yang, 1991). *Actumnus dorsipes* is characterized by the dorsal surface of the carapace being distinctly divided into regions and covered only with soft tomentum (fig. 6E). It is different from *A. uformis* in having the undivided protogastric region entire (fig. 6E) rather than U-shaped (fig. 4A, C).

Another species superficially resembling *A. uformis* is *A. granotuberosus* Garth & Kim, 1983, known only by a male from the Sulu Archipelago, the Philippines, from a depth of 45 m. In the original description (Garth & Kim, 1983), the carapace was described as having a very dense, short and smooth tomentum, and the species was compared with *A. setifer* and *A. dorsipes*. The most important criterion to distinguish *A. granotuberosus* from the congeners is the tuberculation of the cheliped palm – the tubercles are usually surrounded by smaller granules (Garth & Kim, 1983: fig. 7c). In *A. uformis*, the palms are armed only with simple conical granules or tubercles (fig. 5D). The general shape of the carapace of *A. uformis* resembles *A. granotuberosus* figured by Garth & Kim (1983: fig. 7a), but the protogastric region is distinctly U-shaped.

ACKNOWLEDGEMENTS

This study is based on the specimens collected by Dr. Norikazu Shikatani (Shikatani Nature Guide) at Nakagusuku Bay, Okinawa-jima Island and Dr. Masayuki Osawa (Shimane University) on board the TR/V "Toyoshio Maru" of Hiroshima University, to whom the authors' cordial thanks are tendered. Dr. Tohru Naruse (University of the Ryukyus) arranged the registration of the type specimens in the University Museum, Fujukan. We are much indebted to Dr. Peter K. L. Ng (National University of Singapore) for his advice on the taxonomic problems of the genus *Actumnus* and help with the manuscript.

REFERENCES

BELL, T., 1835. Observations on the genus *Cancer* of Dr. Leach (*Platycarcinus*, Latr.), with descriptions of three new species. Transactions of the Zoological Society of London, **1**: 335-342, pls. 42-46.

BOUVIER, E. J., 1915. Décapodes marcheurs (Reptantia) et stomatopodes recueillis à l'île Maurice par M. Paul Carié. Bulletin Scientifique de la France et de la Belgique fondé par Alfred Giuard, **48**: 178-318, pls. 4-7.

CANO, G., 1889. Crostacei Brachyiuri ed Anomuri raccolti nel viaggio della « Vettor Pisani », intorno al glabo. Bollettino della Società de Naturalisti in Napoli, (1) **3**: 79-105.

DAI, A. & S. YANG, 1991. Crabs of the China Seas. (China Ocean Press, Beijing).

DANA, J. D., 1851. On the classification of the Cancroidea. American Journal of Sciences and Arts, **12**: 121-131.

DANA, J. D., 1852. Crustacea. United States Exploring Expedition during the years 1838, 1839, 1840, 1841, 1842 under the command of Charles Wilkes, U.S.N., 13: 1-685.

DANA, J. D., 1855. Atlas. Crustacea. United States Exploring Expedition during the years 1838, 1839, 1840, 1841, 1842 under the command of Charles Wilkes, U.S.N., 14: 1-27, pls. 1-96.

DE HAAN, W., 1833-1850. Crustacea. In: P. F. VON SIEBOLD (ed.), Fauna Japonica sive Descriptio Animalum, quae in Itinere per Japoniam, Jussu et Auspiciis Superiorum, qui Summum in India Batava Imperium Tenent, Suscepto, Annis 1823-1830 Collegit, Notis, Observationibus et Adumbrationibus Illustravit: i-xvii, i-xxxi, ix-xvi, 1-243, pls. A-J, L-Q, 1-55, tabs. 1, 2. (Leiden).

DE MAN, J. G., 1887-1888. Report on the podophthalmous Crustacea of the Mergui Archipelago, collected for the trustees of the Indian Museum, Calcutta, by Dr. John Anderson, F.R.S., Superintendent of the Museum. Journal of the Linnean Society, Zoology, **22**: 1-312, pls. 1-19.

GARTH, J. S. & H.-S. KIM, 1983. Crabs of the family Xanthidae (Crustacea: Brachyura) from the Philippine Islands and adjacent waters based chiefly on collections of the U.S. Fish Commission steamer Albatross in 1908-1909. Journal of Natural History, **17**: 663-729.

HELLER, C., 1861. Synopsis der im rothen Meere vorkommenden Crustaceen. Verhandlungen der kaiserlich-königlichen Zoologisch-Botanischen Gesellschaft in Wien, **11**: 3-32.

HO, H.-P., H.-P. YU & P. K. L. NG, 2001. Indo-Pacific Pilumnidae XIV. On a new species of *Actumnus* (Crustacea: Decapoda: Brachyura) from Taiwan. Proceedings of the Biological Society of Washington, **114**: 145-151.

LEACH, W. E., 1816. A tabular view of the external characters of four classes of animals, which Linné arranged under Insecta; with the distribution of the genera composing three of these classes into orders &c. and descriptions of several new genera and species. Transactions of the Linnean Society of London, **11**: 306-400.

MIERS, E. J., 1879. On a collection of Crustacea made by Capt. H. C. St. John, R.N., in the Corean and Japanese Seas. Part I. Podophthalmia. With an appendix by Capt. H. C. St. John. Proceedings of the Zoological Society of London, **1879**: 18-61, pls. 1-3.

NADERLOO, R. & P. K. L. NG, 2011. Rediscovery of the rare pilumnid crab, *Actumnus simplex* Rathbun, 1911 (Decapoda, Brachyura) and a new record for the Persian Gulf. Crustaceana, **84**: 1593-1604.

NG, P. K. L., D. GUINOT & P. J. F. DAVIE, 2008. Systema Brachyurorum: Part I. An annotated checklist of extant brachyuran crabs of the world. The Raffles Bulletin of Zoology, Supplement, **17**: 1-286.

RATHBUN, M. J., 1907. The Brachyura. Reports on the scientific results of the expedition to the tropical Pacific, in charge of Alexander Agassiz, by the U.S. Fish Commission steamer "Albatross," from August, 1899, to March, 1900, Commander Jefferson F. Moser, U.S.N., commanding. IX. Reports on the scientific results of the expedition to the eastern tropical Pacific, in charge of Alexander Agassiz, by the U.S. Fish Commission steamer "Albatross," from October, 1904, to March, 1905, Lieut.-Commander L. M. Garrett, U.S.N., commanding. X. Memoirs of the Museum of Comparative Zoology at Harvard College, **35**: 23-74, pls. 1-9.

RATHBUN, M. J., 1911. Marine Brachyura. The Percy Sladen Trust Expedition to the Indian Ocean in 1905, under the leadership of Mr. J. Stanley Gardiner, 3 (11). Transactions of Linnean Society of London, **14**: 191-261, pls. 15-20.

SAKAI, T., 1939. Studies on the crabs of Japan. IV. Brachygnatha, Brachyrhyncha: 365-741, pls. 42-111. (Yokendo, Tokyo).

SAKAI, T., 1965a. On two new genera and five new species of xanthoid crabs from the collection of His Majesty the Emperor of Japan made in Sagami Bay. Crustaceana, **8**: 97-106.

SAKAI, T., 1965b. The crabs of Sagami Bay collected by His Majesty the Emperor of Japan: xvi + 206 + 92 + 32, pls. 100. (Maruzen, Tokyo).

SAKAI, T., 1976. Crabs of Japan and the Adjacent Seas: xxix, 773 (English), pp. 461 (Japanese), pp. 16, pls. 251 (Atlas). (Kodansha, Tokyo).

SAKAI, T., 1983. Description of new genera and species of Japanese crabs, together with systematically and biogeographically interesting species. (I). Researches on Crustacea, **12**: 1-44, pls. 1-8, frontispiece. [In English and Japanese.]

SAMOUELLE, G., 1819. The entomologist's useful compendium, or an introduction to the knowledge of the British insects: 1-496. (Thomas Boys, London).

STIMPSON, W., 1858. Prodromus descriptionis animalium evertebratorum, quae in Expeditione ad Oceanum Pacificum Septentrionalem, a Republia Federata missa, Cadwaladaro Ringgold et Johanne Rodgers ducibus, observavit et descripsit W. Stimpson. Pars IV. Crustacea Cancroidea et Corystoidea. Proceedings of the Academy of Natural Sciences of Philadelphia, **10**: 31-40.

STIMPSON, W., 1907. Report on the Crustacea (Brachyura and Anomura) collected by the North Pacific Exploring Expedition 1853-1856. Smithsonian Miscellaneous Collections, **49**: 1-240, pls. 1-26.

TAKEDA, M. & S. MIYAKE, 1969. Pilumnid crabs of the family Xanthidae from the west Pacific. II. Twenty-one species of four genera with descriptions of four new species. Ohmu, **2**: 93-156.

TAKEDA, M. & N. SHIKATANI, 1900. Crabs of the genus *Calappa* from the Ryukyu Islands, with description of a new species. Zoological Science, Japan, **7**: 477-484.

WARD, M., 1941. New Brachyura from the Gulf of Davao, Mindanao, Philippine Islands. American Museum Novitates, **1104**: 1-15.

WARD, M., 1942. A new genus and eight new species of Brachyura from Mauritius and the Chagos Archipelago. Mauritius Institute Bulletin, **2**: 39-48, pls. 2-4.

First received 27 August 2016.
Final version accepted 26 April 2017.

[When citing this volume, please refer to Crustaceana 90 (2017) 771-1288]

Michael Türkay Memorial Issue

A NEW SPECIES OF THE GENUS *LEPTOCHELA* (DECAPODA, CARIDEA, PASIPHAEIDAE) FROM THE YELLOW SEA

BY

YAQIN WANG[1,2,3]), ZHIBIN GAN[1,3,4]) and XINZHENG LI[1,3,4])

[1]) Institute of Oceanology, Chinese Academy of Sciences, 7 Nanhai Road, Qingdao 266071, P.R. China

[2]) University of Chinese Academy of Sciences, Beijing 100049, P.R. China

[3]) Laboratory for Marine Biology and Biotechnology, Qingdao National Laboratory for Marine Science and Technology, 1 Wenhai Road, Qingdao 266071, P.R. China

ABSTRACT

Based on samples collected from the Yellow Sea in October 2014, a new species of pasiphaeid shrimp, *Leptochela* (*Leptochela*) *tuerkayi* sp. nov., is described. The new species is closely related to *L.* (*L.*) *gracilis* Stimpson, 1860 in morphology but differs from the latter and other congeners by the following combination of characters: three longitudinal dorsal carinae in ovigerous females, third maxilliped with the rather short exopod slightly overreaching the distal margin of the antepenultimate segment and the fourth pereiopod with the exopod reaching beyond the end of the merus. Morphological data and a partial sequence of the 16S rRNA gene of the new species are provided to show its systematic position.

Key words. — Crustacea, Pasiphaeidae, *Leptochela*, new species, Yellow Sea

ZUSAMMENFASSUNG

Basierend auf Proben, welche im Oktober 2014 Gelben Meer genommen wurden, wird hier eine neue Art einer pasiphaeiden Garnele, *Leptochela* (*Leptochela*) *tuerkayi* sp. nov., beschrieben. Die neue Art ist morphologisch nahe verwandt mit *L.* (*L.*) *gracilis* Stimpson, 1860, unterscheidet sich jedoch von letzterer und anderen Verwandten durch die folgende Merkmalskombination: drei longitudinale, dorsale Carinae bei eiertragenden Weibchen; dem dritten Maxillipeden mit eher kurzem Exopoden, welcher den Distalrand des drittletzten Gliedes des vierten Pereiopoden leicht überragt und dem Exopoden des vierten Pereiopoden, welcher das Merusende überragt. Morphologische Daten und eine partielle Sequenz des 16S rRNS-Gens der neuen Art werden zur Verfügung gestellt, um deren systematische Position zu zeigen.

[4]) Corresponding authors; e-mails: ganzhibin2005@163.com; lixzh@qdio.ac.cn

INTRODUCTION

Leptochela was established by Stimpson (1860) based on the species *Leptochela gracilis* Stimpson, 1860. In the recent Carideorum Catalogus of De Grave & Fransen (2011), *Leptochela* was listed under the family Pasiphaeidae Dana, 1852 together with *Pasiphaea* Savigny, 1816 and six other genera. To date, 15 species have been described in the genus *Leptochela*. The shrimps of this genus are small in body size with the postorbital carapace length of adult specimens ranging from less than 3 mm to nearly 10 mm. They are often abundant in certain areas of the Indo-West-Pacific region (Chace, 1976). *Leptochela* (*Leptochela*) *gracilis* Stimpson, 1860 is a common pasiphaeid species in Chinese waters and was recorded to be collected in the surface layers of the Yellow Sea during night (Li & Huang, 1964). It often occurs together with the palaemonid species *Palaemon gravieri* Yu, 1930 and the penaeid species *Acetes chinensis* Hansen, 1919 during the spring and summer along the Chinese coast of the East China Sea (Dong, 1991).

When we sorted the samples from the Yellow Sea collected in 2014, some specimens were found to belong to *Leptochela*, representing an unknown species, that is herein described as a new species. Its systematic position is discussed with morphological characters and 16S rRNA gene sequence.

MATERIAL AND METHODS

Sample collection.— The material examined in the present study was collected by trawling from the Yellow Sea on 19 October 2014 from a sandy bottom at the depths of 71 and 21.5 m. The tissue samples for DNA extraction were collected from the pleopods of specimens directly on board and preserved in 75% ethanol. All the specimens are deposited in the Marine Biological Museum of the Chinese Academy of Sciences (MBMCAS) in the Institute of Oceanology, Chinese Academy of Sciences, Qingdao, P.R. China (IOCAS).

Molecular analysis.— In this study, four species of the genus *Leptochela*, plus two outgroup species of *Pasiphaea*, i.e., *P. merriami* Schmitt, 1931 and *P. telacantha* Hayashi, 2004 were included (table I). A pleopod of the specimens was used for DNA extraction using the Tissue DNA Kit (Omega, Qingdao, P.R. China) according to the manufacturer's instructions. Finally, the DNA was eluted in 100 μl of sterile distilled H_2O to increase the concentration. The primer pair 5′-CGC CTGTTTATCAAAAACAT-3′ (forward)/5′-AGATAGAAACCAACCTGG-3′ (reverse) (Crandall & Fitzpatrick, 1996) was used to amplify a 500-600 bp segment of the 16S rRNA gene by the following PCR protocol: 5 min at 94°C for initial denaturing followed by 35 cycles of 30 s at 94°C, 30 s at 50°C, 50 s at 72°C, and a final

TABLE I

Details of the specimens of the genera *Leptochela* Stimpson, 1860 and *Pasiphaea* Savigny, 1816 (Pasiphaeidae Dana, 1852) used in this study, and respective GenBank accession numbers

Species	Voucher number	GenBank accession No.
Leptochela carinata Ortmann, 1893	ULLZ 7232	EU868692
Leptochela bermudensis Gurney, 1939	ULLZ 7888	EU868691
Leptochela gracilis Stimpson, 1860	MBM189217	KX395710
Leptochela gracilis Stimpson, 1860	MBM189218	KX395711
Leptochela gracilis Stimpson, 1860	MBM284967	KX395712
Leptochela gracilis Stimpson, 1860	MBM284991	KX395713
Leptochela tuerkayi sp. n.	MBM189216	KX395714
Leptochela tuerkayi sp. n.	MBM189215	KX395715
Pasiphaea merriami Schmitt, 1931	ULLZ 6703	EU868700
Pasiphaea telacantha Hayashi, 2004	MNHN-IU-2008-14048	KP725635

extension for 5 min at 72°C. The total reaction volume was 50 μl, containing 25 μl 2× Ex Taq MasterMix (CWBIO, Beijing, P.R. China), 18 μl free ddH$_2$O, 1 μl of each primer (10 μM) and 5 μl DNA elution. The PCR products were sequenced on an ABI 3730xl Analyzer (Applied Biosystems, Foster City, CA, U.S.A.) using the same forward and reverse primers. The details of specimens and GenBank accession numbers used in this study are listed in table I.

Data analysis.— The obtained sequences were edited in Contig Express system and incorporated in Bioedit version 5.09 (Hall, 1999). Sequence alignments were performed using MUSCLE 3.8 (Edgar, 2004). The base composition, parsimony-informative site, transition/transversion bias and sequence genetic distances were analysed with MEGA 5.05 (Tamura et al., 2011). The Maximum Likelihood (ML) phylogenetic tree was also constructed using MEGA 5.05 based on the Tamura-Nei model.

SYSTEMATICS

Family PASIPHAEIDAE Dana, 1852
Genus *Leptochela* Stimpson, 1860

Leptochela Stimpson, 1860: 42; Chace, 1976: 2.

Diagnosis.— Carapace and rostrum unarmed dorsally. Rostrum moderately small, sharp tooth, dorsal margin straight or sinuous. Sixth abdominal somite with transverse carinate ridge near anterior end of dorsal surface, and pair of curved and distinct teeth on posterior margin. Telson with two or three pairs of dorsolateral spines, one or two posterior, anterior pair mesially, posterior margin usually with five pairs of spiniform setae. Mandibular palp broad, flattened, one segmented.

All pereopods with exopods. Fourth pereopod shorter than third, longer than fifth. Exopods of pleopods not unusually long.

Subgenus *Leptochela* Stimpson, 1860

Leptochela Stimpson, 1860: 42; Chace, 1976: 2; Hanamura, 1987: 15; Hayashi, 1995: 1.

Diagnosis.— Exopod of third pereopod short, not reaching distal end of ischium. Forth pair of pereopods about 1.3 times as long as fifth. Sixth abdominal somite without dorsal lappet.

Leptochela (Leptochela) tuerkayi sp. nov. (figs. 1-3)

Leptochela sp. A — Wang et al., 2017: 609, 610.

Type material.— Holotype: MBM189216, 1 male, postorbital carapace length 7.83 mm, North Yellow Sea (38°01′N 123°59′E), 71 m depth, sandy bottom, collected by Yueyun Wang, Yue Sun, Bingqin Liu, 19 October 2014. Paratypes: MBM189214, 1 ovigerous females, postorbital carapace length 6.60 mm, collection data same as holotype; MBM189215, 1 male, postorbital carapace length 6.30 mm, collection data same as holotype; MBM189219, 4 males, postorbital carapace length 5.80-7.80 mm, 2 ovigerous females, postorbital carapace length 6.92-7.10 mm, collection data same as holotype.

Etymology.— The species is named in honour of the German carcinologist, the late Professor Michael Türkay, for his great contributions to the knowledge of crustacean taxonomy.

Diagnosis.— Small in size. Maximum postorbital carapace length 7.83 mm. Rostrum with dorsal margin variable, straight, convex, concave, or sinuous. Carapace with three longitudinal dorsal ridges in breeding females only, absent in non-breeding females and males. Orbital margin entire, not serrated, without mesially directed tooth on ventral portion, suborbital angle rounded. Dorsal margin of fifth abdominal somite entire, without dorsal elevations but with sharp posterior tooth projecting from posterior margin. Telson with pair of dorsolateral spines in addition to pair of anterior mesial spines; posterior margin with usual 5 pairs of prominent spiniform setae, without minute mesial seta. Length of antennal scale less than 60% of carapace length. Third maxilliped with rather short exopod, slightly overreaching end of antepenultimate segment of third maxilliped. First pereiopod with 47 to 55 spines on cutting margin of movable finger. Second pereiopod with 48 to 61 spines on cutting margin of movable finger. Third pereiopod with exopod not nearly reaching distal end of ischium. Fourth pereiopod with exopod nearly reaching middle of merus. Endopod of first pleopod of male rounded distolaterally, not flared. Appendix masculina (not including setae) not overreaching appendix interna.

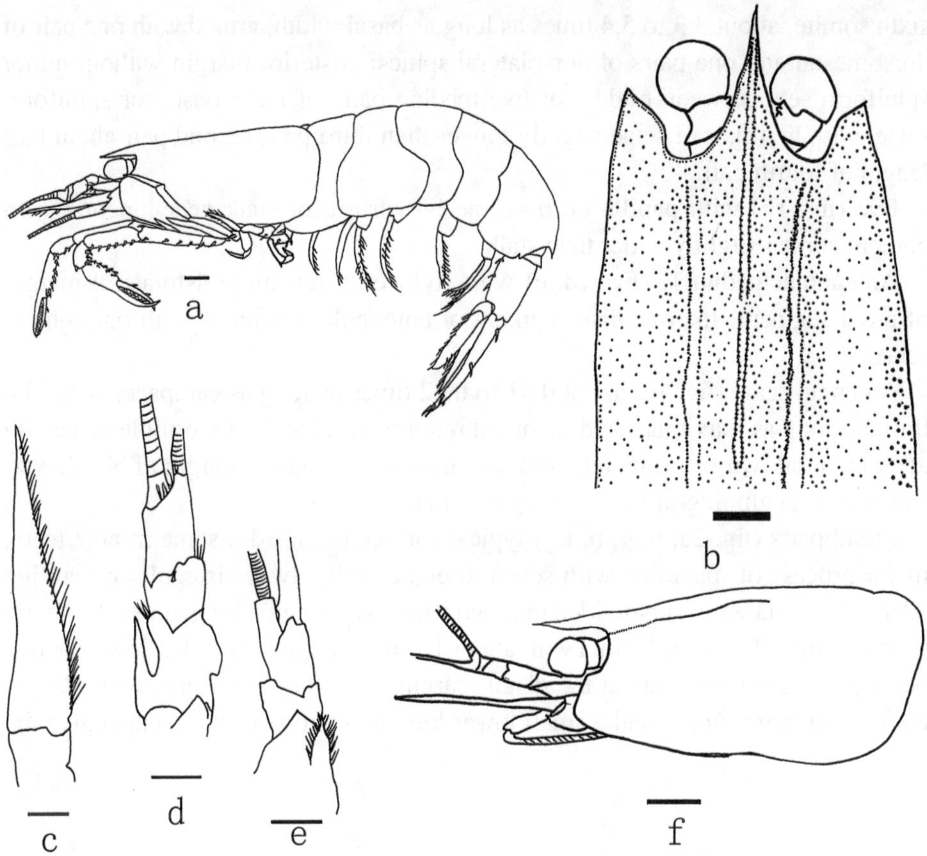

Fig. 1. *Leptochela* (*Leptochela*) *tuerkayi* sp. nov. a, Holotype, MBM189216, lateral view. b-f, paratype, MBM189214: b, dorsal view of eyes; c, left antennal scale; d, dorsal view of left antennula; e, lateral view left antennula; f, lateral view of carapace. Scale = 0.05 mm.

Description.— Small sized *Leptochela* species. Rostrum (fig. 1a) with dorsal margin straight or slightly sinuous, apex sometimes slightly curved downward; in female, rostrum (fig. 1f) always extending beyond eye in female but usually not extending beyond eye in male. Carapace (fig. 1b) usually rounded in males and immature females, but tricarinate in breeding females. Carapace with median dorsal carina on anterior 0.28 to 0.62 of length. Orbital margin entire, not serrated, without mesially directed tooth on ventral portion, suborbital angle rounded.

Abdomen regularly rounded dorsally on anterior 3 somites, usually carinate on posterior part of fourth somite. Fifth somite carinate, without dorsal prominences but with sharp median tooth projecting from posterior margin. Sixth somite with tranverse swelling near anterior end of dorsal surface, posterolateral lobe with distinct acute tooth on posterodorsal margin and two or three additional small spines on ventrolateral surface. Telson (fig. 3a) about 1.2 to 1.6 times as long as sixth somite, about 3.3 to 5.4 times as long as basal width, armed with one pair of dorsomesial and one pairs of dorsolateral spines; posterior margin without minor spiniform seta between middle of five median pairs of large posterior spiniform setae, with first pair of setae slightly longer than third pair, second pair about half length of fourth pair.

Eye (fig. 1b) with papilla on dorsomesial surface of stalk near juncture with cornea; cornea slightly wider than stalk.

Antennular peduncle (fig. 1d, e) with stylocerite reaching distolateral margin of basal segment. Intermediate segment of antennular peduncle with one minute setae.

Antennal scale (fig. 1c) about 0.51 to 0.62 times as long as carapace, about 3.4 times as long as maximum width; lateral margin with barely discernible concavity at about basal two-fifths, distal tooth continuous with mesial margin of blade, with shoulder at beginning of mesial margin of blade.

Mouthparts (fig. 2a, b, g, h, i, j) typical for genus. Besides some minute teeth, molar process of mandible with seven to eight teeth, always irregular on cutting edge. Maxillula with palp divided into two parts, outer region long, inner short with long seta distally, upper lacinia with about 10 slender spines distally, lower lacinia bearing slender long setae at the distal margin. Maxilla with distal endite deeply cleft into a larger upper and a small lower lob, palp without setae, scaphognathite

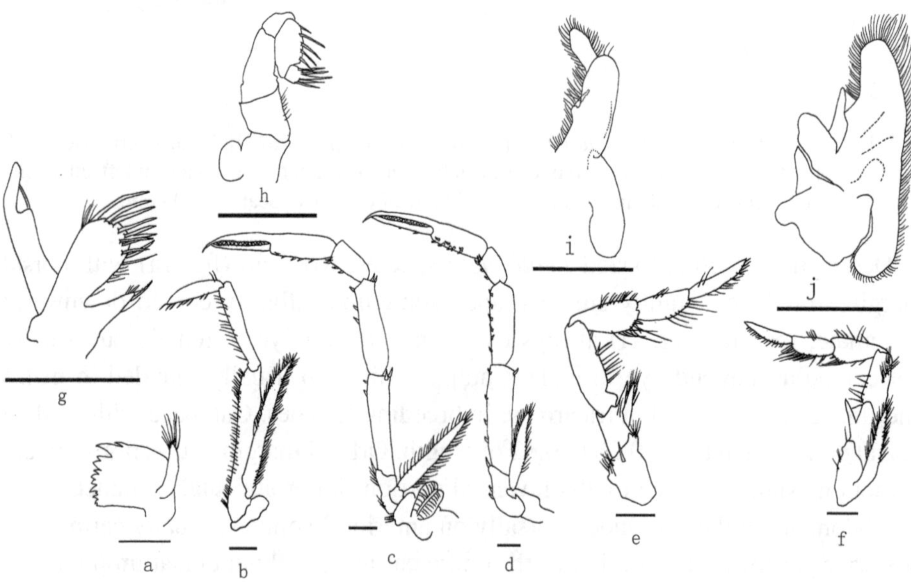

Fig. 2. *Leptochela* (*Leptochela*) *tuerkayi* sp. nov., paratype, MBM189214. a, Mandible; b, third maxilliped; c, left first pereiopod; d, left second pereiopod; e, outer view of right fourth pereiopod; f, inner view of right fourth pereiopod; g, maxillula; h, second maxilliped; i, first maxilliped; j, maxilla. Scale: a-f = 0.05 mm, g-j = 0.5 mm.

with long marginal setae well developed. First maxilliped bearing two-segmented endopod, distal part small and with marginal setae, epipod long, exopod short surrounded by setae. Second maxilliped with oval epipod, no exopod present, distal two segments terminating in strong spines. Third maxilliped exceeding antennular peduncle, with rather short exopod, slightly overreaching distal margin of antepenultimate segment.

First pereiopod (fig. 2c) overreaching antennal scale with propodus and dactylus; dactylus with 47 to 55 spines on cutting margin; propodus with row of four or five stout spines on flexor margin; carpus with three or four spines on flexor margin; merus with about three or four stout spines on flexor margin and two on anterior extensor margin; ischium with several spines near anterior margin; basis with distinct spine on anterior margin. Second pereiopod (fig. 2d) reaching anteriorly nearly as far as first pereiopod; dactylus with 48 to 61 spines on cutting margin, propodus with two rows of four or five stout and movable spines on flexor margin; carpus with three or four spines on flexor margin and several spines on anterior margin; merus with about six stout spines on flexor margin and one on anterior extensor margin; ischium with four or five spines on flexor margin; basis with distinct spine and two minor spines on anterior margin. Third pereiopod overreaching anterior margin of carapace with dactylus; exopod falling far short of distal end of ischium. Fourth pereiopods (fig. 2e, f) short, reaching nearly to distal end of ischium of second pereiopod, with exopod reaching beyond end of merus; carpus with two or three spines on anterior margin; merus with two stout spines on anterior margin; ischium with one stout large spine on posterior margin; basis with one slender spine on anterior margin. Fifth pereiopod similar to fourth pereiopod in shape, reaching nearly to midlength of ischium of second pereiopod when both extended anteriorly; dactyl not longer than propodus, exopod not overreaching merus.

Endopod (fig. 3a) of first pleopod of male rounded distolaterally, not flared. Appendix masculina (fig. 3b) bearing 10 long setae, not overreaching appendix interna, setae not included. Exopod of uropod armed with 10 to 17 movable spines along lateral margin.

Ecology.— Known only from the present contribution, the type specimens of this new species were collected from Yellow Sea, in the subtidal zone and at depths of 71 meters. The Yellow Sea is a marginal sea in the north-western part of the Indo-Pacific Ocean. It is influenced by coastal currents and mixed water mass, as well as the Yellow Sea Cold Water Mass (Li et al., 2012). The North Yellow Sea where the fresh samples were collected is a vital channel for sediment and water exchange between the Bohai Bay and the South Yellow Sea. The average depth of the sea is 38 m and the deepest area can be up to 80 m, while our samples were collected at 71 m, quite near the deepest bottom of the sea (Li et al., 2009). Due to

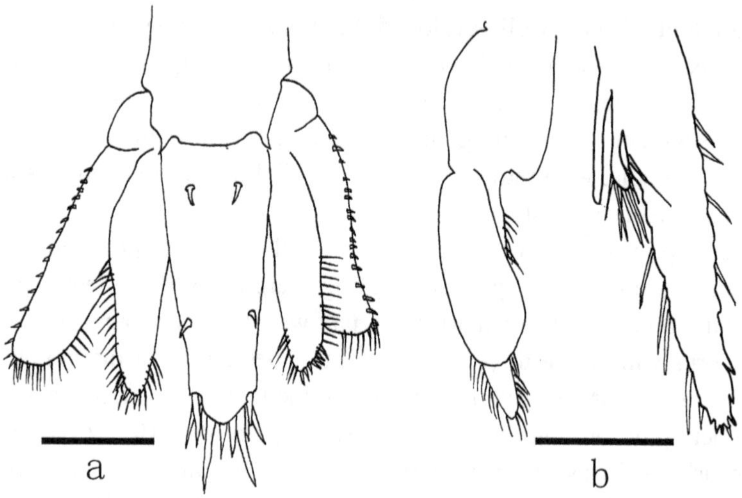

Fig. 3. *Leptochela* (*Leptochela*) *tuerkayi* sp. nov. a, Paratype, MBM189214, telson and uropods; b, paratype, MBM189215, right first pleopod and second pleopod with appendix masculina and appendix interna. Scale = 0.05 mm.

the influence of the cold water mass, the temperature of the bottom stays at 2-10°C throughout the year (Liu & Xu, 1963).

Remarks.— It is very likely that the new species has been confused with and regarded as *Leptochela* (*Leptochela*) *gracilis* Stimpson, 1860 in Chinese coastal waters. Apparently the new species is closely related to *L.* (*L.*) *gracilis* Stimpson, 1860 as both species share the character of the sharp posterior tooth on the fifth abdominal somite. In fact, the immature females of *L.* (*L.*) *tuerkayi* sp. nov. are very hard to be distinguished from *L.* (*L.*) *gracilis* Stimpson, 1860. *L.* (*L.*) *tuerkayi*

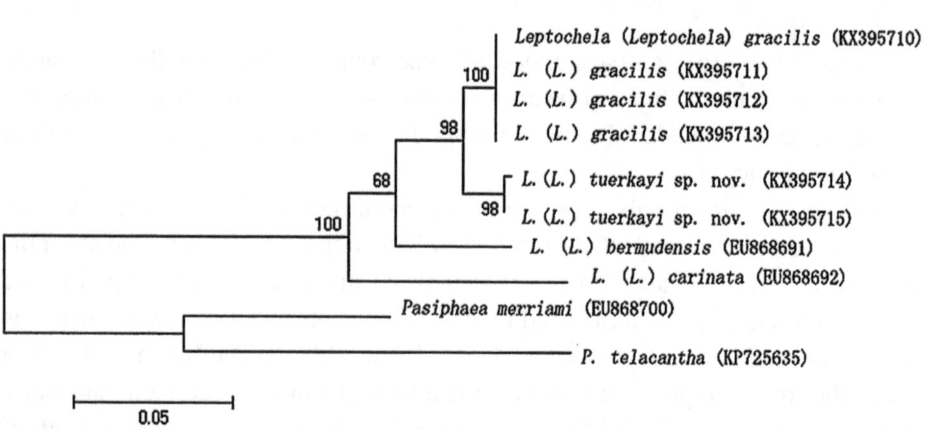

Fig. 4. Phylogenetic tree obtained by maximum likelihood analysis of DNA sequences based on 16S rRNA gene; bootstrap values >50% are shown.

TABLE II

Kimura's 2-parameter pairwise distances of the 16S rRNA between the studied specimens

GenBank accession No.	Specimen	1	2	3	4	5	6	7	8	9
KX395710	1-*Leptochela* (*Leptochela*) *gracilis* Stimpson, 1860									
KX395711	2-*Leptochela* (*Leptochela*) *gracilis*	0.000								
KX395712	3-*Leptochela* (*Leptochela*) *gracilis*	0.000	0.000							
KX395713	4-*Leptochela* (*Leptochela*) *gracilis*	0.000	0.000	0.000						
KX395714	5-*Leptochela* (*Leptochela*) *tuerkayi* sp. nov.	0.040	0.040	0.040	0.040					
KX395715	6-*Leptochela* (*Leptochela*) *tuerkayi* sp. nov.	0.036	0.036	0.036	0.036	0.003				
EU868691	7-*Leptochela* (*Leptochela*) *bermudensis* Gurney, 1939	0.111	0.111	0.111	0.111	0.112	0.116			
EU868692	8-*Leptochela* (*Leptochela*) *carinata* Ortmann, 1893	0.199	0.199	0.199	0.199	0.222	0.217	0.214		
EU868700	9-*Pasiphaea merriami* Schmitt, 1931	0.704	0.704	0.704	0.704	0.677	0.671	0.664	0.773	
KP725635	10-*Pasiphaea telacantha* Hayashi, 2004	0.858	0.858	0.858	0.858	0.908	0.901	0.922	0.939	0.385

sp. nov. is a small sized species, the postorbital carapace length varies from 3.11-7.83 mm; of non-ovigerous females, 3.11-6.70 mm; of ovigerous females, 4.79-7.58 mm. The largest specimen is the male holotype; whereas the specimens of *L.* (*L.*) *gracilis* Stimpson, 1860 we checked in the MBMCAS have a larger body size with the postorbital carapace length usually longer than 8.0 mm, up to 10.0 mm. In addition, the new species differs from *L.* (*L.*) *gracilis* Stimpson, 1860 by the existence of 3 longitudinal dorsal ridges on carapace in ovigerous female specimens. In *L.* (*L.*) *gracilis* Stimpson, 1860, neither males nor females have three longitudinal dorsal ridges on the carapace. The third maxilliped has a rather short exopod, slightly overreaching the distal margin of the antepenultimate segment of the third maxilliped. In *L.* (*L.*) *gracilis* Stimpson, 1860, the exopods is always long, overreaching the distal margin of the antepenultimate segment. The fourth pereiopod has the exopod nearly reaching the middle of the merus, whilst in *L.* (*L.*) *gracilis* Stimpson, 1860, the exopod usually does not overreach the distal margin of the ischium.

Molecular phylogenetic analysis

The 16S rRNA dataset contains 10 sequences from 6 species of the family Pasiphaeidae, including two sequences from the new species *Leptochela* (*Leptochela*) *tuerkayi* sp. nov. (KX395714, KX395715), four from *L.* (*L.*) *gracilis* Stimpson, 1860 (KX395710-KX395713) and one from *L.* (*L.*) *bermudensis* Gurney, 1939 (EU868691), one from *Leptochela* (*Proboloura*) *carinata* Ortmann, 1893 (EU868692), one from *Pasiphaea merriami* Schmitt, 1931 (EU868700) and one from *Pasiphaea telacantha* Hayashi, 2004 (KP725635). The tree with the highest log likelihood (−1509.2688) is shown in fig. 4. The final alignment consists of 452 bp and includes 165 variable sites (approx. 36.50%) and 85 parsimony-informative sites (approx. 18.81%). The estimated transition/transversion bias is 0.95. The pairwise distances are shown in table II. In the maximum likelihood tree, the four species from the genus *Leptochela* Stimpson, 1860 form a robustly supported clade (PP = 1.0, BP = 99%), with the new species *L.* (*L.*) *tuerkayi* sp. nov. In addition, the new species is sister to *L.* (*L.*) *gracilis* Stimpson, 1860 in our phylogenetic tree, in accordance with their great resemblance in morphology.

ACKNOWLEDGEMENTS

The present work was supported by the National Natural Science Foundation of China (No. 41506171) and Strategic Priority Research Program of the Chinese Academy of Sciences (XDA11020303). Many thanks to our colleagues Yueyun Wang, Yue Sun, Bingqin Liu, for their collection of the samples we examined. We thank Dr. Qi Kou, our colleague, for his kind correction to the molecular part of the manuscript. Thanks are due to Mr. Shaoqing Wang and Miss Lianmei Shuai, the managers of the MBMCAS, for their kind help in specimen sorting.

REFERENCES

CHACE, JR., F. A., 1976. Shrimps of the pasiphaeid genus *Leptochela* with descriptions of three new species (Crustacea: Decapoda: Caridea). Smithsonian Contributions to Zoology, **222**: 1-51.

CRANDALL, K. A. & J. F. FITZPATRICK, 1996. Crayfish molecular systematics: using a combination of procedures to estimate phylogeny. Systematic Biology, **45**: 1-26.

DE GRAVE, S. & C. H. J. M. FRANSEN, 2011. Carideorum catalogus: the recent species of the dendrobranchiate, stenopodidean, procarididean and caridean shrimps (Crustacea: Decapoda). Zoologische Mededelingen, **85**: 195-589.

DONG, Y. M., 1991. Fauna of Zhejiang. Hangzhou, Zhengjiang Sciene and Technology, **1**: 1-481.

EDGAR, R. C., 2004. Muscle: multiple sequence alignment with high accuracy and high throughput. Nucleic Acids Research, **32**: 1792-1797.

HALL, T. A., 1999. BioEdit: a user-friendly biological sequence alignment editor and analysis program for Windows 95/98/NT. Nucleic Acids Symposium Series, **41**: 95-98.

HANAMURA, Y., 1987. Caridean shrimps obtained by R.V. "Soela" from north-west Australia, with description of a new species of *Leptochela* (Crustacea: Decapoda: Pasiphaeidae). The Beagle: Records of the Museums and Art Galleries of the Northern Territory, **4**: 15-33.

HAYASHI, K. I., 1995. Brief revision of the genus *Leptochela* with description of two new species (Crustacea, Decapoda, Pasiphaeidae). In: B. RICHER DE FORGES (coord.), Les fonds meubles des lagons de Nouvelle Calédonie (sédimentologie, benthos). Etudes et Thèses, **2**: 83-99. (ORSTOM, Paris).

LI, G. G. & S. Z. HUANG, 1964. The vertical distribution of some zooplankton off Cheefoo and Weihai, China. Journal of Shangdong College of Oceanology, **1**: 106-127.

LI, X. Z., H. F. WANG, L. MA, L. ZHANG & B. L. ZHANG, 2012. Advance of the study on the macrobenthos from the Yellow Sea and East China Sea. Sustainable Development, **2**: 80-102.

LI, Z., X. W. BAO, Y. Z. WANG, N. LI & L. L. QIAO, 2009. Seasonal distribution and relationship of water mass and suspended load in north Yellow Sea. Chinese Journal of Oceanology and Limnology, **27**(4): 907-918.

LIU, R. Y. & F. S. XU, 1963. Preliminary studies on the benthic fauna of the Yellow Sea and the East China Sea. Oceanologia et Limnologia Sinica, **5**: 306-320.

STIMPSON, W., 1860. Prodromus descriptionis animalium evertebratorum, quae in Expeditione ad Oceanum Pacificum Septentrionalem, a Republica Federata missa, Cadwaladore Ringgold et Johanne Rodgers ducibus, observavit et descripsit. Pars VIII, Crustacea Macrura. Proceedings of the Academy of Natural Sciences of Philadelphia, **1860**: 22-47. [Pp. 91-116 on offprint.]

TAMURA, K., D. PETERSON, N. PETERSON, G. STECHER, M. NEI & S. KUMAR, 2011. MEGA5: molecular evolutionary genetics analysis using maximum likelihood, evolutionary distance, and maximum parsimony methods. Molecular Biology and Evolution, **28**: 2731-2739.

WANG, Y., Z. GAN & X. LI, 2017. Two new records of species of *Leptochela* Stimpson, 1860 (Decapoda, Caridea, Pasiphaeidae) from China. Crustaceana, **90**(5): 601-611.

First received 17 September 2016.
Final version accepted 3 December 2016.

[When citing this volume, please refer to Crustaceana 90 (2017) 771-1288]

Michael Türkay Memorial Issue

DESCRIPTION OF A NEW SPECIES OF *PACHYCHELES* (DECAPODA, ANOMURA, PORCELLANIDAE) FROM THE SOUTHERN CARIBBEAN SEA

BY

BERND WERDING[1]) and ALEXANDRA HILLER[2,3])

[1]) Institut für Tierökologie und Spezielle Zoologie der Justus-Liebig-Universität Giessen, Heinrich-Buff-Ring 29 (Tierhaus), D-35392 Giessen, Germany

[2]) Smithsonian Tropical Research Institute, Apartado 0843-03092, Panama, Republic of Panama

ABSTRACT

A new species of porcellanid crab from the southern Caribbean Sea is described. *Pachycheles tuerkayi* n. sp. has been confused with *P. serratus* (Benedict, 1901) since the 1950s because the two species are morphologically and ecologically similar and have overlapping distributions in the southern Caribbean. *P. tuerkayi* n. sp. is restricted to the coasts of Costa Rica, Panamá and Colombia. *P. serratus* ranges from Puerto Rico and the Virgin Islands to the coasts of Panamá, Colombia and Venezuela. Genetic differences based on DNA sequences of the mitochondrial 16S rDNA gene from the two species surpassed those estimated for geminate porcellanids on each side of the Isthmus of Panamá. Field observations where *P. tuerkayi* n. sp. and *P. serratus* overlap indicated that the two species come into contact when sharing the same substrate. The total number of porcellanid species in the western Atlantic rises to 50.

ZUSAMMENFASSUNG

Eine neue Porzellaniden-Art aus dem Karibischen Meer wird beschrieben. *Pachycheles tuerkayi* n. sp. und *P. serratus* (Benedict, 1901), deren Verbreitung sich in der südlichen Karibik überschneidet, sind, seit den 1950er Jahren, wegen ihrer morphologischen Ähnlichkeiten und wegen ihres Vorkommens in der gleichen ökologischen Situation für eine Art gehalten worden. Die Verbreitung von *P. tuerkayi* n. sp. ist auf die Küste von Costa Rica, Panamá und Kolumbien beschränkt, während *P. serratus* in Puerto Rico, den Jungferninseln, sowie in Panamá, Kolumbien und Venezuela vorkommt. Die genetischen Distanzen, die sich aus dem Vergleich von DNS Sequenzen des mitochondrialen 16S rDNS Gens der beiden Arten ergeben, gehen über diejenigen hinaus, die für Zwillingsarten beiderseits des Isthmus von Panamá ermittelt wurden. Feldbeobachtungen im Überlappungsgebiet beider Arten zeigen, dass diese miteinander in Kontakt kommen können, wenn sie das gleiche Substrat bewohnen. Die Gesamtzahl der west atlantischen Porzellaniden beträgt nun 50 Arten.

[3]) Corresponding author; e-mail: hillera@si.edu

INTRODUCTION

The porcellanid fauna of the western Atlantic has been intensively studied since the 1950s (Haig, 1956; Gore, 1970, 1982; Gore & Abele, 1976; Werding, 1977, 1978, 1982; Werding & Hiller, 2002). Werding et al. (2003) reported 48 species (including 2 formerly known only from the East Pacific), and Werding & Hiller (2005) added *Petrolisthes hispaniolensis* Werding & Hiller, 2005, increasing the number to 49 West Atlantic species.

The new species described here as *Pachycheles tuerkayi* n. sp. was first discerned by Haig (1956) as a pubescent form of the tropical *Pachycheles serratus* (Benedict, 1901). Haig highlighted a “thick pile of fine hairs” on the chelipeds of specimens from Bahía Caledonia, Panamá. Later, Gore (1970, 1982) recorded the pubescent and non-pubescent forms from Panamá, although Gore & Abele (1976) only referred to pubescent specimens from the same region. Werding (1977) reported on the two forms in the area of Santa Marta, Colombia, but only on the pubescent form in the Colombian Gulf of Urabá (Werding, 1978) and Islas del Rosario (Werding, 1982).

Here, we describe the pubescent form as *Pachycheles tuerkayi* n. sp. and designate the non-pubescent form as *P. serratus* (fig. 1), as this corresponds to Benedict’s (1901) description (see Discussion). Genetic distances based on DNA sequences of the mitochondrial 16S rDNA gene from the two species were estimated to explore species boundaries and to establish intra- and interspecific divergence. The new species becomes the 50th member of the West Atlantic porcellanid fauna.

MATERIAL AND METHODS

Preserved specimens of *Pachycheles serratus* and *P. tuerkayi* n. sp. collected in Panamá (Bocas del Toro, Playa Diablo), Colombia (Islas del Rosario and Santa Marta) and Venezuela were used for morphological examination and molecular analyses. Material of the new species is deposited in the collections of the Museo de Historia Natural Marina de Colombia (INV CRU) at the Institute of Marine and Coastal Research of Colombia, Santa Marta (INVEMAR), and the Forschungsinstitut und Naturmuseum Senckenberg, Frankfurt am Main (SMF), Germany. Specimens were sexed and measured by using a stereoscope with a micrometer. Measurements correspond to carapace length followed by carapace width.

DNA was extracted from the chelipeds or walking legs using the QIAGEN DNeasy® Kit following the protocol for animal tissue. A region of approximately 540 bp of the mitochondrial 16S rDNA gene was amplified with the primers

16Sar-L (5′-CGCCTGTTTAACAAAAACAT-3′) and 16Sbr-H (5′-CCGGTTTGA ACTCAGATCATGT-3′) (Palumbi et al., 1991). Double-stranded amplifications were performed in 25 μl volume reactions containing 5 μl of *Taq* buffer (5×), 2.5 μl of dNTP mix (8 mM, Promega), 1.2 μl of each primer (10 μM), 2.5 μl of $MgCl_2$ (25 mM), 0.2 μl of GoTaq® Flexi DNA *Taq* Polymerase (Promega), 1 μl of DNA template, and 11.4 μl of ddH_2O. Thermal cycling conditions consisted of an initial denaturation at 96°C for 3 min, followed by 30 cycles of 95°C for 1 min, 52°C for 1 min, and 72°C for 1 min. A final extension step at 72°C for 5 min followed the last cycle. PCR products were cleaned using ExoSAP-IT® (Affymetrix-USB) following the manufacturer's protocol. Cleaned products were sequenced in both directions using the BigDye® Terminator v3.1 Cycle Sequencing Kit (Applied Biosystems) in 10 μl reactions as follows: 1 μl of PCR product was mixed with 4 μl of ddH_20, 3 μl of sequencing buffer (5×), 1 μl of primer and 1 μl of BigDye ready-reaction mix. The mixture was subjected to thermal cycling conditions consisting of an initial denaturation step at 96°C for 1 min, followed by 25 cycles of 96°C for 10 s, 50°C for 5 s, and 60°C for 4 min. Cycle sequencing reactions were run on an ABI PRISM 3130xl Genetic Analyzer (Applied Biosystems). Two GenBank sequences were included in the data set: *Pachycheles serratus* (JF900732) and *P. setimanus* (Lockington, 1878) (JF900730), both from Bocas del Toro, Panamá. The later sequence caught our attention because *P. setimanus* is morphologically close to *P. serratus* (see below), but is restricted to the tropical East Pacific (Haig, 1960). Sequences were aligned using MAFFT v.7 (Katoh et al., 2002) with the L-Ins-i algorithm and default gap opening/extension penalties. Genetic distances within and between *Pachycheles serratus* and *P. tuerkayi* n. sp. were estimated using the Kimura-2-parameter model (K2P; Kimura, 1980) implemented in MEGA v.6.06 (Tamura et al., 2013). DNA sequences obtained for this study were given GenBank accession numbers KY860741-KY860763.

RESULTS

Molecular results

The alignment of 23 partial sequences of the 16S rDNA gene was 535 bp long. Average percent K2P distances within *Pachycheles tuerkayi* n. sp. (14 sequences in total) averaged at 0.17%, and within *P. serratus* (9 sequences in total) at 0.04%. Distances between the two species averaged 6.8%. These distance estimations excluded the GenBank sequence labeled as *P. serratus* from Bocas del Toro, Panamá (JF900732), because it matched DNA sequences of the West Atlantic *P. cristobalensis* Gore, 1970 (A.H., unpublished data). The GenBank sequence labeled as *Pachycheles setimanus* from Bocas del Toro (JF900730) matched sequences of *Pachycheles tuerkayi* n. sp. and, therefore, was included in our distance estimation.

Systematic account

Family PORCELLANIDAE Haworth, 1825

Pachycheles tuerkayi n. sp.

(figs. 1a, 2, 3)

Pachycheles serratus Haig, 1956 (part, pp. 8-9); Gore, 1970 (part, p. 962), 1982 (part, p. 9); Werding, 1977 (part, p. 194-195); Gore & Abele, 1976 (p. 20); Werding, 1978 (p. 219), 1982 (pp. 441-442); Vargas & Cortés, 2006 (p. 485).

Material examined.— Holotype, male (5.1 × 5.4 mm), INV CRU 8408, Colombia, Magdalena, Santa Marta, El Morro, under boulders, 2-3 m, leg. B. Werding, 9 Sep 1974. Paratypes: 3 males (3.6 mm × 3.8 mm, 4.0 × 4.3 mm, 4.8 mm × 5.1 mm), 2 females (4.8 mm × 5.5 mm, 4.2 mm × 4.7 mm), 2 ov. females (4.3 mm × 5.0 mm, 5.2 mm × 5.8 mm), INV CRU 8409, same data as holotype; 3 males (4.7 × 5.0 mm, 4.5 × 4.9 mm, 4.0 × 4.3 mm), 3 ov. females (5.0 × 5.8 mm, 4.8 × 5.5 mm, 4.0 × 4.3 mm), SMF 50188, Colombia, Gulf of Urabá, Napú, under boulders, 2-4 m, leg. B. Werding, 1977; 4 males (4.5 × 4.8 mm, 4.4 × 4.7 mm, 4.3 × 4.7 mm, 4.2 × 4.5 mm), 1 female (4.7 × 5.2 mm), 1 ov. females (4.8 × 5.2 mm), SMF 59189, Colombia, Magdalena, Santa Marta, under boulders, 0.5-2 m, leg. B. Werding, 1974.

Description.— Carapace somewhat broader than long in both sexes, dorsal surface smooth, metabranchial regions plicate, regions not marked. No epibranchial spine, epibranchial border crested. Front narrow, convex or slightly trilobate, scarcely produced, orbits deep, inner orbital angle rounded, outer orbital angle produced into a forwardly directed prominent tooth. Lateral walls with a trapezoid plate in anterior half and a large sub-quadrate plate frequently followed by a small fragment in posterior half.

First movable segment of antenna with a blunt conical tooth; second granulate and third cylindrical, smooth; flagellum two times as long as carapace. Basal segment of antennules forwardly directed, forming a rectangular, serrated lobe along the frontal margin.

Chelipeds thick, one distinctly larger, inner distal edge of merus produced into a large tuberculate lobe, upper surface with continuous transversal ridges; carpus about as long as broad, surface with transversal scale like ridges and round granules, anterior margin armed with three large, irregular teeth, decreasing in size distally; dorsal surface covered distally with long, plumose setae which hide most of the frontal and distal borders. Posterior border slightly convex. Chelae extremely large, the bigger one exceeding by far the size of cephalothorax in large males (figs. 1a, 2). Distal border of chelae somewhat depressed at the level of the base of the pollex. Upper surface of chelae covered with round granules, most of the surface covered with thick setae, except on inner border and fingers. Fingers of

Fig. 1. a, *Pachycheles tuerkayi* n. sp. from Santa Marta, Colombia, large male, INV CRU 8409; scale = 4.5 mm. b, *P. serratus* (Benedict, 1901) from the Gulf of Morrosquillo, Colombia, large male. Scale = 4.5 mm.

chelipeds slender, widely gaping in large specimens, and closing at entire length in smaller ones.

Walking legs slender, with scattered plumose setae more dense on carpus and propodus; dactylus slender, with three movable spines.

Males with pleopods; telson with seven plates.

Coloration.— The smooth carapace of *Pachycheles tuerkayi* n. sp. is brown, sometimes grayish, spotted with irregular light marks. The chelipeds are uniformly brown, darker in the distal parts of the fingers, whose tips are white. Setae on

Fig. 2. *Pachycheles tuerkayi* n. sp., male holotype, INV CRU8408, dorsal view. Scale = 5.4 mm.

carpus and manus are beige. Walking legs are of the same colour as the carapace, a bit paler near articulations.

Ecology.— *Pachycheles tuerkayi* n. sp. lives in crevices between stones and other hard substrates in moderately exposed shallow water from intertidal areas to waters 6 m deep, often where there are fine sediments and some brackish water influence. It occurs with *P. serratus* in the same habitat in the Santa Marta region in Colombia.

Distribution.— Atlantic coast of Costa Rica, Panamá and Colombia, from Puerto Vargas (Costa Rica) to Islas del Rosario and Santa Marta (Colombia).

Etymology.— The new species is named after Michael Türkay, long-time friend and colleague.

Remarks.— *Pachycheles tuerkayi* n. sp. (figs. 1a, 2, 3) can be distinguished from *P. serratus* (figs. 1b, 4) by having (1) a carapace with smooth, shiny surface, (2) most of the cheliped's dorsal surface densely covered with setae, and (3) the cheliped's distal margin somewhat depressed on the level of the base of the pollex. In *P. serratus* the surface of the carapace is coarser, the dorsal surface of the chelipeds bears no setae or only a few scattered ones, and the distal margin of the cheliped is not depressed. *P. tuerkayi* n. sp. and *P. serratus* belong to a

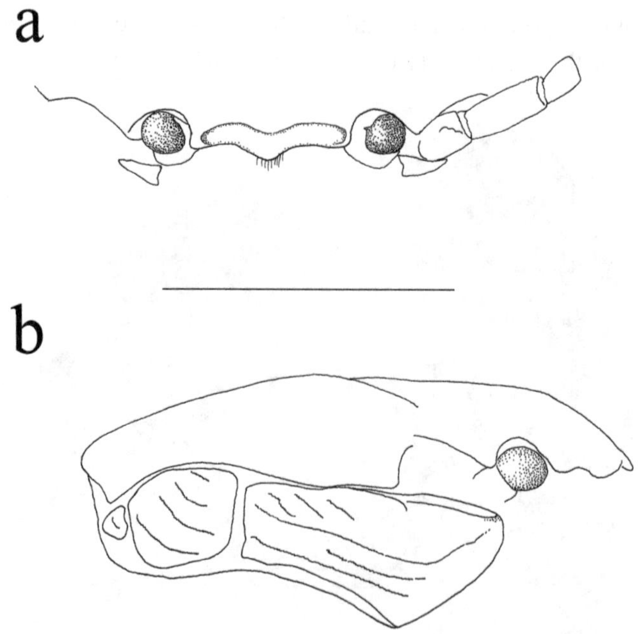

Fig. 3. *Pachycheles tuerkayi* n. sp., male holotype, INV CRU8408. a, carapace, frontal view; b, carapace, lateral view. Scale = 2.8 mm.

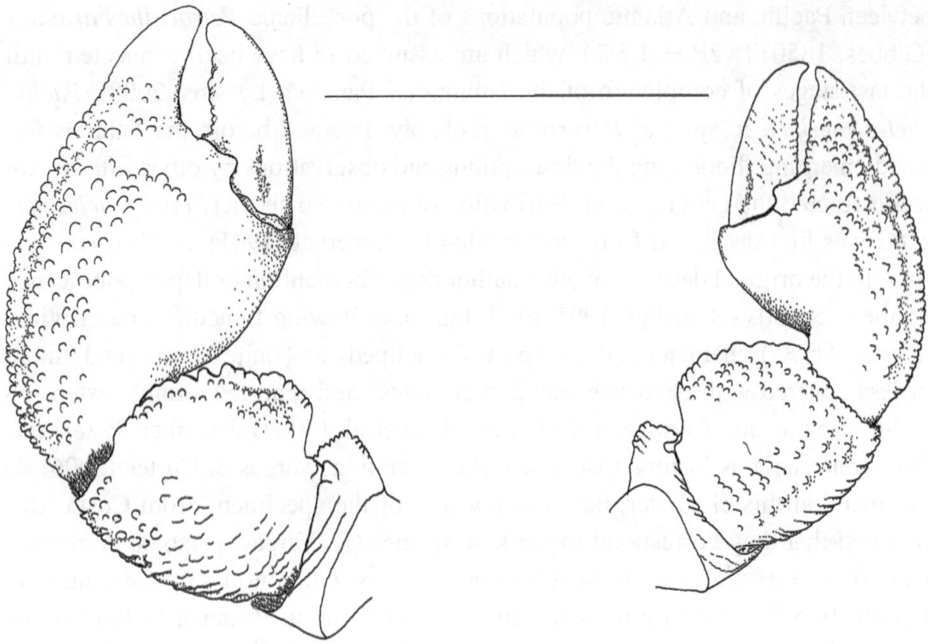

Fig. 4. *Pachycheles serratus* (Benedict, 1901), male, SMF 50196, chelipeds, dorsal view. Scale = 4.5 mm.

morphological group of *Pachycheles* species from the tropical Atlantic and eastern Pacific, characterized by having the carapace with scarce setae, and the carpus and manus of chelipeds covered with granules. This group is represented by *P. chacei* Haig, 1956 and *P. cristobalensis* in the Caribbean, by *P. greeleyi* (Rathbun, 1900) in Brazil, and by *P. sahariensis* Monod, 1933 in the eastern Atlantic. In the eastern Pacific the group is represented by *P. calculosus* Haig, 1960, *P. setimanus* and *P. chacei* (present on both sides of the Isthmus of Panamá). All species have a telson with 7 plates, except *P. chacei*, which has a 5-plated telson. *P. tuerkayi* n. sp., *P. serratus* and *P. setimanus* have the presence of male pleopods in common. Males in the other species in this *Pachycheles* group do not have pleopods.

DISCUSSION

Pachycheles tuerkayi n. sp. poses one more example of cryptic diversity within Porcellanidae. Other cases of species complexes in this family also comprise closely related species with overlapping distributions and ecologies (e.g., Hiller et al., 2006; Hiller & Werding, 2007). The new species and *P. serratus* diverge with average percent genetic distances (K2P = 6.9%) larger than those estimated between Pacific and Atlantic populations of the porcellanid *Petrolisthes armatus* (Gibbes, 1850) (K2P = 1.5%), which are assumed to have been connected until the last stages of completion of the Isthmus of Panamá (Lessios, 2008). *Pachycheles tuerkayi* n. sp. and *P. serratus* probably diverged before the Isthmus finished emerging. Following the descriptions and observations by other authors, we reconstructed the geographical distribution of these two species. *Pachycheles serratus* was first described from Puerto Rico by Benedict (1901) as *Pisosoma serrata*. In the original description, this author does not mention or depict pubescence on the chelipeds (Benedict, 1901, pl. 3, fig. 7). Following Benedict's description, Schmitt (1935) characterized the species' chelipeds as "quite naked" and "more or less covered with bead-like rounded granules" and reported a range extension to St. Thomas and Cartagena, Colombia. Lira et al. (2007) described *P. serratus* from Venezuela as having sparse or lacking setation. Vargas & Cortés (2006) do not mention this character, but later revision of the specimens from Costa Rica revealed that they correspond to the new species (R. Vargas, personal communication). *P. serratus* seems to be a common species with a northern distribution in Puerto Rico and the Virgin Islands and a coastal one in Panamá, Colombia and Venezuela. *Pachycheles tuerkayi* n. sp. seems to be the predominant species on the Caribbean coasts of Costa Rica, Panamá and Colombia, where it co-occurs with *P. serratus*, occupying the same habitats. More ecological observations are needed to establish how the two species share very similar or the same habitat.

ACKNOWLEDGEMENTS

We thank H. A. Lessios, L. Geyer, L. Rivera and A. Calderón (STRI) for support in the lab. We are grateful to S. Kalscheid for the drawings and to R. Vargas (Universidad de Costa Rica) for sharing useful observations of specimens from Costa Rica. We also thank E. Montoya, G. Borrero and M. Martelo (INVEMAR), and B. Trautwein (SMF) for their help with material curation and deposition. This study was supported by a Smithsonian grant to A. H. and by the Centre of Excellence in Marine Sciences (CEMarin).

REFERENCES

BENEDICT, J. E., 1901. The anomuran collections made by the Fish Hawk Expedition to Porto Rico. Bull. U.S. Fish Comm., **20**: 139-148.

GORE, R. H., 1970. *Pachycheles cristobalensis* sp. nov. with notes on the porcellanid crabs of the southwestern Caribbean. Bull. Mar. Sci., **20**: 957-970.

GORE, R. H., 1982. Porcellanid crabs from the coasts of Mexico and central America (Crustacea: Decapoda: Anomura). Smithson. Contrib. Zool., **363**: 1-34.

GORE, R. H. & L. G. ABELE, 1976. Shallow water porcelain crabs from the Pacific coast of Panama and adjacent Caribbean waters (Crustacea: Anomura: Porcellanidae). Smithson. Contrib. Zool., **237**: 1-30.

HAIG, J., 1956. The Galatheidea (Crustacea Anomura) of the Allan Hancock Atlantic Expedition with a review of the Porcellanidae of the western North Atlantic. Allan Hancock Atlantic Expedition Report, **8**: 1-45.

HAIG, J., 1960. The Porcellanidae (Crustacea Anomura) of the eastern Pacific. Allan Hancock Pacific Expeditions, **24**: 1-440.

HILLER, A., H. KRAUS, M. ALMON & B. WERDING, 2006. The *Petrolisthes galathinus* complex: species boundaries based on color pattern, morphology and molecules, and evolutionary interrelationships between the complex and other Porcellanidae (Crustacea: Decapoda: Anomura). Mol. Phylogenet. Evol., **40**: 547-569.

HILLER, A. & B. WERDING, 2007. Redescription of *Petrolisthes edwardsii* (de Saussure) and description of a new, sibling species from the eastern Pacific based on different colour, morphology and genetic identity (Crustacea: Anomura: Porcellanidae). Org. Divers. Evol., **7**: 181-194.

KATOH, K., K. MISAWA, K. KUMA & T. MIYATA, 2002. MAFFT: a novel method for rapid multiple sequence alignment based on fast Fourier transform. Nucleic Acids Res., **30**: 3059-3066.

KIMURA, M., 1980. A simple method of estimating evolutionary rates of base substitutions through comparative studies of nucleotide sequences. J. Mol. Evol., **16**: 111-120.

LESSIOS, H. A., 2008. The great American schism: divergence of marine organisms after the rise of the central American isthmus. Annu. Rev. Ecol. Evol. Syst., **39**: 63-91.

LIRA, C., G. HERNÁNDEZ, J. BOLAÑOS, K. GRATEROL & M. PIÑATE, 2007. Cangrejos porcelánidos (Decapoda: Anomura) de las islas nororientales de Venezuela. II. El género *Pachycheles* Stimpson, 1858. Bol. Inst. Oceanogr. Venez., **46**: 37-50.

PALUMBI, S. R., A. MARTIN, S. ROMANO, W. O. MCMILLAN, L. STICE & G. GRABOWSKI, 1991. The Simple Fool's Guide to PCR, ver. 2: 1-45. (Department of Zoology and Kewalo Marine Laboratory, University of Hawaii, Honolulu, HI).

SCHMITT, W. L., 1935. Crustacea Macrura and Anomura of Porto Rico and the Virgin Islands. Scientific Survey of Porto Rico and the Virgin Islands, **15**: 125-227.

TAMURA, K., G. STECHER, D. PETERSON, A. FILIPSKI & S. KUMAR, 2013. MEGA6: molecular evolutionary genetics analysis version 6.0. Mol. Biol. Evol., **30**: 2725-2729.

VARGAS, R. & J. CORTÉS, 2006. Biodiversidad marina de Costa Rica: Crustacea: infraorden Anomura. Rev. Biol. Trop., **54**: 461-488.

WERDING, B., 1977. Los porcelánidos (Crustacea, Anomura, Porcellanidae) de la región de Santa Marta, Colombia. An. Ist. Invest. Mar. Punta de Betín, **9**: 173-214.

WERDING, B., 1978. Los porcelánidos (Crustacea: Anomura: Porcellanidae) de la región de Acandí (Golfo de Urabá) con algunos encuentros nuevos de la región de Santa Marta (Colombia). An. Ist. Invest. Mar. Punta de Betín, **10**: 213-221.

WERDING, B., 1982. Porcellanid crabs of the Rosario Islands, Caribbean coast of Colombia, with a description of *Petrolisthes rosariensis* new species (Crustacea: Anomura). Bull. Mar. Sci., **32**: 439-447.

WERDING, B. & A. HILLER, 2002. A new species of the *Petrolisthes galathinus* complex from the southern Caribbean Sea, with a discussion on the identity of *P. galathinus* (Bosc, 1802) (Decapoda, Porcellanidae). Crustaceana, **75**: 849-857.

WERDING, B. & A. HILLER, 2005. A new porcellanid species from the Caribbean Sea, with a discussion on the American "*Petrolisthes lewisi*-group" (Decapoda, Anomura, Porcellanidae). Crustaceana, **78**: 1185-1190.

WERDING, B., A. HILLER & R. LEMAITRE, 2003. Geographic and depth distributional patterns of western Atlantic Porcellanidae (Crustacea: Decapoda: Anomura) with an updated list of species. Proc. ICC5-Anomuran Symposium. Mem. Mus. Victoria, **60**: 77-85.

First received 28 September 2016.
Final version accepted 19 April 2017.